Springer Proceedings in Mathematics & Statistics

Volume 58

For further volumes:
http://www.springer.com/series/10533

Springer Proceedings in Mathematics & Statistics

This book series features volumes composed of selected contributions from workshops and conferences in all areas of current research in mathematics and statistics, including OR and optimization. In addition to an overall evaluation of the interest, scientific quality, and timeliness of each proposal at the hands of the publisher, individual contributions are all refereed to the high quality standards of leading journals in the field. Thus, this series provides the research community with well-edited, authoritative reports on developments in the most exciting areas of mathematical and statistical research today.

Toke M. Carlsen • Søren Eilers
Gunnar Restorff • Sergei Silvestrov
Editors

Operator Algebra and Dynamics

Nordforsk Network Closing Conference, Faroe Islands, May 2012

Springer

Editors
Toke M. Carlsen
Department of Mathematical Sciences
Norwegian University of Science and Technology
Trondheim
Norway

Søren Eilers
Department of Mathematical Sciences
University of Copenhagen
Copenhagen
Denmark

Gunnar Restorff
Department of Science and Technology
University of the Faroe Islands
Tórshavn
Faroe Islands

Sergei Silvestrov
Division of Applied Mathematics, The School of Education, Culture and Communication (UKK)
Mälardalen University
Västerås
Sweden

ISSN 2194-1009 ISSN 2194-1017 (electronic)
ISBN 978-3-662-51241-8 ISBN 978-3-642-39459-1 (eBook)
DOI 10.1007/978-3-642-39459-1
Springer Heidelberg New York Dordrecht London

Math. Subject Classification (2010): 46Lxx, 46L55, 47Lxx, 46L55, 54H20, 54H15, 37-XX

Printed on acid-free paper

Springer is part of Springer Science+Business Media (www.springer.com)

Preface

In May 2012, 42 mathematicians congregated in the beautiful seclusion offered by Gjógv, a small community on the Faroese island Eysturoy, to create a productive scientific event centered between the mathematical fields of operator algebras and dynamical systems. Experiencing both the force of the Atlantic Ocean and the tranquil beauty of sunny pastures, the participants enjoyed the hospitality of the Gjáargarður guest house as well as of the University of the Faroe Islands for almost a week, exchanging mathematical ideas and results at lectures as well as at more informal occasions.

The conference marked the conclusion of a 3-year program made possible by the generous support of the NordForsk program funded by the Nordic Research Council, involving 60 mathematicians in Denmark, the Faroe Islands, Norway, and Sweden. The Gjógv meeting was also generously supported by the Faroese Research Council and the University of the Faroe Islands as well as by the Centre for Symmetry and Deformation at the University of Copenhagen. The organizers of the conference were the group leaders of the nodes of the NordForsk network, namely, Toke Meier Carlsen [Trondheim], Søren Eilers [Copenhagen], Nadia Larsen [Oslo], Gunnar Restorff [Tórshavn], Sergei Silvestrov [Västerås/Lund], Wojciech Szymański [Odense], Klaus Thomsen [Aarhus], and Lyudmila Turowska [Gothenburg], with Restorff acting as the local organizer in charge of the rather nontrivial logistics for this memorable event.

Apart from members of the network, senior and junior alike, the organizing committee invited five external speakers:

- Claire Anantharaman-Delaroche [Orléans]
- Siegfried Echterhoff [Münster]
- Wolfgang Krieger [Heidelberg]
- Efren Ruiz [Hilo, Hawaii]
- Dana Williams [Dartmouth]

Fig. 1 The grass-roofed guesthouse Gjáargarður

The interplay between operator algebras and dynamical systems, the scientific focus of both the NordForsk network and the conference, is a topic of dramatic current interest. These two areas benefited from the genius of John von Neumann in their early days, but have developed independently over the decades following World War II. The network aimed to steer the force resulting from the leading international position of Nordic mathematics in the area of operator algebras in the direction of the exciting cross field at the boundary of dynamics and functional analysis, the main goal being to understand and analyze C^*-algebras and von Neumann algebras associated to dynamics, as well as to develop the relevant concepts in dynamics.

This volume documents some of the substantial progress made by the network, which existed for almost 3 years prior to the closing conference. However, the network's impact on Nordic mathematics will be felt for some time, in particular due to the strong scientific ties forged between the NordForsk network members and the eight nodes as a result of conferences such as the one in Gjógv and the many personal visits by researchers in the network to other nodes.

There are many ways in which operator algebra and dynamics interact and during the existence of the NordForsk network several or perhaps even most of these interactions were explored at meetings or focused visits. The individual chapters of this proceedings volume illustrate several of these interactions. Chapter 1 deals with von Neumann algebras arising from discrete measured groupoids, Chap. 2 with purely infinite Cuntz-Krieger algebras, and Chap. 3 with filtered K-theory over finite topological spaces, whereas C^*-algebras associated to shift spaces (or subshifts) is the topic of Chap. 4. Graph C^*-algebras are studied in Chaps. 5 and 7, and in Chap. 6 irrational extended rotation algebras are shown to be C^*-alloys. Chapter 8 deals with free probability and Chap. 9 with renewal systems, whereas KMS-states of Cuntz-algebras are used in Chap. 10 to give a new proof of the Grothendieck

Fig. 2 Group photo of the participants.
Back row: Wolfgang Krieger [Heidelberg], Søren Eilers [Copenhagen], Hannes Thiel [Copenhagen/Münster], Tron Omland [Trondheim], Johan Öinert [Copenhagen], Sigurd Segtnan [Oslo], Søren Knudby [Copenhagen], Sören Möller [Odense], Nadia Larsen [Oslo], Klaus Thomsen [Aarhus], James Gabe [Copenhagen], Tim de Laat [Copenhagen], Efren Ruiz [Hawaii Hilo], Jonas Andersen Seebach [Aarhus], Sara E. Arklint [Copenhagen], Fredrik Ekström [Lund], Johan Richter [Lund].
Middle row: Maria Ramirez-Solano [Copenhagen], Gunnar Restorff [Tórshavn], Dana Williams [Dartmouth], Sergei Silvestrov [Västerås], Magnus Landstad [Trondheim], Toke Meier Carlsen [Trondheim], Eduard Ortega [Trondheim], Rasmus Bentmann [Copenhagen], Rune Johansen [Copenhagen], Adam P.W. Sørensen [Copenhagen/Wollongong], Steven Deprez [Copenhagen], Alexander Stolin [Gothenburg].
Front row: George A. Elliott [Copenhagen/Toronto], Erling Størmer [Oslo], Rui Palma [Oslo], Siegfried Echterhoff [Münster], Claire Ananthataman-Delaroche [Orléans], Martin Wanvik [Trondheim], Wojciech Szymański [Odense], Jesper With Mikkelsen [Odense], Lyudmila Turowska [Gothenburg], Jyotishman Bhowmick [Oslo], Asger Törnquist [Copenhagen]

theorem for jointly completely bounded bilinear forms on C^*-algebras. In Chap. 11, Cuntz-Li algebras associated with the a-adic numbers are constructed as crossed products, and in Chap. 12, crossed products of injective endomorphisms (the so-called Stacey crossed products) are studied. In Chap. 13, another type of operator algebras associated to dynamical systems, namely, C^*-completions of the Hecke algebra of a Hecke pair, is studied, whereas Chap. 14 gives an overview on how operator algebras can be used to study wavelets. Finally, Chap. 15 deals with semiprojective C^*-algebras, and in Chap. 16, the topological dimension of type I C^*-algebras is studied.

Fig. 3 View from the pass Skúvadalsskarð south of the village Gjógv

Fig. 4 View of the village Gjógv

We extend our deep-felt thanks to all the people who made this volume possible—authors, referees, and the technical staff at Springer—as well as the Nordic Research Council, the Faroese Research Council, the University of the Faroe

Fig. 5 The sail-vessel Norðlýsið seen out through the opening of a grotto on the west shore of the island Hestur during a boat trip. After the boat trip, an official reception was held by the University of the Faroe Islands in Tórshavn followed up by a public lecture about mathematics by Søren Eilers

Islands, and the Centre for Symmetry and Deformation in Copenhagen for providing essential funding for the closing conference.

Trondheim, Norway — Toke Meier Carlsen
Copenhagen, Denmark — Søren Eilers
Tórshavn, Faroe Islands — Gunnar Restorff
Västerås, Sweden — Sergei Silvestrov
June 2013

Contents

List of Contributors

Claire Anantharaman-Delaroche Laboratoire MAPMO – CNRS/Université d'Orléans, Orléans, France

Sara E. Arklint Department of Mathematical Sciences, University of Copenhagen, Copenhagen Ø, Denmark

Rasmus Bentmann Department of Mathematical Sciences, University of Copenhagen, Copenhagen Ø, Denmark

Toke Meier Carlsen Department of Mathematical Sciences, Norwegian University of Science and Technology (NTNU), Trondheim, Norway

Søren Eilers Department of Mathematical Sciences, University of Copenhagen, Copenhagen Ø, Denmark

George A. Elliott Department of Mathematics, University of Toronto, Toronto, ON, Canada

James Gabe Department of Mathematical Sciences, University of Copenhagen, Copenhagen Ø, Denmark

Uffe Haagerup Department of Mathematical Sciences, University of Copenhagen, Copenhagen Ø, Denmark

Rune Johansen Department of Mathematical Sciences, University of Copenhagen, Copenhagen Ø, Denmark

Tim de Laat Department of Mathematical Sciences, University of Copenhagen, Copenhagen Ø, Denmark

Sören Möller Department of Mathematics and Computer Science, University of Southern Denmark, Odense M, Denmark

Zhuang Niu Department of Mathematics, University of Wyoming, Laramie, WY, USA

Tron Omland Department of Mathematical Sciences, Norwegian University of Science and Technology (NTNU), Trondheim, Norway

Eduard Ortega Department of Mathematical Sciences, Norwegian University of Science and Technology (NTNU), Trondheim, Norway

Rui Palma University of Oslo, Oslo, Norway

Enrique Pardo Departamento de Matemáticas, Facultad de Ciencias, Universidad de Cádiz, Puerto Real (Cádiz), Spain

Gunnar Restorff Department of Science and Technology, University of the Faroe Islands, Tórshavn, Faroe Islands

Efren Ruiz Department of Mathematics, University of Hawaii, Hilo, HI, USA

Sergei Silvestrov Division of Applied Mathematics, The School of Education, Culture and Communication, Mälardalen University, Västerås, Sweden

Adam P. W. Sørensen School of Mathematics and Applied Statistics, University of Wollongong, Wollongong, NSW, Australia

Hannes Thiel Mathematisches Institut, Universität Münster, Münster, Germany

Chapter 1
The Haagerup Property for Discrete Measured Groupoids

Claire Anantharaman-Delaroche

Abstract We define the Haagerup property in the general context of countable groupoids equipped with a quasi-invariant measure. One of our objectives is to complete an article of Jolissaint devoted to the study of this property for probability measure preserving countable equivalence relations. Our second goal, concerning the general situation, is to provide a definition of this property in purely geometric terms, whereas this notion had been introduced by Ueda in terms of the associated inclusion of von Neumann algebras. Our equivalent definition makes obvious the fact that treeability implies the Haagerup property for such groupoids and that it is not compatible with Kazhdan's property (T).

Keywords Haagerup property • Groupoid • Kazhdan property (T) • The von Neumann algebra of a measured groupoid

Mathematics Subject Classification (2010): 46L55, 37A15, 22A22, 20F65.

1.1 Introduction

Since the seminal paper of Haagerup [15], showing that free groups have the (now so-called) Haagerup property, or property (H), this notion plays an increasingly important role in group theory (see the book [10]). A similar property (H) has been introduced for finite von Neumann algebras [11, 12] and it was proved in [11] that a

C. Anantharaman-Delaroche (✉)
Laboratoire MAPMO – UMR6628, Fédération Denis Poisson (FDP – FR2964), CNRS/Université d'Orléans, B. P. 6759, F-45067 Orléans Cedex 2, France
e-mail: claire.anantharaman@univ-orleans.fr

T.M. Carlsen et al. (eds.), *Operator Algebra and Dynamics*, Springer Proceedings in Mathematics & Statistics 58, DOI 10.1007/978-3-642-39459-1_1,

countable group Γ has property (H) if and only if its von Neumann algebra $L(\Gamma)$ has property (H).

Later, given a von Neumann subalgebra A of a finite von Neumann algebra M, a property (H) for M relative to A has been considered [9,25] and proved to be very useful. It is in particular one of the crucial ingredients used by Popa [25], to provide the first example of a II_1 factor with trivial fundamental group.

A discrete (also called countable) measured groupoid (G, μ) with set of units X (see Sect. 1.2.1) gives rise to an inclusion $A \subset M$, where $A = L^\infty(X, \mu)$ and $M = L(G, \mu)$ is the von Neumann algebra of the groupoid. This inclusion is canonically equipped with a conditional expectation $E_A : M \to A$. Although M is not always a finite von Neumann algebra, there is still a notion of property (H) relative to A and E_A (see [32]). However, to our knowledge, this property has not been translated in terms only involving (G, μ), as in the group case. A significant exception concerns the case where $G = \mathscr{R}$ is a countable equivalence relation on X, preserving the probability measure μ, i.e. a type II_1 equivalence relation [18].

Our first goal is to extend the work of Jolissaint [18] in order to cover the general case of countable measured groupoids, and in particular the case of group actions leaving a probability measure quasi-invariant. Although it is not difficult to guess the right definition of property (H) for (G, μ) (see Definition 8), it is more intricate to prove the equivalence of this notion with the fact that $L(G, \mu)$ has property (H) relative to $L^\infty(X, \mu)$.

We begin in Sect. 1.2 by introducing the basic notions and notation relative to countable measured groupoids. In particular we discuss the Tomita-Takesaki theory for their von Neumann algebras. This is essentially a reformulation of the pioneering results of P. Hahn [16] in a way that fits better for our purpose. In Sect. 1.3 we discuss in detail several facts about the von Neumann algebra of the Jones' basic construction for an inclusion $A \subset M$ of von Neumann algebras, assuming that A is abelian. We also recall here the notion of relative property (H) in this setting.

In Sects. 1.4 and 1.5, we study the relations between positive definite functions on our groupoids and completely positive maps on the corresponding von Neumann algebras. These results are extensions of well known results for groups and of results obtained by Jolissaint in [18] for equivalence relations, but additional difficulties must be overcome. After this preliminary work, it is immediate (Sect. 1.6) to show the equivalence of our definition of property (H) for groupoids with the definition involving operator algebras (Theorem 1).

Our main motivation originates from the reading of Ueda's paper [32] and concerns treeable groupoids. This notion was introduced by Adams for probability measure preserving countable equivalence relations [1]. Treeable groupoids may be viewed as the groupoid analogue of free groups. So a natural question, raised by C.C. Moore in his survey [22, p. 277] is whether a treeable equivalence relation must have the Haagerup property. In fact, this problem is solved in [32] using operator algebras techniques. In Ueda's paper, the notion of treeing is translated in an operator algebra framework regarding the inclusion $L^\infty(X, \mu) \subset L(G, \mu)$, and it is proved that this condition implies that $L(G, \mu)$ has the Haagerup property relative to $L^\infty(X, \mu)$.

Our approach is opposite. For us, it seems more natural to compare these two notions, treeability and property (H), purely at the level of the groupoid. Indeed, the definition of treeability is more nicely read at this level: roughly speaking, it means that there is a measurable way to endow each fibre of the groupoid with a structure of tree (see Definition 13). The direct proof that treeability implies property (H) is given in Sect. 1.7 (Theorem 3).

Using our previous work [6] on groupoids with property (T), we prove in Sect. 1.8 that, under an assumption of ergodicity, this property is incompatible with the Haagerup property (Theorem 5). As a consequence, we recover the result of Jolissaint [18, Proposition 3.2] stating that if Γ is a Kazhdan countable group which acts ergodically on a Lebesgue space (X, μ) and leaves the probability measure μ invariant, then the orbit equivalence relation $(\mathscr{R}_\Gamma, \mu)$ has not the Haagerup property (Corollary 3). A fortiori, $(\mathscr{R}_\Gamma, \mu)$ is not treeable, a result due to Adams and Spatzier [2, Theorem 18] and recovered in a different way by Ueda.

This text is an excerpt from the survey [7] which was not intended for publication.

1.2 The von Neumann Algebra of a Measured Groupoid

1.2.1 Preliminaries on Countable Measured Groupoids

Our references for measured groupoids are [8, 16, 27]. Let us recall that a *groupoid* is a set G endowed with a product $(\gamma, \gamma') \mapsto \gamma\gamma'$ defined on a subset $G^{(2)}$ of $G \times G$, called the set of *composable* elements, and with an inverse map $\gamma \mapsto \gamma^{-1}$, satisfying the natural properties expected for a product and an inverse such as associativity. For every $\gamma \in G$, we have that $(\gamma^{-1}, \gamma) \in G^{(2)}$ and if $(\gamma, \gamma') \in G^{(2)}$, then $\gamma^{-1}(\gamma\gamma') = (\gamma^{-1}\gamma)\gamma' = \gamma'$ (and similarly $(\gamma\gamma')\gamma'^{-1} = \gamma(\gamma'\gamma'^{-1}) = \gamma$). We set $r(\gamma) = \gamma\gamma^{-1}$, $s(\gamma) = \gamma^{-1}\gamma$ and $G^{(0)} = r(G) = s(G)$. The maps r and s are called respectively the *range* and the *source* map. The pair (γ, γ') is composable if and only if $s(\gamma) = r(\gamma')$ and then we have $r(\gamma\gamma') = r(\gamma)$ and $s(\gamma\gamma') = s(\gamma')$. The set $G^{(0)}$ is called the *unit space* of G. Indeed, its elements are units in the sense that $\gamma\, s(\gamma) = \gamma$ and $r(\gamma)\gamma = \gamma$.

The fibres corresponding to $r, s : G \to G^{(0)}$ are denoted respectively by $G^x = r^{-1}(x)$ and $G_x = s^{-1}(x)$. Given a subset A of $G^{(0)}$, the *reduction* of G to A is the groupoid $G|_A = r^{-1}(A) \cap s^{-1}(A)$. Two elements x, y of $G^{(0)}$ are said to be equivalent if $G^x \cap G_y \neq \emptyset$. We denote by $\mathscr{R}_G$ this equivalence relation. Given $A \subset G^{(0)}$, its *saturation* $[A]$ is the set $s(r^{-1}(A))$ of all elements in $G^{(0)}$ that are equivalent to some element of A. When $A = [A]$, we say that A is *invariant*.

A *Borel groupoid* is a groupoid G endowed with a standard Borel structure such that the range, source, inverse and product are Borel maps, where $G^{(2)}$ has the Borel structure induced by $G \times G$ and $G^{(0)}$ has the Borel structure induced by G. We say that G is *countable* (or *discrete*) if the fibres G^x (or equivalently G_x) are countable.

In the sequel, we only consider such groupoids. We *always denote by X the set $G^{(0)}$ of units of G*. A *bisection* S is a Borel subset of G such that the restrictions of r and s to S are injective. A useful fact, consequence of a theorem of Lusin-Novikov, states that, since r and s are countable-to-one Borel maps between standard Borel spaces, there exists a countable partition of G into bisections (see [20, Theorem 18.10]).

Let μ be a probability measure on $X = G^{(0)}$. We define a σ-finite measure ν on G by the formula

$$\int_G F \,\mathrm{d}\nu = \int_X \Big(\sum_{\{\gamma \in G : s(\gamma) = x\}} F(\gamma) \Big) \,\mathrm{d}\mu(x) \tag{1.1}$$

for every non-negative Borel function F on G. The fact that $x \mapsto \sum_{\{\gamma : s(\gamma)=x\}} F(\gamma)$ is Borel is proved as in [14, Theorem 2] by using a countable partition of G by Borel subsets on which s is injective.

We say that μ is *quasi-invariant* if ν is equivalent to it image ν^{-1} under $\gamma \mapsto \gamma^{-1}$. In other terms, for every bisection S, one has $\mu(s(S)) = 0$ if and only if $\mu(r(S)) = 0$. This notion only depends on the measure class of μ. We set $\delta = \frac{\mathrm{d}\nu^{-1}}{\mathrm{d}\nu}$. Whenever $\nu = \nu^{-1}$, we say that μ is *invariant*.

Definition 1. A *countable (or discrete) measured groupoid*[1] (G, μ) is a countable Borel groupoid G with a quasi-invariant probability measure μ on $X = G^{(0)}$.

In the rest of this paper, G will always be equipped with the corresponding σ-finite measure ν defined in (1.1).

Example 1. (a) Let $\Gamma \curvearrowright X$ be a (right) action of a countable group Γ on a standard Borel space X, and assume that the action preserves the class of a probability measure μ. Let $G = X \rtimes \Gamma$ be the *semi-direct product groupoid*. We have $r(x,t) = x$ and $s(x,t) = xt$. The product is given by the formula $(x,s)(xs,t) = (x,st)$. Equipped with the quasi-invariant measure μ, (G,μ) is a countable measured groupoid. As a particular case, we find the group $G = \Gamma$ when X is reduced to a point.

(b) Another important family of examples concerns *countable measured equivalence relations*. A countable Borel equivalence relation $\mathcal{R} \subset X \times X$ on a standard Borel space X is a Borel subset of $X \times X$ whose equivalence classes are finite or countable. It has an obvious structure of Borel groupoid with $r(x,y) = x$, $s(x,y) = y$ and $(x,y)(x,z) = (x,z)$. When equipped with a quasi-invariant probability measure μ, we say that $(\mathcal{R}, \mu)$ is a countable measured equivalence relation. Here, quasi-invariance also means that for every Borel subset $A \subset X$, we have $\mu(A) = 0$ if and only if the measure of the saturation $s(r^{-1}(A))$ of A is still 0.

---

[1]In [8], a countable measured groupoid is called r-discrete. Another difference is that we have swapped here the definitions of ν and ν^{-1}.

The orbit equivalence relation associated with an action $\Gamma \curvearrowright (X, \mu)$ is denoted $(\mathscr{R}_\Gamma, \mu)$.

A general groupoid is a combination of an equivalence relation and groups. Indeed, let (G, μ) be a countable measured groupoid. Let $c = (r, s)$ be the map $\gamma \mapsto (r(\gamma), s(\gamma))$ from G into $X \times X$. The range of c is the graph $\mathscr{R}_G$ of the equivalence relation induced on X by G. Moreover $(\mathscr{R}_G, \mu)$ is a countable measured equivalence relation. The kernel of the groupoid homomorphism c is the subgroupoid $\{\gamma \in G : s(\gamma) = r(\gamma)\}$. For every $x \in X$, the subset $G(x) = s^{-1}(x) \cap r^{-1}(x)$, endowed with the induced operations, is a group, called the *isotropy group* at x. So the kernel of c is the bundle of groups $x \mapsto G(x)$ over X, called the *isotropy bundle*, and G appears as an extension of the equivalence relation $\mathscr{R}_G$ by this bundle of groups.

A reduction $(G_{|U}, \mu_{|U})$ such that U is conull in X is called *inessential*. Since we are working in the setting of measured spaces, it will make no difference to replace (G, μ) by any of its inessential reductions.

1.2.2 The von Neumann Algebra of (G, μ)

If $f : G \to \mathbb{C}$ is a Borel function, we set

$$\|f\|_I = \max \left\{ \left\| x \mapsto \sum_{r(\gamma)=x} |f(\gamma)| \right\|_\infty , \left\| x \mapsto \sum_{s(\gamma)=x} |f(\gamma)| \right\|_\infty \right\}.$$

Let $I(G)$ be the set of functions such that $\|f\|_I < +\infty$. It only depends on the measure class of μ. We endow $I(G)$ with the (associative) convolution product

$$(f * g)(\gamma) = \sum_{\gamma_1\gamma_2=\gamma} f(\gamma_1)g(\gamma_2) = \sum_{s(\gamma)=s(\gamma_2)} f(\gamma\gamma_2^{-1})g(\gamma_2) = \sum_{r(\gamma_1)=r(\gamma)} f(\gamma_1)g(\gamma_1^{-1}\gamma).$$

and the involution $f^*(\gamma) = \overline{f(\gamma^{-1})}$.

We have $I(G) \subset L^1(G, \nu) \cap L^\infty(G, \nu) \subset L^2(G, \nu)$, with $\|f\|_1 \le \|f\|_I$ when $f \in I(G)$. Therefore $\|\cdot\|_I$ is a norm on $I(G)$, where two functions which coincide ν-almost everywhere are identified. It is easily checked that $I(G)$ is complete for this norm. Moreover for $f, g \in I(G)$ we have $\|f * g\|_I \le \|f\|_I \|g\|_I$. Therefore $(I(G), \|\cdot\|_I)$ is a Banach $*$-algebra.

This variant of the Banach algebra $I(G)$ introduced by Hahn [16] has been considered by Renault in [29, p. 50]. Its advantage is that it does not involve the Radon-Nikodym derivative δ.

For $f \in I(G)$ we define a bounded operator $L(f)$ on $L^2(G, \nu)$ by

$$(L(f)\xi)(\gamma) = (f * \xi)(\gamma) = \sum_{\gamma_1\gamma_2=\gamma} f(\gamma_1)\xi(\gamma_2). \tag{1.2}$$

We have $\|L(f)\| \leq \|f\|_I$, $L(f)^* = L(f^*)$ and $L(f)L(g) = L(f*g)$. Hence, L is a representation of $I(G)$, called the *left regular representation*.

Definition 2. The *von Neumann algebra of the countable measured groupoid* (G, μ) is the von Neumann subalgebra $L(G, \mu)$ of $\mathscr{B}(L^2(G, \nu))$ generated by $L(I(G))$. It will also be denoted by M in the rest of the paper.

Note that $L^2(G, \nu)$ is a direct integral of Hilbert spaces :

$$L^2(G, \nu) = \int_X^{\oplus} \ell^2(G_x)\,\mathrm{d}\mu(x).$$

We define on $L^2(G, \nu)$ a structure of $L^\infty(X)$-module by $(f\xi)(\gamma) = f \circ s(\gamma)\xi(\gamma)$, where $f \in L^\infty(X)$ and $\xi \in L^2(G, \nu)$. In fact $L^\infty(X)$ is the algebra of diagonalizable operators with respect to the above disintegration of $L^2(G, \nu)$.

Obviously, the representation L commutes with this action of $L^\infty(X)$. It follows that the elements of $L(G, \mu)$ are decomposable operators ([13, Theorem 1, p. 164]). We have $L(f) = \int_X^{\oplus} L_x(f)\,\mathrm{d}\mu(x)$, where $L_x(f) : \ell^2(G_x) \to \ell^2(G_x)$ is defined as in (1.2), but for $\xi \in \ell^2(G_x)$.

Let $C_n = \{1/n \leq \delta \leq n\}$. Then (C_n) is an increasing sequence of measurable subsets of G with $\cup_n C_n = G$ (up to null sets). We denote by $I_n(G)$ the set of elements in $I(G)$ taking value 0 outside C_n and we set $I_\infty(G) = \cup_n I_n(G)$. Obviously, $I_\infty(G)$ is an involutive subalgebra of $I(G)$. It is easily checked that $I_\infty(G)$ is dense into $L^2(G, \nu)$ and that $L(G, \mu)$ is generated by $L(I_\infty(G))$.

The von Neumann algebra $L^\infty(X)$ is isomorphic to a subalgebra of $I_\infty(G)$, by giving to $f \in L^\infty(X)$ the value 0 outside $X \subset G$. Note that, for $\xi \in L^2(G, \nu)$,

$$(L(f)\xi)(\gamma) = f \circ r(\gamma)\xi(\gamma).$$

In this way, $A = L^\infty(X)$ appears as a von Neumann subalgebra of M.

Obviously, the pair $A \subset M$ only depends on the measure class of μ, up to unitary equivalence.

We view $I(G)$ as a subspace of $L^2(G, \nu)$. The characteristic function $\mathbf{1}_X$ of $X \subset G$ is a norm one vector in $L^2(G, \nu)$. Let φ be the normal state on M defined by

$$\varphi(T) = \langle \mathbf{1}_X, T\mathbf{1}_X \rangle_{L^2(G,\nu)}.$$

For $f \in I(G)$, we have $\varphi(L(f)) = \int_X f(x)\,\mathrm{d}\mu(x)$, and therefore, for $f, g \in I(G)$,

$$\varphi(L(f)^* L(g)) = \langle f, g \rangle_{L^2(G,\nu)}. \tag{1.3}$$

Lemma 1. *Let g be a Borel function on G such that $\delta^{-1/2} g = f \in I(G)$ (for instance $g \in I_\infty(G)$). Then $\xi \mapsto \xi * g$ is a bounded operator on $L^2(G, \nu)$. More precisely, we have*

$$\|\xi * g\|_2 \leq \|f\|_I \|\xi\|_2.$$

Proof. Straightforward. □

We set $R(g)(\xi) = \xi * g$. We have $L(f) \circ R(g) = R(g) \circ L(f)$ for every $g \in I_\infty(G)$ and $f \in I(G)$. We denote by $R(G,\mu)$ the von Neumann algebra generated by $R(I_\infty(G))$.

Lemma 2. *The vector $\mathbf{1}_X$ is cyclic and separating for $L(G,\mu)$, and therefore φ is a faithful state.*

Proof. Immediate from the fact that $L(f)$ and $R(g)$ commute for $f, g \in I_\infty(G)$, with $L(f)\mathbf{1}_X = f$ and $R(g)\mathbf{1}_X = g$, and from the density of $I_\infty(G)$ into $L^2(G,\nu)$. □

The von Neumann algebra $L(G,\mu)$ is on standard form on $L^2(G,\nu)$, canonically identified with $L^2(M,\varphi)$ (see (1.3)). We identify M with a dense subspace of $L^2(G,\nu)$ by $T \mapsto \hat{T} = T(\mathbf{1}_X)$. The modular conjugation J and the one-parameter modular group σ relative to the vector $\mathbf{1}_X$ (and φ) have been computed in [16]. With our notations, we have

$$\forall \xi \in L^2(G,\nu), \quad (J\xi)(\gamma) = \delta(\gamma)^{1/2}\overline{\xi(\gamma^{-1})}$$

and

$$\forall T \in L(G,\mu), \quad \sigma_t(T) = \delta^{it} T \delta^{-it}.$$

Here, for $t \in \mathbb{R}$, the function δ^{it} acts on $L^2(G,\nu)$ by pointwise multiplication and defines a unitary operator. Note that for $f \in L(G,\mu)$, we have $\delta^{it}L(f)\delta^{-it} = L(\delta^{it} f)$. In particular, σ acts trivially on A. Therefore (see [31]), there exists a unique faithful conditional expectation $E_A : M \to A$ such that $\varphi = \varphi \circ E_A$, and for $T \in M$, we have

$$\widehat{E_A(T)} = e_A(\hat{T}),$$

where e_A is the orthogonal projection from $L^2(G,\nu)$ onto $L^2(X,\mu)$. If we view the elements of M as functions on G, then E_A is the restriction map to X. The triple (M, A, E_A) only depends on the class of μ, up to equivalence.

For $f \in I(G)$ and $\xi \in L^2(G,\nu)$ we observe that

$$(JL(f)J)\xi = R(g)\xi = \xi * g \quad \text{with} \quad g = \delta^{1/2} f^*. \tag{1.4}$$

1.3 Basic Facts on the Module $L^2(M)_A$

We consider, in an abstract setting, the situation we have met above. Let $A \subset M$ be a pair of von Neumann algebras, where $A = L^\infty(X,\mu)$ is abelian. We assume the existence of a normal faithful conditional expectation $E_A : M \to A$ and we set $\varphi = \tau_\mu \circ E_A$, where τ_μ is the state on A defined by the probability measure μ. Recall that

M is on standard form on the Hilbert space $L^2(M,\varphi)$ of the Gelfand-Naimark-Segal construction associated with φ. We view $L^2(M,\varphi)$ as a left M-module and a right A-module. Identifying[2] M with a subspace of $L^2(M,\varphi)$, we know that E_A is the restriction to M of the orthogonal projection $e_A : L^2(M,\varphi) \to L^2(A,\tau_\mu)$.

For further use, we make the following observation

$$\forall m \in M, \forall a \in A, \quad \hat{m}a = Ja^* J\hat{m} = \widehat{ma} = m\hat{a}. \tag{1.5}$$

Indeed, if S is the closure of the map $\hat{m} \mapsto \widehat{m^*}$ and if $S = J\Delta^{1/2} = \Delta^{-1/2}J$ is its polar decomposition, then every $a \in A$ commutes with Δ since it is invariant under σ^φ. Then (1.5) follows easily. Note that (1.4) gives a particular case of this remark.

1.3.1 The Commutant $\langle M, e_A\rangle$ of the Right Action

The algebra of all operators which commute with the right action of A is the von Neumann algebra of the basic construction for $A \subset M$. It is denoted $\langle M, e_A\rangle$ since it is generated by M and e_A. The linear span of $\{m_1 e_A m_2 : m_1, m_2 \in M\}$ is a $*$-subalgebra which is weak operator dense in $\langle M, e_A\rangle$. Moreover $\langle M, e_A\rangle$ is a semi-finite von Neumann algebra, carrying a canonical normal faithful semi-finite trace Tr_μ (depending on the choice of μ), defined by

$$\mathrm{Tr}_\mu(m_1 e_A m_2) = \int_X E_A(m_2 m_1)\,\mathrm{d}\mu = \varphi(m_2 m_1).$$

(for these classical results, see [19, 24]). We shall give more information on this trace in Lemma 4 and its proof. We need some preliminaries.

Definition 3. A vector $\xi \in L^2(M,\varphi)$ is *A-bounded* if there exists $c > 0$ such that $\|\xi a\|_2 \le c\tau_\mu(a^*a)^{1/2}$ for every $a \in A$.

We denote by $L^2(M,\varphi)^0$, or $\mathscr{L}^2(M,\varphi)$, the subspace of A-bounded vectors. It contains M. We also recall the obvious fact that $T \mapsto T(1_A)$ is an isomorphism from the space $\mathscr{B}(L^2(A,\tau_\mu)_A, L^2(M,\varphi)_A)$ of bounded (right) A-linear operators $T : L^2(A,\tau_\mu) \to L^2(M,\varphi)$ onto $\mathscr{L}^2(M,\varphi)$. For $\xi \in \mathscr{L}^2(M,\varphi)$, we denote by L_ξ the corresponding operator from $L^2(A,\tau_\mu)$ into $L^2(M,\varphi)$. In particular, for $m \in M$, we have $L_m = m|_{L^2(A,\tau_\mu)}$. It is easy to see that $\mathscr{L}^2(M,\varphi)$ is stable under the actions of $\langle M, e_A\rangle$ and A, and that $L_{T\xi a} = T \circ L_\xi \circ a$ for $T \in \langle M, e_A\rangle$, $\xi \in \mathscr{L}^2(M,\varphi)$, $a \in A$.

[2]When necessary, we shall write $\hat{m}$ the element $m \in M$, when viewed in $L^2(M,\varphi)$, in order to stress this fact.

For $\xi, \eta \in \mathscr{L}^2(M, \varphi)$, the operator $L_\xi^* L_\eta \in \mathscr{B}(L^2(A, \tau_\mu))$ is in A, since it commutes with A. We set $\langle \xi, \eta \rangle_A = L_\xi^* L_\eta$. In particular, we have $\langle m_1, m_2 \rangle_A = E_A(m_1^* m_2)$ for $m_1, m_2 \in M$. The A-valued inner product $\langle \xi, \eta \rangle_A = L_\xi^* L_\eta$ gives to $\mathscr{L}^2(M, \varphi)$ the structure of a self-dual Hilbert right A-module [23]. It is a normed space with respect to the norm $\|\xi\|_{\mathscr{L}^2(M)} = \|\langle \xi, \xi \rangle_A\|_A^{1/2}$. Note that

$$\|\xi\|_{L^2(M)}^2 = \tau_\mu(\langle \xi, \xi \rangle_A) \leq \|\xi\|_{\mathscr{L}^2(M)}^2.$$

On the algebraic tensor product $\mathscr{L}^2(M, \varphi) \odot L^2(A)$ a positive hermitian form is defined by

$$\langle \xi \otimes f, \eta \otimes g \rangle = \int_X \bar{f} g \langle \xi, \eta \rangle_A \, \mathrm{d}\mu.$$

The Hilbert space $\mathscr{L}^2(M, \varphi) \otimes_A L^2(A)$ obtained by separation and completion is isomorphic to $L^2(M, \varphi)$ as a right A-module by $\xi \otimes f \mapsto \xi f$. Moreover the von Neumann algebra $\mathscr{B}(\mathscr{L}^2(M, \varphi)_A)$ of bounded A-linear endomorphisms of $\mathscr{L}^2(M, \varphi)$ is isomorphic to $\langle M, e_A \rangle$ by $T \mapsto T \otimes 1$. We shall identify these two von Neumann algebras (see [23, 30] for details on these facts).

Definition 4. An *orthonormal basis* of the A-module $L^2(M, \varphi)$ is a family (ξ_i) of elements of $\mathscr{L}^2(M, \varphi)$ such that $\overline{\sum_i \xi_i A} = L^2(M, \varphi)$ and $\langle \xi_i, \xi_j \rangle_A = \delta_{i,j} p_j$ for all i, j, where the p_j are projections in A.

It is easily checked that $L_{\xi_i} L_{\xi_i}^*$ is the orthogonal projection on $\overline{\xi_i A}$, and that these projections are mutually orthogonal with $\sum_i L_{\xi_i} L_{\xi_i}^* = 1$.

Using a generalization of the Gram-Schmidt orthonormalization process, one shows the existence of orthonormal bases (see [23]).

Lemma 3. *Let (ξ_i) be an orthonormal basis of the A-module $L^2(M, \varphi)$. For every $\xi \in \mathscr{L}^2(M, \varphi)$, we have (weak* convergence)*

$$\langle \xi, \xi \rangle_A = \sum_i \langle \xi, \xi_i \rangle_A \langle \xi_i, \xi \rangle_A. \tag{1.6}$$

Proof. Indeed

$$\langle \xi, \xi \rangle_A = L_\xi^* L_\xi = L_\xi^* (\sum_i L_{\xi_i} L_{\xi_i}^*) L_\xi = \sum_i (L_\xi^* L_{\xi_i})(L_{\xi_i}^* L_\xi) = \sum_i \langle \xi, \xi_i \rangle_A \langle \xi_i, \xi \rangle_A.$$

□

Lemma 4. *Let $(\xi_i)_{i \in I}$ be an orthonormal basis of the A-module $L^2(M, \varphi)$.*

1. For every $x \in \langle M, e_A \rangle_+$ we have

$$\mathrm{Tr}_\mu(x) = \sum_i \tau_\mu(\langle \xi_i, x \xi_i \rangle_A) = \sum_i \langle \xi_i, x \xi_i \rangle_{L^2(M)}. \tag{1.7}$$

2. span $\left\{L_\xi L_\eta^* : \xi, \eta \in \mathscr{L}^2(M,\varphi)\right\}$ *is contained in the ideal of definition of* Tr_μ *and we have, for* $\xi, \eta \in \mathscr{L}^2(M,\varphi)$,

$$\mathrm{Tr}_\mu(L_\xi L_\eta^*) = \tau_\mu(L_\eta^* L_\xi) = \tau_\mu(\langle \eta, \xi \rangle_A). \tag{1.8}$$

Proof. 1. The map $U : L^2(M,\varphi) = \oplus_i \overline{\xi_i A} \to \oplus_i p_i L^2(A)$ defined by $U(\xi_i a) = p_i a$ is an isomorphism which identifies $L^2(M,\varphi)$ to the submodule $p(\ell^2(I) \otimes L^2(A))$ of $\ell^2(I) \otimes L^2(A)$, with $p = \oplus_i p_i$. The canonical trace on $\langle M, e_A \rangle$ is transfered to the restriction to $p\big(\mathscr{B}(\ell^2(I))\overline{\otimes} A\big)p$ of the trace $\mathrm{Tr} \otimes \tau_\mu$, defined on $T = [T_{i,j}] \in (\mathscr{B}(\ell^2(I)\overline{\otimes} A))_+$ by $(\mathrm{Tr} \otimes \tau_\mu)(T) = \sum_i \tau_\mu(T_{ii})$. It follows that

$$\mathrm{Tr}_\mu(x) = \sum_i \tau_\mu((UxU^*)_{ii}) = \sum_i \langle \xi_i, x\xi_i \rangle_{L^2(M)} = \sum_i \tau_\mu(\langle \xi_i, x\xi_i \rangle_A).$$

2. Taking $x = L_\xi L_\xi^*$ in *1*, the equality $\mathrm{Tr}_\mu(L_\xi L_\xi^*) = \tau_\mu(\langle \xi, \xi \rangle_A)$ follows from Eqs. (1.6) and (1.7). Formula (1.8) is deduced by polarization. □

1.3.2 Compact Operators

In a semi-finite von Neumann algebra N, there is a natural notion of ideal of compact operators, namely the norm-closed ideal $\mathscr{I}(N)$ generated by its finite projections (see [25, Sect. 1.3.2] or [26]).

Concerning $N = \langle M, e_A \rangle$, there is another natural candidate for the space of compact operators. First, we observe that given $\xi, \eta \in \mathscr{L}^2(M,\varphi)$, the operator $L_\xi L_\eta^* \in \langle M, e_A \rangle$ plays the role of a rank one operator in ordinary Hilbert spaces: indeed, if $\alpha \in \mathscr{L}^2(M,\varphi)$, we have $(L_\xi L_\eta^*)(\alpha) = \xi \langle \eta, \alpha \rangle_A$. In particular, for $m_1, m_2 \in M$, we note that $m_1 e_A m_2$ is a "rank one operator" since $m_1 e_A m_2 = L_{m_1} L_{m_2^*}^*$. We denote by $\mathscr{K}(\langle M, e_A \rangle)$ the norm closure into $\langle M, e_A \rangle$ of

$$\mathrm{span}\left\{L_\xi L_\eta^* : \xi, \eta \in \mathscr{L}^2(M,\varphi)\right\}.$$

It is a two-sided ideal of $\langle M, e_A \rangle$.

For every $\xi \in \mathscr{L}^2(M,\varphi)$, we have $L_\xi e_A \in \langle M, e_A \rangle$. Since

$$L_\xi L_\eta^* = (L_\xi e_A)(L_\eta e_A)^*$$

we see that $\mathscr{K}(\langle M, e_A \rangle)$ is the norm closed two-sided ideal generated by e_A in $\langle M, e_A \rangle$. The projection e_A being finite (because $\mathrm{Tr}_\mu(e_A) = 1$), we have

$$\mathscr{K}(\langle M, e_A \rangle) \subset \mathscr{I}(\langle M, e_A \rangle).$$

The subtle difference between $\mathscr{K}(\langle M, e_A\rangle)$ and $\mathscr{I}(\langle M, e_A\rangle)$ is studied in [25, Sect. 1.3.2]. We recall in particular that for every $T \in \mathscr{I}(\langle M, e_A\rangle)$ and every $\varepsilon > 0$, there is a projection $p \in A$ such that $\tau_\mu(1-p) \le \varepsilon$ and $TJpJ \in \mathscr{K}(\langle M, e_A\rangle)$ (see [25, Proposition 1.3.3 (3)]).[3]

1.3.3 The Relative Haagerup Property

Let Φ be a unital completely positive map from M into M such that $E_A \circ \Phi = E_A$. Then for $m \in M$, we have

$$\|\Phi(m)\|_2^2 = \varphi(\Phi(m)^*\Phi(m)) \le \varphi(\Phi(m^*m)) = \varphi(m^*m) = \|m\|_2^2.$$

It follows that Φ extends to a contraction $\hat{\Phi}$ of $L^2(M,\varphi)$. Whenever Φ is A-bimodular, $\hat{\Phi}$ commutes with the right action of A (due to (1.5)) and so belongs to $\langle M, e_A\rangle$. It also commutes with the left action of A and so belongs to $A' \cap \langle M, e_A\rangle$.

Definition 5. We say that M has the *Haagerup property (or property* (H)) *relative to A and E_A* if there exists a net (Φ_i) of unital A-bimodular completely positive maps from M to M such that

1. $E_A \circ \Phi_i = E_A$ for all i ;
2. $\widehat{\Phi_i} \in \mathscr{K}(\langle M, e_A\rangle)$ for all i ;
3. $\lim_i \|\Phi_i(x) - x\|_2 = 0$ for every $x \in M$.

This notion is due to Boca [9]. In [25], Popa uses a slightly different formulation.

Lemma 5. *In the previous definition, we may equivalently assume that, for every i, $\widehat{\Phi_i} \in \mathscr{I}(\langle M, e_A\rangle)$*

Proof. This fact is explained in [25]. Let Φ be a unital A-bimodular completely positive map from M to M such that $E_A \circ \Phi = E_A$ and $\hat{\Phi} \in \mathscr{I}(\langle M, e_A\rangle)$. As already said, by [25, Proposition 1.3.3 (3)], for every $\varepsilon > 0$, there is a projection p in A with $\tau_\mu(1-p) < \varepsilon$ and $\hat{\Phi}JpJ \in \mathscr{K}(\langle M, e_A\rangle)$. Thus we have $p\hat{\Phi}JpJ \in \mathscr{K}(\langle M, e_A\rangle)$. Moreover, this operator is associated with the completely positive map $\Phi_p : m \in M \mapsto \Phi(pmp)$, since

$$(p\hat{\Phi}JpJ)(\hat{m}) = p\hat{\Phi}(\widehat{mp}) = p\hat{\Phi}(\widehat{mp}) = p\widehat{\Phi(mp)} = \widehat{\Phi(pmp)}.$$

Then, $\Phi' = \Phi_p + (1-p)E_A$ is unital, satisfies $E_A \circ \Phi' = E_A$ and still provides an element of $\mathscr{K}(\langle M, e_A\rangle)$. This modification allows to prove that if Definition 5 holds with $\mathscr{K}(\langle M, e_A\rangle)$ replaced by $\mathscr{I}(\langle M, e_A\rangle)$, then the relative Haagerup property is satisfied (see [25, Proposition 2.2 (1)]). □

[3] In [25], $\mathscr{K}(\langle M, e_A\rangle)$ is denoted $\mathscr{I}_0(\langle M, e_A\rangle)$.

1.3.4 Back to $L^2(G,\nu)_A$

We apply the facts just reminded to $M = L(G,\mu)$, which is on standard form on $L^2(G,\nu) = L^2(M,\varphi)$. This Hilbert space is viewed as a right A-module: for $\xi \in L^2(G,\mu)$ and $f \in A$, the action is given by $\xi f \circ s$.

It is easily seen that $\mathscr{L}^2(M,\varphi)$ is the space of $\xi \in L^2(G,\nu)$ such that $x \mapsto \sum_{s(\gamma)=x} |\xi(\gamma)|^2$ is in $L^\infty(X)$. Moreover, for $\xi,\eta \in \mathscr{L}^2(M,\varphi)$ we have

$$\langle \xi,\eta\rangle_A = \sum_{s(\gamma)=x} \overline{\xi(\gamma)}\eta(\gamma).$$

For simplicity of notation, we shall often identify $f \in I(G) \subset L^2(G,\nu)$ with the operator $L(f)$.[4] For instance, for $f,g \in I(G)$, the operator $L(f)\circ L(g)$ is also written $f * g$, and for $T \in \mathscr{B}(L^2(G,\mu))$, we write $T \circ f$ instead of $T \circ L(f)$.

Let $S \subset G$ be a bisection. Its characteristic function $\mathbf{1}_S$ is an element of $I(G)$ and a partial isometry in M since

$$\mathbf{1}_S^* * \mathbf{1}_S = \mathbf{1}_{s(S)}, \quad \text{and} \quad \mathbf{1}_S * \mathbf{1}_S^* = \mathbf{1}_{r(S)}.$$

Let $G = \sqcup S_n$ be a countable partition of G into Borel bisections. Another straightforward computation shows that $(\mathbf{1}_{S_n})_n$ is an orthonormal basis of the right A-module $L^2(M,\varphi)$.

By Lemma 4, for $x \in \langle M, e_A\rangle_+$ we have

$$\mathrm{Tr}_\mu(x) = \sum_n \langle \mathbf{1}_{S_n}, x\mathbf{1}_{S_n}\rangle_{L^2(M)}.$$

In particular, whenever x is the multiplication operator $\mathfrak{m}(f)$ by some bounded non-negative Borel function f, we get

$$\mathrm{Tr}_\mu(\mathfrak{m}(f)) = \int_G f \,\mathrm{d}\nu. \tag{1.9}$$

1.4 From Completely Positive Maps to Positive Definite Functions

Recall that if G is a countable group, and $\Phi : L(G) \to L(G)$ is a completely positive map, then $t \mapsto F_\Phi(t) = \tau(\Phi(u_t)u_t^*)$ is a positive definite function on G, where τ is the canonical trace on $L(G)$ and u_t, $t \in G$, are the canonical

[4]The reader should not confuse $L(f) : L^2(G,\nu) \to L^2(G,\nu)$ with its restriction $L_f : L^2(A,\tau_\mu) \to L^2(G,\nu)$.

unitaries in $L(G)$. We want to extend this classical fact to the groupoid case. This was achieved by Jolissaint [18] for countable probability measure preserving equivalence relations.

Let (G, μ) be a countable measured groupoid and $M = L(G, \mu)$. Let $\Phi : M \to M$ be a normal A-bimodular unital completely positive map. Let $G = \sqcup S_n$ be a partition into Borel bisections. We define $F_\Phi : G \to \mathbb{C}$ by

$$F_\Phi(\gamma) = E_A(\Phi(\mathbf{1}_{S_n}) \circ \mathbf{1}_{S_n}^*) \circ r(\gamma), \tag{1.10}$$

where S_n is the bisection which contains γ.

That F_Φ does not depend (up to null sets) on the choice of the partition is a consequence of the following lemma.

Lemma 6. *Let S_1 and S_2 be two Borel bisections. Then*

$$E_A(\Phi(\mathbf{1}_{S_1}) \circ \mathbf{1}_{S_1}^*) = E_A(\Phi(\mathbf{1}_{S_2}) \circ \mathbf{1}_{S_2}^*)$$

almost everywhere on $r(S_1 \cap S_2)$.

Proof. Denote by e the characteristic function of $r(S_1 \cap S_2)$. Then $e * \mathbf{1}_{S_1} = e * \mathbf{1}_{S_2} = \mathbf{1}_{S_1 \cap S_2}$. Thus we have

$$\begin{aligned} eE_A(\Phi(\mathbf{1}_{S_1}) \circ \mathbf{1}_{S_1}^*)e &= E_A\big(\Phi(e * \mathbf{1}_{S_1}) \circ (\mathbf{1}_{S_1}^* * e)\big) \\ &= E_A\big(\Phi(e * \mathbf{1}_{S_2}) \circ (\mathbf{1}_{S_2}^* * e)\big) = eE_A(\Phi(\mathbf{1}_{S_2}) \circ \mathbf{1}_{S_2}^*)e. \end{aligned}$$

□

We now want to show that F_Φ is a positive definite function in the following sense. We shall need some preliminary facts.

Definition 6. A Borel function $F : G \to \mathbb{C}$ is said to be *positive definite* if there exists a μ-null subset N of $X = G^{(0)}$ such that for every $x \notin N$, and every $\gamma_1, \dots, \gamma_k \in G^x$, the $k \times k$ matrix $[F(\gamma_i^{-1}\gamma_j)]$ is non-negative.

Definition 7. We say that a *Borel bisection S is admissible* if there exists a constant $c > 0$ such that $1/c \le \delta(\gamma) \le c$ almost everywhere on S.

In other terms, $\mathbf{1}_S \in I_\infty(G)$ and so the convolution to the right by $\mathbf{1}_S$ defines a bounded operator $R(\mathbf{1}_S)$ on $L^2(M, \varphi)$, by (1.4).

Lemma 7. *Let S be a Borel bisection and let $T \in M$. We have $\widehat{\mathbf{1}_S \circ T} = \mathbf{1}_S * \hat{T}$. Moreover, if S is admissible, we have $\widehat{T \circ \mathbf{1}_S} = \hat{T} * \mathbf{1}_S$.*

Proof. First, we observe that $\widehat{\mathbf{1}_S \circ T} = \mathbf{1}_S \circ T(\mathbf{1}_X) = \mathbf{1}_S * \hat{T}$.

On the other hand, given $f \in I(G)$, we have $L(f)(\widehat{\mathbf{1}_S}) = f * 1_S$. So, if (f_n) is a sequence in $I(G)$ such that $\lim_n L(f_n) = T$ in the strong operator topology, we have

$$\widehat{T \circ \mathbf{1}_S} = T(\widehat{\mathbf{1}_S}) = \lim_n L(f_n)(\widehat{\mathbf{1}_S}) = \lim_n f_n * \mathbf{1}_S$$

in $L^2(G,\nu)$. But, when S is admissible, the convolution to the right by $\mathbf{1}_S$ is the bounded operator $R(\mathbf{1}_S)$. Noticing that $\lim_n \left\| f_n - \hat{T} \right\|_2 = 0$, it follows that

$$\widehat{T \circ \mathbf{1}_S} = \lim_n f_n * \mathbf{1}_S = \hat{T} * \mathbf{1}_S. \qquad \square$$

Lemma 8. *Let $T \in M$, and let S be an admissible bisection. Then*

$$\mathbf{1}_S(\gamma) E_A(T \circ \mathbf{1}_S)(s(\gamma)) = \mathbf{1}_S(\gamma) E_A(\mathbf{1}_S \circ T)(r(\gamma))$$

for almost every γ.

Proof. We have

$$(\widehat{T \circ \mathbf{1}_S})(x) = \sum_{\gamma_1\gamma_2 = x} \hat{T}(\gamma_1)\mathbf{1}_S(\gamma_2) = \hat{T}(\gamma_2^{-1})$$

whenever $x \in s(S)$, where γ_2 is the unique element of S with $s(\gamma_2) = x$. Otherwise $(\widehat{T \circ \mathbf{1}_S})(x) = 0$.

On the other hand,

$$(\widehat{\mathbf{1}_S \circ T})(x) = \hat{T}(\gamma_1^{-1})$$

whenever $x \in r(S)$, where γ_1 is the unique element of S with $r(\gamma_1) = x$. Otherwise $(\widehat{\mathbf{1}_S \circ T})(x) = 0$. Our statement follows immediately. $\square$

Lemma 9. *F_Φ is a positive definite function.*

Proof. We assume that F_Φ is defined by Eq. (1.10) through a partition under admissible bisections. We set $S_{ij} = S_i^{-1}S_j = \{\gamma^{-1}\gamma' : \gamma \in S_i, \gamma' \in S_j\}$. Note that $\mathbf{1}_{S_i}^* * \mathbf{1}_{S_j} = \mathbf{1}_{S_{ij}}$. Moreover, the S_{ij} are admissible bisections. We set

$$Z_{ijm} = \left\{x \in r(S_{ij} \cap S_m) : E_A(\Phi(\mathbf{1}_{S_{ij}}) \circ \mathbf{1}_{S_{ij}}^*)(x) \neq E_A(\Phi(\mathbf{1}_{S_m}) \circ \mathbf{1}_{S_m}^*)(x)\right\}$$

and $Z = \cup_{i,j,m} Z_{ijm}$. It is a null set by Lemma 6.

By Lemma 8, for every i there is a null set $E_i \subset r(S_i)$ such that for $\gamma \in S_i$ with $r(\gamma) \notin E_i$ and for every j, we have

$$E_A(\Phi(\mathbf{1}_{S_{ij}}) \circ \mathbf{1}_{S_j}^* \circ \mathbf{1}_{S_i})(s(\gamma)) = E_A(\mathbf{1}_{S_i} \circ \Phi(\mathbf{1}_{S_{ij}}) \circ \mathbf{1}_{S_j}^*)(r(\gamma))$$

We set $E = \cup_i E_i$. Let Y be the saturation of $Z \cup E$. It is a null set, since μ is quasi-invariant.

Let $x \notin Y$, and $\gamma_1, \dots, \gamma_k \in G^x$. Assume that $\gamma_i^{-1}\gamma_j \in S_{n_i}^{-1}S_{n_j} \cap S_m$. We have $r(\gamma_i^{-1}\gamma_j) = s(\gamma_i) \notin Y$ since $r(\gamma_i) = x \notin Y$. Therefore,

$$F_\Phi(\gamma_i^{-1}\gamma_j) = E_A(\Phi(\mathbf{1}_{S_{n_i n_j}}) \circ \mathbf{1}_{S_{n_j}}^* \circ \mathbf{1}_{S_{n_i}})(s(\gamma_i)).$$

But $\gamma_i \in S_{n_i}$ with $r(\gamma_i) = x \notin Y$, so

$$E_A(\Phi(\mathbf{1}_{S_{n_i n_j}}) \circ \mathbf{1}^*_{S_{n_j}} \circ \mathbf{1}_{S_{n_i}})(s(\gamma_i)) = E_A(\mathbf{1}_{S_{n_i}} \circ \Phi(\mathbf{1}_{S_{n_i n_j}}) \circ \mathbf{1}^*_{S_{n_j}})(r(\gamma_i)).$$

Given $\lambda_1, \dots, \lambda_k \in \mathbb{C}$, we have

$$\sum_{i,j=1}^{k} \lambda_i \overline{\lambda_j} F_\Phi(\gamma_i^{-1}\gamma_j) = E_A\Big(\sum_{i=1}^{k}(\lambda_i \mathbf{1}_{S_{n_i}}) \circ \Phi(\mathbf{1}^*_{S_{n_i}} \circ \mathbf{1}_{S_{n_j}}) \circ \sum_{j=1}^{k}(\lambda_j \mathbf{1}_{S_{n_j}})^*\Big)(x) \geq 0.$$

□

Obviously, if Φ is unital, F_Φ takes value 1 almost everywhere on X.

Proposition 1. *We now assume that Φ is unital, with $E_A \circ \Phi = E_A$ and $\hat{\Phi} \in \mathcal{K}(\langle M, e_A\rangle)$. Then, for every $\varepsilon > 0$, we have*

$$\nu(\{|F_\Phi| > \varepsilon\}) < +\infty.$$

Proof. Let (S_n) be a partition of G into Borel bisections. Given $\varepsilon > 0$ we choose $\xi_1, \dots, \xi_k, \eta_1, \dots, \eta_k \in \mathcal{L}^2(M, \varphi)$ such that

$$\Big\| \hat{\Phi} - \sum_{i=1}^{k} L_{\xi_i} L^*_{\eta_i} \Big\| \leq \varepsilon/2.$$

We view $\hat{\Phi} - \sum_{i=1}^{k} L_{\xi_i} L^*_{\eta_i}$ as an element of $\mathcal{B}(\mathcal{L}^2(M, \varphi)_A)$ and we apply it to $\mathbf{1}_{S_n} \in \mathcal{L}^2(M, \varphi)$. Then

$$\Big\| \Phi(\mathbf{1}_{S_n}) - \sum_{i=1}^{k} \xi_i \langle \eta_i, \mathbf{1}_{S_n}\rangle_A \Big\|_{\mathcal{L}^2(M)} \leq \varepsilon/2 \|\mathbf{1}_{S_n}\|_{\mathcal{L}^2(M)} \leq \varepsilon/2.$$

Using the Cauchy-Schwarz inequality $\langle \xi, \eta\rangle_A^* \langle \xi, \eta\rangle_A \leq \|\xi\|^2_{\mathcal{L}^2(M)} \langle \eta, \eta\rangle_A$, we get

$$\Big\| \Big\langle \mathbf{1}_{S_n}, \Phi(\mathbf{1}_{S_n}) - \sum_{i=1}^{k} \xi_i \langle \eta_i, \mathbf{1}_{S_n}\rangle_A \Big\rangle_A \Big\| \leq \Big\| \Phi(\mathbf{1}_{S_n}) - \sum_{i=1}^{k} \xi_i \langle \eta_i, \mathbf{1}_{S_n}\rangle_A \Big\|_{\mathcal{L}^2(M)} \leq \varepsilon/2.$$

We have, for almost every $\gamma \in S_n$ and $x = s(\gamma)$,

$$\begin{aligned} |F_\Phi(\gamma)| &= \big|E_A(\Phi(\mathbf{1}_{S_n}) \circ \mathbf{1}^*_{S_n})(r(\gamma))\big| = \big|E_A(\mathbf{1}^*_{S_n} \circ \Phi(\mathbf{1}_{S_n}))(x)\big| = \big|\langle \mathbf{1}_{S_n}, \Phi(\mathbf{1}_{S_n})\rangle_A(x)\big| \\ &\leq \Big| \Big\langle \mathbf{1}_{S_n}, \Phi(\mathbf{1}_{S_n}) - \sum_{i=1}^{k} \xi_i \langle \eta_i, \mathbf{1}_{S_n}\rangle_A \Big\rangle_A (x) \Big| + \sum_{i=1}^{k} \Big| \big\langle \mathbf{1}_{S_n}, \xi_i \langle \eta_i, \mathbf{1}_{S_n}\rangle_A \big\rangle_A (x) \Big|. \end{aligned}$$

The first term is $\leq \varepsilon/2$ for almost every $x \in s(S_n)$. As for the second term, we have, almost everywhere,

$$\left|\langle \mathbf{1}_{S_n}, \xi_i\rangle_A(x)\langle \eta_i, \mathbf{1}_{S_n}\rangle_A(x)\right| \leq \|\xi_i\|_{\mathscr{L}^2(M)}\left|\langle \eta_i, \mathbf{1}_{S_n}\rangle_A(x)\right|.$$

Hence, we get

$$|F_\Phi(\gamma)| \leq \varepsilon/2 + \sum_{i=1}^k \|\xi_i\|_{\mathscr{L}^2(M)}\left|\langle \eta_i, \mathbf{1}_{S_n}\rangle_A(s(\gamma))\right|$$

for almost every $\gamma \in S_n$.

We want to estimate

$$\nu(\{|F_\Phi| > \varepsilon\}) = \sum_n \nu(\{\gamma \in S_n : |F_\Phi(\gamma)| > \varepsilon\}).$$

For almost every $\gamma \in S_n$ such that $|F_\Phi(\gamma)| > \varepsilon$, we see that

$$\sum_{i=1}^k \|\xi_i\|_{\mathscr{L}^2(M)}\left|\langle \eta_i, \mathbf{1}_{S_n}\rangle_A(s(\gamma))\right| > \varepsilon/2.$$

Therefore

$$\begin{aligned}\nu(\{|F_\Phi| > \varepsilon\}) &\leq \sum_n \nu\Big(\Big\{\gamma \in S_n : \sum_{i=1}^k \|\xi_i\|_{\mathscr{L}^2(M)}\left|\langle \eta_i, \mathbf{1}_{S_n}\rangle_A(s(\gamma))\right| > \varepsilon/2\Big\}\Big)\\ &\leq \sum_n \mu\Big(\Big\{x \in s(S_n) : \sum_{i=1}^k \|\xi_i\|_{\mathscr{L}^2(M)}\left|\langle \eta_i, \mathbf{1}_{S_n}\rangle_A(x)\right| > \varepsilon/2\Big\}\Big).\end{aligned}$$

Now,

$$\sum_{i=1}^k \|\xi_i\|_{\mathscr{L}^2(M)}\left|\langle \eta_i, \mathbf{1}_{S_n}\rangle_A(x)\right| \leq \Big(\sum_{i=1}^k \|\xi_i\|^2_{\mathscr{L}^2(M)}\Big)^{1/2}\Big(\sum_{i=1}^k \left|\langle \eta_i, \mathbf{1}_{S_n}\rangle_A(x)\right|^2\Big)^{1/2}.$$

We set $c = \sum_{i=1}^k \|\xi_i\|^2_{\mathscr{L}^2(M)}$ and $f_n(x) = \sum_{i=1}^k \left|\langle \eta_i, \mathbf{1}_{S_n}\rangle_A(x)\right|^2$. We have

$$\begin{aligned}\sum_n f_n(x) &= \sum_n\sum_{i=1}^k \left|\langle \eta_i, \mathbf{1}_{S_n}\rangle_A(x)\right|^2 = \sum_{i=1}^k\sum_n \left|\langle \eta_i, \mathbf{1}_{S_n}\rangle_A(x)\right|^2\\ &= \sum_{i=1}^k \langle \eta_i, \eta_i\rangle_A(x) \leq \sum_{i=1}^k \|\eta_i\|^2_{\mathscr{L}^2(M)},\end{aligned}$$

since, by Lemma 3 (or directly here),

$$\langle \eta_i, \eta_i \rangle_A = \sum_k \langle \eta_i, \mathbf{1}_{S_k} \rangle_A \langle \mathbf{1}_{S_k}, \eta_i \rangle_A = \sum_k \left| \langle \eta_i, \mathbf{1}_{S_k} \rangle_A \right|^2.$$

We set $d = \sum_{i=1}^{k} \|\eta_i\|^2_{\mathscr{L}^2(M)}$.
We have

$$\nu(\{|F_\Phi| > \varepsilon\}) \le \sum_n \mu(\{x \in s(S_n) : c f_n(x) > (\varepsilon/2)^2\}).$$

We set $\alpha = c^{-1}(\varepsilon/2)^2$. Denote by $i(x)$ the number of indices n such that $f_n(x) > \alpha$. Then $i(x) \le N$, where N is the integer part of d/α. We denote by $\mathscr{P} = \{P_n\}$ the set of subsets of $\mathbb{N}$ whose cardinal is $\le N$. Then there is a partition $X = \sqcup_m B_m$ into Borel subsets such that

$$\forall x \in B_m, \quad P_m = \{n \in \mathbb{N} : f_n(x) > \alpha\}.$$

We have

$$\begin{aligned}
\nu(\{|F_\Phi| > \varepsilon\}) &\le \sum_{n,m} \mu(\{x \in B_m \cap s(S_n) : f_n(x) > \alpha\}) \\
&\le \sum_m \Big(\sum_n \mu(\{x \in B_m \cap s(S_n) : f_n(x) > \alpha\}) \Big) \\
&\le \sum_m \sum_{n \in P_m} \mu(\{x \in B_m \cap s(S_n) : f_n(x) > \alpha\}) \le \sum_m N \mu(B_m) = N
\end{aligned}$$

□

1.5 From Positive Definite Functions to Completely Positive Maps

Again, we want to extend a well known result in the group case, namely that, given a positive definite function F on a countable group G, there is a normal completely positive map $\Phi : L(G) \to L(G)$, well defined by the formula $\Phi(u_t) = F(t)u_t$ for every $t \in G$.

We need some preliminaries. For the notion of representation used below, see for instance [6, Sect. 3.1].

Lemma 10. *Let F be a positive definite function on (G, μ). There exists a representation π of G on a measurable field $\mathscr{K} = \{\mathscr{K}(x)\}_{x \in X}$ of Hilbert spaces, and a measurable section $\xi : x \mapsto \xi(x) \in \mathscr{K}(x)$ such that*

$$F(\gamma) = \langle \xi \circ r(\gamma), \pi(\gamma) \xi \circ s(\gamma) \rangle$$

almost everywhere, that is F is the coefficient of the representation π, associated with ξ.

Proof. This classical fact may be found in [28]. The proof is straightforward, and similar to the classical GNS construction in the case of groups. Let $V(x)$ the space of finitely supported complex-valued functions on G^x, endowed with the semi-definite positive hermitian form

$$\langle f, g\rangle_x = \sum_{\gamma_1,\gamma_2\in G^x} \overline{f(\gamma_1)} g(\gamma_2) F(\gamma_1^{-1}\gamma_2).$$

We denote by $\mathcal{H}(x)$ the Hilbert space obtained by separation and completion of $V(x)$, and $\pi(\gamma) : \mathcal{H}(s(\gamma)) \to \mathcal{H}(r(\gamma))$ is defined by $(\pi(\gamma)f)(\gamma_1) = f(\gamma^{-1}\gamma_1)$. The Borel structure on the field $\{\mathcal{H}(x)\}_{x\in X}$ is provided by the Borel functions on G whose restriction to the fibres G^x are finitely supported. Finally, ξ is the characteristic function of X, viewed as a Borel section. □

Now we assume that $F(x) = 1$ for almost every $x \in X$, and thus ξ is a unit section. We consider the measurable field $\{\ell^2(G_x) \otimes \mathcal{H}(x)\}_{x\in X}$. Note that

$$\ell^2(G_x) \otimes \mathcal{H}(x) = \ell^2(G_x, \mathcal{H}(x)).$$

Let $f \in \ell^2(G_x)$. We define $S_x(f) \in \ell^2(G_x, \mathcal{H}(x))$ by

$$S_x(f)(\gamma) = f(\gamma)\pi(\gamma)^*\xi \circ r(\gamma)$$

for $\gamma \in G_x$. Then

$$\sum_{s(\gamma)=x} \|S_x(f)(\gamma)\|^2_{\mathcal{H}(x)} = \|f\|_{\ell^2(G_x)}.$$

The field $(S_x)_{x\in X}$ of operators defines an isometry

$$S : L^2(G, \nu) \to \int_X^{\oplus} \ell^2(G_x, \mathcal{H}(x))\,\mathrm{d}\mu(x),$$

by

$$S(f)(\gamma) = f(\gamma)\pi(\gamma)^*\xi \circ r(\gamma).$$

Note that $\int_X^{\oplus} \ell^2(G_x, \mathcal{H}(x))\,\mathrm{d}\mu(x)$ is a right A-module, by

$$(\eta a)_x = \eta_x a(x) : \gamma \in G_x \mapsto \eta(\gamma) a \circ s(\gamma).$$

Of course, S commutes with the right actions of A. We also observe that, as a right A-module, $\mathscr{L}^2(M,\varphi) \otimes_A \int_X^\oplus \mathscr{K}(x)\,\mathrm{d}\mu(x)$ and $\int_X^\oplus \ell^2(G_x, \mathscr{K}(x))\,\mathrm{d}\mu(x)$ are canonically isomorphic under the map

$$\zeta \otimes \eta \mapsto \zeta\eta \circ s, \quad \forall \zeta \in \mathscr{L}^2(M,\varphi), \forall \eta \in \int_X^\oplus \mathscr{K}(x)\,\mathrm{d}\mu(x),$$

where $(\zeta\eta \circ s)_x$ is the function $\gamma \in G_x \mapsto \zeta(\gamma)\eta \circ s(\gamma)$ in $\ell^2(G_x, \mathscr{K}(x))$. It follows that M acts on $\int_X^\oplus \ell^2(G_x, \mathscr{K}(x))\,\mathrm{d}\mu(x)$ by $m \mapsto m \otimes \mathrm{Id}$. In particular, for $f \in I(G)$, we see that $L(f) \otimes \mathrm{Id}$, viewed as an operator on $\int_X^\oplus \ell^2(G_x, \mathscr{K}(x))\,\mathrm{d}\mu(x)$, is acting as

$$((L(f) \otimes \mathrm{Id})\eta)(\gamma) = \sum_{\gamma_1\gamma_2=\gamma} f(\gamma_1)\eta(\gamma_2) \in \mathscr{K}(s(\gamma)).$$

Lemma 11. *For $f \in I(G)$, we have $S^*\big(L(f) \otimes \mathrm{Id}\big)S = L(Ff)$.*

Proof. A straightforward computation shows that for $\eta \in \int_X^\oplus \ell^2(G_x, \mathscr{K}(x))\,\mathrm{d}\mu(x)$, we have

$$(S^*\eta)(\gamma) = \langle \pi(\gamma)^*\xi \circ r(\gamma), \eta(\gamma)\rangle_{\mathscr{K}(s(\gamma))}.$$

Moreover, given $h \in L^2(G,\nu)$, we have

$$\big((L(f)\otimes\mathrm{Id})S\,h\big)(\gamma) = \sum_{\gamma_1\gamma_2=\gamma} f(\gamma_1)(S\,h)(\gamma_2) = \sum_{\gamma_1\gamma_2=\gamma} f(\gamma_1)h(\gamma_2)\pi(\gamma_2)^*\xi\circ r(\gamma_2).$$

Hence,

$$\begin{aligned}
\big(S^*(L(f) \otimes \mathrm{Id})S\,h\big)(\gamma) &= \Big\langle \pi(\gamma)^*\xi \circ r(\gamma), \sum_{\gamma_1\gamma_2=\gamma} f(\gamma_1)h(\gamma_2)\pi(\gamma_2)^*\xi \circ r(\gamma_2)\Big\rangle \\
&= \Big\langle \xi \circ r(\gamma), \sum_{\gamma_1\gamma_2=\gamma} f(\gamma_1)h(\gamma_2)\pi(\gamma_1)\xi \circ r(\gamma_2)\Big\rangle \\
&= \sum_{\gamma_1\gamma_2=\gamma} f(\gamma_1)h(\gamma_2)\langle \xi \circ r(\gamma_1), \pi(\gamma_1)\xi \circ r(\gamma_2)\rangle \\
&= \sum_{\gamma_1\gamma_2=\gamma} f(\gamma_1)F(\gamma_1)h(\gamma_2) = (L(Ff)h)(\gamma).
\end{aligned}$$

□

Proposition 2. *Let $F : G \to \mathbb{C}$ be a Borel positive definite function on G such that $F_{|X} = 1$. Then there exists a unique normal completely positive map Φ from M into*

M such that $\Phi(L(f)) = L(Ff)$ for every $f \in I(G)$. Morever, Φ is A-bimodular, unital and $E_A \circ \Phi = E_A$.

Proof. The uniqueness is a consequence of the normality of Φ, combined with the density of $L(I(G))$ into M. With the notation of the previous lemma, for $m \in M$ we put $\Phi(m) = S^*\big(m \otimes \mathrm{Id}\big)S$. Obviously, Φ satisfies the required conditions. □

Remark 1. We keep the notation of the previous proposition. A straightforward computation shows that F is the positive definite function F_Φ constructed from Φ.

Proposition 3. *Let F be a Borel positive definite function on G such that $F_{|X} = 1$. We assume that for every $\varepsilon > 0$, we have $\nu(\{|F| > \varepsilon\}) < +\infty$. Let Φ be the completely positive map defined by F. Then $\hat{\Phi}$ belongs to the norm closed ideal $\mathscr{I}(\langle M, e_A\rangle)$ generated by the finite projections of $\langle M, e_A\rangle$.*

Proof. We observe that $T = \hat{\Phi}$ is the multiplication operator $\mathfrak{m}(F)$ by F. We need to show that for every $t > 0$, the spectral projection $e_t(|T|)$ of $|T|$ relative to $]t, +\infty[$ is finite. This projection is the multiplication operator by $f_t = \mathbf{1}_{]t,+\infty[} \circ |F|$. By (1.9), we have

$$\mathrm{Tr}_\mu(\mathfrak{m}(f_t)) = \nu(f_t) = \nu(\{|F| > t\}) < +\infty. \qquad \square$$

1.6 Characterizations of the Relative Haagerup Property

We keep the same notation as in the previous section.

Theorem 1. *The following conditions are equivalent:*

1. *M has the Haagerup property relative to A and E_A.*
2. *There exists a sequence (F_n) of positive definite functions on G such that*
   - *(a)* $(F_n)_{|X} = 1$ *almost everywhere ;*
   - *(b) For every* $\varepsilon > 0$, $\nu(\{|F_n| > \varepsilon\}) < +\infty$ *;*
   - *(c)* $\lim_n F_n = 1$ *almost everywhere.*

Proof. *1* ⇒ 2. Let (Φ_n) a sequence of unital completely positive maps $M \to M$ satisfying conditions *1, 2, 3* of Definition 5. We set $F_n = F_{\Phi_n}$. By Proposition 1 we know that condition *b* of 2 above is satisfied. It remains to check *c*. For $m \in M$, we have

$$\|\Phi_n(m) - m\|_2^2 = \int_X E_A((\Phi_n(m) - m)^*(\Phi_n(m) - m))(x)\, \mathrm{d}\mu(x).$$

Let $G = \sqcup_n S_n$ be a partition of G by Borel bisections. There is a null subset Y of X such that, for every k and for $\gamma \in S_k \cap r^{-1}(X \setminus Y)$ we have

$$F_n(\gamma) - 1 = E_A(\mathbf{1}_{S_k}^* \circ \Phi_n(\mathbf{1}_{S_k}))(s(\gamma)) - E_A(\mathbf{1}_{S_k}^* \circ \mathbf{1}_{S_k})(s(\gamma)).$$

Thus

$$|F_n(\gamma) - 1|^2 = \left|E_A(\mathbf{1}_{S_k}^* \circ (\Phi_n(\mathbf{1}_{S_k}) - \mathbf{1}_{S_k}))(s(\gamma))\right|^2$$
$$\leq E_A((\Phi_n(\mathbf{1}_{S_k}) - \mathbf{1}_{S_k})^*(\Phi_n(\mathbf{1}_{S_k}) - \mathbf{1}_{S_k}))(s(\gamma)).$$

It follows that

$$\int_G |F_n - 1|^2 \mathbf{1}_{S_k} \,\mathrm{d}\nu \leq \int_{s(S_k)} E_A((\Phi_n(\mathbf{1}_{S_k}) - \mathbf{1}_{S_k})^*(\Phi_n(\mathbf{1}_{S_k}) - \mathbf{1}_{S_k}))(x)\,\mathrm{d}\mu(x)$$
$$\leq \|\Phi_n(\mathbf{1}_{S_k}) - \mathbf{1}_{S_k}\|_2^2 \to 0.$$

So there is a subsequence of $(|F_n(\gamma) - 1|)_n$ which goes to 0 almost everywhere on S_k. Using the Cantor diagonal process, we get the existence of a subsequence $(F_{n_k})_k$ of $(F_n)_n$ such that $\lim_k F_{n_k} = 1$ almost everywhere, which is enough for our purpose.

2 $\Rightarrow$ *1*. Assume the existence of a sequence $(F_n)_n$ of positive definite functions on G, satisfying the three conditions of *2*. Let Φ_n be the completely positive map defined by F_n. Let us show that for every $m \in M$, we have

$$\lim_n \|\Phi_n(m) - m\|_2 = 0.$$

We first consider the case $m = L(f)$ with $f \in I(G)$. Then we have

$$\|\Phi_n(L(f)) - L(f)\|_2 = \|L((F_n - 1)f)\|_2 = \|(F_n - 1)f\|_2 \to 0$$

by the Lebesgue dominated convergence theorem.

Let now $m \in M$. Then

$$\|\Phi_n(m) - m\|_2 \leq \|\Phi_n(m - L(f))\|_2 + \|\Phi_n(L(f)) - L(f)\|_2 + \|L(f) - m\|_2.$$

We conclude by a classical approximation argument, since

$$\|\Phi_n(m - L(f))\|_2 \leq \|L(f) - m\|_2.$$

Together with Propositions 2, 3 and Lemma 5, this proves *1*. □

This theorem justifies the following definition.

Definition 8. We say that a countable measured groupoid (G, μ) has the *Haagerup property* (or has *property* (H)) if there exists a sequence (F_n) of positive definite functions on G such that

1. $(F_n)_{|X} = 1$ almost everywhere ;
2. For every $\varepsilon > 0$, $\nu(\{|F_n| > \varepsilon\}) < +\infty$;
3. $\lim_n F_n = 1$ almost everywhere.

We observe that, by Theorem 1, this notion only involves the conditional expectation E_A and therefore only depends on the measure class of μ. This fact does not seem to be obvious directly from the above Definition 8.

Of course, we get back the usual definition for a countable group. The other equivalent definitions for groups also extend to groupoids as we shall see now.

Definition 9. A *real conditionally negative definite function* on G is a Borel function $\psi : G \to \mathbb{R}$ such that

1. $\psi(x) = 0$ for every $x \in G^{(0)}$;
2. $\psi(\gamma) = \psi(\gamma^{-1})$ for every $\gamma \in G$;
3. For every $x \in G^{(0)}$, every $\gamma_1, \dots, \gamma_n \in G^x$ and every real numbers $\lambda_1, \dots, \lambda_n$ with $\sum_{i=1}^n \lambda_i = 0$, then

$$\sum_{i,j=1}^{n} \lambda_i \lambda_j \psi(\gamma_i^{-1}\gamma_j) \le 0.$$

Such a function is non-negative.

Definition 10. Let (G, μ) be a countable measured groupoid. A *real conditionally negative definite function* on (G, μ) is a Borel function $\psi : G \to \mathbb{R}$ such that there exists a co-null subset U of $G^{(0)}$ with the property that the restriction of ψ to the inessential reduction $G_{|U}$ satisfies the conditions of the previous definition.

We say that ψ is *proper* if for every $c > 0$, we have $\nu(\{\psi \le c\}) < +\infty$.

Theorem 2. *The groupoid (G, μ) has the Haagerup property if and only if there exists a real conditionally negative definite function ψ on (G, μ) such that*

$$\forall c > 0, \quad \nu(\{\psi \le c\}) < +\infty.$$

Proof. We follow the steps of the proof given by Jolissaint [18] for equivalence relations and previously by Akemann-Walter [3] for groups. Let ψ be a proper conditionally negative definite function. We set $F_n = \exp(-\psi/n)$. Then (F_n) is a sequence of positive definite functions which goes to 1 pointwise. Moreover, we have $F_n(\gamma) > c$ if and only if $\psi(\gamma) < -n \ln c$. Therefore (G, μ) has the Haagerup property.

Conversely, let (F_n) be a sequence of positive definite functions on G satisfying conditions a, b, c of Theorem 1. We choose sequences (α_n) and (ε_n) of positive numbers such (α_n) is increasing with $\lim_n \alpha_n = +\infty$, (ε_n) is decreasing with $\lim_n \varepsilon_n = 0$, and such that $\sum_n \alpha_n (\varepsilon_n)^{1/2} < +\infty$.

Let $G = \sqcup S_n$ be any partition of G into Borel bisections. Taking if necessary a subsequence of (F_n), we may assume that for every n,

$$\sum_{1 \le k \le n} \int_G |1 - F_n|^2 \mathbf{1}_{S_k} \, \mathrm{d}\nu \le \varepsilon_n^2.$$

It follows that

$$\int_G \Re(1 - F_n)^2 \mathbf{1}_{\cup_{1 \leq k \leq n} S_k} \, d\nu \leq \int_G |1 - F_n|^2 \mathbf{1}_{\cup_{1 \leq k \leq n} S_k} \, d\nu$$
$$\leq \varepsilon_n^2.$$

We set $E_n = \{\gamma \in \cup_{1 \leq k \leq n} S_k : |\Re(1 - F_n(\gamma))| \geq (\varepsilon_n)^{1/2}\}$ and $E = \cap_l \cup_{n \geq l} E_n$. Since $\nu(E_n) \leq \varepsilon_n$ and $\sum_n \varepsilon_n < +\infty$, we see that $\nu(E) = 0$.

Let us set $\psi = \sum_n \alpha_n \Re(1 - F_n)$ on $G \setminus E$ and $\psi = 0$ on E. We claim that the series converges pointwise. Indeed, let $\gamma \in G \setminus E$. There exists m such that

$$\gamma \in (\cup_{1 \leq i \leq m} S_i) \cap (\cap_{n \geq m} E_n^c).$$

Thus, $|\Re(1 - F_n(\gamma))| \leq (\varepsilon_n)^{1/2}$ for $n \geq m$, which shows our claim.

It remains to show that ψ is proper. Let $c > 0$, and let $\gamma \in G \setminus E$ with $\psi(\gamma) \leq c$. Then we have $\Re(1 - F_n(\gamma)) \leq c/\alpha_n$ for every n and therefore $\Re F_n(\gamma) \geq 1 - c/\alpha_n$. Let n be large enough such that $1 - c/\alpha_n \geq 1/2$. Then, if $\psi(\gamma) \leq c$, we have

$$|F_n(\gamma)| \geq \Re F_n(\gamma) \geq 1 - c/\alpha_n \geq 1/2$$

and thus

$$\nu(\{\psi \leq c\}) \leq \nu(\{|F_n| \geq 1/2\}) < +\infty. \qquad \square$$

Definition 11. Let π be a representation of (G, μ) on a measurable field $\mathcal{K} = \{\mathcal{K}(x)\}_{x \in X}$ of Hilbert spaces. A *π-cocycle* is a Borel section b of the pull-back bundle $r : r^*\mathcal{K} = \{(\gamma, \xi) : \xi \in \mathcal{K}(r(\gamma))\} \to G$ for which there exists an inessential reduction $(G_{|U}, \mu_{|U})$ such that we have, for composable $\gamma_1, \gamma_2 \in G_{|U}$,

$$b(\gamma_1\gamma_2) = b(\gamma_1) + \pi(\gamma_1)b(\gamma_2).$$

We say that b is *proper* if for every $c > 0$, we have $\nu(\{\|b\| \leq c\}) < +\infty$, where $\|b(\gamma)\|$ denotes the Hilbert norm of $b(\gamma)$ in $\mathcal{K}(r(\gamma))$.

Let b be a π-cocycle. It is easily seen that $\gamma \mapsto \|b(\gamma)\|^2$ is conditionally negative definite. Moreover, every real conditionally negative definite is of this form (see [6, Proposition 5.21]).

Corollary 1. *The groupoid (G, μ) has the Haagerup property if and only if it admits a proper π-cocycle for some representation π.*

Example 2. Let $\Gamma \curvearrowright (X, \mu)$ be an action of a countable group Γ which leaves quasi-invariant the probability measure μ. If Γ has the property (H), then $(X \rtimes \Gamma, \mu)$ inherits this property. Indeed, if $\psi : \Gamma \to \mathbb{R}$ is a proper conditionally negative definite function, then $\tilde{\psi} : (x, s) \mapsto \psi(s)$ is a proper conditionally negative definite function on $(X \rtimes \Gamma, \mu)$. Conversely, when the action is free, preserves μ and is such that $(X \rtimes \Gamma, \mu)$ has property (H), then Γ has property (H) [18, Proposition 3.3]. However, free non-singular actions of groups not having property (H) can generate semi-direct product groupoids with this property. Such actions can even be amenable (see for instance [8, Examples 5.2.2]).

Interesting examples are provided by treeable groupoids, as we shall see now. For instance, the free product of the type II_1 hyperfinite equivalence relation by itself, being treeable [5, Proposition 2.4], has the Haagerup property. Note also that property (H) passes to subgroupoids.

1.7 Treeable Countable Measured Groupoids Have Property (H)

The notion of treeable countable measured equivalence relation has been introduced by Adams in [1]. Its obvious extension to the case of countable measured groupoids is exposed in [6]. We recall here the main definitions. Let Q be a Borel subset of a countable Borel groupoid G. We set $Q^0 = X$ and for $n \geq 1$, we set

$$Q^n = \{\gamma \in G : \exists \gamma_1, \dots, \gamma_n \in Q, \gamma = \gamma_1 \cdots \gamma_n\}.$$

Definition 12. A *graphing* of G is a Borel subset Q of G such that $Q = Q^{-1}$, $Q \cap X = \emptyset$ and $\cup_{n \geq 0} Q^n = G$.

A graphing defines a structure of G-bundle of graphs on X: the set of vertices is G and

$$\mathbb{E} = \{(\gamma_1, \gamma_2) \in G \times G : r(\gamma_1) = r(\gamma_2), \gamma_1^{-1}\gamma_2 \in Q\}$$

is the set of edges. In particular, for every $x \in X$, the fibre G^x is a graph, its set or edges being $\mathbb{E} \cap (G^x \times G^x)$. Moreover, for $\gamma \in G$, the map $\gamma_1 \mapsto \gamma\gamma_1$ induces an isomorphism of graphs from $G^{s(\gamma)}$ onto $G^{r(\gamma)}$. Thus, a graphing is an equivariant Borel way of defining a structure of graph on each fibre G^x. These graphs are connected since $\cup_{n \geq 0} Q^n = G$.

When the graphs G^x are trees for every $x \in X$, the graphing Q is called a *treeing*.

Definition 13. A countable Borel groupoid G is said to be *treeable* if there is a graphing which gives to $r : G \to X$ a structure of G-bundle of trees.

A countable measured groupoid (G, μ) is said to be *treeable* if there exists an inessential reduction $G_{|U}$ which is a treeable Borel groupoid in the above sense.

Equipped with such a structure, (G, μ) is said to be a *treed measured groupoid.*

Consider the case where G is a countable group and Q is a symmetric set of generators. The corresponding graph structure on G is the Cayley graph defined by Q. If $Q = S \cup S^{-1}$ with $S \cap S^{-1} = \emptyset$, then Q is a treeing if and only if S is a free subset of generators of G (and thus G is a free group).

As made precise in [4, Proposition 3.9], treeable groupoids are the analogue of free groups and therefore the following theorem is no surprise.

Theorem 3 (Ueda). *Let (G, μ) be a countable measured groupoid which is treeable. Then (G, μ) has the Haagerup property.*

Let Q be a treeing of (G, μ). We endow G^x with the length metric d_x defined by

$$d_x(\gamma_1, \gamma_2) = \min \{n \in \mathbb{N} : \gamma_1^{-1}\gamma_2 \in Q^n\}.$$

The map $(\gamma_1, \gamma_2) \in \{(\gamma_1, \gamma_2) : r(\gamma_1) = r(\gamma_2)\} \mapsto d_{r(\gamma_1)}(\gamma_1, \gamma_2)$ is Borel.

We set $\psi(\gamma) = d_{r(\gamma)}(r(\gamma), \gamma)$. It is a real conditionally negative definite function on G. Indeed, given $\gamma_1, \dots, \gamma_n \in G^x$ and $\lambda_1, \dots, \lambda_n \in \mathbb{R}$ such that $\sum_{i=1}^n \lambda_i = 0$, we have

$$\sum_{i,j=1}^n \lambda_i\lambda_j \psi(\gamma_i^{-1}\gamma_j) = \sum_{i,j=1}^n \lambda_i\lambda_j d_{s(\gamma_i)}(s(\gamma_i), \gamma_i^{-1}\gamma_j) = \sum_{i,j=1}^n \lambda_i\lambda_j d_x(\gamma_i, \gamma_j) \le 0,$$

since the length metric on a tree is conditionally negative definite (see [17, p. 69] for instance).

We begin by proving Theorem 3 in the case where Q is bounded, i.e. there exists $k > 0$ such that $\sharp Q^x \le k$ for almost every $x \in X$.

Lemma 12. *Assume that Q is bounded. For every $c > 0$ we have $\nu(\{\psi \le c\}) < +\infty$.*

Proof. We have

$$\nu(\{\psi \le c\}) = \int_X \sharp\{\gamma : s(\gamma) = x, d_x(x, \gamma^{-1}) \le c\} \, \mathrm{d}\mu(x).$$

If k is such that $\sharp Q^x \le k$ for almost every $x \in X$, the cardinal of the ball in G^x of center x and radius c is smaller than k^c. It follows that $\nu(\{\psi \le c\}) \le k^c$. □

In view of the proof in the general case, we make a preliminary observation. Whenever Q is bounded, G is the union of the increasing sequence $(\{\psi \le k\})_{k\in\mathbb{N}}$ of Borel subsets with $\nu(\{\psi \le k\}) < +\infty$. Moreover, setting $F_n = \exp(-\psi/n)$, we have $\lim_n F_n = 1$ uniformly on each subset $\{\psi \le k\}$. Indeed, if $\psi(\gamma) \le k$, we get

$$0 \le 1 - F_n(\gamma) \le \sum_{j\ge 1} \frac{1}{n^j}\frac{\psi(\gamma)^j}{j!} \le \frac{k}{n} \exp(k/n).$$

Proof of Theorem 3. The treeing Q is no longer supposed to be bounded. Let $G = \sqcup S_k$ be a partition of G into Borel bisections. For every integer n we set

$$Q_n' = \cup_{k\le n}(Q \cap S_k) \quad \text{and} \quad Q_n = Q_n' \cup (Q_n')^{-1}.$$

Note that (Q_n) is an increasing sequence of Borel symmetric and bounded subsets of Q with $\cup_n Q_n = Q$. Let G_n be the subgroupoid of G generated by Q_n, that is $G_n = \cup_{k\geq 0} Q_n^k$, where we put $Q_n^0 = X$.

We observe that Q_n is a treeing for G_n. Denote by ψ_n the associated conditionally negative definite function on G_n. Since $Q_{n-1} \subset Q_n$, we have

$$(\psi_n)_{|G_{n-1}} \leq \psi_{n-1}.$$

Given two integers k and N, we set

$$A_{k,N} = \{\gamma \in G_k : \psi_k(\gamma) \leq N\}.$$

Then, obviously we have

$$A_{k,N} \subset A_{k+1,N} \quad \text{and} \quad A_{k,N} \subset A_{k,N+1}.$$

In particular, $(A_{k,k})_k$ is an increasing sequence of Borel subsets of G with $\cup_k A_{k,k} = G$.

We fix k. We set $F_{k,n}(\gamma) = \exp(-\psi_k(\gamma)/n)$ if $\gamma \in G_k$ and $F_{k,n}(\gamma) = 0$ if $\gamma \notin G_k$. By Lemma 13 to follow, $F_{k,n}$ is positive definite on G. Since Q_k is bounded, Lemma 12 implies that for every $\varepsilon > 0$, and for every n, we have $\nu(\{F_{k,n} \geq \varepsilon\}) < +\infty$. Moreover, $\lim_n F_{k,n} = 1$ uniformly on each $A_{k,N}$, $N \geq 1$, as previously noticed.

We choose, step by step, a strictly increasing sequence $(n_i)_{i\geq 1}$ of integers such that for every k,

$$\sup_{\gamma \in A_{k,k}} 1 - F_{k,n_k}(\gamma) \leq 1/k.$$

Then the sequence $(F_{k,n_k})_k$ of positive definite functions satisfies the required conditions showing that (G, μ) has property (H). □

Lemma 13. *Let H be a subgroupoid of a groupoid G with $G^{(0)} = H^{(0)}$. Let F be a positive definite function on H and extend F to G by setting $F(\gamma) = 0$ if $\gamma \notin H$. Then F is positive definite on G.*

Proof. Let $\gamma_1, \ldots, \gamma_n \in G^x$ and let $\lambda_1, \ldots, \lambda_n \in \mathbb{C}$. We want to show that

$$\sum_{i,j=1}^{n} \lambda_i \overline{\lambda_j} F(\gamma_i^{-1}\gamma_j) \geq 0.$$

We assume that this inequality holds for every number $k < n$ of indices. For $k = n$, this inequality is obvious if for every $i \neq j$ we have $\gamma_i^{-1}\gamma_j \notin H$. Otherwise, up to a permutation of indices, we take $j = 1$ and we assume that $2, \ldots, l$ are the indices i

such that $\gamma_i^{-1}\gamma_1 \in H$. Then, if $1 \leq i, j \leq l$ we have $\gamma_i^{-1}\gamma_j = (\gamma_i^{-1}\gamma_1)(\gamma_1^{-1}\gamma_j) \in H$ and for $i \leq l < j$ we have $\gamma_i^{-1}\gamma_j \notin H$. It follows that

$$\sum_{i,j=1}^{n} \lambda_i \overline{\lambda_j} F(\gamma_i^{-1}\gamma_j) = \sum_{i,j=1}^{l} \lambda_i \overline{\lambda_j} F(\gamma_i^{-1}\gamma_j) + \sum_{i,j>l} \lambda_i \overline{\lambda_j} F(\gamma_i^{-1}\gamma_j),$$

where the first term of the right hand side is ≥ 0. As for the second term, it is also ≥ 0 by the induction assumption. □

1.8 Properties (T) and (H) Are Not Compatible

Property (T) for group actions and equivalence relations has been introduced by Zimmer in [33]. Its extension to measured groupoids is immediate. We say that (G,μ) has *property (T)* if whenever a representation of (G,μ) almost has unit invariant sections, it actually has a unit invariant section (see [6, Definitions 4.2, 4.3] for details). We have proved in [6, Theorem 5.22] the following characterization of property (T).

Theorem 4. *Let (G,μ) be an ergodic countable measured groupoid. The following conditions are equivalent:*

1. *(G,μ) has property (T) ;*
2. *For every real conditionally negative definite function ψ on G, there exists a Borel subset E of X, with $\mu(E) > 0$, such that the restriction of ψ to $G_{|E}$ is bounded.*

Theorem 5. *Let (G,μ) be an ergodic countable measured groupoid. We assume that $(G^{(0)},\mu)$ is a diffuse standard probability space. Then (G,μ) cannot have simultaneously properties (T) and (H).*

Proof. Assume that (G,μ) has both properties (H) and (T). There exists a Borel conditionally negative definite function ψ such that for every $c > 0$, we have $\nu(\{\psi \leq c\}) < +\infty$. Moreover, there exists a Borel subset E of X, with $\mu(E) > 0$, such that the restriction of ψ to $G_{|E}$ is bounded. Then, we have

$$\int_E \sharp\{\gamma : s(\gamma) = x, r(\gamma) \in E\} \, \mathrm{d}\mu(x) < +\infty.$$

Therefore, for almost every $x \in E$, we have $\sharp\{\gamma : s(\gamma) = x, r(\gamma) \in E\} < +\infty$. Replacing if necessary E by a smaller subset we may assume the existence of $N > 0$ such that for every $x \in E$,

$$\sharp\{\gamma : s(\gamma) = x, r(\gamma) \in E\} \leq N.$$

Since $(G_{|E}, \mu_{|E})$ is ergodic, we may assume that all the fibres of this groupoid have the same finite cardinal. Therefore, this groupoid is proper and so the quotient Borel space $E/(G_{|E})$ is countably separated (see [8, Lemma 2.1.3]). A classical argument (see [34, Proposition 2.1.10]) shows that $\mu_{|E}$ is supported by an equivalence class, that is by a finite subset of E. But this contradicts the fact that the measure is diffuse. □

In the following corollaries, we always assume that (X, μ) is a diffuse standard probability measure space.

Corollary 2. *Let (G, μ) be a countable ergodic measured groupoid with the property that $(\mathscr{R}_G, \mu)$ has property (H) (e.g. is treeable). Then (G, μ) has not property (T).*

Proof. If (G, μ) had property (T) then $(\mathscr{R}_G, \mu)$ would have the same property by [6, Theorem 5.18]. But this is impossible by Theorem 5. □

This allows to retrieve results of Jolissaint [18, Proposition 3.2] and Adams-Spatzier [2, Theorem 1.8].

Corollary 3. *Let $\Gamma \curvearrowright (X, \mu)$ be an ergodic probability measure preserving action of a countable group Γ having property (T). Then $(\mathscr{R}_\Gamma, \mu)$ has not property (H) and in particular is not treeable.*

Proof. Indeed, under the assumptions of the corollary, the semi-direct product groupoid $(X \rtimes \Gamma, \mu)$ has property (T) by [33, Proposition 2.4], and we apply the previous corollary. □

Corollary 4. *Let $(\mathscr{R}, \mu)$ be a type II_1 equivalence relation on X having property (H). Then its full group $[\mathscr{R}]$ does not contain any countable subgroup Γ which acts ergodically on (X, μ) and has property (T).*

Proof. If $[\mathscr{R}]$ contains such a subgroup, then $(\mathscr{R}_\Gamma, \mu)$ has property (T), and also property (H) as a subequivalence relation of $\mathscr{R}$, in contradiction with Corollary 2.

□

Problem 1. Since by Dye's theorem $(\mathscr{R}, \mu)$ is entirely determined by its full group [21, Theorem 4.1], it would be interesting to characterize property (H) in terms of this full group.

References

1. Adams, S.: Trees and amenable equivalence relations. Ergodic Theory Dynam. Systems **10**(1), 1–14 (1990). DOI 10.1017/S0143385700005368. URL http://dx.doi.org/10.1017/S0143385700005368
2. Adams, S.R., Spatzier, R.J.: Kazhdan groups, cocycles and trees. Amer. J. Math. **112**(2), 271–287 (1990). DOI 10.2307/2374716. URL http://dx.doi.org/10.2307/2374716

3. Akemann, C.A., Walter, M.E.: Unbounded negative definite functions. Canad. J. Math. **33**(4), 862–871 (1981). DOI 10.4153/CJM-1981-067-9. URL http://dx.doi.org/10.4153/CJM-1981-067-9
4. Alvarez, A.: Une théorie de Bass-Serre pour les relations d'équivalence et les groupoïdes boréliens. Ph.D. thesis, Ecole Normale Supérieure de Lyon (2008). URL http://www.univ-orleans.fr/mapmo/membres/alvarez/MAPMO/accueil.html
5. Alvarez, A., Gaboriau, D.: Free products, orbit equivalence and measure equivalence rigidity. Groups Geom. Dyn. **6**(1), 53–82 (2012). DOI 10.4171/GGD/150. URL http://dx.doi.org/10.4171/GGD/150
6. Anantharaman-Delaroche, C.: Cohomology of property T groupoids and applications. Ergodic Theory Dynam. Systems **25**(4), 977–1013 (2005). DOI 10.1017/S0143385704000884. URL http://dx.doi.org/10.1017/S0143385704000884
7. Anantharaman-Delaroche, C.: Old and new about treeability and the Haagerup property for measured groupoids (2011), arXiv:1105.5961v1
8. Anantharaman-Delaroche, C., Renault, J.: Amenable groupoids, *Monographies de L'Enseignement Mathématique [Monographs of L'Enseignement Mathématique]*, vol. 36. L'Enseignement Mathématique, Geneva (2000). With a foreword by Georges Skandalis and Appendix B by E. Germain
9. Boca, F.: On the method of constructing irreducible finite index subfactors of Popa. Pacific J. Math. **161**(2), 201–231 (1993). URL http://projecteuclid.org/getRecord?id=euclid.pjm/1102623229
10. Cherix, P.A., Cowling, M., Jolissaint, P., Julg, P., Valette, A.: Groups with the Haagerup property. Gromov's a-T-menability, *Progress in Mathematics*, vol. 197. Birkhäuser Verlag, Basel (2001). DOI 10.1007/978-3-0348-8237-8. URL http://dx.doi.org/10.1007/978-3-0348-8237-8
11. Choda, M.: Group factors of the Haagerup type. Proc. Japan Acad. Ser. A Math. Sci. **59**(5), 174–177 (1983). URL http://projecteuclid.org/getRecord?id=euclid.pja/1195515589
12. Connes, A.: Classification des facteurs. In: Operator algebras and applications, Part 2 (Kingston, Ont., 1980), *Proc. Sympos. Pure Math.*, vol. 38, pp. 43–109. Amer. Math. Soc., Providence, R.I. (1982)
13. Dixmier, J.: von Neumann algebras, *North-Holland Mathematical Library*, vol. 27. North-Holland Publishing Co., Amsterdam (1981). With a preface by E. C. Lance, Translated from the second French edition by F. Jellett
14. Feldman, J., Moore, C.C.: Ergodic equivalence relations, cohomology, and von Neumann algebras. I. Trans. Amer. Math. Soc. **234**(2), 289–324 (1977)
15. Haagerup, U.: An example of a nonnuclear C^*-algebra, which has the metric approximation property. Invent. Math. **50**(3), 279–293 (1978/79). DOI 10.1007/BF01410082. URL http://dx.doi.org/10.1007/BF01410082
16. Hahn, P.: The regular representations of measure groupoids. Trans. Amer. Math. Soc. **242**, 35–72 (1978). DOI 10.2307/1997727. URL http://dx.doi.org/10.2307/1997727
17. de la Harpe, P., Valette, A.: La propriété (T) de Kazhdan pour les groupes localement compacts (avec un appendice de Marc Burger). Astérisque (175), 158 (1989). With an appendix by M. Burger
18. Jolissaint, P.: The Haagerup property for measure-preserving standard equivalence relations. Ergodic Theory Dynam. Systems **25**(1), 161–174 (2005). DOI 10.1017/S0143385704000562. URL http://dx.doi.org/10.1017/S0143385704000562
19. Jones, V.F.R.: Index for subfactors. Invent. Math. **72**(1), 1–25 (1983). DOI 10.1007/BF01389127. URL http://dx.doi.org/10.1007/BF01389127
20. Kechris, A.S.: Classical descriptive set theory, *Graduate Texts in Mathematics*, vol. 156. Springer-Verlag, New York (1995). DOI 10.1007/978-1-4612-4190-4. URL http://dx.doi.org/10.1007/978-1-4612-4190-4
21. Kechris, A.S.: Global aspects of ergodic group actions, *Mathematical Surveys and Monographs*, vol. 160. American Mathematical Society, Providence, RI (2010)

22. Moore, C.C.: Virtual groups 45 years later. In: Group representations, ergodic theory, and mathematical physics: a tribute to George W. Mackey, *Contemp. Math.*, vol. 449, pp. 263–300. Amer. Math. Soc., Providence, RI (2008). DOI 10.1090/conm/449/08716. URL http://dx.doi.org/10.1090/conm/449/08716
23. Paschke, W.L.: Inner product modules over B^*-algebras. Trans. Amer. Math. Soc. **182**, 443–468 (1973)
24. Popa, S.: Classification of subfactors and their endomorphisms, *CBMS Regional Conference Series in Mathematics*, vol. 86. Published for the Conference Board of the Mathematical Sciences, Washington, DC (1995)
25. Popa, S.: On a class of type II_1 factors with Betti numbers invariants. Ann. of Math. (2) **163**(3), 809–899 (2006). DOI 10.4007/annals.2006.163.809. URL http://dx.doi.org/10.4007/annals.2006.163.809
26. Popa, S., Rădulescu, F.: Derivations of von Neumann algebras into the compact ideal space of a semifinite algebra. Duke Math. J. **57**(2), 485–518 (1988). DOI 10.1215/S0012-7094-88-05722-5. URL http://dx.doi.org/10.1215/S0012-7094-88-05722-5
27. Ramsay, A.: Virtual groups and group actions. Advances in Math. **6**, 253–322 (1971) (1971)
28. Ramsay, A., Walter, M.E.: Fourier-Stieltjes algebras of locally compact groupoids. J. Funct. Anal. **148**(2), 314–367 (1997). DOI 10.1006/jfan.1996.3083. URL http://dx.doi.org/10.1006/jfan.1996.3083
29. Renault, J.: A groupoid approach to C^*-algebras, *Lecture Notes in Mathematics*, vol. 793. Springer, Berlin (1980)
30. Rieffel, M.A.: Morita equivalence for C^*-algebras and W^*-algebras. J. Pure Appl. Algebra **5**, 51–96 (1974)
31. Takesaki, M.: Conditional expectations in von Neumann algebras. J. Functional Analysis **9**, 306–321 (1972)
32. Ueda, Y.: Notes on treeability and costs for discrete groupoids in operator algebra framework. In: Operator Algebras: The Abel Symposium 2004, *Abel Symp.*, vol. 1, pp. 259–279. Springer, Berlin (2006). DOI 10.1007/978-3-540-34197-0_13. URL http://dx.doi.org/10.1007/978-3-540-34197-0_13
33. Zimmer, R.J.: On the cohomology of ergodic actions of semisimple Lie groups and discrete subgroups. Amer. J. Math. **103**(5), 937–951 (1981). DOI 10.2307/2374253. URL http://dx.doi.org/10.2307/2374253
34. Zimmer, R.J.: Ergodic theory and semisimple groups, *Monographs in Mathematics*, vol. 81. Birkhäuser Verlag, Basel (1984)

Chapter 2
Do Phantom Cuntz-Krieger Algebras Exist?

Sara E. Arklint

Abstract If phantom Cuntz-Krieger algebras do not exist, then purely infinite Cuntz-Krieger algebras can be characterized by outer properties. In this survey paper, a summary of the known results on non-existence of phantom Cuntz-Krieger algebras is given.

Keywords Cuntz-Krieger algebras • Graph C^*-algebras • Purely infinite C^*-algebras • K-theory • Filtered K-theory • Primitive ideal space

Mathematics Subject Classification (2010): 46L55, 46L35, 46L80, 46M15.

2.1 Introduction

The Cuntz-Krieger algebras were introduced by Joachim Cuntz and Wolfgang Krieger in 1980, cf. [8], and are a generalization of the Cuntz algebras. Given a nondegenerate $n \times n$ matrix A with entries in $\{0, 1\}$, its associated Cuntz-Krieger algebra O_A is defined as the universal C^*-algebra generated by n partial isometries $s_1, \dots, s_n$ satisfying the relations

$$1 = s_1 s_1^* + \cdots + s_n s_n^*,$$

$$s_i^* s_i = \sum_{j=1}^{n} A_{ij} s_j s_j^* \text{ for all } i = 1, \dots n.$$

S.E. Arklint (✉)
Department of Mathematical Sciences, University of Copenhagen, Universitetsparken 5, DK-2100 Copenhagen Ø, Denmark
e-mail: arklint@math.ku.dk

T.M. Carlsen et al. (eds.), *Operator Algebra and Dynamics*, Springer Proceedings in Mathematics & Statistics 58, DOI 10.1007/978-3-642-39459-1_2,

The Cuntz-Krieger algebras arise from shifts of finite type, and it has been shown that the Cuntz-Krieger algebras are exactly the graph algebras $C^*(E)$ arising from finite directed graphs E with no sinks, [3].

Neither of the two equivalent definitions of Cuntz-Krieger algebras give an outer characterization of Cuntz-Krieger algebras; i.e., neither give a way of determining whether a C^*-algebra is a Cuntz-Krieger algebra, unless it is constructed from a graph or a shift of finite type.

A Cuntz-Krieger algebra is purely infinite if and only if it has real rank zero, cf. [11], and in the following we will mainly restrict to real rank zero Cuntz-Krieger algebras since we will rely on classification results that only hold in the purely infinite case. The Cuntz-Krieger algebra O_A is purely infinite if and only if A satisfies Cuntz's condition (II), and equivalently the Cuntz-Krieger algebra $C^*(E)$ is purely infinite if and only if the graph E satisfies condition (K), cf. [11].

The notion of C^*-algebras over a topological space is useful for defining phantom Cuntz-Krieger algebras and for defining filtered K-theory, and in [12], Eberhard Kirchberg proved some very powerful classification results for O_∞-absorbing C^*-algebras over a space X using $KK(X)$-theory. A C^*-algebra A over the finite T_0-space X is a C^*-algebra equipped with a lattice-preserving map from the open sets of X to the ideals in A, denoted $U \mapsto A(U)$, satisfying $A(\emptyset) = 0$ and $A(X) = A$, and extended to locally closed subsets as $A(U \setminus V) = A(U)/A(V)$. In particular, a C^*-algebra with finitely many ideals is a C^*-algebra over its primitive ideal space.

Definition 1. A C^*-algebra A with primitive ideal space X *looks like a purely infinite Cuntz-Krieger algebra* if

1. A is unital, purely infinite, nuclear, separable, and of real rank zero,
2. X is finite,
3. For all $x \in X$, the group $K_*(A(x))$ is finitely generated, the group $K_1(A(x))$ is free, and $\operatorname{rank} K_0(A(x)) = \operatorname{rank} K_1(A(x))$,
4. For all $x \in X$, $A(x)$ is in the bootstrap class of Rosenberg and Schochet.

A C^*-algebra that looks like a purely infinite Cuntz-Krieger algebra but is not isomorphic to a Cuntz-Krieger algebra, is called a *phantom Cuntz-Krieger algebra*.

All purely infinite Cuntz-Krieger algebras look like purely infinite Cuntz-Krieger algebras. It is not known whether all C^*-algebra s that look like purely infinite Cuntz-Krieger algebras are Cuntz-Krieger algebras. If it is established that they are, i.e., that phantom Cuntz-Krieger algebras do not exist, then the above definition gives a characterization of the purely infinite Cuntz-Krieger algebras.

An example to point out the relevance of such a characterization is given by Proposition 1. If phantom Cuntz-Krieger algebras do not exist, the proposition determines exactly when an extension of purely infinite Cuntz-Krieger algebras is a purely infinite Cuntz-Krieger algebra.

By a result of Lawrence G. Brown and Gert K. Pedersen, Theorem 3.14 of [7], an extension of real rank zero C^*-algebras has real rank zero if and only if projections in the quotient lift to projections in the extension. Hence, if a C^*-algebra A with primitive ideal space X has real rank zero, then $K_0(A(Y \setminus U)) \to K_1(A(U))$ vanishes for all Y and U where Y is a locally closed subsets of X and U is an open subsets of Y. Using this, an induction argument shows that for a C^*-algebra that looks like a purely infinite Cuntz-Krieger algebra, (3) and (4) of Definition 1 hold for all locally closed subsets Y of X.

Proposition 1. *Consider a unital extension $0 \to I \to A \to A/I \to 0$ and assume that A/I is a purely infinite Cuntz-Krieger algebra and that I is stably isomorphic to a purely infinite Cuntz-Krieger algebra. Then A looks like a purely infinite Cuntz-Krieger algebra if and only if the induced map $K_0(A/I) \to K_1(I)$ vanishes.*

Proof. By Theorem 3.14 of [7], the C^*-algebra A is of real rank zero if and only if the induced map $K_0(A/I) \to K_1(I)$ vanishes. It is well-known or easy to check that the other properties stated in Definition 1 are closed under extensions. □

2.2 Special Cases

One of the first places one would look for phantom Cuntz-Krieger algebras is among the matrix algebras over purely infinite Cuntz-Krieger algebras. Clearly, if O_A is a purely infinite Cuntz-Krieger algebra, then $M_n(O_A)$ looks like a purely infinite Cuntz-Krieger algebra for all n. Since $M_n(O_A)$ is a graph algebra, one then immediately asks if a graph algebra can be a phantom Cuntz-Krieger algebra. It turns out that it cannot.

Theorem 1 ([3]). *Let E be a directed graph and assume that its graph algebra $C^*(E)$ is unital and satisfies* $\operatorname{rank} K_0(C^*(E)) = \operatorname{rank} K_1(C^*(E))$. *Then $C^*(E)$ is isomorphic to a Cuntz-Krieger algebra.*

Theorem 2 ([3]). *Let A be a unital C^*-algebra and assume that A is stably isomorphic to a Cuntz-Krieger algebra. Then A is isomorphic to a Cuntz-Krieger algebra.*

As a small corollary to the work of Eberhard Kirchberg on $KK(X)$-theory, phantom Cuntz-Krieger algebras cannot have vanishing K-theory.

Theorem 3 ([12]). *Let A and B be unital, nuclear, separable C^*-algebras with primitive ideal space X. Then $A \otimes O_2$ and $B \otimes O_2$ are isomorphic.*

Corollary 1. *Let A be a C^*-algebra that looks like a purely infinite Cuntz-Krieger algebra, and assume that $K_*(A) = 0$. Then A is a Cuntz-Krieger algebra.*

Proof. Let X denote the finite primitive ideal space of A. Since $K_*(A) = 0$ and A looks like a purely infinite Cuntz-Krieger algebra, $K_*(A(x)) = 0$ for all $x \in X$. So for all $x \in X$, $A(x)$ is O_2-absorbing since it is a UCT Kirchberg algebra with vanishing K-theory. By applying Theorem 4.3 of [15] finitely many times, we see that A itself is O_2-absorbing. Let O_B be a Cuntz-Krieger algebra with primitive ideal space X and with $O_B(x)$ (stably) isomorphic to O_2 for all $x \in X$. Then by Theorem 3, A is isomorphic to O_B. □

2.3 Using Filtered K-Theory

Via K-theoretic classification results it can be established that a phantom Cuntz-Krieger algebra cannot have a so-called accordion space as its primitive ideal space. We will first restrict to the cases where the primitive ideal space has at most two points in order to describe the historical development and due to the importance and powerfulness of the results needed. The most crucial result is by Eberhard Kirchberg who showed in [12] that for stable, purely infinite, nuclear, separable C^*-algebras A and B with finite primitive ideal space X, any $KK(X)$-equivalence between A and B lifts to a $*$-isomorphism.

Simple C^*-algebras that look like purely infinite Cuntz-Krieger algebras are UCT Kirchberg algebras, hence the classification result by Eberhard Kirchberg and N. Christoffer Phillips applies. For a unital C^*-algebra A with unit 1_A, denote by $[1_A]$ the class of 1_A in $K_0(A)$. For unital C^*-algebras A and B an isomorphism from $(K_*(A), [1_A])$ to $(K_*(B), [1_B])$ is defined as a pair (ϕ_0, ϕ_1) of group isomorphisms $\phi_i\colon K_i(A) \to K_i(B)$, $i = 0, 1$, for which $\phi_0([1_A]) = \phi_0([1_B])$.

Theorem 4 ([13]). *Let A and B be unital, simple, purely infinite, nuclear, separable C^*-algebras in the bootstrap class. If $(K_*(A), [1_A])$ and $(K_*(B), [1_B])$ are isomorphic, then A and B are isomorphic.*

The range of K_* for graph algebras has been determined by Wojciech Szymański, and his result has been extended by Søren Eilers, Takeshi Katsura, Mark Tomforde, and James West to include the class of the unit.

Theorem 5 ([9]). *Let G and F be finitely generated groups, let $g \in G$, and assume that F is free and that* $\operatorname{rank} G = \operatorname{rank} F$. *Then there exists a simple, purely infinite Cuntz-Krieger algebra O_A realising $(G \oplus F, g)$ as $(K_*(O_A), [1_{O_A}])$.*

Corollary 2. *Simple phantom Cuntz-Krieger algebras do not exist.*

Proof. Let A be a simple C^*-algebra that looks like a purely infinite Cuntz-Krieger algebra. By Theorem 5, there exists a Cuntz-Krieger algebra O_B of real rank zero for which $(K_*(A), [1_A]) \cong (K_*(O_B), [1_{O_B}])$. Since A and O_B are UCT Kirchberg algebras, it follows from Theorem 4 that A and O_B are isomorphic. □

For C^*-algebras with exactly one nontrivial ideal, the suitable invariant seems to be the induced six-term exact sequence in K-theory.

Definition 2. Let X_{six} denote the space $\{1, 2\}$ with $\{2\}$ open and $\{1\}$ not open. For a C^*-algebra A with primitive ideal space X_{six}, $K_{\text{six}}(A)$ is defined as the cyclic six-term exact sequence

$$\begin{array}{ccccc} K_0(A(2)) & \xrightarrow{i} & K_0(A) & \xrightarrow{r} & K_0(A(1)) \\ {\scriptstyle\delta}\uparrow & & & & \downarrow{\scriptstyle\delta} \\ K_1(A(1)) & \xleftarrow[r]{} & K_1(A) & \xleftarrow[i]{} & K_1(A(2)) \end{array}$$

induced by the extension $0 \to A(2) \to A \to A(1) \to 0$. For unital C^*-algebras A and B with primitive ideal space X_{six}, an isomorphism from $(K_{\text{six}}(A), [1_A])$ to $(K_{\text{six}}(B), [1_B])$ is defined as a triple $(\phi_*^{\{2\}}, \phi_*^{X_{\text{six}}}, \phi_*^{\{1\}})$ of graded isomorphisms $\phi_*^Y \colon K_*(A(Y)) \to K_*(B(Y))$, $Y \in \{\{2\}, X_{\text{six}}, \{1\}\}$, that commute with the maps i, r, and δ and satisfies $\phi_0^{X_{\text{six}}}([1_A]) = [1_B]$.

This invariant was originally introduced by Mikael Rørdam to classify stable extensions of UCT Kirchberg algebras. Alexander Bonkat established a UCT for K_{six} (that was later generalized by Ralf Meyer and Ryszard Nest), and by combining his UCT with the result of Eberhard Kirchberg (and a meta theorem by Søren Eilers, Gunnar Restorff and Efren Ruiz in [10] to achieve unital and not stable isomorphism) one obtains the following theorem.

Theorem 6 ([6,12]). *Let A and B be unital, purely infinite, nuclear, separable C^*-algebras with primitive ideal space X_{six}, and assume that $A(x)$ and $B(x)$ are in the bootstrap class for all $x \in \{1, 2\}$. Then $(K_{\text{six}}(A), [1_A]) \cong (K_{\text{six}}(B), [1_B])$ implies $A \cong B$.*

The range of K_{six} for graph algebras has been determined by Søren Eilers, Takeshi Katsura, Mark Tomforde, and James West.

Theorem 7 ([9]). *Let a six-term exact sequence*

$$\mathscr{E}: \quad \begin{array}{ccccc} G_1 & \longrightarrow & G_2 & \longrightarrow & G_3 \\ \uparrow & & & & \downarrow{\scriptstyle 0} \\ F_3 & \longleftarrow & F_2 & \longleftarrow & F_1. \end{array}$$

be given with G_1, G_2, G_3 and F_1, F_2, F_3 finitely generated groups, and let $g \in G_2$. Assume that the groups F_1, F_2, F_3 are free, and that $\operatorname{rank} G_i = \operatorname{rank} F_i$ *for all $i = 1, 2, 3$. Then there exists a purely infinite Cuntz-Krieger algebra O_A with primitive ideal space X_{six} realising $(\mathscr{E}, g)$ as $(K_{\text{six}}(O_A), [1_{O_A}])$.*

Corollary 3. *Phantom Cuntz-Krieger algebras with exactly one nontrivial ideal do not exist.*

The generalization of the invariant K_{six} to larger primitive ideal spaces is called filtered K-theory or filtrated K-theory and was introduced by Gunnar Restorff and by Ralf Meyer and Ryszard Nest. Filtered K-theory consists of the six-term exact sequences induced by all extensions of subquotients. A smaller invariant, the reduced filtered K-theory $\mathrm{FK}_{\mathcal{R}}$ originally defined by Gunnar Restorff to classify purely infinite Cuntz-Krieger algebras, has so far proven suitable for classifying C^*-algebras that look like purely infinite Cuntz-Krieger algebras.

Let X be a finite T_0-space. For $x \in X$, we denote by $\widetilde{\{x\}}$ the smallest open subset of X containing x, and we define $\tilde{\partial}(x)$ as $\widetilde{\{x\}} \setminus \{x\}$. For $x, y \in X$ we write $y \to x$ when $y \in \tilde{\partial}(x)$ and there is no $z \in \tilde{\partial}(x)$ for which $y \in \tilde{\partial}(z)$.

Definition 3. For a C^*-algebra A with primitive ideal space X, its *reduced filtered K-theory* $\mathrm{FK}_{\mathcal{R}}(A)$ consists of the groups and maps

$$K_1(A(x)) \xrightarrow{\delta} K_0(A(\tilde{\partial}(x))) \xrightarrow{i} K_0(A(\widetilde{\{x\}}))$$

induced by the extension $0 \to A(\tilde{\partial}(x)) \to A(\widetilde{\{x\}}) \to A(x) \to 0$, for all $x \in X$, together with the groups and maps

$$K_0(A(\widetilde{\{y\}})) \xrightarrow{i} K_0(A(\tilde{\partial}(x)))$$

induced by the extension $0 \to A(\widetilde{\{y\}}) \to A(\tilde{\partial}(x)) \to A(\tilde{\partial}(x) \setminus \widetilde{\{y\}}) \to 0$, for all $x, y \in X$ with $y \to x$.

Example 1. Let $X = \{1,2,3\}$ be given the topology $\{\emptyset, \{3\}, \{3,2\}, \{3,1\}, X\}$. Then for a C^*-algebra A with primitive ideal space X, its reduced filtered K-theory $\mathrm{FK}_{\mathcal{R}}(A)$ consists of the groups and maps

$$\begin{array}{ccccc} K_1(A(2)) & & & & K_0(A(\{3,1\})) \\ & \overset{\delta}{\searrow} & & \overset{i}{\nearrow} & \\ & & K_0(A(3)) & & \\ & \overset{\delta}{\nearrow} & & \overset{i}{\searrow} & \\ K_1(A(1)) & & & & K_0(A(\{3,2\})) \end{array}$$

together with the group $K_1(A(3))$.

It is shown in [2] that if A is a C^*-algebra of real rank zero with primitive ideal space X, then the sequence

$$\bigoplus_{\substack{y \to x,\, y \to x' \\ x, x' \in X}} K_0(A(\widetilde{\{y\}})) \xrightarrow{(ii - ii)} \bigoplus_{x \in X} K_0(A(\widetilde{\{x\}})) \xrightarrow{(i)} K_0(A) \longrightarrow 0$$

is exact.

Definition 4. For a unital C^*-algebra A of real rank zero with primitive ideal space X, $1(A)$ is defined as the unique element in

$$\bigoplus_{x\in X} K_0(A(\widetilde{\{x\}})) \Big/ \bigoplus_{\substack{y\to x, y\to x' \\ x,x'\in X}} K_0(A(\widetilde{\{y\}}))$$

that is mapped to $[1_A]$. For A and B unital C^*-algebras of real rank zero with primitive ideal space X, an isomorphism from $(\mathrm{FK}_{\mathcal{R}}(A), 1(A))$ to $(\mathrm{FK}_{\mathcal{R}}(B), 1(B))$ is defined as a family of isomorphisms

$$\phi^1_{\{x\}} \colon K_1(A(x)) \to K_1(B(x))$$
$$\phi^0_{\tilde{\partial}(x)} \colon K_0(A(\tilde{\partial}(x)) \to K_0(B(\tilde{\partial}(x))$$
$$\phi^0_{\widetilde{\{x\}}} \colon K_0(A(\widetilde{\{x\}})) \to K_0(B(\widetilde{\{x\}}))$$

(where $\phi^0_{\tilde{\partial}(x)} = \phi^0_{\widetilde{\{y\}}}$ when $\tilde{\partial}(x) = \widetilde{\{y\}}$) for all $x \in X$ that commute with the maps i and δ and maps $1(A)$ to $1(B)$.

Using Theorem 7, Rasmus Bentmann, Takeshi Katsura, and the author have established the range of reduced filtered K-theory $\mathrm{FK}_{\mathcal{R}}$ for graph algebras, cf. [2]. They also show in [2] that if a C^*-algebra has as its primitive ideal space an accordion space, cf. Definition 5, or one of the spaces defined in Definition 6, and looks like a purely infinite Cuntz-Krieger algebra, then there exists a purely infinite Cuntz-Krieger algebra with the same primitive ideal space and filtered K-theory. It is not known if this holds for other types of primitive ideal spaces.

Theorem 8 ([2]). *Let B be a C^*-algebra that looks like a purely infinite Cuntz-Krieger algebra. Then there exists a Cuntz-Krieger algebra O_A of real rank zero with* $\mathrm{Prim}(O_A) \cong \mathrm{Prim}(B)$ *for which* $(\mathrm{FK}_{\mathcal{R}}(O_A), [1_{O_A}])$ *is isomorphic to* $(\mathrm{FK}_{\mathcal{R}}(B), [1_B])$.

Definition 5. A finite, connected T_0-space X is called an *accordion space* if the following holds:

- For all $x \in X$ there are at most two elements $y \in X$ for which $y \to x$,
- There is at least two elements $x \in X$ for which there is exactly one element $y \in X$ for which $y \to x$.

The notion of accordion spaces was introduced by Rasmus Bentmann in [4]. Intuitively, a space is an accordion space if and only if the Hasse diagram of the ordering defined by $y \leq x$ when $y \in \widetilde{\{x\}}$, looks like an accordion. All finite, linear spaces are accordion spaces, and the following five spaces are examples of connected spaces that are not accordion spaces.

Definition 6. Define a topology on the space $\mathscr{X} = \{1, 2, 3, 4\}$ by defining $U \subseteq \mathscr{X}$ to be open if U is empty or $4 \in U$. Define $\mathscr{X}^{\text{op}}$ as having the opposite topology. Then $\mathscr{X}$ and $\mathscr{X}^{\text{op}}$ have Hasse diagrams

respectively. Define a topology on the space $\mathscr{Y} = \{1, 2, 3, 4\}$ by defining $U \subseteq \mathscr{Y}$ to be open if $U \in \{\emptyset, \{4\}\}$ or if $\{3, 4\} \subseteq U$. Define $\mathscr{Y}^{\text{op}}$ as having the opposite topology. Then $\mathscr{Y}$ and $\mathscr{Y}^{\text{op}}$ have Hasse diagrams

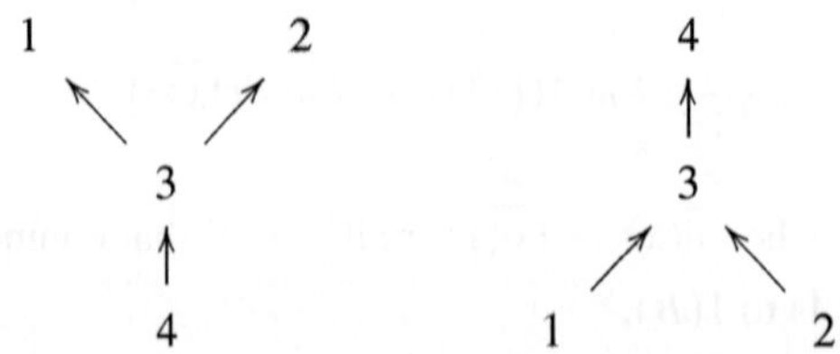

respectively. Finally, define a topology on the space $\mathscr{D} = \{1, 2, 3.4\}$ as the open sets being $\{\emptyset, \{4\}, \{3, 4\}, \{2, 4\}, \{2, 3, 4\}, \mathscr{D}\}$. Then $\mathscr{D}$ has Hasse diagram

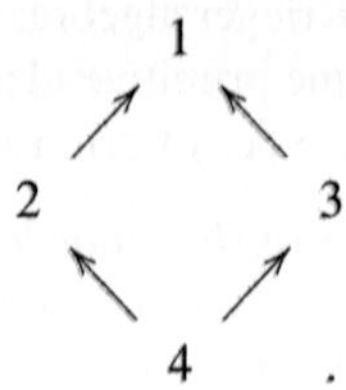

Ralf Meyer and Ryszard Nest showed in [14] that if X is a finite, linear space, then filtered K-theory is a complete invariant for all stable, purely infinite, nuclear, separable C^*-algebras A with primitive ideal space X that satisfy that $A(x)$ are in the bootstrap class for all $x \in X$. They also gave a counter-example to completeness of filtered K-theory for the space $\mathscr{X}$. Using their methods, Rasmus Bentmann and Manuel Köhler showed in [5] that filtered K-theory is a complete invariant for such C^*-algebras exactly when their primitive ideal space X is an accordion space.

However, Gunnar Restorff, Efren Ruiz, and the author showed in [1] that for the spaces $\mathscr{X}$, $\mathscr{X}^{\text{op}}$, $\mathscr{Y}$, and $\mathscr{Y}^{\text{op}}$, filtered K-theory is a complete invariant for such C^*-algebras if one adds the assumption of real rank zero. And in [2], Rasmus Bentmann, Takeshi Katsura, and the author showed that for the space $\mathscr{D}$, reduced filtered K-theory is a complete invariant for C^*-algebras that look like purely infinite Cuntz-Krieger algebras. It is also shown in [2] that for C^*-algebras that look like purely infinite Cuntz-Krieger algebras and have either an accordion space or one of the spaces defined in Definition 6 as primitive ideal space, any isomorphism on reduced filtered K-theory can be lifted to an isomorphism on filtered K-theory.

The five spaces of Definition 6 are so far the only non-accordion spaces for which such results have been achieved. However, combining these results with Theorem 3.3 of [10] gives the following theorem, cf. [2].

Theorem 9 ([1, 2, 5, 14]). *Let X be either an accordion space or one of the spaces defined in Definition 6. Let A and B be C^*-algebras that look like purely infinite Cuntz-Krieger algebras and both have X as primitive ideal space. Then $(\mathrm{FK}_{\mathcal{R}}(A), 1(A)) \cong (\mathrm{FK}_{\mathcal{R}}(B), 1(B))$ implies $A \cong B$.*

Corollary 4. *Let X be either an accordion space or one of the spaces defined in Definition 6. Then phantom Cuntz-Krieger algebras with primitive ideal space X do not exist.*

2.4 Summary

The results stated in this article, are recaptured in the following theorem.

Theorem 10. *Let A be a C^*-algebra that looks like a purely infinite Cuntz-Krieger algebra. If A satisfies either of the following conditions,*

- *A is a graph algebra,*
- $K_*(A) = 0$,
- $\mathrm{Prim}(A)$ *is an accordion space,*
- $\mathrm{Prim}(A)$ *is one of the five four-point spaces of Definition 6,*

then A is isomorphic to a Cuntz-Krieger algebra.

In general, it is unknown whether phantom Cuntz-Krieger algebras exist.

Acknowledgements This research was supported by the Danish National Research Foundation through the Centre for Symmetry and Deformation (DNRF92), and by the NordForsk Research Network "Operator Algebra and Dynamics" (grant #11580).

References

1. Arklint, S., Restorff, G., Ruiz, E.: Filtrated K-theory for real rank zero C^*-algebras. Internat. J. Math. **23**(8), 1250,078, 19 (2012). DOI 10.1142/S0129167X12500784. URL http://dx.doi.org/10.1142/S0129167X12500784
2. Arklint, S.E., Bentmann, R., Katsura, T.: Reduction of filtered K-theory and a characterization of Cuntz-Krieger algebras (2013), arXiv:1301.7223v1
3. Arklint, S.E., Ruiz, E.: Corners of Cuntz-Krieger algebras (2012), arXiv:1209.4336v2

4. Bentmann, R.: Filtrated K-theory and classification of C^*-algebras. Diplom thesis, Georg-August-Universität Göttingen (2010). URL http://www.math.ku.dk/~bentmann/thesis.pdf
5. Bentmann, R., Köhler, M.: Universal coefficient theorems for C^*-algebras over finite topological spaces (2011), arXiv:1101.5702
6. Bonkat, A.: Bivariante K-Theorie für Kategorien projektiver Systeme von C^*-Algebren. Ph.D. thesis, Westfälische Wilhelms-Universität Münster (2002). URL http://deposit.ddb.de/cgi-bin/dokserv?idn=967387191
7. Brown, L.G., Pedersen, G.K.: C^*-algebras of real rank zero. J. Funct. Anal. **99**(1), 131–149 (1991). DOI 10.1016/0022-1236(91)90056-B. URL http://dx.doi.org/10.1016/0022-1236(91)90056-B
8. Cuntz, J., Krieger, W.: A class of C^*-algebras and topological Markov chains. Invent. Math. **56**(3), 251–268 (1980). DOI 10.1007/BF01390048. URL http://dx.doi.org/10.1007/BF01390048
9. Eilers, S., Katsura, T., Tomforde, M., West, J.: The ranges of K-theoretic invariants for nonsimple graph algebras (2012), arXiv:1202.1989v1
10. Eilers, S., Restorff, G., Ruiz, E.: Strong classification of extensions of classifiable C^*-algebras (2013), arXiv:1301.7695v1
11. Hong, J.H., Szymański, W.: Purely infinite Cuntz-Krieger algebras of directed graphs. Bull. London Math. Soc. **35**(5), 689–696 (2003). DOI 10.1112/S0024609303002364. URL http://dx.doi.org/10.1112/S0024609303002364
12. Kirchberg, E.: Das nicht-kommutative Michael-Auswahlprinzip und die Klassifikation nicht-einfacher Algebren. In: C^*-algebras (Münster, 1999), pp. 92–141. Springer, Berlin (2000)
13. Kirchberg, E., Phillips, N.C.: Embedding of exact C^*-algebras in the Cuntz algebra $\mathcal{O}_2$. J. Reine Angew. Math. **525**, 17–53 (2000). DOI 10.1515/crll.2000.065. URL http://dx.doi.org/10.1515/crll.2000.065
14. Meyer, R., Nest, R.: C*-algebras over topological spaces: filtrated K-theory. Canad. J. Math. **64**(2), 368–408 (2012). DOI 10.4153/CJM-2011-061-x. URL http://dx.doi.org/10.4153/CJM-2011-061-x
15. Toms, A.S., Winter, W.: Strongly self-absorbing C^*-algebras. Trans. Amer. Math. Soc. **359**(8), 3999–4029 (2007). DOI 10.1090/S0002-9947-07-04173-6. URL http://dx.doi.org/10.1090/S0002-9947-07-04173-6

Chapter 3
Projective Dimension in Filtrated K-Theory

Rasmus Bentmann

Abstract Under mild assumptions, we characterise modules with projective resolutions of length $n \in \mathbb{N}$ in the target category of filtrated K-theory over a finite topological space in terms of two conditions involving certain Tor-groups. We show that the filtrated K-theory of any separable C^*-algebra over any topological space with at most four points has projective dimension 2 or less. We observe that this implies a universal coefficient theorem for rational equivariant KK-theory over these spaces. As a contrasting example, we find a separable C^*-algebra in the bootstrap class over a certain five-point space, the filtrated K-theory of which has projective dimension 3. Finally, as an application of our investigations, we exhibit Cuntz-Krieger algebras which have projective dimension 2 in filtrated K-theory over their respective primitive spectrum.

Keywords K-theory • Filtered K-theory • Ideal-related KK-theory • Universal coefficient theorem

Mathematics Subject Classification (2010): 46L80, 19K35, 46M20.

3.1 Introduction

A far-reaching classification theorem in [7] motivates the computation of Eberhard Kirchberg's ideal-related Kasparov groups $\mathrm{KK}(X; A, B)$ for separable C^*-algebras A and B over a non-Hausdorff topological space X by means of

R. Bentmann (✉)
Department of Mathematical Sciences, University of Copenhagen, Universitetsparken 5, DK-2100 Copenhagen Ø, Denmark
e-mail: bentmann@math.ku.dk

T.M. Carlsen et al. (eds.), *Operator Algebra and Dynamics*, Springer Proceedings in Mathematics & Statistics 58, DOI 10.1007/978-3-642-39459-1_3,

K-theoretic invariants. We are interested in the specific case of finite spaces here. In [10, 11], Ralf Meyer and Ryszard Nest laid out a theoretic framework that allows for a generalisation of Jonathan Rosenberg's and Claude Schochet's universal coefficient theorem [16] to the equivariant setting. Starting from a set of generators of the equivariant bootstrap class, they define a homology theory with a certain universality property, which computes $\mathrm{KK}(X)$-theory via a spectral sequence. In order for this *universal coefficient* spectral sequence to degenerate to a short exact sequence, it remains to be checked *by hand* that objects in the range of the homology theory admit projective resolutions of length 1 in the Abelian target category.

Generalising earlier results from [3, 11, 15] the verification of the condition mentioned above for *filtrated K-theory* was achieved in [2] for the case that the underlying space is a disjoint union of so-called accordion spaces. A finite connected T_0-space X is an accordion space if and only if the directed graph corresponding to its specialisation pre-order is a Dynkin quiver of type A. Moreover, it was shown in [2, 11] that, if X is a finite T_0-space which is not a disjoint union of accordion spaces, then the projective dimension of filtrated K-theory over X is *not* bounded by 1 and objects in the equivariant bootstrap class are *not* classified by filtrated K-theory. The assumption of the separation axiom T_0 is not a loss of generality in this context (see [9, §2.5]).

There are two natural approaches to tackle the problem arising for non-accordion spaces: one can either try to refine the invariant—this has been done with some success in [11] and [1]; or one can hold onto the invariant and try to establish projective resolutions of length 1 on suitable subcategories or localisations of the category $\mathfrak{KK}(X)$, in which X-equivariant KK-theory is organised. The latter is the course we pursue in this note. We state our results in the next section.

3.2 Statement of Results

The definition of filtrated K-theory and related notation are recalled in Sect. 3.3.

Proposition 1. *Let X be a finite topological space. Assume that the ideal $\mathcal{NT}_{\mathrm{nil}} \subset \mathcal{NT}^*(X)$ is nilpotent and that the decomposition $\mathcal{NT}^*(X) = \mathcal{NT}_{\mathrm{nil}} \rtimes \mathcal{NT}_{\mathrm{ss}}$ holds. Fix $n \in \mathbb{N}$. For an $\mathcal{NT}^*(X)$-module M, the following assertions are equivalent:*

1. *M has a projective resolution of length n.*
2. *The Abelian group $\mathrm{Tor}_n^{\mathcal{NT}^*(X)}(\mathcal{NT}_{\mathrm{ss}}, M)$ is free and the Abelian group $\mathrm{Tor}_{n+1}^{\mathcal{NT}^*(X)}(\mathcal{NT}_{\mathrm{ss}}, M)$ vanishes.*

The basic idea of this paper is to compute the Tor-groups above by writing down projective resolutions for the fixed right-module $\mathcal{NT}_{\mathrm{ss}}$.

Let Z_m be the $(m+1)$-point space on the set $\{1, 2, \ldots, m+1\}$ such that $Y \subseteq Z_m$ is open if and only if $Y \ni m+1$ or $Y = \emptyset$. A C^*-algebra over Z_m is a C^*-algebra A

with a distinguished ideal such that the corresponding quotient decomposes as a direct sum of m orthogonal ideals. Let S be the set $\{1,2,3,4\}$ equipped with the topology $\{\emptyset, 4, 24, 34, 234, 1234\}$, where we write $24 := \{2,4\}$ etc. A C^*-algebra over S is a C^*-algebra together with two distinguished ideals which need not satisfy any further conditions; see [9, Lemma 2.35].

Proposition 2. *Let X be a topological space with at most 4 points. Let $M = \mathrm{FK}(A)$ for some C^*-algebra A over X. Then M has a projective resolution of length 2 and $\mathrm{Tor}_2^{\mathcal{N}\mathcal{T}^*}(\mathcal{N}\mathcal{T}_{\mathrm{ss}}, M) = 0$.*

Moreover, we can find explicit formulas for $\mathrm{Tor}_1^{\mathcal{N}\mathcal{T}^*}(\mathcal{N}\mathcal{T}_{\mathrm{ss}}, M)$; for instance, $\mathrm{Tor}_1^{\mathcal{N}\mathcal{T}^*(Z_3)}(\mathcal{N}\mathcal{T}_{\mathrm{ss}}, M)$ is isomorphic to the homology of the complex

$$\bigoplus_{j=1}^{3} M(j4) \xrightarrow{\begin{pmatrix} i & -i & 0 \\ -i & 0 & i \\ 0 & i & -i \end{pmatrix}} \bigoplus_{k=1}^{3} M(1234 \setminus k) \xrightarrow{(i\ i\ i)} M(1234)\,. \tag{3.1}$$

A similar formula holds for the space S; see (3.6).

The situation simplifies if we consider *rational* $\mathrm{KK}(X)$-theory, whose morphism groups are given by $\mathrm{KK}(X; A, B) \otimes \mathbb{Q}$; see [6]. This is a $\mathbb{Q}$-linear triangulated category which can be constructed as a localisation of $\mathfrak{KK}(X)$; the corresponding localisation of filtrated K-theory is given by $A \mapsto \mathrm{FK}(A) \otimes \mathbb{Q}$ and takes values in the category of modules over the $\mathbb{Q}$-linear category $\mathcal{N}\mathcal{T}^*(X) \otimes \mathbb{Q}$.

Proposition 3. *Let X be a topological space with at most 4 points. Let A and B be C^*-algebras over X. If A belongs to the equivariant bootstrap class $\mathcal{B}(X)$, then there is a natural short exact universal coefficient sequence*

$$\mathrm{Ext}^1_{\mathcal{N}\mathcal{T}^*(X)\otimes\mathbb{Q}}\big(\mathrm{FK}_{*+1}(A)\otimes\mathbb{Q}, \mathrm{FK}_*(B)\otimes\mathbb{Q}\big) \rightarrowtail \mathrm{KK}_*(X; A, B)\otimes\mathbb{Q}$$
$$\twoheadrightarrow \mathrm{Hom}_{\mathcal{N}\mathcal{T}^*(X)\otimes\mathbb{Q}}\big(\mathrm{FK}_*(A)\otimes\mathbb{Q}, \mathrm{FK}_*(B)\otimes\mathbb{Q}\big)\,.$$

In [6], a long exact sequence is constructed which in our setting, by the above proposition, reduces the computation of $\mathrm{KK}_*(X; A, B)$, up to extension problems, to the computation of a certain torsion theory $\mathrm{KK}_*(X; A, B; \mathbb{Q}/\mathbb{Z})$.

The next proposition says that the upper bound of 2 for the projective dimension in Proposition 2 does not hold for all finite spaces.

Proposition 4. *There is an $\mathcal{N}\mathcal{T}^*(Z_4)$-module M of projective dimension 2 with free entries and $\mathrm{Tor}_2^{\mathcal{N}\mathcal{T}^*}(\mathcal{N}\mathcal{T}_{\mathrm{ss}}, M) \neq 0$. The module $M \otimes_{\mathbb{Z}} \mathbb{Z}/k$ has projective dimension 3 for every $k \in \mathbb{N}_{\geq 2}$. Both M and $M \otimes_{\mathbb{Z}} \mathbb{Z}/k$ can be realised as the filtrated* K*-theory of an object in the equivariant bootstrap class $\mathcal{B}(X)$.*

As an application of Proposition 2 we investigate in Sect. 3.10 the obstruction term $\mathrm{Tor}_1^{\mathcal{N}\mathcal{T}^*}\big(\mathcal{N}\mathcal{T}_{\mathrm{ss}}, \mathrm{FK}(A)\big)$ for certain Cuntz-Krieger algebras with four-point primitive ideal spaces. We find:

Proposition 5. *There is a Cuntz-Krieger algebra with primitive ideal space homeomorphic to Z_3 which fulfills Cuntz's condition (II) and has projective dimension 2 in filtrated* K*-theory over Z_3. The analogous statement for the space S holds as well.*

The relevance of this observation lies in the following: *if* Cuntz-Krieger algebras *had* projective dimension at most 1 in filtrated K-theory over their primitive ideal space, this would lead to a strengthened version of Gunnar Restorff's classification result [14] with a proof avoiding reference to results from symbolic dynamics.

3.3 Preliminaries

Let X be a finite topological space. A subset $Y \subseteq X$ is called *locally closed* if it is the difference $U \setminus V$ of two open subsets U and V of X; in this case, U and V can always be chosen such that $V \subseteq U$. The set of locally closed subsets of X is denoted by $\mathbb{LC}(X)$. By $\mathbb{LC}(X)^*$, we denote the set of *non-empty, connected* locally closed subsets of X.

Recall from [9] that a *C^*-algebra over X* is pair (A, ψ) consisting of a C^*-algebra A and a continuous map $\psi\colon \mathrm{Prim}(A) \to X$. A C^*-algebra (A, ψ) over X is called *tight* if the map ψ is a homeomorphism. A C^*-algebra (A, ψ) over X comes with *distinguished subquotients* $A(Y)$ for every $Y \in \mathbb{LC}(X)$.

There is an appropriate version $\mathrm{KK}(X)$ of bivariant K-theory for C^*-algebras over X (see [7, 9]). The corresponding category, denoted by $\mathfrak{KK}(X)$, is equipped with the structure of a triangulated category (see [12]); moreover, there is an equivariant analogue $\mathscr{B}(X) \subseteq \mathfrak{KK}(X)$ of the bootstrap class [9].

Recall that a triangulated category comes with a class of distinguished candidate triangles. An *anti-distinguished* triangle is a candidate triangle which can be obtained from a distinguished triangle by reversing the sign of one of its three morphisms. Both distinguished and anti-distinguished triangles induce long exact Hom-sequences.

As defined in [11], for $Y \in \mathbb{LC}(X)$, we let $\mathrm{FK}_Y(A) := \mathrm{K}_*\big(A(Y)\big)$ denote the $\mathbb{Z}/2$-graded K-group of the subquotient of A associated to Y. Let $\mathscr{NT}(X)$ be the $\mathbb{Z}/2$-graded pre-additive category whose object set is $\mathbb{LC}(X)$ and whose space of morphisms from Y to Z is $\mathscr{NT}_*(X)(Y, Z)$—the $\mathbb{Z}/2$-graded Abelian group of all natural transformations $\mathrm{FK}_Y \Rightarrow \mathrm{FK}_Z$. Let $\mathscr{NT}^*(X)$ be the full subcategory with object set $\mathbb{LC}(X)^*$. We often abbreviate $\mathscr{NT}^*(X)$ by $\mathscr{NT}^*$.

Every open subset of a locally closed subset of X gives rise to an extension of distinguished subquotients. The corresponding natural maps in the associated six-term exact sequence yield morphisms in the category $\mathscr{NT}$, which we briefly denote by i, r and δ.

A *(left-)module* over $\mathscr{NT}(X)$ is a grading-preserving, additive functor from $\mathscr{NT}(X)$ to the category $\mathfrak{Ab}^{\mathbb{Z}/2}$ of $\mathbb{Z}/2$-graded Abelian groups. A morphism of $\mathscr{NT}(X)$-modules is a natural transformation of functors. Similarly, we define left-modules over $\mathcal{NT}^*(X)$. By $\mathfrak{Mod}\big(\mathcal{NT}^*(X)\big)_{\mathrm{c}}$ we denote the category of countable $\mathcal{NT}^*(X)$-modules.

Filtrated K-*theory* is the functor $\mathfrak{KK}(X) \to \mathfrak{Mod}\big(\mathcal{NT}^*(X)\big)_{\mathrm{c}}$ which takes a C^*-algebra A over X to the collection $\big(\mathrm{K}_*(A(Y))\big)_{Y\in\mathbb{LC}(X)^*}$ equipped with the obvious $\mathcal{NT}^*(X)$-module structure.

Let $\mathcal{NT}_{\mathrm{nil}} \subset \mathcal{NT}^*$ be the ideal generated by all natural transformations between different objects, and let $\mathcal{NT}_{\mathrm{ss}} \subset \mathcal{NT}^*$ be the subgroup spanned by the identity transformations id_Y^Y for objects $Y \in \mathbb{LC}(X)^*$. The subgroup $\mathcal{NT}_{\mathrm{ss}}$ is in fact a subring of $\mathcal{NT}^*$ isomorphic to $\mathbb{Z}^{\mathbb{LC}(X)^*}$. We say that $\mathcal{NT}^*$ decomposes as semi-direct product $\mathcal{NT}^* = \mathcal{NT}_{\mathrm{nil}} \rtimes \mathcal{NT}_{\mathrm{ss}}$ if $\mathcal{NT}^*$ as an Abelian group is the inner direct sum of $\mathcal{NT}_{\mathrm{nil}}$ and $\mathcal{NT}_{\mathrm{ss}}$; see [2, 11]. We do not know if this fails for any finite space.

We define *right-modules* over $\mathcal{NT}^*(X)$ as *contravariant*, grading-preserving, additive functors $\mathcal{NT}^*(X) \to \mathfrak{Ab}^{\mathbb{Z}/2}$. If we do not specify between left and right, then we always mean left-modules. The subring $\mathcal{NT}_{\mathrm{ss}} \subset \mathcal{NT}^*$ is regarded as an $\mathcal{NT}^*$-right-module by the obvious action: The ideal $\mathcal{NT}_{\mathrm{nil}} \subset \mathcal{NT}^*$ acts trivially, while $\mathcal{NT}_{\mathrm{ss}}$ acts via right-multiplication in $\mathcal{NT}_{\mathrm{ss}} \cong \mathbb{Z}^{\mathbb{LC}(X)^*}$. For an $\mathcal{NT}^*$-module M, we set $M_{\mathrm{ss}} := M/\mathcal{NT}_{\mathrm{nil}} \cdot M$.

For $Y \in \mathbb{LC}(X)^*$ we define the *free* $\mathcal{NT}^*$*-left-module on* Y by $P_Y(Z) := \mathcal{NT}(Y,Z)$ for all $Z \in \mathbb{LC}(X)^*$ and similarly for morphisms $Z \to Z'$ in $\mathcal{NT}^*$. Analogously, we define the *free* $\mathcal{NT}^*$*-right-module on* Y by $Q_Y(Z) := \mathcal{NT}(Z,Y)$ for all $Z \in \mathbb{LC}(X)^*$. An $\mathcal{NT}^*$-left/right-module is called *free* if it is isomorphic to a direct sum of degree-shifted free left/right-modules on objects $Y \in \mathbb{LC}(X)^*$. It follows directly from Yoneda's Lemma that free $\mathcal{NT}^*$-left/right-modules are projective.

An $\mathcal{NT}$-module M is called *exact* if the $\mathbb{Z}/2$-graded chain complexes

$$\cdots \to M(U) \xrightarrow{i_U^Y} M(Y) \xrightarrow{r_Y^{Y\setminus U}} M(Y\setminus U) \xrightarrow{\delta_{Y\setminus U}^U} M(U)[1] \to \cdots$$

are exact for all $U, Y \in \mathbb{LC}(X)$ with U open in Y. An $\mathcal{NT}^*$-module M is called *exact* if the corresponding $\mathcal{NT}$-module is exact (see [2]).

We use the notation $C \in\in \mathscr{C}$ to denote that C is an object in a category $\mathscr{C}$.

In [11], the functors FK_Y are shown to be representable, that is, there are objects $\mathscr{R}_Y \in\in \mathfrak{KK}(X)$ and isomorphisms of functors $\mathrm{FK}_Y \cong \mathrm{KK}_*(X;\mathscr{R}_Y,\llcorner\!\lrcorner)$. We let $\widehat{\mathrm{FK}}$ denote the stable *co*homological functor on $\mathfrak{KK}(X)$ represented by the same set of objects $\{\mathscr{R}_Y \mid Y \in \mathbb{LC}(X)^*\}$; it takes values in $\mathcal{NT}^*$-*right*-modules. We warn that $\mathrm{KK}_*(X;A,\mathscr{R}_Y)$ does not identify with the K-homology of $A(Y)$. By Yoneda's lemma, we have $\mathrm{FK}(\mathscr{R}_Y) \cong P_Y$ and $\widehat{\mathrm{FK}}(\mathscr{R}_Y) \cong Q_Y$.

We occasionally use terminology from [10, 11] concerning homological algebra in $\mathfrak{KK}(X)$ relative to the ideal $\mathfrak{J} := \ker(\mathrm{FK})$ of morphisms in $\mathfrak{KK}(X)$ inducing trivial module maps on FK. An object $A \in\in \mathfrak{KK}(X)$ is called $\mathfrak{J}$*-projective* if $\mathfrak{J}(A,B) = 0$ for every $B \in\in \mathfrak{KK}(X)$. We recall from [10] that FK restricts to an equivalence of categories between the subcategories of $\mathfrak{J}$-projective objects in

$\mathfrak{KK}(X)$ and of projective objects in $\mathfrak{Mod}\big(\mathcal{NT}^*(X)\big)_{\mathrm{c}}$. Similarly, the functor $\widehat{\mathrm{FK}}$ induces a contravariant equivalence between the $\mathfrak{J}$-projective objects in $\mathfrak{KK}(X)$ and projective $\mathcal{NT}^*$-*right*-modules.

3.4 Proof of Proposition 1

Recall the following result from [11].

Lemma 1 ([11, Theorem 3.12]). *Let X be a finite topological space. Assume that the ideal $\mathcal{NT}_{\mathrm{nil}} \subset \mathcal{NT}^*(X)$ is nilpotent and that the decomposition $\mathcal{NT}^*(X) = \mathcal{NT}_{\mathrm{nil}} \rtimes \mathcal{NT}_{\mathrm{ss}}$ holds. Let M be an $\mathcal{NT}^*(X)$-module. The following assertions are equivalent:*

1. *M is a free $\mathcal{NT}^*(X)$-module.*
2. *M is a projective $\mathcal{NT}^*(X)$-module.*
3. *M_{ss} is a free Abelian group and $\mathrm{Tor}_1^{\mathcal{NT}^*(X)}(\mathcal{NT}_{\mathrm{ss}}, M) = 0$.*

Now we prove Proposition 1. We consider the case $n = 1$ first. Choose an epimorphism $f\colon P \twoheadrightarrow M$ for some projective module P, and let K be its kernel. M has a projective resolution of length 1 if and only if K is projective. By Lemma 1, this is equivalent to K_{ss} being a free Abelian group and $\mathrm{Tor}_1^{\mathcal{NT}^*}(\mathcal{NT}_{\mathrm{ss}}, K) = 0$. We have $\mathrm{Tor}_1^{\mathcal{NT}^*}(\mathcal{NT}_{\mathrm{ss}}, K) = 0$ if and only if $\mathrm{Tor}_2^{\mathcal{NT}^*}(\mathcal{NT}_{\mathrm{ss}}, M) = 0$ because these groups are isomorphic. We will show that K_{ss} is free if and only if $\mathrm{Tor}_1^{\mathcal{NT}^*}(\mathcal{NT}_{\mathrm{ss}}, M)$ is free. The extension $K \rightarrowtail P \twoheadrightarrow M$ induces the following long exact sequence:

$$0 \to \mathrm{Tor}_1^{\mathcal{NT}^*}(\mathcal{NT}_{\mathrm{ss}}, M) \to K_{\mathrm{ss}} \to P_{\mathrm{ss}} \to M_{\mathrm{ss}} \to 0\,.$$

Assume that K_{ss} is free. Then its subgroup $\mathrm{Tor}_1^{\mathcal{NT}^*}(\mathcal{NT}_{\mathrm{ss}}, M)$ is free as well. Conversely, if $\mathrm{Tor}_1^{\mathcal{NT}^*}(\mathcal{NT}_{\mathrm{ss}}, M)$ is free, then K_{ss} is an extension of free Abelian groups and thus free. Notice that P_{ss} is free because P is projective. The general case $n \in \mathbb{N}$ follows by induction using an argument based on syzygies as above. This completes the proof of Proposition 1.

3.5 Free Resolutions for $\mathcal{NT}_{\mathrm{ss}}$

The $\mathcal{NT}^*$-right-module $\mathcal{NT}_{\mathrm{ss}}$ decomposes as a direct sum $\bigoplus_{Y\in\mathbb{LC}(X)^*} S_Y$ of the simple submodules S_Y which are given by $S_Y(Y) \cong \mathbb{Z}$ and $S_Y(Z) = 0$ for $Z \neq Y$. We obtain

$$\mathrm{Tor}_n^{\mathcal{NT}^*}(\mathcal{NT}_{\mathrm{ss}}, M) = \bigoplus_{Y\in\mathbb{LC}(X)^*} \mathrm{Tor}_n^{\mathcal{NT}}(S_Y, M)\,.$$

Our task is then to write down projective resolutions for the $\mathscr{N}\mathscr{T}^*$-right-modules S_Y. The first step is easy: we map Q_Y onto S_Y by mapping the class of the identity in $Q_Y(Y)$ to the generator of $S_Y(Y)$. Extended by zero, this yields an epimorphism $Q_Y \twoheadrightarrow S_Y$.

In order to surject onto the kernel of this epimorphism, we use the indecomposable transformations in $\mathscr{N}\mathscr{T}^*$ whose range is Y. Denoting these by $\eta_i \colon W_i \to Y$, $1 \leq i \leq n$, we obtain the two step resolution

$$\bigoplus_{i=1}^{n} Q_{W_i} \xrightarrow{(\eta_1\ \eta_2\ \cdots\ \eta_n)} Q_Y \twoheadrightarrow S_Y\ .$$

In the notation of [11], the map $\bigoplus_{i=1}^{n} Q_{W_i} \to Q_Y$ corresponds to a morphism $\phi \colon \mathscr{R}_Y \to \bigoplus_{i=1}^{n} \mathscr{R}_{W_i}$ of $\mathfrak{I}$-projectives in $\mathfrak{KK}(X)$. If the mapping cone C_ϕ of ϕ is again $\mathfrak{I}$-projective, the distinguished triangle $\Sigma C_\phi \to \mathscr{R}_Y \overset{\phi}{\to} \bigoplus_{i=1}^{n} \mathscr{R}_{W_i} \to C_\phi$ yields the projective resolution

$$\cdots \to Q_Y \to Q_\phi[1] \to \bigoplus_{i=1}^{n} Q_{W_i}[1] \to Q_Y[1] \to Q_\phi \to \bigoplus_{i=1}^{n} Q_{W_i} \to Q_Y \twoheadrightarrow S_Y\ ,$$

where $Q_\phi = \mathrm{FK}(C_\phi)$. We denote periodic resolutions like this by

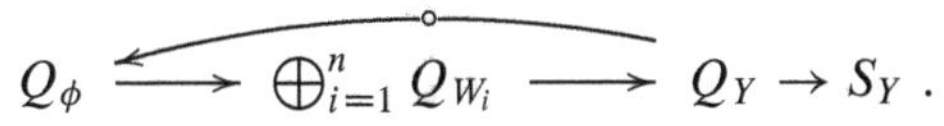

If the mapping cone C_ϕ is not $\mathfrak{I}$-projective, the situation has to be investigated individually. We will see examples of this in Sects. 3.7 and 3.9. The resolutions we construct in these cases exhibit a certain six-term periodicity as well. However, they begin with a finite number of "non-periodic steps" (one in Sect. 3.7 and two in Sect. 3.9), which can be considered as a symptom of the deficiency of the invariant filtrated K-theory over non-accordion spaces from the homological viewpoint. We remark without proof that the mapping cone of the morphism $\phi \colon \mathscr{R}_Y \to \bigoplus_{i=1}^{n} \mathscr{R}_{W_i}$ is $\mathfrak{I}$-projective for every $Y \in \mathbb{LC}(X)^*$ if and only if X is a disjoint union of accordion spaces.

3.6 Tensor Products with Free Right-Modules

Lemma 2. *Let M be an $\mathscr{N}\mathscr{T}^*$-left-module. There is an isomorphism $Q_Y \otimes_{\mathscr{N}\mathscr{T}^*} M \cong M(Y)$ of $\mathbb{Z}/2$-graded Abelian groups which is natural in $Y \in\in \mathscr{N}\mathscr{T}^*$.*

Proof. This is a simple consequence of Yoneda's lemma and the tensor-hom adjunction.

Lemma 3. *Let*

$$\Sigma\mathscr{R}_{(3)} \xrightarrow{\gamma} \mathscr{R}_{(1)} \xrightarrow{\alpha} \mathscr{R}_{(2)} \xrightarrow{\beta} \mathscr{R}_{(3)}$$

be a distinguished or anti-distinguished triangle in $\mathfrak{KK}(X)$, *where*

$$\mathscr{R}_{(i)} = \bigoplus_{j=1}^{m_i} \mathscr{R}_{Y_j^i} \oplus \bigoplus_{k=1}^{n_i} \Sigma\mathscr{R}_{Z_k^i}$$

for $1 \leq i \leq 3$, $m_i, n_i \in \mathbb{N}$ *and* $Y_j^i, Z_k^i \in \mathbb{LC}(X)^*$. *Set* $Q_{(i)} = \widehat{\mathrm{FK}}(\mathscr{R}_{(i)})$. *If* $M = \mathrm{FK}(A)$ *for some* $A \in\in \mathfrak{KK}(X)$, *then the induced sequence*

$$\begin{array}{ccccc}
Q_{(1)} \otimes_{\mathscr{N}\mathscr{T}^*} M & \xrightarrow{\alpha^*\otimes\mathrm{id}_M} & Q_{(2)} \otimes_{\mathscr{N}\mathscr{T}^*} M & \xrightarrow{\beta^*\otimes\mathrm{id}_M} & Q_{(3)} \otimes_{\mathscr{N}\mathscr{T}^*} M \\
{\scriptstyle \gamma^*\otimes\mathrm{id}_M[1]}\uparrow & & & & \downarrow{\scriptstyle \gamma^*\otimes\mathrm{id}_M} \\
Q_{(3)} \otimes_{\mathscr{N}\mathscr{T}^*} M[1] & \xleftarrow[\beta^*\otimes\mathrm{id}_M[1]]{} & Q_{(2)} \otimes_{\mathscr{N}\mathscr{T}^*} M[1] & \xleftarrow[\alpha^*\otimes\mathrm{id}_M[1]]{} & Q_{(1)} \otimes_{\mathscr{N}\mathscr{T}^*} M[1]
\end{array} \tag{3.2}$$

is exact.

Proof. Using the previous lemma and the representability theorem, we naturally identify $Q_{(i)} \otimes_{\mathscr{N}\mathscr{T}^*} M \cong \mathrm{KK}_*(X; \mathscr{R}_{(i)}, A)$. Since, in triangulated categories, distinguished or anti-distinguished triangles induce long exact Hom-sequences, the sequence (3.2) is thus exact.

3.7 Proof of Proposition 2

We may restrict to connected T_0-spaces. In [9], a list of isomorphism classes of connected T_0-spaces with three or four points is given. If X is a disjoint union of accordion spaces, then the assertion follows from [2]. The remaining spaces fall into two classes:

1. All connected non-accordion four-point T_0-spaces except for the pseudocircle;
2. The pseudocircle (see Sect. 3.7.2).

The spaces in the first class have the following in common: If we fix two of them, say X, Y, then there is an ungraded isomorphism $\Phi\colon \mathscr{N}\mathscr{T}^*(X) \to \mathscr{N}\mathscr{T}^*(Y)$ between the categories of natural transformations on the respective filtrated K-theories such that the induced equivalence of ungraded module categories

$$\Phi^*\colon \mathfrak{Mod}^{\mathrm{ungr}}\big(\mathcal{NT}^*(Y)\big)_{\mathrm{c}} \to \mathfrak{Mod}^{\mathrm{ungr}}\big(\mathcal{NT}^*(X)\big)_{\mathrm{c}}$$

restricts to a bijective correspondence between exact ungraded $\mathcal{NT}^*(Y)$-modules and exact ungraded $\mathcal{NT}^*(X)$-modules. Moreover, the isomorphism Φ restricts to an isomorphism from $\mathcal{NT}_{\mathrm{ss}}(X)$ onto $\mathcal{NT}_{\mathrm{ss}}(Y)$ and one from $\mathcal{NT}_{\mathrm{nil}}(X)$ onto $\mathcal{NT}_{\mathrm{nil}}(Y)$. In particular, the assertion holds for X if and only if it holds for Y.

The above is a consequence of the investigations in [1, 2, 11]; the same kind of relation was found in [2] for the categories of natural transformations associated to accordion spaces with the same number of points. As a consequence, it suffices to verify the assertion for one representative of the first class—we choose Z_3—and for the pseudocircle.

3.7.1 *Resolutions for the Space Z_3*

We refer to [11] for a description of the category $\mathcal{NT}^*(Z_3)$, which in particular implies, that the space Z_3 satisfies the conditions of Proposition 1. Using the extension triangles from [11, (2.5)], the procedure described in Sect. 3.5 yields the following projective resolutions induced by distinguished triangles as in Lemma 3:

$$Q_1[1] \longrightarrow Q_4 \longrightarrow Q_{14} \to S_{14}\,, \qquad \text{and similarly for } S_{24},\, S_{34};$$

$$Q_{1234}[1] \longrightarrow Q_1[1] \oplus Q_2[1] \oplus Q_3[1] \longrightarrow Q_4 \to S_4\,;$$

$$Q_{234} \longrightarrow Q_{1234} \longrightarrow Q_1 \to S_1\,, \qquad \text{and similarly for } S_2,\, S_3.$$

Next we will deal with the modules S_{jk4}, where $1 \le j < k \le 3$. We observe that there is a Mayer-Vietoris type exact sequence of the form

$$Q_4 \longrightarrow Q_{j4} \oplus Q_{k4} \longrightarrow Q_{jk4}\,. \tag{3.3}$$

Lemma 4. *The candidate triangle $\Sigma\mathcal{R}_4 \to \mathcal{R}_{jk4} \to \mathcal{R}_{j4} \oplus \mathcal{R}_{k4} \to \mathcal{R}_4$ corresponding to the periodic part of the sequence* (3.3) *is distinguished or anti-distinguished (depending on the choice of signs for the maps in* (3.3)*).*

Proof. We give the proof for $j = 1$ and $k = 2$. The other cases follow from cyclicly permuting the indices 1, 2 and 3. We denote the morphism $\mathcal{R}_{124} \to \mathcal{R}_{14} \oplus \mathcal{R}_{24}$ by φ and the corresponding map $Q_{14} \oplus Q_{24} \to Q_{124}$ in (3.3) by φ^*. It suffices to check that $\widehat{\mathrm{FK}}(\mathrm{Cone}_\varphi)$ and Q_4 correspond, possibly up to a sign, to the same element in $\mathrm{Ext}^1_{\mathcal{NT}^*(Z_3)^{\mathrm{op}}}\big(\ker(\varphi^*), \mathrm{coker}(\varphi^*)[1]\big)$. We have $\mathrm{coker}(\varphi^*) \cong S_{124}$ and an extension $S_{124}[1] \rightarrowtail Q_4 \twoheadrightarrow \ker(\varphi^*)$. Since $\operatorname{Hom}(Q_4, S_{124}[1]) \cong S_{124}(4)[1] = 0$ and $\operatorname{Ext}^1(Q_4, S_{124}[1]) = 0$ because Q_4 is projective, the long exact Ext-sequence yields $\operatorname{Ext}^1\big(\ker(\varphi^*), \operatorname{coker}(\varphi^*)[1]\big) \cong \operatorname{Hom}(S_{124}[1], S_{124}[1]) \cong \mathbb{Z}$. Considering the sequence of transformations $3 \xrightarrow{\delta} 124 \xrightarrow{i} 1234 \xrightarrow{r} 3$, it is straight-forward to check that such an extension corresponds to one of the generators $\pm 1 \in \mathbb{Z}$ if and only if its underlying module is exact. This concludes the proof because both $\widehat{\mathrm{FK}}(\mathrm{Cone}_\varphi)$ and Q_4 are exact.

Hence we obtain the following projective resolutions induced by distinguished or anti-distinguished triangles as in Lemma 3:

$$Q_4 \longrightarrow Q_{j4} \oplus Q_{k4} \longrightarrow Q_{jk4} \to S_{jk4} \ .$$

To summarize, by Lemma 3, $\operatorname{Tor}_n^{\mathcal{N}\mathcal{T}^*}(S_Y, M) = 0$ for $Y \neq 1234$ and $n \geq 1$.

As we know from [11], the subset 1234 of Z_3 plays an exceptional role. In the notation of [11] (with the direction of the arrows reversed because we are dealing with *right*-modules), the kernel of the homomorphism $Q_{124} \oplus Q_{134} \oplus Q_{234} \xrightarrow{(i\ i\ i)} Q_{1234}$ is of the form

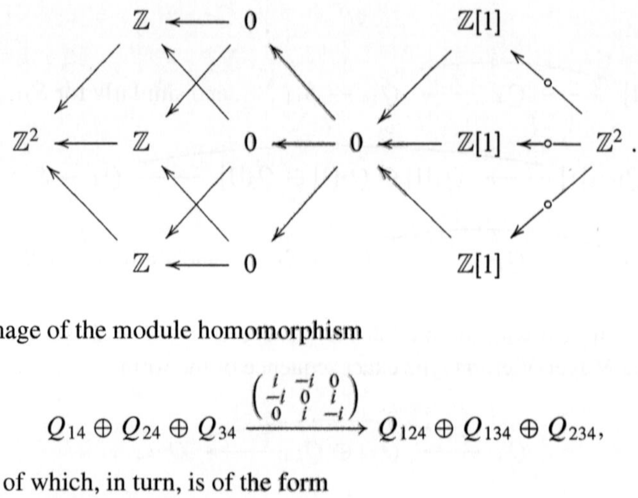

It is the image of the module homomorphism

$$Q_{14} \oplus Q_{24} \oplus Q_{34} \xrightarrow{\begin{pmatrix} i & -i & 0 \\ -i & 0 & i \\ 0 & i & -i \end{pmatrix}} Q_{124} \oplus Q_{134} \oplus Q_{234}, \tag{3.4}$$

the kernel of which, in turn, is of the form

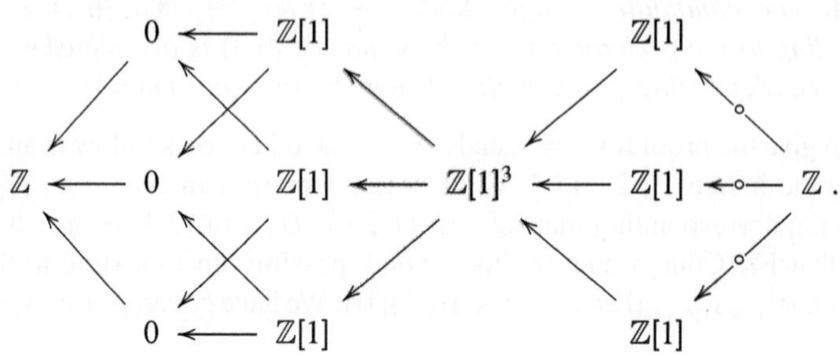

A surjection from $Q_4 \oplus Q_{1234}[1]$ onto this module is given by $\left(\begin{smallmatrix} i & i & i \\ \delta^{14}_{1234} & 0 & 0 \end{smallmatrix}\right)$, where $\delta^{14}_{1234} := \delta^{14}_{3} \circ r^{3}_{1234}$. The kernel of this homomorphism has the form

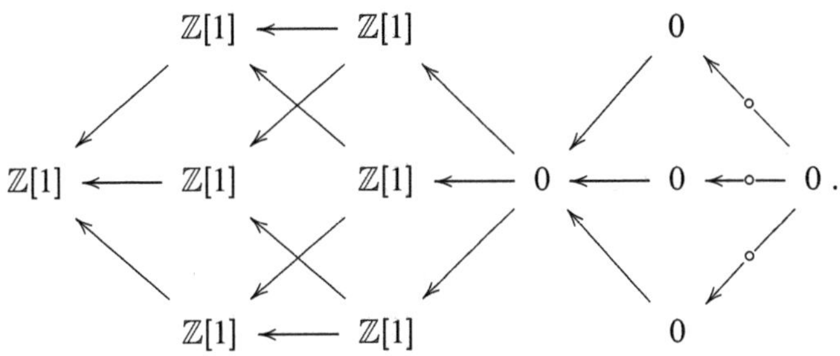

This module is isomorphic to $\mathrm{Syz}_{1234}[1]$, where $\mathrm{Syz}_{1234} := \ker(Q_{1234} \twoheadrightarrow S_{1234})$. Therefore, we end up with the projective resolution

$$Q_4 \oplus Q_{1234}[1] \longrightarrow Q_{14} \oplus Q_{24} \oplus Q_{34} \longrightarrow Q_{124} \oplus Q_{134} \oplus Q_{234} \longrightarrow Q_{1234} \to S_{1234}\,. \tag{3.5}$$

The homomorphism from $Q_{124} \oplus Q_{134} \oplus Q_{234}$ to $Q_4 \oplus Q_{1234}[1]$ is given by

$$\begin{pmatrix} 0 & 0 & -\delta^{4}_{234} \\ i & i & i \end{pmatrix},$$

where $\delta^{4}_{234} := \delta^{4}_{2} \circ r^{2}_{234}$.

Lemma 5. *The candidate triangle in $\mathfrak{KK}(X)$ corresponding to the periodic part of the sequence* (3.5) *is distinguished or anti-distinguished (depending on the choice of signs for the maps in* (3.5)*).*

Proof. The argument is analogous to the one in the proof of Lemma 4. Again, we consider the group $\mathrm{Ext}^1_{\mathcal{N}\mathcal{T}^*(Z_3)^{\mathrm{op}}}\big(\ker(\varphi^*), \mathrm{coker}(\varphi^*)[1]\big)$ where φ^* now denotes the map (3.4). We have $\mathrm{coker}(\varphi^*) \cong \mathrm{Syz}_{1234}$ and an extension $Q_4 \rightarrowtail \ker(\varphi^*) \twoheadrightarrow S_{1234}[1]$. Using long exact sequences, we obtain

$$\begin{aligned} \mathrm{Ext}^1\big(\ker(\varphi^*), \mathrm{coker}(\varphi^*)[1]\big) &\cong \mathrm{Ext}^1(S_{1234}[1], \mathrm{Syz}_{1234}[1]) \\ &\cong \mathrm{Hom}(S_{1234}[1], S_{1234}[1]) \cong \mathbb{Z}. \end{aligned}$$

Again, an extension corresponds to a generator if and only if its underlying module is exact.

By the previous lemma and Sect. 3.6, computing the tensor product of this complex with M and taking homology shows that $\mathrm{Tor}_n^{\mathcal{NT}^*}(\mathcal{NT}_{\mathrm{ss}}, M) = 0$ for $n \geq 2$ and that $\mathrm{Tor}_1^{\mathcal{NT}^*}(\mathcal{NT}_{\mathrm{ss}}, M)$ is equal to $\mathrm{Tor}_1^{\mathcal{NT}^*}(S_{1234}, M)$ and isomorphic to the homology of the complex (3.1).

Example 1. For the filtrated K-module with projective dimension 2 constructed in [11, §5] we get $\mathrm{Tor}_1^{\mathcal{NT}^*}(\mathcal{NT}_{\mathrm{ss}}, M) \cong \mathbb{Z}/k$.

Remark 1. As explicated in the beginning of this section, the category $\mathcal{NT}^*(S)$ corresponding to the four-point space S defined in the introduction is isomorphic in an appropriate sense to the category $\mathcal{NT}^*(Z_3)$. As has been established in [1], the indecomposable morphisms in $\mathcal{NT}^*(S)$ are organised in the diagram

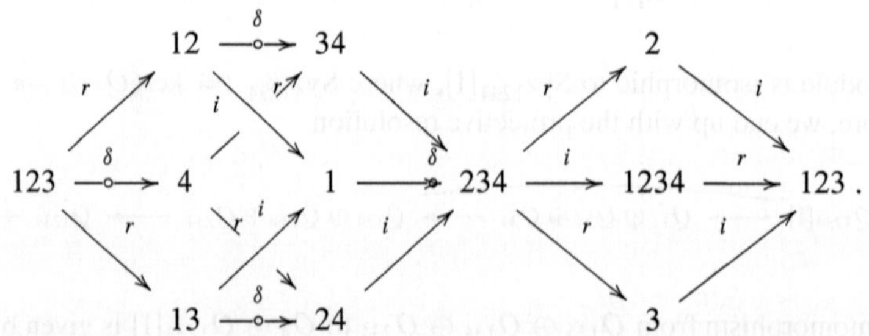

In analogy to (3.1), we have that $\mathrm{Tor}_1^{\mathcal{NT}^*(S)}(\mathcal{NT}_{\mathrm{ss}}, M)$ is isomorphic to the homology of the complex

$$M(12)[1] \oplus M(4) \oplus M(13)[1] \xrightarrow{\begin{pmatrix} \delta & -r & 0 \\ -i & 0 & i \\ 0 & r & -\delta \end{pmatrix}} M(34) \oplus M(1)[1] \oplus M(24) \xrightarrow{(i\ \delta\ i)} M(234)\,, \quad (3.6)$$

where $M = \mathrm{FK}(A)$ for some separable C^*- algebra A over X.

3.7.2 Resolutions for the Pseudocircle

Let $C_2 = \{1, 2, 3, 4\}$ with the partial order defined by $1 < 3$, $1 < 4$, $2 < 3$, $2 < 4$. The topology on C_2 is thus given by $\{\emptyset, 3, 4, 34, 134, 234, 1234\}$. Hence the non-empty, connected, locally closed subsets are

$$\mathbb{LC}(C_2)^* = \{3, 4, 134, 234, 1234, 13, 14, 23, 24, 124, 123, 1, 2\}\,.$$

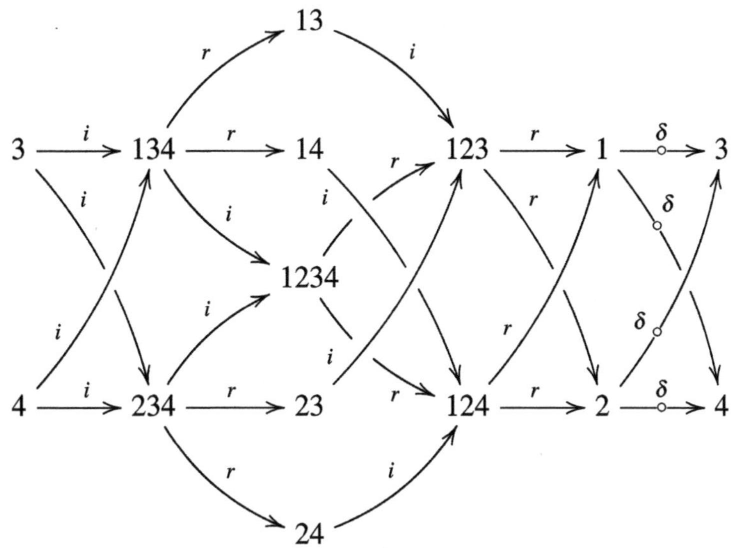

Fig. 3.1 Indecomposable natural transformations in $\mathscr{N}\mathscr{T}^*(C_2)$

The partial order on C_2 corresponds to the directed graph

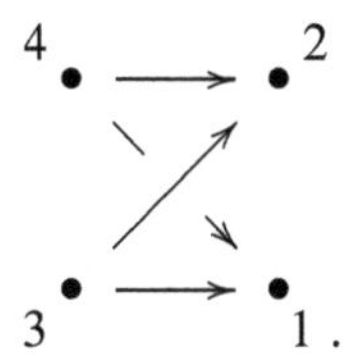

The space C_2 is the only T_0-space with at most four points with the property that its order complex (see [11, Definition 2.6]) is not contractible; in fact, it is homeomorphic to the circle $\mathbb{S}^1$. Therefore, by the representability theorem [11, §2.1] we find

$$\mathscr{N}\mathscr{T}_*(C_2, C_2) \cong \mathrm{KK}_*(X; \mathscr{R}_{C_2}, \mathscr{R}_{C_2}) \cong \mathrm{K}_*\big(\mathscr{R}_{C_2}(C_2)\big) \cong \mathrm{K}^*\left(\mathbb{S}^1\right) \cong \mathbb{Z} \oplus \mathbb{Z}[1]\,,$$

that is, there are non-trivial odd natural transformations $\mathrm{FK}_{C_2} \Rightarrow \mathrm{FK}_{C_2}$. These are generated, for instance, by the composition $C_2 \xrightarrow{r} 1 \xrightarrow{\delta} 3 \xrightarrow{i} C_2$. This follows from the description of the category $\mathscr{N}\mathscr{T}^*(C_2)$ below. Note that $\delta_{C_2}^{C_2} \circ \delta_{C_2}^{C_2}$ vanishes because it factors through $r_{13}^1 \circ i_3^{13} = 0$.

Figure 3.1 displays a set of indecomposable transformations generating the category $\mathscr{N}\mathscr{T}^*(C_2)$ determined in [1, §6.3.2], where also a list of relations generating the relations in the category $\mathscr{N}\mathscr{T}^*(C_2)$ can be found. From this, it is straight-forward to verify that the space C_2 satisfies the conditions of Proposition 1.

Proceeding as described in Sect. 3.5, we find projective resolutions of the following form (we omit explicit descriptions of the boundary maps):

$$Q_{123}[1] \longrightarrow Q_1[1] \oplus Q_2[1] \longrightarrow Q_3 \to S_3\,, \qquad \text{and similarly for } S_4;$$

$$Q_1[1] \longrightarrow Q_3 \oplus Q_4 \longrightarrow Q_{134} \to S_{134}\,, \qquad \text{and similarly for } S_{234};$$

$$Q_4 \longrightarrow Q_{134} \longrightarrow Q_{13} \to S_{13}\,, \qquad \text{and similarly for } S_{14}, S_{23}, S_{24};$$

$$Q_3 \oplus Q_4 \longrightarrow Q_{134} \oplus Q_{234} \longrightarrow Q_{1234} \to S_{1234}\,;$$

$$Q_4 \oplus Q_{123}[1] \longrightarrow Q_{134} \oplus Q_{234} \longrightarrow Q_{1234} \oplus Q_{13} \oplus Q_{23} \to Q_{123} \to S_{123}\,,$$

and similarly for S_{124};

$$Q_{234} \oplus Q_1[1] \longrightarrow Q_{1234} \oplus Q_{23} \oplus Q_{24} \longrightarrow Q_{123} \oplus Q_{124} \to Q_1 \to S_1\,,$$

and similarly for S_2. Again, the periodic part of each of these resolutions is induced by an extension triangle, a Mayer-Vietoris triangle as in Lemma 4 or a more exotic (anti-)distinguished triangle as in Lemma 5 (we omit the analogous computation here).

We get $\mathrm{Tor}_1^{\mathcal{NT}^*}(S_Y, M) = 0$ for every $Y \in \mathbb{LC}(C_2)^* \setminus \{123, 124, 1, 2\}$, and further $\mathrm{Tor}_n^{\mathcal{NT}^*}(S_Y, M) = 0$ for all $Y \in \mathbb{LC}(C_2)^*$ and $n \geq 2$. Therefore,

$$\mathrm{Tor}_1^{\mathcal{NT}^*}(\mathcal{NT}_{\mathrm{ss}}, M) \cong \bigoplus_{Y \in \{123,124,1,2\}} \mathrm{Tor}_1^{\mathcal{NT}^*}(S_Y, M)\,.$$

The four groups $\mathrm{Tor}_1^{\mathcal{NT}^*}(S_Y, M)$ with $Y \in \{123, 124, 1, 2\}$ can be described explicitly as in Sect. 3.7.1 using the above resolutions. This finishes the proof of Proposition 2.

3.8 Proof of Proposition 3

We apply the Meyer-Nest machinery to the homological functor $\mathrm{FK} \otimes \mathbb{Q}$ on the triangulated category $\mathfrak{KK}(X) \otimes \mathbb{Q}$. We need to show that every $\mathcal{NT}^* \otimes \mathbb{Q}$ module of the form $M = \mathrm{FK}(A) \otimes \mathbb{Q}$ has a projective resolution of length 1. It is easy to see that analogues of Propositions 1 and 2 hold. In particular, the term $\mathrm{Tor}_2^{\mathcal{NT}^* \otimes \mathbb{Q}}(\mathcal{NT}_{\mathrm{ss}} \otimes \mathbb{Q}, M)$ always vanishes. Here we use that $\mathbb{Q}$ is a flat

$\mathbb{Z}$-module, so that tensoring with $\mathbb{Q}$ turns projective $\mathcal{NT}^*$-module resolutions into projective $\mathcal{NT}^* \otimes \mathbb{Q}$-module resolutions. Moreover, the freeness condition for the $\mathbb{Q}$-module $\mathrm{Tor}_1^{\mathcal{NT}^*\otimes\mathbb{Q}}(\mathcal{NT}_{\mathrm{ss}} \otimes \mathbb{Q}, M)$ is empty since $\mathbb{Q}$ is a field.

3.9 Proof of Proposition 4

The computations to determine the category $\mathcal{NT}^*(Z_4)$ are very similar to those for the category $\mathcal{NT}^*(Z_3)$ which were carried out in [11]. We summarise its structure in Fig. 3.2. The relations in $\mathcal{NT}^*(Z_4)$ are generated by the following:

- The hypercube with vertices $5, 15, 25, \ldots, 12345$ is a commuting diagram;
- The following compositions vanish:

$$1235 \xrightarrow{i} 12345 \xrightarrow{r} 4\,, \quad 1245 \xrightarrow{i} 12345 \xrightarrow{r} 3\,,$$

$$1345 \xrightarrow{i} 12345 \xrightarrow{r} 2\,, \quad 2345 \xrightarrow{i} 12345 \xrightarrow{r} 1\,,$$

$$1 \xrightarrow{\delta} 5 \xrightarrow{i} 15\,, \quad 2 \xrightarrow{\delta} 5 \xrightarrow{i} 25\,, \quad 3 \xrightarrow{\delta} 5 \xrightarrow{i} 35\,, \quad 4 \xrightarrow{\delta} 5 \xrightarrow{i} 45\,;$$

- The sum of the four maps $12345 \to 5$ via 1, 2, 3, and 4 vanishes.

This implies that the space Z_4 satisfies the conditions of Proposition 1.

In the following, we will define an exact $\mathcal{NT}^*$-left-module M and compute $\mathrm{Tor}_2^{\mathcal{NT}^*}(S_{12345}, M)$. By explicit computation, one finds a projective resolution of the simple $\mathcal{NT}^*$-right-module S_{12345} of the following form (again omitting explicit formulas for the boundary maps):

$$\cdots \to Q_5 \oplus \bigoplus_{1\le i\le 4} Q_{12345\setminus i}[1] \longrightarrow \bigoplus_{1\le l\le 4} Q_{l5} \oplus Q_{12345}[1] \longrightarrow \bigoplus_{1\le j<k\le 4} Q_{jk5} \to$$
$$\to \bigoplus_{1\le i\le 4} Q_{12345\setminus i} \longrightarrow Q_{12345} \longrightarrow S_{12345}.$$

Notice that this sequence is periodic as a cyclic six-term sequence except for the first *two* steps.

Consider the exact $\mathcal{NT}^*$-left-module M defined by the exact sequence

$$0 \to P_{12345} \xrightarrow{\begin{pmatrix} i\\ i\\ i\\ i \end{pmatrix}} \bigoplus_{1\le i\le 4} P_{12345\setminus i} \xrightarrow{\begin{pmatrix} i & -i & 0 & 0\\ -i & 0 & i & 0\\ 0 & i & -i & 0\\ i & 0 & 0 & -i\\ 0 & -i & 0 & i\\ 0 & 0 & i & -i \end{pmatrix}} \bigoplus_{1\le j<k\le 4} P_{jk5} \twoheadrightarrow M\,. \tag{3.7}$$

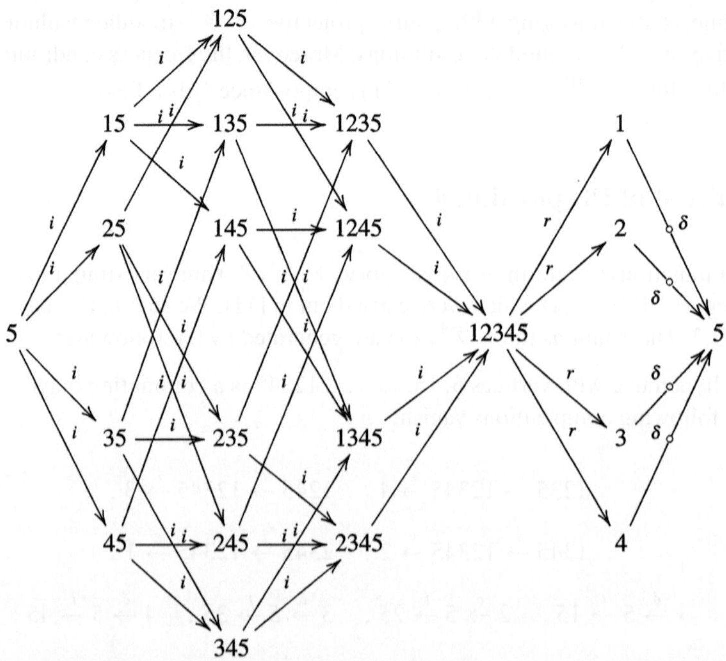

Fig. 3.2 Indecomposable natural transformations in $\mathcal{N}\mathcal{T}^*(Z_4)$

We have $\bigoplus_{1\leq l\leq 4} M(l5) \oplus M(12345)[1] \cong 0 \oplus \mathbb{Z}^3$, $\bigoplus_{1\leq j<k\leq 4} M(jk5) \cong \mathbb{Z}^6$, and $M(5) \oplus \bigoplus_{1\leq i\leq 4} M(12345\setminus i)[1] \cong \mathbb{Z}[1] \oplus \mathbb{Z}[1]^8$. Since

$$\begin{array}{ccc} \bigoplus\limits_{1\leq l\leq 4} M(l5) \oplus M(12345)[1] & \longrightarrow & \bigoplus\limits_{1\leq j<k\leq 4} M(jk5) \\ & \nwarrow & \big\downarrow \\ & & M(5) \oplus \bigoplus\limits_{1\leq i\leq 4} M(12345\setminus i)[1] \end{array}$$

is exact, a rank argument shows that the map

$$\bigoplus_{1\leq l\leq 4} M(l5) \oplus M(12345)[1] \to \bigoplus_{1\leq j<k\leq 4} M(jk5)$$

is zero. On the other hand, the kernel of the map

$$\bigoplus_{1\leq j<k\leq 4} M(jk5) \xrightarrow{\begin{pmatrix} i & -i & 0 & i & 0 & 0 \\ -i & 0 & i & 0 & -i & 0 \\ 0 & i & -i & 0 & 0 & i \\ 0 & 0 & 0 & -i & i & -i \end{pmatrix}} \bigoplus_{1\leq i\leq 4} M(12345\setminus i)$$

is non-trivial; it consists precisely of the elements in

$$\bigoplus_{1\leq j<k\leq 4} M(jk5) \cong \bigoplus_{1\leq j<k\leq 4} \mathbb{Z}[\mathrm{id}_{jk5}^{jk5}]$$

which are multiples of $([\mathrm{id}_{jk5}^{jk5}])_{1\leq j<k\leq 4}$. This shows $\mathrm{Tor}_2^{\mathcal{NT}^*}(S_{12345}, M) \cong \mathbb{Z}$. Hence, by Proposition 1, the module M has projective dimension at least 2. On the other hand, (3.7) is a resolution of length 2. Therefore, the projective dimension of M is exactly 2.

Let $k \in \mathbb{N}_{\geq 2}$ and define $M_k = M \otimes_{\mathbb{Z}} \mathbb{Z}/k$. Since $\mathrm{Tor}_2^{\mathcal{NT}^*}(S_{12345}, M_k) \cong \mathbb{Z}/k$ is non-free, Proposition 1 shows that M_k has at least projective dimension 3. On the other hand, if we abbreviate the resolution (3.7) for M by

$$0 \to P^{(5)} \xrightarrow{\alpha} P^{(4)} \xrightarrow{\beta} P^{(3)} \twoheadrightarrow M\ , \tag{3.8}$$

a projective resolution of length 3 for M_k is given by

$$0 \to P^{(5)} \xrightarrow{\binom{k}{\alpha}} P^{(5)} \oplus P^{(4)} \xrightarrow{\begin{pmatrix} \alpha & -k \\ 0 & \beta \end{pmatrix}} P^{(4)} \oplus P^{(3)} \xrightarrow{(\beta\ k)} P^{(3)} \twoheadrightarrow M_k\ ,$$

where k denotes multiplication by k.

It remains to show that the modules M and M_k can be realised as the filtrated K-theory of objects in $\mathcal{B}(X)$. It suffices to prove this for the module M since tensoring with the Cuntz algebra $\mathcal{O}_{k+1}$ then yields a separable C^*- algebra with filtrated K-theory M_k by the Künneth Theorem.

The projective resolution (3.8) can be written as

$$0 \to \mathrm{FK}(P^2) \xrightarrow{\mathrm{FK}(f_2)} \mathrm{FK}(P^1) \xrightarrow{\mathrm{FK}(f_1)} \mathrm{FK}(P^0) \twoheadrightarrow M,$$

because of the equivalence of the category of projective $\mathcal{NT}^*$-modules and the category of $\mathfrak{J}$-projective objects in $\mathfrak{KK}(X)$. Let N be the cokernel of the module map $\mathrm{FK}(f_2)$. Using [11, Theorem 4.11], we obtain an object $A \in\in \mathcal{B}(X)$ with $\mathrm{FK}(A) \cong N$. We thus have a commutative diagram of the form

$$\begin{array}{ccccccccc} 0 & \longrightarrow & \mathrm{FK}(P^2) & \xrightarrow{\mathrm{FK}(f_2)} & \mathrm{FK}(P^1) & \xrightarrow{\qquad\mathrm{FK}(f_1)\qquad} & \mathrm{FK}(P^0) & \twoheadrightarrow & M\ . \\ & & & & & \searrow \qquad \nearrow_{\gamma} & & & \\ & & & & & \mathrm{FK}(A) & & & \end{array}$$

Since A belongs to the bootstrap class $\mathcal{B}(X)$ and $\mathrm{FK}(A)$ has a projective resolution of length 1, we can apply the universal coefficient theorem to lift the homomorphism

γ to an element $f \in \mathrm{KK}(X; A, P^0)$. Now we can argue as in the proof of [11, Theorem 4.11]: since f is $\mathfrak{J}$-monic, the filtrated K-theory of its mapping cone is isomorphic to $\operatorname{coker}(\gamma) \cong M$. This completes the proof of Proposition 4.

3.10 Cuntz-Krieger Algebras with Projective Dimension 2

In this section we exhibit a Cuntz-Krieger algebra A which is a tight C^*-algebra over the space Z_3 and for which the odd part of $\mathrm{Tor}_1^{\mathcal{N}\mathcal{T}^*(Z_3)}(\mathcal{N}\mathcal{T}_{\mathrm{ss}}, \mathrm{FK}(A))$—denoted $\mathrm{Tor}_1^{\mathrm{odd}}$ in the following—is not free. By Proposition 2 this C^*-algebra has projective dimension 2 in filtrated K-theory.

In the following we will adhere to the conventions for graph algebras and adjacency matrices from [4]. Let E be the finite graph with vertex set $E^0 = \{v_1, v_2, \ldots, v_8\}$ and edges corresponding to the adjacency matrix

$$\begin{pmatrix} B_4 & 0 & 0 & 0 \\ X_1 & B_1 & 0 & 0 \\ X_2 & 0 & B_2 & 0 \\ X_3 & 0 & 0 & B_3 \end{pmatrix} := \begin{pmatrix} \begin{pmatrix} 3 & 2 \\ 2 & 3 \end{pmatrix} & 0 & 0 & 0 \\ \begin{pmatrix} 1 & 1 \\ 1 & 1 \end{pmatrix} & \begin{pmatrix} 3 & 2 \\ 1 & 2 \end{pmatrix} & 0 & 0 \\ \begin{pmatrix} 1 & 1 \\ 1 & 1 \end{pmatrix} & 0 & \begin{pmatrix} 3 & 2 \\ 1 & 2 \end{pmatrix} & 0 \\ \begin{pmatrix} 1 & 1 \\ 1 & 1 \end{pmatrix} & 0 & 0 & \begin{pmatrix} 3 & 2 \\ 1 & 2 \end{pmatrix} \end{pmatrix}. \tag{3.9}$$

Since this is a finite graph with no sinks and no sources, the associated graph C^*- algebra $C^*(E)$ is in fact a Cuntz-Krieger algebra (we can replace E with its *edge graph*; see [13, Remark 2.8]). Moreover, the graph E is easily seen to fulfill condition (K) because every vertex is the base of two or more simple cycles. As a consequence, the adjacency matrix of the edge graph of E fulfills condition (II) from [5]. In fact, condition (K) is designed as a generalisation of condition (II): see, for instance, [8].

Applying [13, Theorem 4.9]—and carefully translating between different graph algebra conventions—we find that the ideals of $C^*(E)$ correspond bijectively and in an inclusion-preserving manner to the open subsets of the space Z_3. By [9, Lemma 2.35], we may turn A into a tight C^*- algebra over Z_3 by declaring $A(\{4\}) = I_{\{v_1,v_2\}}$, $A(\{1,4\}) = I_{\{v_1,v_2,v_3,v_4\}}$, $A(\{2,4\}) = I_{\{v_1,v_2,v_5,v_6\}}$ as well as $A(\{3,4\}) = I_{\{v_1,v_2,v_7,v_8\}}$, where I_S denotes the ideal corresponding to the saturated hereditary subset S.

It is known how to compute the six-term sequence in K-theory for an extension of graph C^*- algebras: see [4]. Using this and Proposition 2, $\mathrm{Tor}_1^{\mathrm{odd}}$ is the homology of the complex

$$\ker(\phi_0) \xrightarrow{\left(\begin{smallmatrix} i & -i & 0 \\ -i & 0 & i \\ 0 & i & -i \end{smallmatrix}\right)} \ker(\phi_1) \xrightarrow{(i\ i\ i)} \ker(\phi_2)\,, \tag{3.10}$$

where $\phi_0 = \operatorname{diag}\left(\left(\begin{smallmatrix} B_4' & X_1^t \\ 0 & B_1' \end{smallmatrix}\right), \left(\begin{smallmatrix} B_4' & X_2^t \\ 0 & B_2' \end{smallmatrix}\right), \left(\begin{smallmatrix} B_4' & X_3^t \\ 0 & B_3' \end{smallmatrix}\right)\right)$, $\quad \phi_2 = \begin{pmatrix} B_4' & X_1^t & X_2^t & X_3^t \\ 0 & B_1' & 0 & 0 \\ 0 & 0 & B_2' & 0 \\ 0 & 0 & 0 & B_3' \end{pmatrix}$,

$$\phi_1 = \operatorname{diag}\left(\begin{pmatrix} B_4' & X_1^t & X_2^t \\ 0 & B_1' & 0 \\ 0 & 0 & B_2' \end{pmatrix}, \begin{pmatrix} B_4' & X_1^t & X_3^t \\ 0 & B_1' & 0 \\ 0 & 0 & B_3' \end{pmatrix}, \begin{pmatrix} B_4' & X_2^t & X_3^t \\ 0 & B_2' & 0 \\ 0 & 0 & B_3' \end{pmatrix}\right),$$

and $B_4' = B_4^t - \left(\begin{smallmatrix} 1 & 0 \\ 0 & 1 \end{smallmatrix}\right) = \left(\begin{smallmatrix} 2 & 2 \\ 2 & 2 \end{smallmatrix}\right)$ and $B_j' = B_j^t - \left(\begin{smallmatrix} 1 & 0 \\ 0 & 1 \end{smallmatrix}\right) = \left(\begin{smallmatrix} 2 & 1 \\ 2 & 1 \end{smallmatrix}\right)$ for $1 \le j \le 3$. We obtain a commutative diagram

$$\begin{array}{ccccc}
\ker(\phi_0) & \rightarrowtail & (\mathbb{Z}^{\oplus 2})^{\oplus(2\cdot 3)} & \xrightarrow{\phi_0} & \operatorname{im}(\phi_0) \\
\downarrow{\scriptstyle f_K} & & \downarrow{\scriptstyle f} & & \downarrow{\scriptstyle f_I} \\
\ker(\phi_1) & \rightarrowtail & (\mathbb{Z}^{\oplus 2})^{\oplus(3\cdot 3)} & \xrightarrow{\phi_1} & \operatorname{im}(\phi_1) \\
\downarrow{\scriptstyle g_K} & & \downarrow{\scriptstyle g} & & \downarrow{\scriptstyle g_I} \\
\ker(\phi_2) & \rightarrowtail & (\mathbb{Z}^{\oplus 2})^{\oplus(4\cdot 1)} & \xrightarrow{\phi_2} & \operatorname{im}(\phi_2)\,,
\end{array} \tag{3.11}$$

where f and g have the block forms

$$f = \begin{pmatrix} \mathrm{id} & 0 & -\mathrm{id} & 0 & 0 & 0 \\ 0 & \mathrm{id} & 0 & 0 & 0 & 0 \\ 0 & 0 & 0 & -\mathrm{id} & 0 & 0 \\ -\mathrm{id} & 0 & 0 & 0 & \mathrm{id} & 0 \\ 0 & -\mathrm{id} & 0 & 0 & 0 & 0 \\ 0 & 0 & 0 & 0 & 0 & \mathrm{id} \\ 0 & 0 & \mathrm{id} & 0 & -\mathrm{id} & 0 \\ 0 & 0 & 0 & \mathrm{id} & 0 & 0 \\ 0 & 0 & 0 & 0 & 0 & -\mathrm{id} \end{pmatrix}, \qquad g = \begin{pmatrix} \mathrm{id} & 0 & 0 & \mathrm{id} & 0 & 0 & \mathrm{id} & 0 & 0 \\ 0 & \mathrm{id} & 0 & 0 & \mathrm{id} & 0 & 0 & 0 & 0 \\ 0 & 0 & \mathrm{id} & 0 & 0 & 0 & 0 & \mathrm{id} & 0 \\ 0 & 0 & 0 & 0 & 0 & \mathrm{id} & 0 & 0 & \mathrm{id} \end{pmatrix},$$

and $f_K := f|_{\ker(\phi_0)}$, $f_I := f|_{\operatorname{im}(\phi_0)}$, $g_K := g|_{\ker(\phi_1)}$, $g_I := g|_{\operatorname{im}(\phi_1)}$. Notice that f and g are defined in a way such that the restrictions $f|_{\ker(\phi_0)}$ and $g|_{\ker(\phi_1)}$ are exactly the maps from (3.10) in the identification made above.

We abbreviate the above short exact sequence of cochain complexes (3.11) as $K_\bullet \rightarrowtail Z_\bullet \twoheadrightarrow I_\bullet$. The part $\mathrm{H}^0(Z_\bullet) \to \mathrm{H}^0(I_\bullet) \to \mathrm{H}^1(K_\bullet) \to \mathrm{H}^1(Z_\bullet)$ in the corresponding long exact homology sequence can be identified with

$$\ker(f) \xrightarrow{\phi_0} \ker(f_I) \to \frac{\ker(g_K)}{\operatorname{im}(f_K)} \to 0\,.$$

Hence

$$\mathrm{Tor}_1^{\mathrm{odd}} \cong \frac{\ker(g_K)}{\mathrm{im}(f_K)} \cong \frac{\ker(f_I)}{\phi_0\big(\ker(f)\big)} \cong \frac{\ker(f) \cap \mathrm{im}(\phi_0)}{\phi_0\big(\ker(f)\big)} .$$

We have $\ker(f) = \{(v, 0, v, 0, v, 0) \mid v \in \mathbb{Z}^2\} \subset (\mathbb{Z}^{\oplus 2})^{\oplus(2\cdot 3)}$.

From the concrete form (3.9) of the adjacency matrix, we find that $\ker(f) \cap \mathrm{im}(\phi_0)$ is the free cyclic group generated by $(1, 1, 0, 0, 1, 1, 0, 0, 1, 1, 0, 0)$, while $\phi_0\big(\ker(f)\big)$ is the subgroup generated by $(2, 2, 0, 0, 2, 2, 0, 0, 2, 2, 0, 0)$. We see that $\mathrm{Tor}_1^{\mathrm{odd}} \cong \mathbb{Z}/2$ is not free.

Now we briefly indicate how to construct a similar counterexample for the space S. Consider the integer matrix

$$\begin{pmatrix} B_4 & 0 & 0 & 0 \\ X_{43} & B_3 & 0 & 0 \\ X_{42} & 0 & B_2 & 0 \\ X_{41} & X_{31} & X_{21} & B_1 \end{pmatrix} := \begin{pmatrix} (3) & 0 & 0 & 0 \\ (2) & (3) & 0 & 0 \\ (2) & 0 & (3) & 0 \\ \begin{pmatrix} 2 \\ 0 \end{pmatrix} & \begin{pmatrix} 1 \\ 0 \end{pmatrix} & \begin{pmatrix} 1 \\ 0 \end{pmatrix} & \begin{pmatrix} 2 & 1 \\ 1 & 2 \end{pmatrix} \end{pmatrix} .$$

The corresponding graph F fulfills condition (K) and has no sources or sinks. The associated graph C^*- algebra $C^*(F)$ is therefore a Cuntz-Krieger algebra satisfying condition (II). It is easily read from the block structure of the edge matrix that the primitive ideal space of $C^*(F)$ is homeomorphic to S. We are going to compute the even part of $\mathrm{Tor}_1^{\mathcal{N}\mathcal{T}^*(S)}\big(\mathcal{N}\mathcal{T}_{\mathrm{ss}}, \mathrm{FK}(C^*(F))\big)$. Since the nice computation methods from the previous example do not carry over, we carry out a more ad hoc calculation.

By Remark 1, the even part of our Tor-term is isomorphic to the homology of the complex

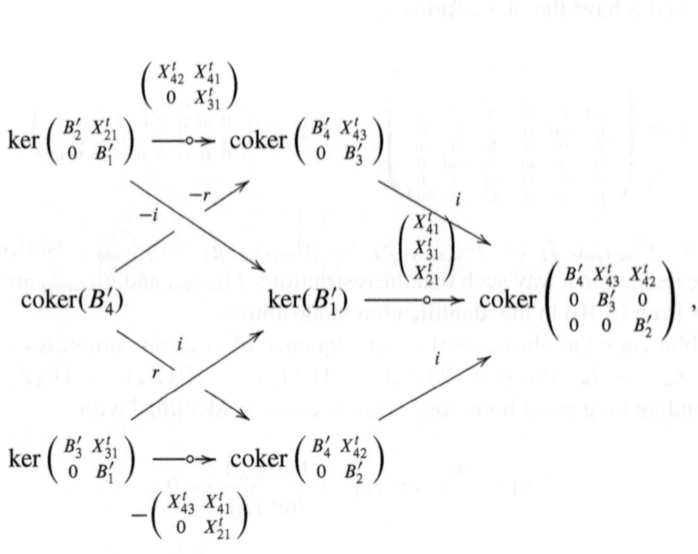

where column-wise direct sums are taken. Here $B'_1 = B^t_1 - \left(\begin{smallmatrix}1&0\\0&1\end{smallmatrix}\right) = \left(\begin{smallmatrix}1&1\\1&1\end{smallmatrix}\right)$ and $B'_j = B^t_j - \big(1\big) = \big(2\big)$ for $2 \leq j \leq 4$. This complex can be identified with

$$\mathbb{Z} \oplus \mathbb{Z}/2 \oplus \mathbb{Z} \xrightarrow{\begin{pmatrix} 0 & 1 & 0 \\ 0 & 0 & 0 \\ -2 & 0 & 2 \\ 0 & 1 & 0 \\ 0 & 0 & 0 \end{pmatrix}} (\mathbb{Z}/2)^2 \oplus \mathbb{Z} \oplus (\mathbb{Z}/2)^2 \xrightarrow{\begin{pmatrix} 1 & 0 & 0 & 1 & 0 \\ 0 & 1 & 1 & 0 & 0 \\ 0 & 0 & 1 & 0 & 1 \end{pmatrix}} (\mathbb{Z}/2)^3 ,$$

the homology of which is isomorphic to $\mathbb{Z}/2$; a generator is given by the class of $(0, 1, 1, 0, 1) \in (\mathbb{Z}/2)^2 \oplus \mathbb{Z} \oplus (\mathbb{Z}/2)^2$. This concludes the proof of Proposition 5.

Acknowledgements The author was supported by the Danish National Research Foundation through the Centre for Symmetry and Deformation (DNRF92) at the University of Copenhagen and by the Marie Curie Research Training Network EU-NCG. Parts of this paper are based on the author's Diploma thesis [1] which was supervised by Ralf Meyer at the University of Göttingen. I would like to thank my PhD-supervisors, Søren Eilers and Ryszard Nest, for helpful advice, Takeshi Katsura for pointing out a mistake in an earlier version of the paper and the anonymous referee for the suggested improvements.

References

1. Bentmann, R.: Filtrated K-theory and classification of C^*-algebras. Diplom thesis,Georg-August-Universität Göttingen (2010). URL http://www.math.ku.dk/~bentmann/thesis.pdf
2. Bentmann, R., Köhler, M.: Universal coefficient theorems for C^*-algebras over finite topological spaces (2011), arXiv:1101.5702
3. Bonkat, A.: Bivariante K-Theorie für Kategorien projektiver Systeme von C^*-Algebren. Ph.D. thesis, Westfälische Wilhelms-Universität Münster (2002). URL http://deposit.ddb.de/cgi-bin/dokserv?idn=967387191
4. Carlsen, T.M., Eilers, S., Tomforde, M.: Index maps in the K-theory of graph algebras. J. K-Theory **9**(2), 385–406 (2012). DOI 10.1017/is011004017jkt156. URL http://dx.doi.org/10.1017/is011004017jkt156
5. Cuntz, J.: A class of C^*-algebras and topological Markov chains. II. Reducible chains and the Ext-functor for C^*-algebras. Invent. Math. **63**(1), 25–40 (1981). DOI 10.1007/BF01389192. URL http://dx.doi.org/10.1007/BF01389192
6. Inassaridze, H., Kandelaki, T., Meyer, R.: Localisation and colocalisation of KK-theory. Abh. Math. Semin. Univ. Hambg. **81**(1), 19–34 (2011). DOI 10.1007/s12188-011-0050-7. URL http://dx.doi.org/10.1007/s12188-011-0050-7
7. Kirchberg, E.: Das nicht-kommutative Michael-Auswahlprinzip und die Klassifikation nicht-einfacher Algebren. In: C^*-algebras (Münster, 1999), pp. 92–141. Springer, Berlin (2000)
8. Kumjian, A., Pask, D., Raeburn, I., Renault, J.: Graphs, groupoids, and Cuntz-Krieger algebras. J. Funct. Anal. **144**(2), 505–541 (1997). DOI 10.1006/jfan.1996.3001. URL http://dx.doi.org/10.1006/jfan.1996.3001
9. Meyer, R., Nest, R.: C^*-algebras over topological spaces: the bootstrap class. Münster J. Math. **2**, 215–252 (2009)
10. Meyer, R., Nest, R.: Homological algebra in bivariant K-theory and other triangulated categories. I. In: Triangulated categories, *London Math. Soc. Lecture Note Ser.*, vol. 375, pp. 236–289. Cambridge Univ. Press, Cambridge (2010)
11. Meyer, R., Nest, R.: C*-algebras over topological spaces: filtrated K-theory. Canad. J. Math. **64**(2), 368–408 (2012). DOI 10.4153/CJM-2011-061-x. URL http://dx.doi.org/10.4153/CJM-2011-061-x

12. Neeman, A.: Triangulated categories, *Annals of Mathematics Studies*, vol. 148. Princeton University Press, Princeton, NJ (2001)
13. Raeburn, I.: Graph algebras, *CBMS Regional Conference Series in Mathematics*, vol. 103. Published for the Conference Board of the Mathematical Sciences, Washington, DC (2005)
14. Restorff, G.: Classification of Cuntz-Krieger algebras up to stable isomorphism. J. Reine Angew. Math. **598**, 185–210 (2006). DOI 10.1515/CRELLE.2006.074. URL http://dx.doi.org/10.1515/CRELLE.2006.074
15. Restorff, G.: Classification of non-simple C^*-algebras. Ph.D. thesis, University of Copenhagen (2008). URL http://www.math.ku.dk/~restorff/papers/thesis.pdf
16. Rosenberg, J., Schochet, C.: The Künneth theorem and the universal coefficient theorem for Kasparov's generalized K-functor. Duke Math. J. **55**(2), 431–474 (1987). DOI 10.1215/S0012-7094-87-05524-4. URL http://dx.doi.org/10.1215/S0012-7094-87-05524-4

Chapter 4
An Introduction to the C^*-Algebra of a One-Sided Shift Space

Toke Meier Carlsen

Abstract This paper gives an introduction to the C^*-algebra of a one-sided shift space. Focus will be given to the fundamental structure of the C^*-algebra of a one-sided shift space, but some of the most important results about C^*-algebras associated to shift spaces will also be presented.

Keywords C^*-algebras of shift spaces • C^*-algebras of subshifts

Mathematics Subject Classification (2010): 46L05, 37B10, 46L80, 46L55.

4.1 Introduction

I will in this paper give an introduction to C^*-algebras associated to shift spaces (also called subshifts). The paper also contains an appendix about C^*-algebras, Morita equivalence and K-theory of C^*-algebras which hopefully will provide a reader without any knowledge of operator algebra with the necessary background for reading this paper.

C^*-algebras associated to shift spaces were introduced by Kengo Matsumoto in [25] as a generalization of Cuntz-Krieger algebras (cf. [17]), and all the major results about them are essentially due to him. C^*-algebras associated to shift spaces have been studied by Matsumoto and his collaborators in [20, 22, 23, 26–28, 30–35, 37], and I have together with various collaborators contributed in [2, 7, 10–16].

T.M. Carlsen (✉)
Department of Mathematical Sciences, Norwegian University of Science and Technology (NTNU), NO-7491 Trondheim, Norway
e-mail: toke.meier.carlsen@math.ntnu.no

T.M. Carlsen et al. (eds.), *Operator Algebra and Dynamics*, Springer Proceedings in Mathematics & Statistics 58, DOI 10.1007/978-3-642-39459-1_4,

The approach I will take in this paper, is a little bit different from Matsumoto's original approach. One notable difference is that I will associate C^*-algebras to *one-sided* shift spaces, whereas Matsumoto associated C^*-algebras to *two-sided* shift spaces, but there are other differences as well. See [15, Sect. 7] for a discussion of the relationship between the different C^*-algebras that have been associated to shift spaces.

The focus of this paper is the fundamental structure of the C^*-algebra of a one-sided shift space, but I will also describe some of the most important results about C^*-algebras associated to shift spaces. In Sect. 4.2 one-sided shift spaces are defined and representations of one-sided shift spaces are introduced and studied. In Sect. 4.3 the C^*-algebra of a one-sided shift space is introduced and studied, and in Sect. 4.4 the gauge action of the C^*-algebra of a one-sided shift space is studied. In Sect. 4.5 it is proved that the C^*-algebra of a one-sided shift space is a conjugacy invariant. Finally, it is in Sect. 4.6 briefly explained that the Morita equivalence class of the C^*-algebra of a one-sided shift space is a flow invariant, and in Sect. 4.7 the K-theory of the C^*-algebra of a one-sided shift space is described. As mentioned above, the paper contains an appendix which hopefully will provide a reader without any knowledge of operator algebra with the necessary background for reading this paper. This appendix contains a section which very briefly introduces C^*-algebras, a section which very briefly introduce Morita equivalence of C^*-algebras, and a section which very briefly introduce the K-theory of C^*-algebras.

It should be noticed that parts of this paper are taken from the notes [9] which I wrote for the summer school "Symbolic dynamics and homeomorphisms of the Cantor set" at the University of Copenhagen, 23–27 June 2008.

4.2 Representations of One-Sided Shift Spaces

Let $\mathfrak{a}$ be a finite set endowed with the discrete topology. We will call this set the *alphabet* and its elements *letters*. Let $\mathfrak{a}^{\mathbb{N}_0}$ be the infinite product space $\prod_{n=0}^{\infty} \mathfrak{a}$ endowed with the product topology. The transformation σ on $\mathfrak{a}^{\mathbb{N}_0}$ given by

$$\big(\sigma(x)\big)_i = x_{i+1}, \; i \in \mathbb{N}_0,$$

is called the *(one-sided) shift*. Let X be a shift invariant closed subset of $\mathfrak{a}^{\mathbb{N}_0}$ (by shift invariant we mean that $\sigma(\mathsf{X}) \subseteq \mathsf{X}$, not necessarily $\sigma(\mathsf{X}) = \mathsf{X}$). The topological dynamical system $(\mathsf{X}, \sigma_{|\mathsf{X}})$ is called a *one-sided shift space* (or a *one-sided subshift*).

Example 1. If $\mathfrak{a}$ is an alphabet, then $\mathfrak{a}^{\mathbb{N}_0}$ itself is a shift space. We call $\mathfrak{a}^{\mathbb{N}_0}$ the full one-sided $\mathfrak{a}$-shift.

We will denote $\sigma_{|\mathsf{X}}$ by σ_{X} or σ for simplicity, and on occasion the alphabet $\mathfrak{a}$ by $\mathfrak{a}_{\mathsf{X}}$. We denote the n-fold composition of σ with itself by σ^n, and we denote the preimage of a set X under σ^n by $\sigma^{-n}(X)$.

A finite sequence $u = (u_1, \ldots, u_k)$ of elements $u_i \in \mathfrak{a}$ is called a finite *word*. The *length* of u is k and is denoted by $|u|$. For each $k \in \mathbb{N}$, we let $\mathfrak{a}^k$ be the set of all words with length k, and we let $\mathsf{L}^k(\mathsf{X})$ be the set of all words with length k appearing in some $x \in \mathsf{X}$. We let $\mathsf{L}^0(\mathsf{X}) = \mathfrak{a}^0$ denote the set $\{\epsilon\}$ consisting of the empty word ϵ which has length 0. We set $\mathsf{L}_l(\mathsf{X}) = \bigcup_{k=0}^{l} \mathsf{L}^k(\mathsf{X})$ and $\mathsf{L}(\mathsf{X}) = \bigcup_{k=0}^{\infty} \mathsf{L}^k(\mathsf{X})$, and likewise $\mathfrak{a}_l = \bigcup_{k=0}^{l} \mathfrak{a}^k$ and $\mathfrak{a}^* = \bigcup_{k=0}^{\infty} \mathfrak{a}^k$. The set $\mathsf{L}(\mathsf{X})$ is called the *language* of X. Note that $\mathsf{L}(\mathsf{X}) \subseteq \mathfrak{a}^*$ for every shift space.

If $u \in \mathfrak{a}^*$ with $|u| > 0$, then we will by u_1 denote the first letter (the leftmost) letter of u, by u_2 the second letter of u, and so on till $u_{|u|}$ which denotes the last (the rightmost) letter of u. Thus $u = u_1 u_2 \cdots u_{|u|}$.

We will often denote an element $x = (x_n)_{n \in \mathbb{N}_0}$ of $\mathfrak{a}^{\mathbb{N}_0}$ by

$$x_0 x_1 \cdots ,$$

and if $u \in \mathfrak{a}^*$, then we will by ux denote the sequence

$$u_1 u_2 \cdots u_{|u|} x_0 x_1 \cdots .$$

We will also often for a sequence x belonging to either $\mathfrak{a}^{\mathbb{N}_0}$ or $\mathfrak{a}^{\mathbb{Z}}$ and for integers $k < l$ belonging to the appropriate index set denote $x_k x_{k+1} \cdots x_{l-1}$ by $x_{[k,l[}$ and regard it as an element of $\mathfrak{a}^*$. Similarly, $x_{[k,\infty[}$ will denote the element

$$x_k x_{k+1} \cdots$$

of $\mathfrak{a}^{\mathbb{N}_0}$.

Definition 1. Let X be a one-sided shift space. We let $l^\infty(\mathsf{X})$ be the C^*-algebra of bounded functions on X. We define two maps $\alpha : l^\infty(\mathsf{X}) \to l^\infty(\mathsf{X})$ and $\mathscr{L} : l^\infty(\mathsf{X}) \to l^\infty(\mathsf{X})$ by for $f \in l^\infty(\mathsf{X})$ and $x \in \mathsf{X}$ letting

$$\alpha(f)(x) = f(\sigma(x)) \text{ and } \mathscr{L}(f)(x) = \begin{cases} \frac{1}{\#\sigma^{-1}(\{x\})} \sum_{y \in \sigma^{-1}(\{x\})} f(y) & \text{if } x \in \sigma(\mathsf{X}), \\ 0 & \text{if } x \notin \sigma(\mathsf{X}), \end{cases}$$

where $\#\sigma^{-1}(\{x\})$ denotes the number of elements of $\sigma^{-1}(\{x\})$ (which is finite).

Definition 2. Let X be a one-sided shift space over the alphabet $\mathfrak{a}$. For every pair (u, v) of words in $\mathfrak{a}^*$, we let $C(u, v)$ denote the subset

$$\{vx \in \mathsf{X} \mid x, ux \in \mathsf{X}\}$$

of X which consists of those elements which begins with a v and which satisfies that the element obtained by replacing the beginning v with u also is an element of X.

We let $\mathscr{D}_{\mathsf{X}}$ be the C^*-subalgebra of $l^\infty(\mathsf{X})$ generated by $\{1_{C(u,v)} \mid u, v \in \mathfrak{a}^*\}$ where $1_{C(u,v)}$ denotes the characteristic function of $C(u, v)$.

Proposition 1. *Let* X *be a one-sided shift space over the alphabet* $\mathfrak{a}$. *Then we have:*

1. $C(\mathsf{X}) \subseteq \mathscr{D}_{\mathsf{X}}$.
2. $\mathscr{D}_{\mathsf{X}}$ *is the closure of*

$$\operatorname{span}\left\{\prod_{i=1}^{n} 1_{C(u_i,v_i)} \mid u_1, \ldots, u_n, v_1, \ldots, v_n \in \mathfrak{a}^*\right\}.$$

3. $\mathscr{D}_{\mathsf{X}}$ *is closed under* α *and* $\mathscr{L}$ *(i.e.,* $f \in \mathscr{D}_{\mathsf{X}} \implies \alpha(f), \mathscr{L}(f) \in \mathscr{D}_{\mathsf{X}}$*).*
4. *If* $\mathscr{X}$ *is a* C^**-subalgebra of* $l^\infty(\mathsf{X})$ *that is closed under* α *and* $\mathscr{L}$ *and contains* $C(\mathsf{X})$*, then* $\mathscr{D}_{\mathsf{X}} \subseteq \mathscr{X}$.

Proof. 1. For $u \in \mathfrak{a}^*$ we have that $Z(u) := C(\epsilon, u) = \{ux \in \mathsf{X} \mid x \in \mathsf{X}\}$ is a clopen subset of X, and thus that $1_{Z(u)} \in C(\mathsf{X})$. Since $\{Z(u) \mid u \in \mathfrak{a}^*\}$ separates the points of X, it follows from the Stone-Weierstrass Theorem that the C^*-subalgebra of $l^\infty(\mathsf{X})$ generated by $\{1_{Z(u)} \mid u \in \mathfrak{a}^*\}$ is equal to $C(X)$. Thus $C(X) \subseteq \mathscr{D}_{\mathsf{X}}$.

2. By definition $\mathscr{D}_{\mathsf{X}}$ is the smallest C^*-subalgebra of $l^\infty(\mathsf{X})$ which contains $\{1_{C(u,v)} \mid u, v \in \mathfrak{a}^*\}$. It is not difficult to check that the closure of

$$\operatorname{span}\left\{\prod_{i=1}^{n} 1_{C(u_i,v_i)} \mid u_1, \ldots, u_n, v_1, \ldots, v_n \in \mathfrak{a}^*\right\}.$$

satisfies this condition.

3. Since α is a $*$-homomorphism, and $\mathscr{L}$ is linear and continuous, it is enough to prove that $\alpha(1_{C(u,v)}) \in \mathscr{D}_{\mathsf{X}}$ for all $u, v \in \mathfrak{a}^*$, and that $\mathscr{L}(\prod_{i=1}^{n} 1_{C(u_i,v_i)}) \in \mathscr{D}_{\mathsf{X}}$ for all $u_1, \ldots, u_n, v_1, \ldots, v_n \in \mathfrak{a}^*$, so let us do that:

If $u, v \in \mathfrak{a}^*$, then we have

$$\alpha(1_{C(u,v)}) = \sum_{a \in \mathfrak{a}} 1_{C(u,va)} \in \mathscr{D}_{\mathsf{X}}.$$

If $A, B \subseteq \mathsf{X}$ such that $1_A, 1_B \in \mathscr{D}_{\mathsf{X}}$, then $1_{A \cup B} = 1_A + 1_B - 1_A 1_B \in \mathscr{D}_{\mathsf{X}}$. Thus $1_{\sigma^n(\mathsf{X})} = 1_{\bigcup_{u \in \mathfrak{a}^n} C(u,\epsilon)} \in \mathscr{D}_{\mathsf{X}}$. It follows that the function $1 - 1_{\sigma(\mathsf{X})} + \sum_{a \in \mathfrak{a}} 1_{C(a,\epsilon)}$ also belongs to $\mathscr{D}_{\mathsf{X}}$. Let us denote it by h. We have for $x \in \mathsf{X}$ that

$$h(x) = \begin{cases} \#\sigma^{-1}(\{x\}) & \text{if } x \in \sigma(\mathsf{X}), \\ 1 & \text{if } x \notin \sigma(\mathsf{X}). \end{cases}$$

Thus h is invertible, and it follows from Fact 8 that $h^{-1} \in \mathscr{D}_{\mathsf{X}}$. So the function $1_{\sigma(\mathsf{X})} - 1 + h^{-1}$ belongs to $\mathscr{D}_{\mathsf{X}}$. Let us denote it by d. We have for $x \in \mathsf{X}$ that

$$d(x) = \begin{cases} \frac{1}{\#\sigma^{-1}(\{x\})} & \text{if } x \in \sigma(\mathsf{X}), \\ 0 & \text{if } x \notin \sigma(\mathsf{X}). \end{cases}$$

If $u_1, \ldots, u_n, v_1, \ldots, v_n \in \mathfrak{a}^*$, then either $\prod_{i=1}^n 1_{C(u_i,v_i)} = 0$, all the v_i's are equal to the empty word, or all the non-empty v_i's begin with the same letter a'. In the first case $\mathscr{L}(\prod_{i=1}^n 1_{C(u_i,v_i)}) = 0$, in the second case we have

$$\mathscr{L}\left(\prod_{i=1}^n 1_{C(u_i,v_i)}\right) = d\left(\sum_{a\in\mathfrak{a}}\prod_{i=1}^n 1_{C(u_ia,\epsilon)}\right) \in \mathscr{D}_{\mathsf{X}},$$

and in the third case we have

$$\mathscr{L}\left(\prod_{i=1}^n 1_{C(u_i,v_i)}\right) = d 1_{C(a',\epsilon)} \prod_{i\in I} 1_{C(u_i,(v_i)_2(v_i)_3\ldots(v_i)_{|v_i|})} \prod_{i\in I'} 1_{C(u_ia,\epsilon)} \in \mathscr{D}_{\mathsf{X}}$$

where $I = \{i \in \{1,2,\ldots,n\} \mid v_i \neq \epsilon\}$ and $I' = \{i \in \{1,2,\ldots,n\} \mid v_i = \epsilon\}$.

4. Let $\mathscr{X}$ be a C^*-subalgebra of $l^\infty(\mathsf{X})$ that is closed under α and $\mathscr{L}$ and contains $C(\mathsf{X})$. For $n \in \mathbb{N}_0$ let g_n be the function

$$1 - \mathscr{L}^n(1) + \sum_{u\in\mathfrak{a}^n} \left(\mathscr{L}^n(1_{z(u)})\right)^2.$$

Then $g_n \in \mathscr{X}$, and for every $x \in \mathsf{X}$ we have

$$g_n(x) = \begin{cases} \frac{1}{\#\sigma^{-n}(\{x\})} & \text{if } x \in \sigma^n(\mathsf{X}), \\ 1 & \text{if } x \notin \sigma^n(\mathsf{X}). \end{cases}$$

Thus, g_n is invertible. It follows from Fact 8 that g_n^{-1} and hence $f_n := g_n^{-1} + \mathscr{L}^n(1) - 1$ belong to $\mathscr{X}$. For every $x \in \mathsf{X}$ we have $f_n(x) = \#\sigma^{-n}(\{x\})$. Thus if $u, v \in \mathfrak{a}^*$, then $1_{C(u,v)} = 1_{Z(v)}\alpha^{|v|}(f_{|u|}\mathscr{L}^{|u|}(1_{z(u)})) \in \mathscr{X}$. Since $\mathscr{D}_{\mathsf{X}}$ is generated by $\{1_{C(u,v)} \mid u, v \in \mathfrak{a}^*\}$, it follows that $\mathscr{D}_{\mathsf{X}} \subseteq \mathscr{X}$. □

Remark 1. It follows from [19, Theorem 1] that $C(\mathsf{X}) = \mathscr{D}_{\mathsf{X}}$ if and only if X is of finite type.

Definition 3. Let X be a one-sided shift space over the alphabet $\mathfrak{a}$. For $w \in \mathfrak{a}^*$ we let λ_w be the map from $l^\infty(\mathsf{X})$ to $l^\infty(\mathsf{X})$ given by

$$\lambda_w(f)(x) = \begin{cases} f(wx) & \text{if } wx \in \mathsf{X}, \\ 0 & \text{if } wx \notin \mathsf{X}, \end{cases}$$

for $f \in l^\infty(\mathsf{X})$ and $x \in \mathsf{X}$.

Lemma 1. *Let* X *be a one-sided shift space over the alphabet* $\mathfrak{a}$ *and let* $w \in \mathfrak{a}^*$. *Then* λ_w *is a* $*$*-homomorphism and* $\lambda_w(\mathscr{D}_\mathsf{X}) \subseteq \mathscr{D}_\mathsf{X}$.

Proof. It is easy to check that λ_w is a $*$-homomorphism. Since $\mathscr{D}_\mathsf{X}$ is generated by $\{1_{C(u,v)} \mid u, v \in \mathfrak{a}^*\}$ and λ_w is a $*$-homomorphism, it is enough to check that $\lambda_w(1_{C(u,v)}) \in \mathscr{D}_\mathsf{X}$ for all $u, v \in \mathfrak{a}^*$, and this follows from the fact that

$$\lambda_w(1_{C(u,v)}) = \begin{cases} 1_{C(w,\epsilon)}1_{C(uw',\epsilon)} & \text{if } w = vw', \\ 1_{C(w,\epsilon)}1_{C(u,v')} & \text{if } wv' = v, \\ 0 & \text{otherwise.} \end{cases}$$

□

Definition 4. Let X be a one-sided shift space over the alphabet $\mathfrak{a}$. By a *representation* of X on a C^*-algebra $\mathscr{X}$ we mean a pair $(\phi, (t_u)_{u\in\mathfrak{a}^*})$ where ϕ is a $*$-homomorphism from $\mathscr{D}_\mathsf{X}$ to $\mathscr{X}$ and $(t_u)_{u\in\mathfrak{a}^*}$ is a family of elements of $\mathscr{X}$ such that

1. $t_u t_v = t_{uv}$,
2. $\phi(1_{C(u,v)}) = t_v t_u^* t_u t_v^*$

for all $u, v \in \mathfrak{a}^*$.

We denote by $C^*((\phi, (t_u)_{u\in\mathfrak{a}^*}))$ the C^*-subalgebra of $\mathscr{X}$ generated by $\{t_u \mid u \in \mathfrak{a}^*\}$.

Let X be a one-sided shift space over the alphabet $\mathfrak{a}$ and let H be a Hilbert space with an orthonormal basis $(e_x)_{x\in\mathsf{X}}$ with the same cardinality as X (we can for example let H be $l^2(\mathsf{X})$ and $e_x = \delta_x$).

For every $u \in \mathfrak{a}^*$, let T_u be the bounded operator on H defined by

$$T_u(e_x) = \begin{cases} e_{ux} & \text{if } ux \in \mathsf{X}, \\ 0 & \text{if } ux \notin \mathsf{X}, \end{cases} \tag{4.1}$$

and let $\phi : \mathscr{D}_\mathsf{X} \to \mathscr{B}(\mathsf{H})$ be the $*$-homomorphism defined by

$$\phi(f)(e_x) = f(x)e_x. \tag{4.2}$$

It is easy to check that $(\phi, (T_u)_{u\in\mathfrak{a}^*})$ is a representation of X. Thus we have:

Proposition 2. *Let* X *be a one-sided shift space over the alphabet* $\mathfrak{a}$ *and let* H *be a Hilbert space with an orthonormal basis* $(e_x)_{x\in\mathsf{X}}$ *with the same cardinality as* X. *For every* $u \in \mathfrak{a}^*$, *let* T_u *be the bounded operator on* H *defined by* (4.1), *and let* $\phi : \mathscr{D}_\mathsf{X} \to \mathscr{B}(\mathsf{H})$ *be the* $*$*-homomorphism defined by* (4.2). *Then* $(\phi, (T_u)_{u\in\mathfrak{a}^*})$ *is a representation of* X *on* $\mathscr{B}(\mathsf{H})$.

4.3 The C^*-Algebra of a One-Sided Shift Space

In this section, the *C^*-algebra of a one-sided shift space* will be introduced. This will be done via the following theorem which says that every one-sided shift space has a *universal representation*.

Theorem 1 (cf. [10, Remark 7.3] and [15, Theorem 10]). *Let X be a one-sided shift space over the alphabet $\mathfrak{a}$. There exists a C^*-algebra $\mathcal{O}_\mathsf{X}$ and a representation $(\iota, (s_u)_{u\in\mathfrak{a}^*})$ of X on $\mathcal{O}_\mathsf{X}$ satisfying:*

1. *$C^*(\iota, (s_u)_{u\in\mathfrak{a}^*}) = \mathcal{O}_\mathsf{X}$.*
2. *If $(\phi, (t_u)_{u\in\mathfrak{a}^*})$ is a representation of X on a C^*-algebra $\mathscr{X}$, then there exist a $*$-homomorphism $\psi_{(\phi,(t_u)_{u\in\mathfrak{a}^*})} : \mathcal{O}_\mathsf{X} \to \mathscr{X}$ such that $\psi_{(\phi,(t_u)_{u\in\mathfrak{a}^*})} \circ \iota = \phi$ and $\psi_{(\phi,(t_u)_{u\in\mathfrak{a}^*})}(s_u) = t_u$ for every $u \in \mathfrak{a}^*$.*

The C^*-algebra $\mathcal{O}_\mathsf{X}$, which is called *the C^*-algebra of* X, can be constructed in different ways, for example as the C^*-algebra of a groupoid (see [8]), as the C^*-algebra of a C^*-correspondence (see [10] and [16]), or as one of Ruy Exel's crossed product C^*-algebras of an endomorphism and a transfer operator (see [15]).

Throughout these notes we will let $(\iota, (s_u)_{u\in\mathfrak{a}^*})$ denote the representation of X on $\mathcal{O}_\mathsf{X}$ mentioned in Theorem 1.

Remark 2. Since $\mathcal{O}_\mathsf{X}$ is generated by a countable family, it is separable.

Lemma 2. *Let X be a one-sided shift space over the alphabet $\mathfrak{a}$. The $*$-homomorphism $\iota : \mathscr{D}_\mathsf{X} \to \mathcal{O}_\mathsf{X}$ is injective.*

Proof. Notice that the $*$-homomorphism $\phi : \mathscr{D}_\mathsf{X} \to \mathscr{B}(\mathsf{H})$ from Proposition 2 is injective. It follows from Theorem 1 that there exists a $*$-homomorphism $\psi : \mathcal{O}_\mathsf{X} \to \mathscr{B}(\mathsf{H})$ such that $\psi \circ \iota = \phi$. It follows that ι is injective. □

We will from now on view $\mathscr{D}_\mathsf{X}$ as a subalgebra of $\mathcal{O}_\mathsf{X}$ and suppress ι. This allows us to state and prove the following lemma and propositions about the fundamental structure of $\mathcal{O}_\mathsf{X}$.

Lemma 3. *Let X be a one-sided shift space over the alphabet $\mathfrak{a}$. We then have:*

1. *$s_\epsilon = s_\epsilon^* = s_\epsilon^2 = 1_\mathsf{X}$ is a unit for $\mathcal{O}_\mathsf{X}$.*
2. *If $u \in \mathfrak{a}^*$, then $s_u s_u^* = 1_{C(\epsilon,u)}$.*
3. *If $u \in \mathfrak{a}^*$, then s_u is a partial isometry (i.e., $s_u s_u^* s_u = s_u$ and $s_u^* s_u s_u^* = s_u^*$).*
4. *If $u, v \in \mathfrak{a}^*$ and $|u| = |v|$, then we have*

$$s_u^* s_v = \begin{cases} 1_{C(u,\epsilon)} & \text{if } u = v, \\ 0 & \text{if } u \neq v. \end{cases}$$

Proof. 1. Since $\epsilon\epsilon = \epsilon$, it follows from 1 of Definition 4 that $s_\epsilon^2 = s_\epsilon$. It follows from 2 of Definition 4 that $s_\epsilon s_\epsilon^* s_\epsilon s_\epsilon^* = 1_{C(\epsilon,\epsilon)}$ which is a projection, so $s_\epsilon s_\epsilon^* = s_\epsilon s_\epsilon^* s_\epsilon s_\epsilon^*$ according to Fact 7. It follows that $(s_\epsilon - s_\epsilon s_\epsilon^*)(s_\epsilon - s_\epsilon s_\epsilon^*)^* = 0$ and thus that $s_\epsilon = s_\epsilon s_\epsilon^* = s_\epsilon s_\epsilon^* s_\epsilon s_\epsilon^* = 1_{C(\epsilon,\epsilon)} = 1_\mathsf{X}$ which is a projection.

Let $u \in \mathfrak{a}^*$. Since $\epsilon u = u\epsilon = u$, it follows from 1 of Definition 4 that $s_\epsilon s_u = s_u s_\epsilon = s_u$. Since s_ϵ is self-adjoint, it follows that it is a unit for $\mathcal{O}_\mathsf{X}$.

2. If $u \in \mathfrak{a}^*$, then $s_u s_u^* = s_u s_\epsilon^* s_\epsilon s_u^* = 1_{C(\epsilon,u)}$.

3. It follows from 2 that $s_u s_u^*$ is a projection and thus that s_u is a partial isometry.

4. Let $u, v \in \mathfrak{a}^*$ with $|u| = |v|$. If $u \neq v$ then $C(\epsilon, u) \cap C(\epsilon, v) = \emptyset$ and so $s_u^* s_v = s_u^* s_u s_u^* s_v s_v^* s_v = s_u^* 1_{C(\epsilon,u)} 1_{C(\epsilon,v)} s_v = 0$. If $u = v$, then $s_u^* s_v = s_u^* s_u = s_\epsilon s_u^* s_u s_\epsilon^* = 1_{C(u,\epsilon)}$. □

Proposition 3. *Let* X *be a one-sided shift space over the alphabet* $\mathfrak{a}$, *let* $n \in \mathbb{N}_0$, *let* $w \in \mathfrak{a}^n$ *and let* $f \in \mathcal{D}_\mathsf{X}$. *Then we have:*

1. $\lambda_w(f) = s_w^* f s_w$.
2. $s_w^* f = \lambda_w(f) s_w^*$.
3. $\alpha^n(f) = \sum_{u \in \mathfrak{a}^n} s_u f s_u^*$.
4. $s_w f = \alpha^n(f) s_w$.
5. $\sum_{u,v \in \mathfrak{a}^n} s_u s_v^* s_v s_u^*$ *is equal to the function* $x \mapsto \#\sigma^{-n}(\{\sigma^n(x)\})$ *and is thus invertible.*
6. $\mathcal{L}^n(f) = (\sum_{u \in \mathfrak{a}^n} s_u)^* (\sum_{u,v \in \mathfrak{a}^n} s_u s_v^* s_v s_u^*)^{-1} f (\sum_{u \in \mathfrak{a}^n} s_u)$.

Proof. *1.* It is clear that the map $f \mapsto s_w^* f s_w$ is linear and $*$-preserving. If $f, g \in \mathcal{D}_\mathsf{X}$, then it follows from Lemma 3 and the fact that $\mathcal{D}_\mathsf{X}$ is commutative that we have

$$s_w^* f s_w s_w^* g s_w = s_w^* f 1_{C(\epsilon,w)} g s_w = s_w^* 1_{C(\epsilon,w)} f g s_w = s_w^* s_w s_w^* f g s_w = s_w^* f g s_w,$$

which shows that the map $f \mapsto s_w^* f s_w$ is also multiplicative and thus is a $*$-homomorphism. According to Lemma 1, λ_w is a $*$-homomorphism, and since $\mathcal{D}_\mathsf{X}$ is generated by $\{1_{C(u,v)} \mid u, v \in \mathfrak{a}^*\}$, it therefore suffices to check that $\lambda_w(1_{C(u,v)}) = s_w^* 1_{C(u,v)} s_w$ for $u, v \in \mathfrak{a}^*$, so let us do that:

It is easy to check that

$$\lambda_w(1_{C(u,v)}) = \begin{cases} 1_{C(w,\epsilon)} 1_{C(uw',\epsilon)} & \text{if } w = vw', \\ 1_{C(w,\epsilon)} 1_{C(u,v')} & \text{if } wv' = v, \\ 0 & \text{otherwise.} \end{cases}$$

It follows from Lemma 3 that if $s_w^* s_v \neq 0$, then either $w = vw'$ for some $w' \in \mathfrak{a}^*$, or $v = wv'$ for some $v' \in \mathfrak{a}^*$. In the first case we have

$$\begin{aligned} s_w^* s_v s_u^* s_u s_v^* s_w &= s_{w'}^* 1_{C(v,\epsilon)} 1_{C(u,\epsilon)} 1_{C(v,\epsilon)} s_{w'} = s_{w'}^* 1_{C(v,\epsilon)} 1_{C(w',\epsilon)} 1_{C(u,\epsilon)} s_{w'} \\ &= s_{w'}^* s_v^* s_v s_{w'}^* s_{w'} s_u^* s_u s_{w'} = s_{vw'}^* s_{vw'} s_{uw'}^* s_{uw'} = 1_{C(w,\epsilon)} 1_{C(uw',\epsilon)}, \end{aligned}$$

and the second case we have

$$s_w^* s_v s_u^* s_u s_v^* s_w = s_w^* s_w s_{v'} s_u^* s_u s_{v'}^* s_w^* s_w = 1_{C(w,\epsilon)} 1_{C(u,v')}.$$

Thus $\lambda_w(1_{C(u,v)}) = s_w^* 1_{C(u,v)} s_w$ as wanted.

2. It follows from 1, Lemma 3 and the fact that $\mathscr{D}_{\mathsf{X}}$ is commutative that we have

$$s_w^* f = s_w^* s_w s_w^* f = s_w^* 1_{C(\epsilon,w)} f = s_w^* f 1_{C(\epsilon,w)} = s_w^* f s_w s_w^* = \lambda_w(f) s_w^*.$$

3. The map $f \mapsto \sum_{u\in\mathfrak{a}^n} s_u f s_u^*$ is clearly linear and $*$-preserving. If $f, g \in \mathscr{D}_{\mathsf{X}}$, then it follows from Lemma 3 and the fact that $\mathscr{D}_{\mathsf{X}}$ is commutative that we have

$$\left(\sum_{u\in\mathfrak{a}^n} s_u f s_u^*\right)\left(\sum_{v\in\mathfrak{a}^n} s_v g s_v^*\right) = \sum_{u\in\mathfrak{a}^n} s_u f 1_{C(u,\epsilon)} g s_u^* = \sum_{u\in\mathfrak{a}^n} s_u f g 1_{C(u,\epsilon)} s_u^* = \sum_{u\in\mathfrak{a}^n} s_u f g s_u^*,$$

which proves that the map $f \mapsto \sum_{u\in\mathfrak{a}^n} s_u f s_u^*$ is multiplicative, and thus a $*$-homomorphism. Since α^n is also a $*$-homomorphism and $\mathscr{D}_{\mathsf{X}}$ is generated by $\{1_{C(u,v)} \mid u, v \in \mathfrak{a}^*\}$, it therefore suffices to check that $\alpha^n(1_{C(u',v')}) = \sum_{u\in\mathfrak{a}^n} s_u 1_{C(u',v')} s_u^*$ for $u', v' \in \mathfrak{a}^*$, and that can be done in the following way:

$$\begin{aligned}\sum_{u\in\mathfrak{a}^n} s_u 1_{C(u',v')} s_u^* &= \sum_{u\in\mathfrak{a}^n} s_u s_{v'} s_{u'}^* s_{u'} s_{v'}^* s_u^* \\ &= \sum_{u\in\mathfrak{a}^n} s_{uv'} s_{u'}^* s_{u'} s_{uv'}^* = \sum_{u\in\mathfrak{a}^n} 1_{C(u',uv')} = \alpha^n(1_{C(u',v')}).\end{aligned}$$

4. It follows from 3, Lemma 3 and the fact that $\mathscr{D}_{\mathsf{X}}$ is commutative that we have

$$s_w f = s_w s_w^* s_w f = s_w 1_{C(w,\epsilon)} f = s_w f 1_{C(w,\epsilon)} = s_w f s_w^* s_w = \sum_{u\in\mathfrak{a}^n} s_u f s_u^* s_w = \alpha^n(f) s_w.$$

5. Follows from Lemma 3 and 3.

6. If $u, v \in \mathfrak{a}^n$ and $u \neq v$, then it follows from Lemma 3 and the fact that $\mathscr{D}_{\mathsf{X}}$ is commutative, that we have

$$\begin{aligned}s_u^* \left(\sum_{u,v\in\mathfrak{a}^n} s_u s_v^* s_v s_u^*\right)^{-1} f s_v &= s_u^* 1_{C(\epsilon,u)} \left(\sum_{u,v\in\mathfrak{a}^n} s_u s_v^* s_v s_u^*\right)^{-1} f 1_{C(\epsilon,v)} s_v \\ &= s_u^* \left(\sum_{u,v\in\mathfrak{a}^n} s_u s_v^* s_v s_u^*\right)^{-1} f 1_{C(\epsilon,u)} 1_{C(\epsilon,v)} s_v = 0.\end{aligned}$$

Hence it follows from 1 to 5 that we have

$$\left(\sum_{u\in\mathfrak{a}^n} s_u\right)^* \left(\sum_{u,v\in\mathfrak{a}^n} s_u s_v^* s_v s_u^*\right)^{-1} f \left(\sum_{u\in\mathfrak{a}^n} s_u\right)$$

$$= \sum_{w\in\mathfrak{a}^n} \left(s_w^* \left(\sum_{u,v\in\mathfrak{a}^n} s_u s_v^* s_v s_u^*\right)^{-1} f s_w \right)$$

$$= \sum_{w\in\mathfrak{a}^n} \lambda_w \left(\sum_{u,v\in\mathfrak{a}^n} s_u s_v^* s_v s_u^*\right)^{-1} f = \mathscr{L}^n(f).$$

□

Proposition 4. *Let* X *be a one-sided shift space over the alphabet* $\mathfrak{a}$. *Then* $\mathscr{O}_{\mathsf{X}}$ *is the closure of*

$$\operatorname{span}\{s_u f s_v^* \mid u, v \in \mathfrak{a}^*,\ f \in \mathscr{D}_{\mathsf{X}}\}.$$

Proof. Let us by $\mathscr{X}$ denote $\operatorname{span}\{s_u f s_v^* \mid u, v \in \mathfrak{a}^*,\ f \in \mathscr{D}_{\mathsf{X}}\}$. Since $\{s_u \mid u \in \mathfrak{a}^*\} \subseteq \mathscr{X}$, it suffices to prove that $\mathscr{X}$ is a $*$-subalgebra of $\mathscr{O}_{\mathsf{X}}$. It is obvious that $\mathscr{X}$ is closed under addition and conjugation, so it suffices to prove that if $u, v, u', v' \in \mathfrak{a}^*$ and $f, f' \in \mathscr{D}_{\mathsf{X}}$, then $s_u f s_v^* s_{u'} f' s_{v'}^* \in \mathscr{X}$, so let us do that:

Let us first assume that $|v| \geq |u'|$. It follows from Lemma 3 that if $s_u f s_v^* s_{u'} f' s_{v'}^* \neq 0$, then there exists a $w \in \mathfrak{a}^*$ such that $v = u'w$, and in that case it follows from Proposition 3 that we have

$$s_u f s_v^* s_{u'} f' s_{v'}^* = s_u f s_w^* s_{u'}^* s_{u'} f' s_{v'}^* = s_u f s_w^* 1_{C(u',\epsilon)} f' s_{v'}^* = s_u f \lambda_w (1_{C(u',\epsilon)} f') s_w^* s_{v'}^* \in \mathscr{X}.$$

That $s_u f s_v^* s_{u'} f' s_{v'}^* \in \mathscr{X}$ if $|v| \leq |u'|$, then follows by taking the adjoint. □

4.4 The Gauge Action

In this section, the *gauge action* of $\mathscr{O}_{\mathsf{X}}$ will be introduced and studied, and the *gauge invariant uniqueness theorem* for $\mathscr{O}_{\mathsf{X}}$ will be described.

An *automorphism* of a C^*-algebra $\mathscr{X}$ is a $*$-isomorphism from $\mathscr{X}$ onto itself. We will by $\operatorname{Aut}(\mathscr{X})$ denote the set of automorphisms of $\mathscr{X}$. The set $\operatorname{Aut}(\mathscr{X})$ becomes a group when equipped with composition. An *action* of a group G on a C^*-algebra $\mathscr{X}$ is a homomorphism from G to $\operatorname{Aut}(\mathscr{X})$. We say that an action $\alpha : G \to \operatorname{Aut}(\mathscr{X})$ of a topological group G is *strongly continuous* if for every convergent sequence $(g_n)_{n\in\mathbb{N}}$ in G and every $x \in \mathscr{X}$, the sequence $\alpha(g_n)(x)$ converges to $\alpha(\lim_{n\to\infty} g_n)(x)$.

We will by $\mathbb{T}$ denote the group $\{z \in \mathbb{C} \mid |z| = 1\}$. The following lemma will be useful for checking if an action of $\mathbb{T}$ on a C^*-algebra is strongly continuous.

Lemma 4. *Let* $\mathfrak{a}$ *be an alphabet. If* $\mathscr{X}$ *is a* C^**-algebra generated by a family* $(x_u)_{u\in\mathfrak{a}^*}$ *and* $\alpha : \mathbb{T} \to \mathscr{X}$ *is an action such that* $\alpha(z)(x_u) = z^{|u|}x_u$ *for every* $z \in \mathbb{T}$ *and every* $u \in \mathfrak{a}^*$*, then* α *is strongly continuous.*

Proof. Let X be the set of elements x of $\mathscr{X}$ such that if $(z_n)_{n\in\mathbb{N}}$ converges to z in $\mathbb{T}$, then $\alpha(z_n)(x)$ converges to $\alpha_z(x)$ in $\mathscr{X}$. It is straightforward to check that X is a C^*-subalgebra of $\mathscr{X}$, and since we for every $u \in \mathfrak{a}^*$ have $x_u \in X$, it follows that $X = \mathscr{X}$. □

Proposition 5 (cf. [10, Remark 3.2], [15, Sect. 9] and [25, p. 361]). *Let* X *be a one-sided shift space over the alphabet* $\mathfrak{a}$*. Then there exists a strongly continuous action* $z \mapsto \gamma_z$ *of* $\mathbb{T}$ *on* $\mathscr{O}_{\mathsf{X}}$ *such that* $\gamma_z(s_u) = z^{|u|}s_u$ *and* $\gamma_z(f) = f$ *for every* $z \in \mathbb{T}$*,* $u \in \mathfrak{a}^*$ *and* $f \in \mathscr{D}_{\mathsf{X}}$*.*

Proof. Let $z \in \mathbb{T}$. It is easy to check that $(\iota, (z^{|u|}s_u)_{u\in\mathfrak{a}^*})$ is a representation of X on $\mathscr{O}_{\mathsf{X}}$. Thus there exists a $*$-homomorphism $\gamma_z : \mathscr{O}_{\mathsf{X}} \to \mathscr{O}_{\mathsf{X}}$ such that $\gamma_z(s_u) = z^{|u|}s_u$ and $\gamma_z(f) = f$ for every $u \in \mathfrak{a}^*$ and every $f \in \mathscr{D}_{\mathsf{X}}$.

If $z_1, z_2 \in \mathbb{T}$ and $u \in \mathfrak{a}^*$, then we have

$$\gamma_{z_1}\big(\gamma_{z_2}(s_u)\big) = \gamma_{z_1}(z_2^{|u|}s_u) = z_1^{|u|}z_2^{|u|}s_u = (z_1z_2)^{|u|}s_u = \gamma_{z_1z_2}(s_u).$$

Since $\mathscr{O}_{\mathsf{X}}$ is generated by $\{s_u \mid u \in \mathfrak{a}^*\}$, it follows that $\gamma_{z_1} \circ \gamma_{z_2} = \gamma_{z_1z_2}$. We have in particular that $\gamma_z \circ \gamma_{z^{-1}} = \gamma_{z^{-1}} \circ \gamma_z = \gamma_1 = \mathrm{Id}_{\mathscr{O}_{\mathsf{X}}}$ for every $z \in \mathbb{T}$, so $\gamma_z \in \mathrm{Aut}(\mathscr{O}_{\mathsf{X}})$, and $z \mapsto \gamma_z$ is an action of $\mathbb{T}$ on $\mathscr{O}_{\mathsf{X}}$. That this action is strongly continuous follows from Lemma 4. □

The action of $\mathbb{T}$ on $\mathscr{O}_{\mathsf{X}}$ from Proposition 5 is called *the gauge action* of $\mathscr{O}_{\mathsf{X}}$. Since γ is strongly continuous, it follows that we for every $x \in \mathscr{O}_{\mathsf{X}}$ have that the function $z \mapsto \gamma_z(x)$ is a continuous function from $\mathbb{T}$ to $\mathscr{O}_{\mathsf{X}}$. Thus we can make sense out of the integral

$$\int_{\mathbb{T}} \gamma_z(x)dz$$

(cf. [40, Lemma C.3.]).

Proposition 6 (cf. [15, Sect. 9] and [25, p. 361]). *Let* X *be a one-sided shift space over the alphabet* $\mathfrak{a}$*. If we for every* $x \in \mathscr{O}_{\mathsf{X}}$ *let*

$$E(x) = \int_{\mathbb{T}} \gamma_z(x)dz,$$

then E is a linear contraction (i.e., $||E(x)|| \leq ||x||$ for all $x \in \mathcal{O}_{\mathsf{X}}$) from $\mathcal{O}_{\mathsf{X}}$ to itself such that

$$E(s_u f s_v^*) = \begin{cases} s_u f s_v^* & \text{if } |u| = |v|, \\ 0 & \text{if } |u| \neq |v| \end{cases}$$

for $u, v \in \mathfrak{a}^*$ and $f \in \mathcal{D}_{\mathsf{X}}$.

Proof. It is clear that E is linear. If $\phi : \mathbb{T} \to \mathcal{O}_{\mathsf{X}}$ is continuous, then $\|\int_{\mathbb{T}} f(z)dz\| \leq \int_{\mathbb{T}} \|f(z)\|dz$ (see [40, Lemma C.3.]), and if $x \in \mathcal{O}_{\mathsf{X}}$, then $\|\gamma_z(x)\| = \|x\|$ for every $z \in \mathbb{T}$ since γ_z is an automorphism, so we have

$$\|E(x)\| = \left\| \int_{\mathbb{T}} \gamma_z(x)dz \right\| \leq \int_{\mathbb{T}} \|\gamma_z(x)\|dz = \int_{\mathbb{T}} \|x\|dz = \|x\|.$$

Let $u, v \in \mathfrak{a}^*$ and $f \in \mathcal{D}_{\mathsf{X}}$. If $|u| = |v|$, then $\gamma_z(s_u f s_v^*) = s_u f s_v^*$ for every $z \in \mathbb{T}$, so $E(s_u f s_v^*) = s_u f s_v^*$. If $|u| \neq |v|$, then we have for every $z \in \mathbb{T}$ that $\gamma_z(s_u f s_v^*) = z^n s_u f s_v^*$ where $n = |u| - |v| \neq 0$, and since $\int_{\mathbb{T}} z^n dz = 0$, it follows that $E(s_u f s_v^*) = 0$. □

The map E from Proposition 6 is a so-called *faithful conditional expectation.*

Definition 5. Let X be a one-sided shift space. We let $\mathcal{F}_{\mathsf{X}}$ denote the *fixed-point algebra*

$$\{x \in \mathcal{O}_{\mathsf{X}} \mid \forall z \in \mathbb{T} : \gamma_z(x) = x\}$$

of the gauge action γ of $\mathcal{O}_{\mathsf{X}}$.

Notice that $\mathcal{F}_{\mathsf{X}}$ is a C^*-subalgebra of $\mathcal{O}_{\mathsf{X}}$.

Proposition 7. *Let X be a one-sided shift space over the alphabet $\mathfrak{a}$. Then we have that $\mathcal{F}_{\mathsf{X}}$ is equal to the closure of*

$$\operatorname{span}\{s_u f s_v^* \mid u, v \in \mathfrak{a}^*,\ |u| = |v|,\ f \in \mathcal{D}_{\mathsf{X}}\},$$

and that $E(\mathcal{O}_{\mathsf{X}}) = \mathcal{F}_{\mathsf{X}}$.

Proof. Let $\mathcal{X}$ denote the closure of $\operatorname{span}\{s_u f s_v^* \mid u, v \in \mathfrak{a}^*,\ |u| = |v|,\ f \in \mathcal{D}_{\mathsf{X}}\}$. It is clear that $\mathcal{X} \subseteq \mathcal{F}_{\mathsf{X}}$, and that $x = E(x) \in E(\mathcal{O}_{\mathsf{X}})$ for every $x \in \mathcal{F}_{\mathsf{X}}$. It follows from Propositions 4 and 6 that $E(\mathcal{O}_{\mathsf{X}}) = \mathcal{X}$. Thus we have $E(\mathcal{O}_{\mathsf{X}}) = \mathcal{X} \subseteq \mathcal{F}_{\mathsf{X}} \subseteq E(\mathcal{O}_{\mathsf{X}})$ from which the conclusion follows. □

The following *gauge-invariant uniqueness theorem* will be proved in [6].

Theorem 2. *Let X be a one-sided shift space, $\mathcal{X}$ a C^*-algebra and $\phi : \mathcal{O}_{\mathsf{X}} \to \mathcal{X}$ a surjective $*$-homomorphism. Then the following two statements are equivalent:*

1. *The $*$-homomorphism $\phi : \mathcal{O}_{\mathsf{X}} \to \mathcal{X}$ is a $*$-isomorphism.*
2. *The restriction of ϕ to $\mathcal{D}_{\mathsf{X}}$ is injective and there exists an action $\tilde{\gamma} : \mathbb{T} \to \operatorname{Aut}(\mathcal{X})$ such that $\tilde{\gamma}_z(\phi(s_u)) = z^{|u|}\phi(s_u)$ for every $z \in \mathbb{T}$ and every $u \in \mathfrak{a}^*$.*

4.5 One-Sided Conjugation

In this section, it will be proved that the C^*-algebra $\mathcal{O}_{\mathsf{X}}$ of a one-sided shift space X is invariant under conjugacy. This has been proved in [10] and [15], and in a special case in [28], but the technique used to prove this statement here is different from the one used in [10] and [15] and will allow us to get a slightly stronger result than the one presented in [10].

Definition 6. Let X_1 and X_2 be one-sided shift spaces. We say that X_1 and X_2 are *conjugate* if there exists a homeomorphism $\phi : \mathsf{X}_1 \to \mathsf{X}_2$ such that $\phi \circ \sigma_{\mathsf{X}_1} = \sigma_{\mathsf{X}_2} \circ \phi$. We call such a homeomorphism a *conjugacy*.

Definition 7. Let X be a one-sided shift space over the alphabet $\mathfrak{a}$. We will by λ_{X} denote the map

$$x \mapsto \left(\sum_{a \in \mathfrak{a}} s_a^*\right) x \left(\sum_{b \in \mathfrak{a}} s_b\right)$$

from $\mathcal{F}_{\mathsf{X}}$ to $\mathcal{F}_{\mathsf{X}}$.

Theorem 3 (cf. [10, Theorem 8.6], [15, Theorem 23], [25, Proposition 5.8] and [28, Lemma 4.5]). *Let X_1 and X_2 be one-sided shift spaces. If X_1 and X_2 are conjugate, then there exists a $*$-isomorphism ψ from $\mathcal{O}_{\mathsf{X}_1}$ to $\mathcal{O}_{\mathsf{X}_2}$ such that*

1. $\psi(C(\mathsf{X}_1)) = C(\mathsf{X}_2)$,
2. $\psi(\mathcal{D}_{\mathsf{X}_1}) = \mathcal{D}_{\mathsf{X}_2}$,
3. $\psi(\mathcal{F}_{\mathsf{X}_1}) = \mathcal{F}_{\mathsf{X}_2}$,
4. $\psi \circ \alpha = \alpha \circ \psi$,
5. $\psi \circ \mathcal{L} = \mathcal{L} \circ \psi$,
6. $\psi \circ \gamma_z = \gamma_z \circ \psi$ *for every* $z \in \mathbb{T}$,
7. $\psi \circ \lambda_{\mathsf{X}_1} = \lambda_{\mathsf{X}_2} \circ \psi$.

Proof. Let ϕ be a conjugacy between X_2 and X_1, and let Φ be the map between the bounded functions on X_1 and the bounded functions on X_2 defined by

$$f \mapsto f \circ \phi.$$

Then $\Phi(C(\mathsf{X}_1)) = C(\mathsf{X}_2)$, $\Phi \circ \alpha = \alpha \circ \Phi$ and $\Phi \circ \mathcal{L} = \mathcal{L} \circ \Phi$, so it follows from Proposition 1 that $\Phi(\mathcal{D}_{\mathsf{X}_1}) = \mathcal{D}_{\mathsf{X}_2}$.

Let $\mathfrak{a}_1$ be the alphabet of X_1 and $\mathfrak{a}_2$ the alphabet of X_2. For $u \in \mathfrak{a}_1^*$ and $v \in \mathfrak{a}_2^*$ with $|u| = |v|$ let $D(u,v) = \{x \in \mathsf{X}_2 \mid vx \in \mathsf{X}_2,\ \phi(vx) = u\phi(x)\}$ and $Z(u) = C(\epsilon, u)$. Then we have that $1_{D(u,v)} = \lambda_v(\Phi(1_{Z(u)})) \in \mathcal{D}_{\mathsf{X}_2}$. For $u \in \mathfrak{a}_1^*$ let $t_u = \sum_{v \in \mathfrak{a}_2^{|u|}} s_v 1_{D(u,v)}$. We will show that $(\Phi, (t_u)_{u \in \mathfrak{a}_1^*})$ is a representation of X_1 on $\mathcal{O}_{\mathsf{X}_2}$. If $u_1, u_2 \in \mathfrak{a}_1^*$ and $v_1, v_2 \in \mathfrak{a}_2^*$ with $|u_1| = |v_1|$ and $|u_2| = |v_2|$, then we have

$$1_{D(u_1u_2, v_1v_2)} = \lambda_{v_2}\Big(\lambda_{v_1}(\Phi(1_{Z(u_1)}))\Phi(1_{Z(u_2)})\Big),$$

so it follows from Proposition 3 that we have

$$\begin{aligned}1_{D(u_1,v_1)}s_{v_2}1_{D(u_2,v_2)} &= \lambda_{v_1}\big(\Phi(1_{Z(u_1)})\big)s_{v_2}\lambda_{v_2}\big(\Phi(1_{Z(u_2)})\big)\\ &= s_{v_2}\lambda_{v_2}\Big(\lambda_{v_1}\big(\Phi(1_{Z(u_1)})\big)\Phi(1_{Z(u_2)})\Big)\\ &= s_{v_2}1_{D(u_1u_2,v_1v_2)}.\end{aligned}$$

It follows that if $u_1, u_2 \in \mathfrak{a}_1^*$, then we have

$$t_{u_1}t_{u_2} = \sum_{v_1\in\mathfrak{a}_2^{|u_1|}} s_{v_1}1_{D(u_1,v_1)} \sum_{v_2\in\mathfrak{a}_2^{|u_2|}} s_{v_2}1_{D(u_2,v_2)} = \sum_{v_1\in\mathfrak{a}_2^{|u_1|}}\sum_{v_2\in\mathfrak{a}_2^{|u_2|}} s_{v_1}s_{v_2}1_{D(u_1u_2,v_1v_2)} = t_{u_1u_2}.$$

We also have that

$$\begin{aligned}&t_{u_1}t_{u_2}^*t_{u_2}t_{u_1}^*\\ &= \sum_{v_1\in\mathfrak{a}_2^{|u_1|}} s_{v_1}1_{D(u_1,v_1)} \sum_{v_2\in\mathfrak{a}_2^{|u_2|}} 1_{D(u_2,v_2)}s_{v_2}^* \sum_{v_3\in\mathfrak{a}_2^{|u_2|}} s_{v_3}1_{D(u_2,v_3)} \sum_{v_4\in\mathfrak{a}_2^{|u_1|}} 1_{D(u_1,v_4)}s_{v_4}^*\\ &= \sum_{v_1\in\mathfrak{a}_2^{|u_1|}}\sum_{v_2\in\mathfrak{a}_2^{|u_2|}}\sum_{v_4\in\mathfrak{a}_2^{|u_1|}} s_{v_1}1_{D(u_1,v_1)}1_{D(u_2,v_2)}1_{C(v_2,\epsilon)}1_{D(u_2,v_2)}1_{D(u_1,v_4)}s_{v_4}^*\\ &= \sum_{v_1\in\mathfrak{a}_2^{|u_1|}}\sum_{v_2\in\mathfrak{a}_2^{|u_2|}}\sum_{v_4\in\mathfrak{a}_2^{|u_1|}} s_{v_1}1_{D(u_1,v_1)}1_{D(u_2,v_2)}1_{D(u_1,v_4)}s_{v_4}^*\\ &= \sum_{v_1\in\mathfrak{a}_2^{|u_1|}}\sum_{v_2\in\mathfrak{a}_2^{|u_2|}} s_{v_1}1_{D(u_1,v_1)}1_{D(u_2,v_2)}s_{v_1}^*\\ &= \sum_{v_1\in\mathfrak{a}_2^{|u_1|}}\sum_{v_2\in\mathfrak{a}_2^{|u_2|}} \alpha^{|v_1|}(1_{D(u_1,v_1)}1_{D(u_2,v_2)})s_{v_1}s_{v_1}^*\\ &= \sum_{v_1\in\mathfrak{a}_2^{|u_1|}}\sum_{v_2\in\mathfrak{a}_2^{|u_2|}} \alpha^{|v_1|}(1_{D(u_1,v_1)}1_{D(u_2,v_2)})1_{Z(v_1)}\\ &= \Phi(1_{C(u_2,u_1)}).\end{aligned}$$

Thus $(\Phi, (t_u)_{u\in\mathfrak{a}_1^*})$ is a representation of X_1 on $\mathcal{O}_{\mathsf{X}_2}$. It follows that there exists a $*$-homomorphism ψ from $\mathcal{O}_{\mathsf{X}_1}$ to $\mathcal{O}_{\mathsf{X}_2}$ such that $\psi(s_u) = t_u = \sum_{v\in\mathfrak{a}_2^{|u|}} s_v 1_{D(u,v)}$ for every $u \in \mathfrak{a}_1^*$ and $\psi(f) = \Phi(f)$ for every $f \in \mathcal{D}_{\mathsf{X}_1}$.

For $u \in \mathfrak{a}_1^*$ and $v \in \mathfrak{a}_2^*$, let $\tilde{D}(v,u) = \{x \in \mathsf{X}_1 \mid ux \in \mathsf{X}_1,\ \phi^{-1}(ux) = v\phi^{-1}(x)\}$. In a manner similar to the preceding paragraph, one can prove that there exists a $*$-homomorphism η from $\mathcal{O}_{\mathsf{X}_2}$ to $\mathcal{O}_{\mathsf{X}_1}$ such that $\psi(s_v) = \sum_{u\in\mathfrak{a}_1^{|v|}} s_u 1_{\tilde{D}(v,u)}$ for every $v \in \mathfrak{a}_2^*$ and $\rho(f) = \Phi^{-1}(f)$ for every $f \in \mathcal{D}_{\mathsf{X}_2}$. If $u, u' \in \mathfrak{a}_1^*$ with $|u| = |u'|$, then $\sum_{v\in\mathfrak{a}_2^{|u|}} 1_{\tilde{D}(v,u')}\Phi^{-1}(1_{D(u,v)}) = 0$ if $u \neq u'$, and $\sum_{v\in\mathfrak{a}_2^{|u|}} 1_{\tilde{D}(v,u')}\Phi^{-1}(1_{D(u,v)}) = 1_{C(u,\epsilon)}$ if $u = u'$. It follows that if $u \in \mathfrak{a}_1^*$, then we have

$$\begin{aligned}\rho(\psi(s_u)) &= \rho\Bigg(\sum_{v\in\mathfrak{a}_2^{|u|}} s_v 1_{D(u,v)}\Bigg) = \sum_{v\in\mathfrak{a}_2^{|u|}} \sum_{u'\in\mathfrak{a}_1^{|v|}} s_{u'} 1_{\tilde{D}(v,u')}\Phi^{-1}(1_{D(u,v)}) \\ &= s_u 1_{C(u,\epsilon)} = s_u s_u^* s_u = s_u.\end{aligned}$$

In a similar way, one can show that $\psi(\rho(s_v)) = s_v$ for every $v \in \mathfrak{a}_2^*$. Thus ρ is the inverse of ψ, and ψ is an isomorphism.

Since $\psi(f) = \Phi(f)$ for $f \in \mathcal{D}_{\mathsf{X}_1}$, it follows that ψ has the properties 1, 2, 4 and 5. If $z \in \mathbb{T}$ and $u \in \mathfrak{a}_1^*$, then we have

$$\psi(\gamma_z(s_u)) = \psi(z^{|u|}s_u) = z^{|u|} \sum_{v\in\mathfrak{a}_2^{|u|}} s_v 1_{D(u,v)} = \gamma_z\Bigg(\sum_{v\in\mathfrak{a}_2^{|u|}} s_v 1_{D(u,v)}\Bigg) = \gamma_z(\psi(s_u)).$$

It follows that ψ has property 6. It follows from this and Proposition 5 that ψ also has property 3.

Let $v \in \mathfrak{a}_2^*$. Then we have that $\sum_{u\in\mathfrak{a}_1^{|v|}} 1_{D(u,v)} = 1_{C(v,\epsilon)}$. Thus we have for every $x \in \mathcal{F}_{\mathsf{X}_1}$ that

$$\begin{aligned}\psi(\lambda_{\mathsf{X}_1}(x)) &= \psi\Bigg(\Bigg(\sum_{a\in\mathfrak{a}_1} s_a^*\Bigg)x\Bigg(\sum_{b\in\mathfrak{a}_1} s_b\Bigg)\Bigg) \\ &= \sum_{a\in\mathfrak{a}_1}\sum_{b\in\mathfrak{a}_1}\sum_{c\in\mathfrak{a}_2}\sum_{d\in\mathfrak{a}_2} 1_{D(a,c)}s_c^*\psi(x)s_d 1_{D(b,d)} \\ &= \sum_{c\in\mathfrak{a}_2}\sum_{d\in\mathfrak{a}_2} 1_{C(c,\epsilon)}s_c^*\psi(x)s_d 1_{C(d,\epsilon)} \\ &= \sum_{c\in\mathfrak{a}_2}\sum_{d\in\mathfrak{a}_2} s_c^*\psi(x)s_d = \lambda_{\mathsf{X}_2}(\psi(x)).\end{aligned}$$

This proves that ψ has property 7. □

4.6 Two-Sided Conjugacy and Flow Equivalence

Let $\mathfrak{a}$ be a finite alphabet and let $\mathfrak{a}^{\mathbb{Z}}$ be the infinite product space $\prod_{n\in\mathbb{Z}} \mathfrak{a}$ endowed with the product topology. The transformation τ on $\mathfrak{a}^{\mathbb{Z}}$ given by

$$\big(\tau(x)\big)_i = x_{i+1},\ i \in \mathbb{Z},$$

is called the *two-sided shift*. Let Λ be a closed subset of $\mathfrak{a}^{\mathbb{Z}}$ such that $\tau(\Lambda) = \Lambda$. The topological dynamical system $(\Lambda, \tau_{|\Lambda})$ is called a *two-sided shift space*. We will denote $\tau_{|\Lambda}$ by τ_Λ or just τ for simplicity.

Given a two-sided shift space Λ we can construct a one-sided shift space, namely

$$\{(x_n)_{n\in\mathbb{N}_0} \mid (x_n)_{n\in\mathbb{Z}} \in \Lambda\}.$$

We will denote this one-sided shift space by X_Λ.

Let Λ_1 and Λ_2 be two two-sided shift spaces. We say that Λ_1 and Λ_2 are *(topological) conjugate* if there exists a homeomorphism $\psi : \Lambda_1 \to \Lambda_2$ such that $\psi \circ \tau_{\Lambda_1} = \tau_{\Lambda_2}$.

The following theorem will be proved in [6].

Theorem 4. *Let Λ_1 and Λ_2 be two two-sided shift spaces which are conjugate. Then $\mathcal{O}_{\mathsf{X}_{\Lambda_1}}$ and $\mathcal{O}_{\mathsf{X}_{\Lambda_2}}$ are Morita equivalent (cf. Sect. 4.7).*

This theorem was essentially proved in [33] under the assumption Λ_1 and Λ_2 both satisfy two conditions called (I) and (E), and later in [36] under the assumption of (I).

The assumption in Theorem 4 that Λ_1 and Λ_2 are conjugate, can in fact be weakened to the assumption that Λ_1 and Λ_2 are *flow equivalent* (we refer to [5, 18, 39] and [24, Sect. 13.6] for the definition of flow equivalence). This was first proved by Matsumoto in [35] under the assumption that both the two two-sided shift spaces satisfy condition (I), and will be proved in [6] without this assumption.

4.7 The K-Theory of C^*-Algebras Associated to Shift Spaces

Since $K_0(\mathcal{X})$ and $K_1(\mathcal{X})$ are invariants of a C^*-algebra $\mathcal{X}$, it follows from the previous section that $K_0(\mathcal{O}_\mathsf{X})$, $K_1(\mathcal{O}_\mathsf{X})$ and $K_0(\mathcal{F}_\mathsf{X})$ are invariants of X ($\mathcal{F}_\mathsf{X}$ is an AF-algebra so $K_1(\mathcal{F}_\mathsf{X}) = 0$ for any one-sided shift space X, cf. [6], [25, Lemma 2.1] and [28, Lemma 4.1]). In this section, we will present formulas based on l-past equivalence for these invariants. This was done in [27, 28, 35] by Matsumoto for the case of one-sided shift spaces of the form X_Λ, where Λ

is a two-sided shift space and generalized to the general case in [15] and [16] (it should be noted that [15, Theorem 26] is not correct as stated. The error has been rectified in [16]). I will not here prove the formulas for $K_0(\mathcal{O}_\mathsf{X})$, $K_1(\mathcal{O}_\mathsf{X})$ and $K_0(\mathcal{F}_\mathsf{X})$, but only establish the necessary setup and state the theorems which give the formulas. The interested reader can find proofs of these theorems in the above mentioned references.

From these formulas, one can directly prove that $K_0(\mathcal{O}_\mathsf{X})$, $K_1(\mathcal{O}_\mathsf{X})$ and $K_0(\mathcal{F}_\mathsf{X})$ are invariants of X without involving C^*-algebras. This is done (for one-sided shift spaces of the form X_Λ, where Λ is a two-sided shift space) in Matsumoto's very interesting paper [29], where also other invariants of shift spaces are presented.

Let X be a one-sided shift space. We will for each $l \in \mathbb{N}_0$ define an equivalence relation on X called *l-past equivalence*. These equivalence relations were introduced by Matsumoto in [28]. For $k \in \mathbb{N}_0$ and $x \in \mathsf{X}$ let $\mathcal{P}_k(x) = \{u \in \mathfrak{a}^k \mid ux \in \mathsf{X}\}$. If $x, y \in \mathsf{X}$ and $l \in \mathbb{N}_0$, then we say that *x and y are l-past equivalent* and write $x \sim_l y$ if $\bigcup_{k=0}^{l} \mathcal{P}_k(x) = \bigcup_{k=0}^{l} \mathcal{P}_k(y)$. Notice that since $\mathfrak{a}^k$ is finite for each $k \in \mathbb{N}_0$, we have for each $l \in \mathbb{N}_0$ only finitely many l-past equivalence classes. We let $m(l)$ be this number of l-past equivalence classes, and we denote the l-past equivalence classes by $\mathrm{e}_1^l, \mathrm{e}_2^l, \dots, \mathrm{e}_{m(l)}^l$. For each $l \in \mathbb{N}_0$, $j \in \{1, 2, \dots, m(l)\}$ and $i \in \{1, 2, \dots, m(l+1)\}$, let

$$I_l(i, j) = \begin{cases} 1 \text{ if } \mathrm{e}_i^{l+1} \subseteq \mathrm{e}_j^l \\ 0 \text{ otherwise.} \end{cases}$$

Let F be a finite set and $i_0 \in F$. Then we denote by e_{i_0} the element in $\mathbb{Z}^F$ for which

$$e_{i_0}(i) = \begin{cases} 1 \text{ if } i = i_0 \\ 0 \text{ otherwise.} \end{cases}$$

Let $0 \le k \le l$. Then we have that $x \sim_l y \implies \mathcal{P}_k(x) = \mathcal{P}_k(y)$. We can therefore for $i \in \{1, 2, \dots, m(l)\}$ define $\mathcal{P}_k(E_i^l)$ to be $\mathcal{P}_k(x)$ for some $x \in E_i^l$. Let M_k^l be defined by

$$M_k^l = \{i \in \{1, 2 \dots, m(l)\} \mid \mathcal{P}_k(\mathrm{e}_i^l) \neq \emptyset\}.$$

Notice that if X is of the form X_Λ for some two-sided shift space Λ (this is equivalent to $\sigma(\mathsf{X}) = \mathsf{X}$), then $M_k^l = \{1, 2, \dots, m(l)\}$ for all $0 \le k \le l$.

If $j \in M_k^l$ and $I_l(i, j) = 1$, then $i \in M_k^{l+1}$, so there exists a positive linear map from $\mathbb{Z}^{M_k^l}$ to $\mathbb{Z}^{M_k^{l+1}}$ given by

$$e_j \mapsto \sum_{i \in M_k^{l+1}} I_l(i, j) e_i.$$

We denote this map by I_k^l.

For a subset e of X and a $u \in \mathfrak{a}^*$, let $ue = \{ux \in \mathsf{X} \mid x \in \mathrm{e}\}$. For each $l \in \mathbb{N}_0$, $j \in \{1, 2, \ldots, m(l)\}$, $i \in \{1, 2, \ldots, m(l+1)\}$ and $a \in \mathfrak{a}$, let

$$A_l(i, j, a) = \begin{cases} 1 \text{ if } \emptyset \neq a\mathrm{e}_i^{l+1} \subseteq \mathrm{e}_j^l \\ 0 \text{ otherwise.} \end{cases}$$

Let $0 \leq k \leq l$. If $j \in M_k^l$ and if there exists an $a \in \mathfrak{a}$ such that $A_l(i, j, a) = 1$, then $i \in M_{k+1}^{l+1}$. Thus there exists a positive linear map from $\mathbb{Z}^{M_k^l}$ to $\mathbb{Z}^{M_{k+1}^{l+1}}$ given by

$$e_j \mapsto \sum_{i \in M_{k+1}^{l+1}} \sum_{a \in \mathfrak{a}} A_l(i, j, a) e_i.$$

We denote this map by A_k^l.

Lemma 5. *Let $0 \leq k \leq l$. Then the following diagram commutes:*

$$\begin{array}{ccc} \mathbb{Z}^{M_k^l} & \xrightarrow{I_k^l} & \mathbb{Z}^{M_k^{l+1}} \\ {\scriptstyle A_k^l}\big\downarrow & & \big\downarrow{\scriptstyle A_k^{l+1}} \\ \mathbb{Z}^{M_{k+1}^{l+1}} & \xrightarrow{I_{k+1}^{l+1}} & \mathbb{Z}^{M_{k+1}^{l+2}}. \end{array}$$

Proof. Let $j \in M_k^l$, $h \in M_{k+1}^{l+2}$ and $a \in \mathfrak{a}$. If $\emptyset \neq a\mathrm{e}_h^{l+2} \subseteq \mathrm{e}_j^l$, then there exists exactly one $i \in M_k^{l+1}$ such that $\mathrm{e}_i^{l+1} \subseteq \mathrm{e}_j^l$ and $\emptyset \neq a\mathrm{e}_h^{l+2} \subseteq \mathrm{e}_i^{l+1}$; and there exists exactly one $i' \in M_{k+1}^{l+1}$ such that $\mathrm{e}_h^{l+2} \subseteq \mathrm{e}_{i'}^{l+1}$ and $\emptyset \neq a\mathrm{e}_{i'}^{l+1} \subseteq \mathrm{e}_j^l$. If $a\mathrm{e}_h^{l+2} = \emptyset$ or $a\mathrm{e}_h^{l+2} \not\subseteq \mathrm{e}_j^l$, then there does not exist an $i \in M_k^{l+1}$ such that $\mathrm{e}_i^{l+1} \subseteq \mathrm{e}_j^l$ and $\emptyset \neq a\mathrm{e}_h^{l+2} \subseteq \mathrm{e}_i^{l+1}$; and there does not exist an $i' \in M_{k+1}^{l+1}$ such that $\mathrm{e}_h^{l+2} \subseteq \mathrm{e}_{i'}^{l+1}$ and $\emptyset \neq a\mathrm{e}_{i'}^{l+1} \subseteq \mathrm{e}_j^l$. Hence we have

$$\sum_{i \in M_k^{l+1}} A_{l+1}(h, i, a) I_l(i, j) = \sum_{i \in M_{k+1}^{l+1}} I_{l+1}(h, i) A_l(i, j, a).$$

It follows from this that

$$A_k^{l+1}(I_k^l(e_j)) = A_k^{l+1}\left(\sum_{i \in M_k^{l+1}} I_l(i, j) e_i \right)$$

$$= \sum_{h \in M_{k+1}^{l+2}} \sum_{a \in \mathfrak{a}} A_{l+1}(h, i, a) \sum_{i \in M_k^{l+1}} I_l(i, j) e_h$$

$$
\begin{aligned}
&= \sum_{h \in M_{k+1}^{l+2}} \sum_{i \in M_{k+1}^{l+1}} \sum_{a \in \mathfrak{a}} I_{l+1}(h,i) A_l(i,j,a) e_h \\
&= I_{k+1}^{l+1}\left(\sum_{i \in M_{k+1}^{l+1}} \sum_{a \in \mathfrak{a}} A_l(i,j,a) e_i \right) \\
&= I_{k+1}^{l+1}(A_k^l(e_j))
\end{aligned}
$$

for every $j \in M_k^l$. Thus the diagram commutes. □

For $k \in \mathbb{N}_0$, the inductive limit $\varinjlim(\mathbb{Z}^{M_k^l}, (\mathbb{Z}^+)^{M_k^l}, I_k^l)$ will be denoted $(\mathbb{Z}_{\mathsf{X}_k}, \mathbb{Z}_{\mathsf{X}_k}^+)$. It follows from Lemma 5 that the family $\{A_k^l\}_{l \geq k}$ induces a positive, linear map A_k from $\mathbb{Z}_{\mathsf{X}_k}$ to $\mathbb{Z}_{\mathsf{X}_{k+1}}$.

Let $0 \leq k < l$. Denote by δ_k^l the linear map from $\mathbb{Z}^{M_k^l}$ to $\mathbb{Z}^{M_{k+1}^l}$ given by

$$
e_j \mapsto \begin{cases} e_j & \text{if } j \in M_{k+1}^l, \\ 0 & \text{if } j \notin M_{k+1}^l, \end{cases}
$$

for $j \in M_k^l$. It is easy to check that the following diagram

$$
\begin{array}{ccc}
\mathbb{Z}^{M_k^l} & \xrightarrow{\delta_k^l} & \mathbb{Z}^{M_{k+1}^l} \\
{\scriptstyle I_k^l}\downarrow & & \downarrow{\scriptstyle I_{k+1}^l} \\
\mathbb{Z}^{M_k^{l+1}} & \xrightarrow{\delta_k^{l+1}} & \mathbb{Z}^{M_{k+1}^{l+1}}
\end{array}
$$

commutes.

Thus the family $\{\delta_k^l\}_{l \geq k}$ induces a positive, linear map from $\mathbb{Z}_{\mathsf{X}_k}$ to $\mathbb{Z}_{\mathsf{X}_{k+1}}$ which we denote by δ_k. Since the diagram

$$
\begin{array}{ccc}
\mathbb{Z}^{M_k^l} & \xrightarrow{\delta_k^l} & \mathbb{Z}^{M_{k+1}^l} \\
{\scriptstyle A_k^l}\downarrow & & \downarrow{\scriptstyle A_{k+1}^l} \\
\mathbb{Z}^{M_{k+1}^{l+1}} & \xrightarrow{\delta_{k+1}^{l+1}} & \mathbb{Z}^{M_{k+2}^{l+1}}
\end{array}
$$

commutes for every $0 \le k < l$, the diagram

$$\begin{array}{ccc} \mathbb{Z}_{\mathsf{X}_k} & \xrightarrow{\delta_k} & \mathbb{Z}_{\mathsf{X}_{k+1}} \\ {\scriptstyle A_k}\downarrow & & \downarrow{\scriptstyle A_{k+1}} \\ \mathbb{Z}_{\mathsf{X}_{k+1}} & \xrightarrow{\delta_{k+1}} & \mathbb{Z}_{\mathsf{X}_{k+2}} \end{array}$$

commutes.

We denote the inductive limit $\varinjlim(\mathbb{Z}_{\mathsf{X}_k}, \mathbb{Z}^+_{\mathsf{X}_k}, A_k)$ by $(\Delta_{\mathsf{X}}, \Delta^+_{\mathsf{X}})$. Since the previous diagram commutes, the family $\{\delta_k\}_{k\in\mathbb{N}_0}$ induces a positive, linear map from Δ_{X} to Δ_{X} which we denote by δ_{X}.

We are now ready to describe the K-theory of $\mathscr{F}_{\mathsf{X}}$ and of $\mathscr{O}_{\mathsf{X}}$.

Theorem 5 (cf. [15, Theorem 25], [25, Theorem 3.11] and [28, Theorem 4.11]). *Let* X *be a one-sided shift space. Then there exists an isomorphism* $\phi : K_0(\mathscr{F}_{\mathsf{X}}) \to \Delta_{\mathsf{X}}$ *which satisfies that* $\phi(K_0^+(\mathscr{F}_{\mathsf{X}})) = \Delta^+_{\mathsf{X}}$ *and that* $\phi \circ (\lambda_{\mathsf{X}})_0 = \delta_{\mathsf{X}} \circ \phi$.

For every $l \in \mathbb{N}_0$ denote by B^l the linear map from $\mathbb{Z}^{m(l)}$ to $\mathbb{Z}^{m(l+1)}$ given by

$$e_j \mapsto \sum_{i=1}^{m(l+1)} \left(I_l(i,j) - \sum_{a\in\mathfrak{a}} A_l(i,j,a) \right) e_i .$$

One can easily check that the following diagram commutes for every $l \in \mathbb{N}_0$:

$$\begin{array}{ccc} \mathbb{Z}^{m(l)} & \xrightarrow{B^l} & \mathbb{Z}^{m(l+1)} \\ {\scriptstyle I_0^l}\downarrow & & \downarrow{\scriptstyle I_0^{l+1}} \\ \mathbb{Z}^{m(l+1)} & \xrightarrow{B^{l+1}} & \mathbb{Z}^{m(l+2)} . \end{array}$$

Hence the family $\{B^l\}_{l\in\mathbb{N}_0}$ induces a linear map B from $\mathbb{Z}_{\mathsf{X}_0}$ to $\mathbb{Z}_{\mathsf{X}_0}$.

Theorem 6 (cf. [16, Theorem 1] and [25, Theorem 4.9]). *Let* X *be a one-sided shift space. Then*

$$K_0(\mathscr{O}_{\mathsf{X}}) \cong \mathbb{Z}_{\mathsf{X}_0}/B\mathbb{Z}_{\mathsf{X}_0},$$

and

$$K_1(\mathscr{O}_{\mathsf{X}}) \cong \ker(B).$$

Appendix

I will in this section give a (very short) introduction to C^*-algebras, Morita equivalence of C^*-algebras and K-theory for C^*-algebras which hopefully will provide a reader without any knowledge of operator algebra with the necessary background for reading this paper.

I will not give any proofs at all. The interested reader is referred to for example [1, 3, 4, 38, 40–42] for more details.

C-Algebras*

Definition 8. A *C^*-algebra* is an algebra $\mathscr{X}$ over the complex numbers equipped with a map $x \mapsto x^*$ and a norm $\|\cdot\|$ satisfying:

1. $\mathscr{X}$ is complete with respect to $\|\cdot\|$.
2. $\|xy\| \le \|x\|\|y\|$ for $x, y \in \mathscr{X}$. .
3. $(x^*)^* = x$ for $x \in \mathscr{X}$.
4. $(xy)^* = y^*x^*$ for $x, y \in \mathscr{X}$.
5. $(\lambda x)^* = \overline{\lambda}x^*$ for $\lambda \in \mathbb{C}$ and $x \in \mathscr{X}$.
6. $(x + y)^* = x^* + y^*$ for $x, y \in \mathscr{X}$.
7. $\|x^*\| = \|x\|$ for $x \in \mathscr{X}$.
8. $\|x^*x\| = \|x\|^2$ for $x \in \mathscr{X}$.

The map $x \mapsto x^*$ is called an *involution*. A C^*-algebra is called *unital* if it has a algebraic unit (i.e, $\mathscr{X}$ is unital if there exists a $1 \in \mathscr{X}$ such that $1x = x1 = x$ for all $x \in \mathscr{X}$). All the C^*-algebras we will meet in this paper (except here in the appendix) are unital.

An algebra equipped with a norm satisfying condition 1 and 2 is called a *Banach algebra*. A Banach algebra equipped with an involution satisfying condition 3–7 is called a *Banach $*$-algebra*. Condition 8 is often called the *C^*-identity*. Although this condition at first glance seems to be a mild condition it is in fact very strong because it ties together the algebraic structure of the C^*-algebra and its topology. One can for example show that if $\mathscr{X}$ is an algebra equipped with an involution satisfying condition 3–6, then there is at most one norm which makes $\mathscr{X}$ a C^*-algebra.

A map $\phi : \mathscr{X}_1 \to \mathscr{X}_2$ between C^*-algebras is called a *$*$-homomorphism* if it satisfies

1. $\phi(ax + by) = a\phi(x) + b\phi(y)$ for $x, y \in \mathscr{X}_1$ and $a, b \in \mathbb{C}$,
2. $\phi(xy) = \phi(x)\phi(y)$ for $x, y \in \mathscr{X}$,
3. $\phi(x^*) = (\phi(x))^*$ for $x \in \mathscr{X}$.

A $*$-homomorphism which is invertible is called a *$*$-isomorphism*, and if there exists a $*$-isomorphism between two C^*-algebras, then they are said to be *isomorphic*.

If $\phi : \mathscr{X}_1 \to \mathscr{X}_2$ is a $*$-homomorphism, then $\|\phi(x)\| \leq \|x\|$ for all $x \in \mathscr{X}_1$, and ϕ is injective if and only if $\|\phi(x)\| = \|x\|$ for all $x \in \mathscr{X}_1$ (see for example [38, Theorem 2.1.7] for a proof of this). Thus a $*$-homomorphism is automatically continuous, and a $*$-isomorphism is automatically isometric. This is another example of how the algebraic structure of a C^*-algebra and its topology are closely related.

Example 2. Let H be a Hilbert space. Then the algebra $\mathscr{B}(\mathsf{H})$ of bounded operators is a C^*-algebra where T^* of an bounded operator $T \in \mathscr{B}(\mathsf{H})$ is the adjoint of T, and the norm $\|T\|$ is the operator norm $\sup\{\|T\eta\| \mid \eta \in \mathsf{H},\ \|\eta\| \leq 1\}$.

Definition 9. A *projection* in a C^*-algebra $\mathscr{X}$ is a $p \in \mathscr{X}$ satisfying $p^2 = p^* = p$. A *partial isometry* is a $s \in \mathscr{X}$ satisfying $ss^*s = s$.

It is easy to see that if s is a partial isometry, then ss^* (and s^*s) is a projection. One can prove (see for example [38, Theorem 2.3.3]) that if s is an element of a C^*-algebra such that ss^* is a projection, then s is a partial isometry.

Using functional calculus (see for example [38, p. 43] or [41, Sect. 1.2.4]) and the uniqueness of a positive square root of a positive element in a C^*-algebra (see [38, Theorem 2.2.1]) one can prove the following fact:

Fact 7. Let s be an element of a C^*-algebra $\mathscr{X}$. If ss^*ss^* is a projection, then $ss^* = ss^*ss^*$ and s is a partial isometry.

Definition 10. A *C^*-subalgebra* of a C^*-algebra $\mathscr{X}$ is a closed subalgebra $\mathscr{Y}$ of $\mathscr{X}$ such that $x \in \mathscr{Y} \implies x^* \in \mathscr{Y}$.

A C^*-subalgebra $\mathscr{Y}$ is a C^*-algebra in itself with the operations it inherits from $\mathscr{X}$. It is a famous theorem by Gelfand and Naimark that every C^*-algebra is isomorphic to some C^*-subalgebra of the C^*-algebra of bounded operators on some Hilbert space.

Example 3. Let X be a set. The algebra of bounded functions from X to $\mathbb{C}$ is a C^*-algebra where the involution f^* of an $f \in l^\infty(X)$ is defined by $f^*(x) = \overline{f(x)}$ for all $x \in X$, and the norm $\|f\|$ of f is $\sup\{|f(x)| \mid x \in X\}$. Notice that $l^\infty(X)$ is abelian.

If X is a locally compact Hausdorff space, then the algebra $C_0(X)$ of continuous functions on X vanishing at infinity is a C^*-subalgebra of $l^\infty(X)$.

It is another famous theorem by Gelfand and Naimark that every abelian C-algebra is isomorphic to $C_0(X)$ for some locally compact Hausdorff space X.

We are going to need (in the proof of Proposition 1) the following fact which follows from [38, Theorem 2.1.11]:

Fact 8. Let $\mathscr{X}$ be a unital C^*-algebra. If $\mathscr{Y}$ is a C^*-subalgebra of $\mathscr{X}$ which contains the unit of $\mathscr{X}$, and $y \in \mathscr{Y}$ is invertible in $\mathscr{X}$, then its inverse y^{-1} belongs to $\mathscr{Y}$.

When $\mathscr{X}$ is C^*-algebra and X is some subset of $\mathscr{X}$, then there exists a C^*-subalgebra $\mathscr{Y}$ of $\mathscr{X}$ which contains X and which is contained in any other C^*-subalgebra of $\mathscr{X}$ that contains X. The C^*-subalgebra $\mathscr{Y}$ is just the intersection of every C^*-subalgebra of $\mathscr{X}$ that contains X. We call $\mathscr{Y}$ *the C^*-subalgebra of $\mathscr{X}$ generated by* X and denote it by $C^*(X)$.

Morita Equivalence

By an *ideal* of a C^*-algebra we mean a closed two-sided ideal. I.e., an ideal of a C^*-algebra $\mathscr{X}$ is a closed subset I of $\mathscr{X}$ such that $\lambda a + \gamma b, xa, ax \in I$ for $a, b \in I$, $\lambda, \gamma \in \mathbb{C}$ and $x \in \mathscr{X}$. An ideal I of a C^*-algebra is automatically closed under involution, i.e., if $x \in I$, then $x^* \in I$. Thus every ideal of a C^*-algebra is also a C^*-subalgebra.

A nonzero ideal of a C^*-algebra $\mathscr{X}$ is said to be *essential* if it has nonzero intersection with every other nonzero ideal of $\mathscr{X}$.

There exists for every C^*-algebra $\mathscr{X}$ a, up to isomorphism, unique maximal unital C^*-algebra $M(\mathscr{X})$ which contains $\mathscr{X}$ as an essential ideal. The C^*-algebra $M(\mathscr{X})$ is known as *the multiplier algebra of* $\mathscr{X}$, cf. [38, Theorem 3.1.8] and [40, Theorem 2.47]. If $\mathscr{X}$ itself is unital, then $M(\mathscr{X}) = \mathscr{X}$.

It is easy to check that if p is a projection in the multiplier algebra $M(\mathscr{X})$ of a the C^*-algebra $\mathscr{X}$, then $p\mathscr{X}p := \{pxp \mid x \in \mathscr{X}\}$ is a C^*-subalgebra of $\mathscr{X}$. Such a C^*-subalgebra is called a *corner*. The projection p is said to be *full* and the corner $p\mathscr{X}p$ is said to be a *full corner* if there is no proper ideal of $\mathscr{X}$ which contains p.

Two projections $p, q \in M(\mathscr{X})$ are said to be *complementary* if $p + q = 1$. If p and q are complementary, then $pq = 0$ and thus $p\mathscr{X}p \cap q\mathscr{X}q = \{0\}$. In this situation, the two corners $p\mathscr{X}p$ and $q\mathscr{X}$ are also called complementary.

Morita equivalence is an equivalence relations between C^*-algebras. I will not give the definition of Morita equivalence here, but instead use the following characterization of Morita equivalence.

Theorem 9 (Cf. [40, Theorem 3.19]). *Two C^*-algebras $\mathscr{X}_1$ and $\mathscr{X}_2$ are Morita equivalent if and only if there is a C^*-algebra $\mathscr{X}$ with complementary full corners isomorphic to $\mathscr{X}_1$ and $\mathscr{X}_2$, respectively.*

It follows directly that Morita equivalence is weaker than isomorphism. It is not difficult to show that if $p\mathscr{X}p$ is a full corner of a C^*-algebra, then $p\mathscr{X}p$ and $\mathscr{X}$ are Morita equivalent.

K-Theory for C-Algebras*

K-theory for C^*-algebras is a pair of covariant functors K_0 and K_1 both defined on the category of C^*-algebras. The functor K_0 associates to each C^*-algebra $\mathscr{X}$ a pair $(K_0^+(\mathscr{X}), K_0(\mathscr{X}))$ consisting of an abelian group $K_0(\mathscr{X})$ and a sub-semigroup $K_0^+(\mathscr{X})$ of $K_0(\mathscr{X})$ (i.e., $K_0^+(\mathscr{X}) \subseteq K_0(\mathscr{X})$ and $g, h \in K_0^+(\mathscr{X}) \implies g + h \in K_0^+(\mathscr{X})$), and associates to each $*$-homomorphism $\phi : \mathscr{X}_1 \to \mathscr{X}_2$ a group homomorphism $K_0(\phi) : K_0(\mathscr{X}_1) \to K_0(\mathscr{X}_2)$ satisfying $K_0(\phi)(K_0^+(\mathscr{X}_1)) \subseteq K_0^+(\mathscr{X}_2)$. The functor K_1 associates to each C-algebra $\mathscr{X}$ an abelian group $K_1(\mathscr{X})$ and to each a $*$-homomorphism $\phi : \mathscr{X}_1 \to \mathscr{X}_2$ a group homomorphism $K_1(\phi) : K_1(\mathscr{X}_1) \to K_1(\mathscr{X}_2)$.

That K_0 and K_1 are functors means that $K_0(\operatorname{Id} \mathscr{X}) = \operatorname{Id}_{K_0(\mathscr{X})}$ and $K_1(\operatorname{Id} \mathscr{X}) = \operatorname{Id}_{K_1(\mathscr{X})}$ for every C^*-algebra $\mathscr{X}$, and that $K_0(\phi_1 \circ \phi_2) = K_0(\phi_1) \circ K_0(\phi_2)$ and $K_1(\phi_1 \circ \phi_2) = K_1(\phi_1) \circ K_1(\phi_2)$ for all $*$-homomorphisms $\phi_1 : \mathscr{X}_1 \to \mathscr{X}_2$ and $\phi_2 : \mathscr{X}_2 \to \mathscr{X}_3$. Thus if two C^*-algebras are isomorphic, then $K_0(\mathscr{X}_1)$ and $K_0(\mathscr{X}_2)$ are isomorphic as groups, and so are $K_1(\mathscr{X}_1)$ and $K_1(\mathscr{X}_2)$. In fact, $K_0(\mathscr{X}_1)$ and $K_0(\mathscr{X}_2)$ are isomorphic by an isomorphism which maps $K_0^+(\mathscr{X}_1)$ onto $K_0^+(\mathscr{X}_2)$.

If $p\mathscr{X}p$ is a full corner of a C^*-algebra $\mathscr{X}$ and ι denotes the inclusion of $p\mathscr{X}p$ into $\mathscr{X}$, then $K_0(\iota)$ and $K_1(\iota)$ are both isomorphisms, and the isomorphism $K_0(\iota)$ maps $K_0^+(p\mathscr{X}p)$ onto $K_0^+(\mathscr{X})$, see [21, Proposition B.3]. Thus if two C^*-algebras are Morita equivalent, then $K_1(\mathscr{X}_1)$ and $K_1(\mathscr{X}_2)$ are isomorphic as groups, and $K_0(\mathscr{X}_1)$ and $K_0(\mathscr{X}_2)$ are isomorphic as groups by an isomorphism which maps $K_0^+(\mathscr{X}_1)$ onto $K_0^+(\mathscr{X}_2)$.

Acknowledgements The author was supported by the NordForsk Research Network Operator Algebra and Dynamics (grant #11580) and the Research Council of Norway through project 191195/V30.

References

1. Arveson, W.: An invitation to C^*-algebras. Springer-Verlag, New York (1976). Graduate Texts in Mathematics, No. 39
2. Bates, T., Carlsen, T.M., Eilers, S.: Dimension groups associated to β-expansions. Math. Scand. **100**(2), 198–208 (2007)
3. Blackadar, B.: K-theory for operator algebras, *Mathematical Sciences Research Institute Publications*, vol. 5, second edn. Cambridge University Press, Cambridge (1998)
4. Blackadar, B.: Operator algebras, *Encyclopaedia of Mathematical Sciences*, vol. 122. Springer-Verlag, Berlin (2006). Theory of C^*-algebras and von Neumann algebras, Operator Algebras and Non-commutative Geometry, III
5. Boyle, M., Handelman, D.: Orbit equivalence, flow equivalence and ordered cohomology. Israel J. Math. **95**, 169–210 (1996). DOI 10.1007/BF02761039. URL http://dx.doi.org/10.1007/BF02761039
6. Carlsen, T.M.: A faithful representation of the C^*-algebra associated to a shift space. In preparation

7. Carlsen, T.M.: On C^*-algebras associated with sofic shifts. J. Operator Theory **49**(1), 203–212 (2003)
8. Carlsen, T.M.: Operator algebraic applications in symbolic dynamics. Ph.D. thesis, University of Copenhagen (2004). URL http://www.math.ku.dk/noter/filer/phd04tmc.pdf
9. Carlsen, T.M.: C^*-algebras associated to shift spaces (2008), arXiv:0808.0301v1
10. Carlsen, T.M.: Cuntz-Pimsner C^*-algebras associated with subshifts. Internat. J. Math. **19**(1), 47–70 (2008). DOI 10.1142/S0129167X0800456X. URL http://dx.doi.org/10.1142/S0129167X0800456X
11. Carlsen, T.M., Eilers, S.: Augmenting dimension group invariants for substitution dynamics. Ergodic Theory Dynam. Systems **24**(4), 1015–1039 (2004). DOI 10.1017/S0143385704000057. URL http://dx.doi.org/10.1017/S0143385704000057
12. Carlsen, T.M., Eilers, S.: Matsumoto K-groups associated to certain shift spaces. Doc. Math. **9**, 639–671 (electronic) (2004)
13. Carlsen, T.M., Eilers, S.: Ordered K-groups associated to substitutional dynamics. J. Funct. Anal. **238**(1), 99–117 (2006). DOI 10.1016/j.jfa.2005.12.028. URL http://dx.doi.org/10.1016/j.jfa.2005.12.028
14. Carlsen, T.M., Matsumoto, K.: Some remarks on the C^*-algebras associated with subshifts. Math. Scand. **95**(1), 145–160 (2004)
15. Carlsen, T.M., Silvestrov, S.: C^*-crossed products and shift spaces. Expo. Math. **25**(4), 275–307 (2007). DOI 10.1016/j.exmath.2007.02.004. URL http://dx.doi.org/10.1016/j.exmath.2007.02.004
16. Carlsen, T.M., Silvestrov, S.: On the K-theory of the C^*-algebra associated with a one-sided shift space. Proc. Est. Acad. Sci. **59**(4), 272–279 (2010). DOI 10.3176/proc.2010.4.04. URL http://dx.doi.org/10.3176/proc.2010.4.04
17. Cuntz, J., Krieger, W.: A class of C^*-algebras and topological Markov chains. Invent. Math. **56**(3), 251–268 (1980). DOI 10.1007/BF01390048. URL http://dx.doi.org/10.1007/BF01390048
18. Franks, J.: Flow equivalence of subshifts of finite type. Ergodic Theory Dynam. Systems **4**(1), 53–66 (1984). DOI 10.1017/S0143385700002261. URL http://dx.doi.org/10.1017/S0143385700002261
19. Ito, S., Takahashi, Y.: Markov subshifts and realization of β-expansions. J. Math. Soc. Japan **26**, 33–55 (1974)
20. Katayama, Y., Matsumoto, K., Watatani, Y.: Simple C^*-algebras arising from β-expansion of real numbers. Ergodic Theory Dynam. Systems **18**(4), 937–962 (1998). DOI 10.1017/S0143385798108350. URL http://dx.doi.org/10.1017/S0143385798108350
21. Katsura, T.: On C^*-algebras associated with C^*-correspondences. J. Funct. Anal. **217**(2), 366–401 (2004). DOI 10.1016/j.jfa.2004.03.010. URL http://dx.doi.org/10.1016/j.jfa.2004.03.010
22. Krieger, W., Matsumoto, K.: Shannon graphs, subshifts and lambda-graph systems. J. Math. Soc. Japan **54**(4), 877–899 (2002). DOI 10.2969/jmsj/1191591995. URL http://dx.doi.org/10.2969/jmsj/1191591995
23. Krieger, W., Matsumoto, K.: A lambda-graph system for the Dyck shift and its K-groups. Doc. Math. **8**, 79–96 (electronic) (2003)
24. Lind, D., Marcus, B.: An introduction to symbolic dynamics and coding. Cambridge University Press, Cambridge (1995). DOI 10.1017/CBO9780511626302. URL http://dx.doi.org/10.1017/CBO9780511626302
25. Matsumoto, K.: On C^*-algebras associated with subshifts. Internat. J. Math. **8**(3), 357–374 (1997). DOI 10.1142/S0129167X97000172. URL http://dx.doi.org/10.1142/S0129167X97000172
26. Matsumoto, K.: Interpolated Cuntz algebras from β-expansions of real numbers. Sūrikaisekikenkyūsho Kōkyūroku (1024), 84–86 (1998). Profound development of operator algebras (Japanese) (Kyoto, 1997)
27. Matsumoto, K.: K-theory for C^*-algebras associated with subshifts. Math. Scand. **82**(2), 237–255 (1998)

28. Matsumoto, K.: Dimension groups for subshifts and simplicity of the associated C^*-algebras. J. Math. Soc. Japan **51**(3), 679–698 (1999). DOI 10.2969/jmsj/05130679. URL http://dx.doi.org/10.2969/jmsj/05130679
29. Matsumoto, K.: Presentations of subshifts and their topological conjugacy invariants. Doc. Math. **4**, 285–340 (electronic) (1999)
30. Matsumoto, K.: Relations among generators of C^*-algebras associated with subshifts. Internat. J. Math. **10**(3), 385–405 (1999). DOI 10.1142/S0129167X99000148. URL http://dx.doi.org/10.1142/S0129167X99000148
31. Matsumoto, K.: A simple C^*-algebra arising from a certain subshift. J. Operator Theory **42**(2), 351–370 (1999)
32. Matsumoto, K.: On automorphisms of C^*-algebras associated with subshifts. J. Operator Theory **44**(1), 91–112 (2000)
33. Matsumoto, K.: Stabilized C^*-algebras constructed from symbolic dynamical systems. Ergodic Theory Dynam. Systems **20**(3), 821–841 (2000). DOI 10.1017/S0143385700000444. URL http://dx.doi.org/10.1017/S0143385700000444
34. Matsumoto, K.: Bowen-Franks groups as an invariant for flow equivalence of subshifts. Ergodic Theory Dynam. Systems **21**(6), 1831–1842 (2001). DOI 10.1017/S0143385701001870. URL http://dx.doi.org/10.1017/S0143385701001870
35. Matsumoto, K.: Bowen-Franks groups for subshifts and Ext-groups for C^*-algebras. K-Theory **23**(1), 67–104 (2001). DOI 10.1023/A:1017568715542. URL http://dx.doi.org/10.1023/A:1017568715542
36. Matsumoto, K.: Strong shift equivalence of symbolic dynamical systems and Morita equivalence of C^*-algebras. Ergodic Theory Dynam. Systems **24**(1), 199–215 (2004). DOI 10.1017/S014338570300018X. URL http://dx.doi.org/10.1017/S014338570300018X
37. Matsumoto, K., Watatani, Y., Yoshida, M.: KMS states for gauge actions on C^*-algebras associated with subshifts. Math. Z. **228**(3), 489–509 (1998). DOI 10.1007/PL00004627. URL http://dx.doi.org/10.1007/PL00004627
38. Murphy, G.J.: C^*-algebras and operator theory. Academic Press Inc., Boston, MA (1990)
39. Parry, B., Sullivan, D.: A topological invariant of flows on 1-dimensional spaces. Topology **14**(4), 297–299 (1975)
40. Raeburn, I., Williams, D.P.: Morita equivalence and continuous-trace C^*-algebras, *Mathematical Surveys and Monographs*, vol. 60. American Mathematical Society, Providence, RI (1998)
41. Rørdam, M., Larsen, F., Laustsen, N.: An introduction to K-theory for C^*-algebras, *London Mathematical Society Student Texts*, vol. 49. Cambridge University Press, Cambridge (2000)
42. Wegge-Olsen, N.E.: K-theory and C^*-algebras. A friendly approach. Oxford Science Publications. The Clarendon Press Oxford University Press, New York (1993)

Chapter 5
Classification of Graph C^*-Algebras with No More than Four Primitive Ideals

Søren Eilers, Gunnar Restorff, and Efren Ruiz

Abstract We describe the status quo of the classification problem of graph C^*-algebras with four primitive ideals or less.

Keywords Graph C^*-algebras • Classification • K-theory

Mathematics Subject Classification (2010): 46L80, 19K35.

5.1 Introduction

The class of graph C^*-algebras (cf. [34] and the references therein) has proven to be an important and interesting venue for classification theory by K-theoretical invariants; in particular with respect to C^*-algebras with finitely many ideals, and in 2009, the authors formulated the following *working conjecture*:

S. Eilers (✉)
Department of Mathematical Sciences,University of Copenhagen,Universitetsparken 5, DK-2100 Copenhagen, Denmark
e-mail: eilers@math.ku.dk

G. Restorff
Department of Science and Technology, University of the Faroe Islands, Nóatún 3, FO-100 Tórshavn, Faroe Islands
e-mail: gunnarr@setur.fo

E. Ruiz
Department of Mathematics, University of Hawaii, Hilo, 200 W. Kawili St., Hilo, HI, 96720-4091 USA
e-mail: ruize@hawaii.edu

T.M. Carlsen et al. (eds.), *Operator Algebra and Dynamics*, Springer Proceedings in Mathematics & Statistics 58, DOI 10.1007/978-3-642-39459-1_5,

Conjecture 1. Graph C^*-algebras $C^*(E)$ with finitely many ideals are classified up to stable isomorphism by their filtered, ordered K-theory $\mathsf{FK}^+_{\mathrm{Prim}(C^*(E))}(C^*(E))$.

Here, the filtered, ordered K-theory is simply the collection of all K_0- and K_1-groups of subquotients of the C^*-algebra in question, taking into account all the natural transformations among them (details will be given below). The conjecture addresses the possibility of a classification result that is not strong (cf. [21]) in the sense that we do not expect every possible isomorphism at the level of the invariant to lift to the C^*-algebras.

The conjecture remains open and we are forthwith optimistic about its veracity, although some of the results which have been obtained, as we shall see, seem to indicate that an added condition of finitely generated K-theory could be needed. In the present paper we will discuss the status of this conjecture for graph algebras with four or fewer primitive ideals; if the number is three or fewer we can present a complete classification under the condition of finitely generated K-theory, but for the number four there are many cases still eluding our methods. Adding, in some cases, the condition of finitely generated K-theory – or even stronger, that the graph algebra is unital – we may solve 103 of the 125 cases, leaving less than one fifth of the cases open. Our main contribution in the present paper concerns the class of *fan spaces*, which has not been accessible through the methods we have used earlier, but we will also go through those results in our two papers [15] and [13] which apply here.

5.1.1 Tempered Primitive Ideal Spaces

Invoking an idea from [18] we organize our overview using a *tempered ideal space* of the C^*-algebra in question. This is defined for any C^*-algebra with only finitely many ideals as the pair $(\mathrm{Prim}(\mathfrak{A}), \tau)$ where $\tau : \mathrm{Prim}(\mathfrak{A}) \to \{0, 1\}$ is defined as

$$\tau(\mathfrak{J}) = \begin{cases} 0 & K_0(\mathfrak{J}/\mathfrak{J}_0)_+ \neq K_0(\mathfrak{J}/\mathfrak{J}_0) \\ 1 & K_0(\mathfrak{J}/\mathfrak{J}_0)_+ = K_0(\mathfrak{J}/\mathfrak{J}_0) \end{cases}$$

with $\mathfrak{J}_0$ the maximal proper ideal of $\mathfrak{J}$ (this exists by the fact that $\mathfrak{J}$ is prime and contains only finitely many ideals). We set

$$X_\square = \{x \in X \mid \tau(x) = 0\} \qquad X_\blacksquare = \{x \in X \mid \tau(x) = 1\}.$$

To be able to work systematically with these objects, we now give them a combinatorial description.

Definition 1. Let $\mathfrak{A}$ be a C^*-algebra. We let $\mathrm{Prim}(\mathfrak{A})$ denote the *primitive ideal space* of $\mathfrak{A}$, equipped with the usual hull-kernel topology, also called the

Jacobson topology. We always identify the open sets of $\mathrm{Prim}(\mathfrak{A})$, $\mathbb{O}(\mathrm{Prim}(\mathfrak{A}))$, and the lattice of ideals of $\mathfrak{A}$, $\mathbb{I}(\mathfrak{A})$, using the lattice isomorphism

$$U \mapsto \bigcap_{\mathfrak{p}\in\mathrm{Prim}(\mathfrak{A})\setminus U} \mathfrak{p}.$$

When U is an open set we write $\mathfrak{A}(U)$ for the corresponding ideal of $\mathfrak{A}$. When $U \supset V$ are both open, so that $U \setminus V$ is locally closed, we write $\mathfrak{A}(U \setminus V)$ for the subquotient $\mathfrak{A}(U)/\mathfrak{A}(V)$.

Note that whenever $X_\square$ or $X_\blacksquare$ are locally closed, standard results in graph C^*-algebra theory give that $\mathfrak{A}(X_\square)$ and $\mathfrak{A}(X_\blacksquare)$ are AF algebras and $\mathcal{O}_\infty$-absorbing algebras, respectively.

Definition 2. Let X be a topological space. The *specialization preorder* $\prec$ on X is defined by $x \prec y$ if and only if $x \in \overline{\{y\}}$.

A topological space satisfies the T_0 separation axiom if and only if its specialization preorder is a partial order.

Definition 3. A subset H of a preordered set $(X, \leq)$ is called *hereditary* if $x \leq y \in H$ implies $x \in H$.

Definition 4. Let $(X, \leq)$ be a preordered set. The *Alexandrov topology* of X is the topology with the closed sets being the hereditary sets.

A topological set is called an *Alexandrov space* if it carries the Alexandrov topology of some preordered set. The preorder is necessarily the specialization preorder. A topological space is an Alexandrov space if and only if arbitrary intersections of open sets are open.

Since we are dealing with C^*-algebras with finite primitive ideal spaces, these are all Alexandrov spaces satisfying the T_0 separation axiom. Consequently, we can equivalently consider all partial orders on finite sets. The tempered primitive ideal space for a C^*-algebra with n primitive ideals may hence be uniquely described using a partial order on $\{1, \dots, n\}$ and a map in $\{0, 1\}^{\{1,\dots,n\}}$.

The *transitive reduction* of a relation R on a set X is a minimal relation S on X having the same transitive closure as R. In general neither existence nor uniqueness are guaranteed, but if the transitive closure of R is antisymmetric and finite, there is a unique transitive reduction. We will illustrate our (finite) topological spaces with graphs of the transitive reduction of the specialization order, where we write an arrow $x \rightarrow y$ if and only if x is less than y in the transitive reduction of the specialization order (similar to the Hasse diagram).[1] The value of τ will be indicated by colors of the vertices of the graph; white for 0 and black for 1.

We obtain a unique signature for each tempered ideal space as follows. Consider the adjacency matrix of the graph of the specialization order and recall that

[1] See Remark 1 for a discussion about the direction of the arrows.

(by transitivity and antisymmetry) we can always permute the vertices so that the adjacency matrix becomes an upper triangular matrix. Since the relation is reflexive, we will have ones in the diagonal, so without loss of information we may write the values of τ there. To each such upper triangular matrix

$$A = \begin{bmatrix} t_1 & a_{1,2} & & a_{1,n-1} & a_{1,n} \\ & t_2 & a_{2,3} & & a_{2,n} \\ & & \ddots & \ddots & \\ & & & t_{n-1} & a_{n-1,n} \\ & & & & t_n \end{bmatrix}$$

we associate two binary numbers

$$a = a_{1,2}a_{1,3}\cdots a_{1,n}a_{2,3}a_{2,4}\cdots a_{2,n}\cdots a_{n-1,n}$$

and

$$t = t_1 \cdots t_n$$

In general, there are several such binary numbers associated with a specialization order by means of permuting the vertices. We choose the order of the vertices to obtain the smallest possible pair (a, t) ordered lexicographically as the unique identifier for this specific tempered ideal structure. In the interest of conserving space we write hexadecimal expansion of the numbers when referring to a certain structure. We write n.a and n.a.t to indicate *signatures* and *tempered signatures*, respectively, defined this way (where n and a are numbers written in decimal expansions and t is a number written in hexadecimal expansion).

If a primitive ideal space is disconnected, we may classify the C^*-algebras associated to each component individually. We will hence assume throughout that the C^*-algebras have connected primitive ideal space (when considering graph algebras, a necessary, but not sufficient, condition for this is that the underlying graphs are connected considered as undirected graphs). Determining the number of connected T_0-spaces with n points is hard for most n; the number has been computed up to $n = 16$ in [6]. But for small n even the number of tempered ideal spaces can readily be found by naive enumeration, by first counting all spaces and then performing inverse Euler transform to obtain those that are connected:

$\lvert\mathrm{Prim}(\mathfrak{A})\rvert$	1	2	3	4	5	6
Number of spaces	1	2	5	16	63	318
Number of connected spaces	1	1	3	10	44	238
Number of tempered spaces	2	10	62	510	5,292	69,364
Number of connected tempered spaces	2	4	20	125	1,058	11,549

We will restrict our attention to $|\operatorname{Prim}(\mathfrak{A})| \leq 4$ and hence have 15 (connected) primitive ideal spaces[2] which may be given temperatures in a total of 151 different ways to concern ourselves with:

1.0	❑	[L],[A]
2.1	❑ ⟶ ❑	[L],[A]
3.7	❑ ⟶ ❑ ⟶ ❑	[L],[A]
4.E	❑ ⟶ ❑ ⟵ ❑ ⟶ ❑	[A]
4.F	❑ ⟶ ❑ ⟵ ❑ ⟵ ❑	[A]
4.39	❑ ⟵ ❑ ⟶ ❑ ⟶ ❑	[A]
4.3F	❑ ⟶ ❑ ⟶ ❑ ⟶ ❑	[L],[A]

3.3	❑ ⟶ ❑ ⟵ ❑	[A],[F]
3.6	❑ ⟵ ❑ ⟶ ❑	[A],[F]
4.A	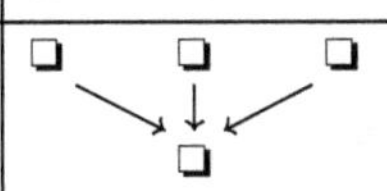	[F]
4.38	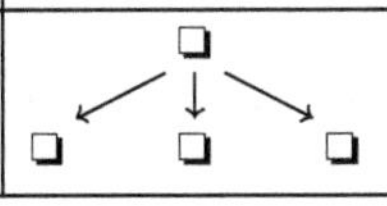	[F]

4.1F	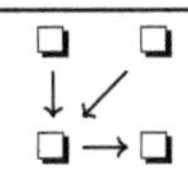	[Y]
4.3E	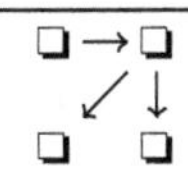	[Y]
4.1E	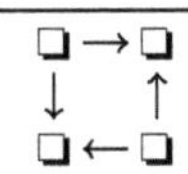	[O]
4.3B	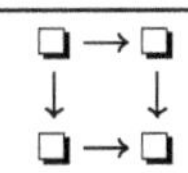	[O]

where ❑ just indicates that it is either □ or ■.

We call a finite T_0 space *linear* ([L]) if its partial order is total. Following [4] we call it an *accordion space* ([A]) if the symmetrization of the space is the symmetrization of a linear space. We call it a *fan space* ([F]) when there is a smallest or largest element in the preorder, so that when this is removed, what remains is a disjoint union of linear spaces. The remaining spaces we organize as [Y]-spaces and [O]-spaces as indicated. In Sect. 5.6 below we summarize our results subject to this organization.

Remark 1. Usually, when representing a relation R with a directed graph, we have an edge from x to y if and only if $x \, R \, y$. This is the convention we use here as well. However, in the literature on filtered K-theory, there are a number of papers choosing the opposite convention. Among these are the papers [1, 2, 4, 29, 30], although it is explicitly mentioned in [1], that it is the Hasse diagram of the opposite relation that is considered.

[2]The space 4.E was forgotten on page 230 of [29]

Apart from being more natural, the convention used in this present paper is in better accordance with graph C^*-algebra theory, as the directed graph used is naturally isomorphic to the graph of the connected components of the underlying graph for, e.g., purely infinite Cuntz-Krieger algebras and fits better with the work of Boyle and Huang. Therefore it is very important to check which convention is being used before applying results.

5.1.2 The Invariant

Let $\mathfrak{A}$ be a C^*-algebra with finitely many ideals and set $\mathsf{X} = \text{Prim}(\mathfrak{A})$. Note that for any locally closed subset $Y = U \setminus V$ of X, we have two groups $K_0(\mathfrak{A}(Y))$ and $K_1(\mathfrak{A}(Y))$. Moreover, for any three open subsets $U \subseteq V \subseteq W$ of X, we have a six term exact sequence

$$\begin{array}{ccccc} K_0(\mathfrak{A}(Y_1)) & \xrightarrow{\iota_0} & K_0(\mathfrak{A}(Y_2)) & \xrightarrow{\pi_0} & K_0(\mathfrak{A}(Y_3)) \\ {\scriptstyle \partial_1}\big\uparrow & & & & \big\downarrow{\scriptstyle \partial_0} \\ K_1(\mathfrak{A}(Y_3)) & \xleftarrow[\pi_1]{} & K_1(\mathfrak{A}(Y_2)) & \xleftarrow[\iota_1]{} & K_1(\mathfrak{A}(Y_1)) \end{array}$$

where $Y_1 = V \setminus U$, $Y_2 = W \setminus U$, and $Y_3 = W \setminus V$. The *filtered, ordered K-theory* $\text{FK}^+_{\mathsf{X}}(\mathfrak{A})$ of $\mathfrak{A}$ is the collection of all K-groups thus occurring, equipped with order on K_0 and the natural transformations $\{\iota_*, \pi_*, \partial_*\}$.

Consequently, if also $\text{Prim}(\mathfrak{B}) = \mathsf{X}$, we write $\text{FK}^+_{\mathsf{X}}(\mathfrak{A}) \cong \text{FK}^+_{\mathsf{X}}(\mathfrak{B})$ if for each locally closed subset Y of X, there exist group isomorphisms

$$\alpha_*^Y : K_*(\mathfrak{A}(Y)) \to K_*(B(Y))$$

preserving all natural transformations in such a way that all α_0^Y are also order isomorphisms. All components of this invariant are readily computable [8], and often, much of it is redundant. We will not pursue that issue here.

The *filtered K-theory* $\text{FK}_{\mathsf{X}}(\mathfrak{A})$ of $\mathfrak{A}$ is defined analogously by disregarding the order structure on K_0. The filtered (ordered) K-theory over a finite T_0-space X can also be used for C^*-algebras over X without being tight.[3]

[3]Although this is not exactly the same definition as the filtrated K-theory in [30], it is known to be the same for all the cases where we have a UCT. For more on this invariant and C^*-algebras over X the reader is referred to [30] and the references therein.

5.1.3 Graph C^*-Algebras

A *graph* (E^0, E^1, r, s) consists of a countable set E^0 of vertices, a countable set E^1 of edges, and maps $r : E^1 \to E^0$ and $s : E^1 \to E^0$ identifying the range and source of each edge. If E is a graph, the *graph C^*-algebra* $C^*(E)$ is the universal C^*-algebra generated by mutually orthogonal projections $\{p_v : v \in E^0\}$ and partial isometries $\{s_e : e \in E^1\}$ with mutually orthogonal ranges satisfying

1. $s_e^* s_e = p_{r(e)}$ for all $e \in E^1$
2. $s_e s_e^* \leq p_{s(e)}$ for all $e \in E^1$
3. $p_v = \sum_{\{e \in E^1 : s(e) = v\}} s_e s_e^*$ for all v with $0 < |s^{-1}(v)| < \infty$.

The countability hypothesis ensures that all our graph C^*-algebras are separable, which is a necessary hypothesis for many of the classification results. We will be mainly interested in graph C^*-algebras with real rank zero. For a graph E, we have that the real rank of $C^*(E)$ is zero if and only if E is satisfying *Condition (K)*, i.e., no vertex of E is the base point of exactly one simple cycle (see Theorem 3.5 of [23]). Moreover, by Proposition 3.3 of [23], every graph C^*-algebra with finitely many ideals has real rank zero. Thus, every graph C^*-algebra with finitely many ideals has a norm-full projection, and by Brown [7], every graph C^*-algebra with finitely many ideals is stably isomorphic to a unital C^*-algebra.

Throughout the paper we will use the following facts about graph C^*-algebras without further mention.

Theorem 1. *Let $C^*(E)$ be a unital graph C^*-algebra with E satisfying Condition (K).*

1. *Every ideal of $C^*(E)$ is stably isomorphic to a unital graph C^*-algebra.*
2. *Every sub-quotient of $C^*(E)$ is stably isomorphic to a unital graph C^*-algebra.*
3. *The K-groups of every sub-quotient of $C^*(E)$ are finitely generated.*
4. *Every non-unital simple sub-quotient of $C^*(E)$ that is an AF-algebra is isomorphic to $\mathbb{K}$.*

Proof. As in the proof of Theorem 5.7 (4) of [37] (see also Proposition 3.4 of [3]), every ideal of a graph C^*-algebra satisfying Condition (K) is Morita equivalent to $C^*(F)$, where $F^0 \subseteq E^0$. Hence, 1 holds since a graph C^*-algebra $C^*(E)$ is unital if and only if E^0 is finite. 2 follows from 1 and Corollary 3.5 of [3]. 3 follows from 2 and Theorem 3.1 of [9].

Suppose $C^*(F)$ is a simple unital AF-algebra. Then F has no cycles. Since $C^*(F)$ is unital, F^0 is finite. Therefore, F has a sink. By Corollary 2.15 of [10], every singular vertex must be reached by any other vertex since $C^*(F)$ is simple. Thus, F must be a finite graph. Hence, $C^*(F) \cong \mathsf{M}_n$. From this observation, 4 follows from 1 and 2 since any non-unital simple C^*-algebra stably isomorphic to $\mathbb{K}$ is isomorphic to $\mathbb{K}$.

See [34] and the references therein for more on graph C^*-algebras.

5.2 General Theory

We first describe the situations in which the graph algebras can be classified using widely applicable results.

5.2.1 *The AF Case*

The *AF* case corresponds to temperatures that are constantly 0. We incur these at the tempered signatures 1.0.0, 2.1.0, 3.3.0, 3.6.0, 3.7.0, 4.A.0, 4.E.0, 4.F.0, 4.1E.0, 4.1F.0, 4.38.0, 4.39.0, 4.3B.0, 4.3E.0, and 4.3F.0. Of course the classification question is resolved by Elliott's theorem:

Theorem 2 ([20]). *AF algebras are classified up to stable isomorphism by their ordered K_0-group.*

5.2.2 *The Purely Infinite Case*

Recall that there are three notions of pure infiniteness for non-simple C^*-algebras, namely *pure infiniteness*, *strong pure infiniteness*, and *$\mathcal{O}_\infty$-absorption*, introduced by E. Kirchberg and M. Rørdam; cf. [26] and [27].

Corollary 1. *For each nuclear, separable C^*-algebra $\mathfrak{A}$ with finite primitive ideal space, the following are equivalent:*

1. *$\mathfrak{A}$ is purely infinite,*
2. *$\mathfrak{A}$ is strongly purely infinite,*
3. *$\mathfrak{A}$ is $\mathcal{O}_\infty$-absorbing, i.e., $\mathfrak{A} \otimes \mathcal{O}_\infty \cong \mathfrak{A}$.*

Proof. It follows from Theorem 9.1 and Corollary 9.2 of [27] that 3 implies 2, that 2 implies 1, and that the three coincide in the simple case. It follows from Proposition 3.5 of [27], that pure infiniteness passes to ideals and subquotients. Thus it follows from [38] that 1 implies 3.

The purely infinite case (the $\mathcal{O}_\infty$-absorbing case) corresponds to temperatures that are constantly 1. We incur these at the tempered signatures 1.0.1, 2.1.3, 3.3.7, 3.6.7, 3.7.7, 4.A.F, 4.E.F, 4.F.F, 4.1E.F, 4.1F.F, 4.38.F, 4.39.F, 4.3B.F, 4.3E.F, and 4.3F.F. As we will outline below, all but the case 4.1E.F are resolved through the recent work of many hands.

The isomorphism result of Kirchberg (cf. [24] and [25]) reduces the classification problem of nuclear and strongly purely infinite C^*-algebras which are also in the bootstrap class to an isomorphism problem in ideal-related *KK*-theory. Since all purely infinite graph C^*-algebras fall in this class we may hence confirm Conjecture 1 in the purely infinite case by providing a universal coefficient theorem which allows the lifting of isomorphisms at the level of filtered K-theory to invertible KK_X-classes. This, however, is not known to be possible in general. Indeed, Meyer and Nest in [30] showed that there are purely infinite C^*-algebras over the space 4.A which fails to have this property, but since the examples provided there cannot possibly come from graph algebras, the question remains open in that setting. The work of Bentmann and Köhler established that general UCTs are available precisely when the space X is an accordion space, and Arklint with the second and third named authors provided UCTs for other spaces, including 4.A, under the added assumption that the C^*-algebra has real rank zero, which is automatic here. Specializing even further, Arklint, Bentmann and Katsura provided a UCT which applies for our space 4.3B under the added assumption that the C^*-algebra has real rank zero and that the K_1 groups of all subquotients are free, which also is automatic here. The space 4.1E remains open. In conclusion:

Theorem 3. *Purely infinite, separable, nuclear C^*-algebras $\mathfrak{A}$ with finite primitive ideal space X in the bootstrap class of Meyer and Nest (i.e., all simple subquotients are in the bootstrap class of Rosenberg and Schochet) are classified up to stable isomorphism by their filtered K-theory $\mathsf{FK}_X(-)$ in the cases*

1. *X is an accordion space [1.0, 2.1, 3.3, 3.6, 3.7, 4.E, 4.F, 4.39, 4.3F] [4, 24, 25, 30, 33, 35, 36]*
2. *X is one of the spaces 4.A, 4.38, 4.1F, 4.3E and* $\operatorname{rr}(\mathfrak{A}) = 0$ *[1]*
3. *X is the space 4.3B,* $\operatorname{rr}(\mathfrak{A}) = 0$, *and $K_1(\mathfrak{J}/\mathfrak{I})$ is free for any $\mathfrak{I} \lhd \mathfrak{J} \trianglelefteq \mathfrak{A}$ [2]*

5.2.3 The Separated Case

The classification problem for the two *mixed* cases with $|\operatorname{Prim}(\mathfrak{A})| = 2$ not covered by the results mentioned above – the tempered signatures 2.1.1 and 2.1.2 – were resolved in [19] drawing heavily on [13]. In [15], we generalized this to more complicated cases having the separation property which is automatic in the two-point case, as detailed below. The idea is to find an ideal $\mathfrak{I}$ such that $\mathfrak{I}$ is AF and $\mathfrak{A}/\mathfrak{I}$ is $\mathcal{O}_\infty$-absorbing, or vice versa. We do not know in general how to prove classification in this case, but under certain added assumptions related to the notion of fullness, this leads to results that may be used to resolve the cases of tempered signature 3.7.1, 3.7.3, 4.F.1, 4.1F.1, 4.1F.3, 4.3B.1, 4.3F.1, 4.3F.3, 4.3F.7 by Proposition 1 below and 3.7.4, 3.7.6, 4.39.8, 4.3B.8, 4.3E.8, 4.3E.C, 4.3F.8, 4.3F.C, 4.3F.E by Proposition 2.

Definition 5. Let $n > 1$ be a given integer. Then we let X_n denote the partially ordered set (actually totally ordered) $\mathsf{X}_n = \{1, 2, \ldots, n\}$ with the usual order. For $a, b \in \mathsf{X}_n$ with $a \leq b$, we let $[a, b]$ denote the set $\{x \in \mathsf{X}_n : a \leq x \leq b\}$.

Proposition 1. *Let $\mathfrak{A}_1$ and $\mathfrak{A}_2$ be separable, nuclear, C^*-algebras over X_n in the bootstrap class of Meyer and Nest (i.e., every simple subquotient is in the bootstrap class of Rosenberg and Schochet). Suppose $\mathfrak{A}_i(\{1\})$ is an AF algebra and $\mathfrak{A}_i([2, n])$*

is a tight stable $\mathcal{O}_\infty$-absorbing C^*-algebra over $[2,n]$, and $\mathfrak{A}_i(\{2\})$ is an essential ideal of $\mathfrak{A}_i([1,2])$. Then $\mathfrak{A}_1 \otimes \mathbb{K} \cong \mathfrak{A}_2 \otimes \mathbb{K}$ if and only if there exists an isomorphism $\alpha : \mathsf{FK}_{\mathsf{X}_n}(\mathfrak{A}_1) \to \mathsf{FK}_{\mathsf{X}_n}(\mathfrak{A}_2)$ such that $\alpha_{\{1\}}$ is positive.

Proposition 2. *Let $\mathfrak{A}_1$ and $\mathfrak{A}_2$ be graph C^*-algebras satisfying Condition (K). Suppose $\mathfrak{A}_i$ is a C^*-algebra over X_n such that $\mathfrak{A}_i(\{n\})$ is an AF algebra, for every ideal $\mathfrak{I}$ of $\mathfrak{A}_i$ we have that $\mathfrak{I} \subseteq \mathfrak{A}_i(\{n\})$ or $\mathfrak{A}_i(\{n\}) \subseteq \mathfrak{I}$, and $\mathfrak{A}_i([1,n-1])$ is a tight, $\mathcal{O}_\infty$-absorbing C^*-algebra over $[1,n-1]$. Then $\mathfrak{A}_1 \otimes \mathbb{K} \cong \mathfrak{A}_2 \otimes \mathbb{K}$ if and only if there exists an isomorphism $\alpha : \mathsf{FK}_{\mathsf{X}_n}(\mathfrak{A}_1) \to \mathsf{FK}_{\mathsf{X}_n}(\mathfrak{A}_2)$ such that $\alpha_{\{n\}}$ is positive.*

5.3 Fan Spaces

In this section, we develop methods to deal mainly with the spaces 3.3, 3.6, 4.A, 4.38. We observe the following in [15]

Lemma 1. *Let E be a graph such that $C^*(E)$ has finitely many ideals and assume that $\mathfrak{I} \lhd \mathfrak{J} \unlhd C^*(E)$ are ideals. Then*

1. *$C^*(E) \otimes \mathbb{K}$ has the corona factorization property.*
2. *$(\mathfrak{J}/\mathfrak{I}) \otimes \mathbb{K}$ is of the form $C^*(F) \otimes \mathbb{K}$ for some graph F.*
3. *$(\mathfrak{J}/\mathfrak{I}) \otimes \mathbb{K}$ has the corona factorization property.*

The graph F above can be chosen as a subgraph of the Drinen-Tomforde desingularization of E [10].

Definition 6. For each C^*-algebra $\mathfrak{A}$, we let $\mathcal{M}(\mathfrak{A})$ and $\mathcal{Q}(\mathfrak{A})$ denote the multiplier algebra and the corona algebra of $\mathfrak{A}$, respectively.

For each extension

$$\mathfrak{e} : 0 \to \mathfrak{B} \to \mathfrak{E} \to \mathfrak{A} \to 0,$$

we let $\eta_\mathfrak{e} : \mathfrak{A} \to \mathcal{Q}(\mathfrak{B})$ denote the Busby map of the extension.

Moreover, for each surjective (or, more generally, proper) $*$-homomorphism $\varphi : \mathfrak{A} \to \mathfrak{B}$, we let $\tilde{\varphi} : \mathcal{M}(\mathfrak{A}) \to \mathcal{M}(\mathfrak{B})$ and $\overline{\varphi} : \mathcal{Q}(\mathfrak{A}) \to \mathcal{Q}(\mathfrak{B})$ denote the unique extension to the multiplier algebras and the induced $*$-homomorphism between the corona algebras, respectively (cf. Sect. 2.1 of [12]).

Lemma 2. *Let $(\mathfrak{B}_i)_{i\in I}$ be a family of C^*-algebras (small enough for direct sums and products to exist). Let $\pi_j : \bigoplus_{i\in I} \mathfrak{B}_i \to \mathfrak{B}_j$ denote the canonical projection, for each $j \in I$. Then there is a canonical isomorphism $\prod_{i\in I} \widetilde{\pi_i} : \mathcal{M}(\bigoplus_{i\in I} \mathfrak{B}_i) \to \prod_{i\in I} \mathcal{M}(\mathfrak{B}_i)$ which has the unique extension $\widetilde{\pi_j} : \mathcal{M}(\bigoplus_{i\in I} \mathfrak{B}_i) \to \mathcal{M}(\mathfrak{B}_j)$ of π_j as the j'th coordinate map.*

Consequently, if I is finite, there is an induced isomorphism

$$\prod_{i\in I} \overline{\pi_i} : \mathcal{Q}(\bigoplus_{i\in I} \mathfrak{B}_i) \to \prod_{i\in I} \mathcal{Q}(\mathfrak{B}_i),$$

and it induces homomorphisms $\overline{\pi_j} : \mathscr{Q}(\bigoplus_{i\in I} \mathfrak{B}_i) \to \mathscr{Q}(\mathfrak{B}_j)$ as the j'th coordinate map. In this case, the direct product coincides with the direct sum.

Proof. Here we view the multiplier algebras as the algebras of double centralizers (cf. pp. 39 and 81–82 in [31]). Let (ρ_1, ρ_2) be a double centralizer on $\bigoplus_{i\in I} \mathfrak{B}_i$ (i.e., an arbitrary element of $\mathscr{M}(\bigoplus_{i\in I} \mathfrak{B}_i)$). Using an approximate unit, it is easy to see that ρ_1 and ρ_2 restricted to $\mathfrak{B}_j$ map into $\mathfrak{B}_j$ itself. In this way we get a canonical $*$-homomorphism from $\mathscr{M}(\bigoplus_{i\in I} \mathfrak{B}_i)$ to $\mathscr{M}(\mathfrak{B}_j)$. By the universal property of the direct product, we get a $*$-homomorphism φ from $\mathscr{M}(\bigoplus_{i\in I} \mathfrak{B}_i)$ to $\prod_{i\in I} \mathscr{M}(\mathfrak{B}_i)$, where the j'th coordinate map clearly is an extension of π_j to the multiplier algebras, and hence it is *the* extension $\widetilde{\pi_j}$ of π_j. Clearly, φ is injective. It is also easy to show that φ is surjective by constructing the preimage.

Therefore, if I is finite, the direct product of the short exact sequences

$$0 \longrightarrow \mathfrak{B}_j \longrightarrow \mathscr{M}(\mathfrak{B}_j) \longrightarrow \mathscr{Q}(\mathfrak{B}_j) \longrightarrow 0$$

is canonically isomorphic to

$$0 \longrightarrow \bigoplus_{i\in I} \mathfrak{B}_i \longrightarrow \mathscr{M}(\bigoplus_{i\in I} \mathfrak{B}_i) \longrightarrow \mathscr{Q}(\bigoplus_{i\in I} \mathfrak{B}_i) \longrightarrow 0 .$$

5.3.1 Primitive Ideal Space with n Maximal Elements

Assumption 1. For this subsection, let $n > 1$ be a fixed integer, and let $X_i = \mathsf{X}_{l_i}$ for $i = 1, 2, \ldots, n$, where $l_1, l_2, \ldots, l_n$ are fixed positive integers. Let, moreover,

$$X = \{m\} \sqcup X_1 \sqcup X_2 \sqcup \cdots \sqcup X_n$$

and define a partial order on X as follows. The element m is the least element of X, and for each $i = 1, 2, \ldots, n$, if $x, y \in X_i$ then $x \leq y$ in X if and only if $x \leq y$ in X_i. There are no other relations between the elements of X.

Lemma 3. *Let $\mathfrak{A}$ be a tight C^*-algebra over X and let $k \in \{1, 2, \ldots, n\}$ be given. Consider the extensions*

$$\mathfrak{e} : 0 \to \mathfrak{A}(X \setminus \{m\}) \to \mathfrak{A} \to \mathfrak{A}(\{m\}) \to 0$$

and

$$\mathfrak{e} \cdot \pi_k : 0 \to \mathfrak{A}(X_k) \to \mathfrak{A}(X_k \cup \{m\}) \to \mathfrak{A}(\{m\}) \to 0,$$

where $\pi_k : \mathfrak{A}(X \setminus \{m\}) \to \mathfrak{A}(X_k)$ is the canonical quotient $*$-homomorphism.

Then $\eta_{\mathfrak{e}\cdot\pi_k} = \overline{\pi}_k \circ \eta_{\mathfrak{e}}$, and $\overline{\pi}_k \circ \eta_{\mathfrak{e}}$ is injective.

Proof. Note that the diagram

$$\begin{array}{ccccccccc} \mathfrak{e} : 0 & \longrightarrow & \mathfrak{A}(X \setminus \{m\}) & \longrightarrow & \mathfrak{A} & \longrightarrow & \mathfrak{A}(\{m\}) & \longrightarrow & 0 \\ & & \downarrow{\scriptstyle \pi_k} & & \downarrow & & \| & & \\ \mathfrak{e} \cdot \pi_k : 0 & \longrightarrow & \mathfrak{A}(X_k) & \longrightarrow & \mathfrak{A}(X_k \cup \{m\}) & \longrightarrow & \mathfrak{A}(\{m\}) & \longrightarrow & 0 \end{array}$$

is commutative. Since π_k is surjective, by Theorem 2.2 of [12], $\overline{\pi}_k \circ \eta_{\mathfrak{e}} = \eta_{\mathfrak{e}\cdot\pi_k}$. Also note, that Corollary 4.3 of [12] justifies the notation $\mathfrak{e} \cdot \pi_k$. Suppose $\overline{\pi}_k \circ \eta_{\mathfrak{e}}$ is not injective, then $\overline{\pi}_k \circ \eta_{\mathfrak{e}} = 0$ since $\mathfrak{A}(\{m\})$ is a simple C^*-algebra. Hence, $\mathfrak{A}(X_k \cup \{m\}) \cong \mathfrak{A}(X_k) \oplus \mathfrak{A}(\{m\})$. Since $\mathfrak{A}(X_k \cup \{m\}) \cong \mathfrak{A}/\mathfrak{A}(X \setminus (X_k \cup \{m\}))$, then there exist proper ideals $\mathfrak{I}$ and $\mathfrak{J}$ of $\mathfrak{A}$ such that $\mathfrak{I} + \mathfrak{J} = \mathfrak{A}$ and $\mathfrak{I} \cap \mathfrak{J} = \mathfrak{A}(X \setminus (X_k \cup \{m\}))$. But this contradicts the fact that $\mathfrak{A}$ is a tight C^*-algebra over X. Hence, $\overline{\pi}_k \circ \eta_{\mathfrak{e}}$ is injective.

Lemma 4. *Let $\mathfrak{A}$ be a tight C^*-algebra over X. Then*

$$\mathfrak{e} : 0 \to \mathfrak{A}(X \setminus \{m\}) \to \mathfrak{A} \to \mathfrak{A}(\{m\}) \to 0$$

is full if and only if $\mathfrak{e} \cdot \pi_k$ is full for all $k = 1, 2, \ldots, n$.

Proof. By Lemma 3, $\eta_{\mathfrak{e}\cdot\pi_k} = \overline{\pi}_k \circ \eta_{\mathfrak{e}}$. Thus, if $\mathfrak{e}$ is a full extension, then $\mathfrak{e} \cdot \pi_k$ is a full extension since $\overline{\pi}_k$ is surjective. Suppose $\mathfrak{e} \cdot \pi_k$ is a full extension for all $k = 1, 2, \ldots, n$. Note that $\mathfrak{A}(X \setminus \{m\})$ is $\bigoplus_{j=1}^n \mathfrak{A}(X_j)$ and thus from Lemma 2 it follows that the j'th coordinate map of $(\bigoplus_{i=1}^n \overline{\pi}_i) \circ \eta_{\mathfrak{e}}$ is exactly $\overline{\pi_j} \circ \eta_{\mathfrak{e}} = \eta_{\mathfrak{e}\cdot\pi_j}$ (according to Lemma 3). Since $\bigoplus_{i=1}^n \overline{\pi}_i$ is an isomorphism and since $\mathfrak{e} \cdot \pi_k$ is a full extension for all $k = 1, 2, \ldots, n$, we have that $\mathfrak{e}$ is a full extension. That this direct sum of full extensions is again full can easily be shown by first cutting down to each coordinate.

The signatures 3.6.1, 3.6.5, 4.39.1, 4.39.3, 4.39.4, 4.39.5, 4.39.7, 4.38.1, 4.38.3, 4.38.7 are covered by the following theorem.

Theorem 4. *Let $\mathfrak{A}$ and $\mathfrak{B}$ be graph C^*-algebras that are tight C^*-algebras over X. Assume that there exists an isomorphism $\alpha : \mathsf{FK}_X^+(\mathfrak{A}) \to \mathsf{FK}_X^+(\mathfrak{B})$. Assume, moreover, that $\mathfrak{A}(\{m\})$ is an AF algebra and that $X_\square$ is hereditary. Then $\mathfrak{A} \otimes \mathbb{K} \cong \mathfrak{B} \otimes \mathbb{K}$.*

Proof. We may assume that $\mathfrak{A}$ and $\mathfrak{B}$ are stable C^*-algebras. Note that for each $x \in X$, $\mathfrak{A}(\{x\})$ is an AF algebra if and only if $\mathfrak{B}(\{x\})$ is an AF algebra, and $\mathfrak{A}(\{x\})$ is $\mathcal{O}_\infty$-absorbing if and only if $\mathfrak{B}(\{x\})$ is $\mathcal{O}_\infty$-absorbing (since there exists a positive isomorphism from $K_0(\mathfrak{A}(\{x\}))$ to $K_0(\mathfrak{B}(\{x\}))$). Specifically, $\mathfrak{B}(\{m\})$ is an AF algebra. First we assume that $X_\blacksquare \neq \emptyset$ and $X_\square \setminus \{m\} \neq \emptyset$.

Note that $\mathfrak{A}(X_\square)$ and $\mathfrak{B}(X_\square)$ are AF algebras. Since

$$\alpha_{X_\square} : K_0(\mathfrak{A}(X_\square)) \to K_0(\mathfrak{B}(X_\square))$$

is a positive isomorphism, there exists an isomorphism $\beta : \mathfrak{A}(X_\square) \to \mathfrak{B}(X_\square)$ such that $K_0(\beta) = \alpha_{X_\square}$ (by Elliott's classification result [20]). Since $\mathfrak{A}(X_\square)$ and $\mathfrak{B}(X_\square)$ are AF algebras and β is an $X_\square$-equivariant isomorphism, we have that $K_0(\beta_Y) = \alpha_Y$ for all $Y \in \mathbb{LC}(X)$ such that $Y \subseteq X_\square$. In particular, $K_0(\beta_{\{x\}}) = \alpha_{\{x\}}$ for all $x \in X_\square$.

Let $X_\blacksquare^{\min}$ be the set of minimal elements of $X_\blacksquare$, and for each $a, b \in X$ let

$$[a, \infty) = \{x \in X \,:\, a \leq x\}\,,$$
$$[a, b) = \{x \in X \,:\, a \leq x < b\}\,.$$

Let $x \in X_\blacksquare^{\min}$ be given. Let $i_x \in \{1, 2, \ldots, n\}$ be the unique number such that $x \in X_{i_x}$. Note that $X_{i_x} \sqcup \{m\} = [m, x) \cup [x, \infty)$, which we will denote by $\tilde{X}_{i_x}$. Let, moreover,

$$\mathfrak{e}_x^{\mathfrak{A}} \,:\, 0 \to \mathfrak{A}([x, \infty)) \to \mathfrak{A}(\tilde{X}_{i_x}) \to \mathfrak{A}([m, x)) \to 0,$$

and

$$\mathfrak{e}_x^{\mathfrak{B}} \,:\, 0 \to \mathfrak{B}([x, \infty)) \to \mathfrak{B}(\tilde{X}_{i_x}) \to \mathfrak{B}([m, x)) \to 0.$$

Since $\alpha : \mathsf{FK}_X^+(\mathfrak{A}) \to \mathsf{FK}_X^+(\mathfrak{B})$ is an isomorphism, we also have an isomorphism $\alpha_{\tilde{X}_{i_x}} : \mathsf{FK}_{\tilde{X}_{i_x}}^+\big(\mathfrak{A}(\tilde{X}_{i_x})\big) \to \mathsf{FK}_{\tilde{X}_{i_x}}^+\big(\mathfrak{B}(\tilde{X}_{i_x})\big)$. So by Theorem 4.14 of [30], Kirchberg [25], and Theorem 3.3 of [15], there exists an isomorphism $\varphi^x : \mathfrak{A}([x, \infty)) \to \mathfrak{B}([x, \infty))$ such that $K_*(\varphi^x) = \alpha_{[x,\infty)}$, and

$$\big[\eta_{\mathfrak{e}_x^{\mathfrak{B}}} \circ \beta_{[m,x)}\big] = \big[\overline{\varphi^x} \circ \eta_{\mathfrak{e}_x^{\mathfrak{A}}}\big]$$

in $KK^1(\mathfrak{A}([m, x)), \mathfrak{B}([x, \infty)))$, since $KK(\beta_{[m,x)})$ is the unique lifting of $\alpha_{[m,x)}$.

As in the proof of Proposition 6.3 of [15], Corollary 5.3 of [15] implies that $\eta_{\mathfrak{e}_x^{\mathfrak{A}}}$ and $\eta_{\mathfrak{e}_x^{\mathfrak{B}}}$ are full extensions, and thus also the extensions with Busby maps $\eta_{\mathfrak{e}_x^{\mathfrak{B}}} \circ \beta_{[m,x)}$ and $\overline{\varphi^x} \circ \eta_{\mathfrak{e}_x^{\mathfrak{A}}}$ are full. Since the extensions are non-unital and $\mathfrak{B}([x, \infty))$ satisfies the corona factorization property, there exists a unitary $u_x \in \mathscr{M}(\mathfrak{B}([x, \infty)))$ such that

$$\eta_{\mathfrak{e}_x^{\mathfrak{B}}} \circ \beta_{[m,x)} = \mathrm{Ad}(\overline{u_x}) \circ \overline{\varphi^x} \circ \eta_{\mathfrak{e}_x^{\mathfrak{A}}}$$

where $\overline{u_x}$ is the image of u_x in the corona algebra (this follows from [22] and [28]). Hence, by Theorem 2.2 of [12], there exists an isomorphism $\eta^x : \mathfrak{A}(\tilde{X}_{i_x}) \to \mathfrak{B}(\tilde{X}_{i_x})$ such that $(\mathrm{Ad}(\overline{u_x}) \circ \varphi^x, \eta^x, \beta_{[m,x)})$ is an isomorphism from $\mathfrak{e}_x^{\mathfrak{A}}$ to $\mathfrak{e}_x^{\mathfrak{B}}$. Let

$$\mathfrak{e}^{\mathfrak{A}} \,:\, 0 \to \mathfrak{A}(X \setminus \{m\}) \to \mathfrak{A} \to \mathfrak{A}(\{m\}) \to 0,$$

and

$$\mathfrak{e}^{\mathfrak{B}} \,:\, 0 \to \mathfrak{B}(X \setminus \{m\}) \to \mathfrak{B} \to \mathfrak{B}(\{m\}) \to 0.$$

Since $\mathfrak{A}(\tilde{X}_{i_x})$ and $\mathfrak{B}(\tilde{X}_{i_x})$ have linear ideal lattices, this induces an isomorphism

$$\begin{array}{rccccccccc} \mathsf{e}^{\mathfrak{A}} \cdot \pi_{i_x}: & 0 & \longrightarrow & \mathfrak{A}(X_{i_x}) & \longrightarrow & \mathfrak{A}(\tilde{X}_{i_x}) & \longrightarrow & \mathfrak{A}(\{m\}) & \longrightarrow & 0, \\ & & & \downarrow \psi^x & & \downarrow & & \downarrow \beta_{\{m\}} & & \\ \mathsf{e}^{\mathfrak{B}} \cdot \pi_{i_x}: & 0 & \longrightarrow & \mathfrak{B}(X_{i_x}) & \longrightarrow & \mathfrak{B}(\tilde{X}_{i_x}) & \longrightarrow & \mathfrak{B}(\{m\}) & \longrightarrow & 0. \end{array}$$

So now by construction,

$$\overline{\psi^x} \circ \eta_{\mathsf{e}^{\mathfrak{A}} \cdot \pi_{i_x}} = \eta_{\mathsf{e}^{\mathfrak{B}} \cdot \pi_{i_x}} \circ \beta_{\{m\}},$$

for all $x \in X_{\blacksquare}^{\min}$, and

$$\overline{\beta_{X_j}} \circ \eta_{\mathsf{e}^{\mathfrak{A}} \cdot \pi_j} = \eta_{\mathsf{e}^{\mathfrak{B}} \cdot \pi_j} \circ \beta_{\{m\}},$$

for all $j = 1, 2, \ldots, n$ satisfying that $\mathfrak{A}(X_j)$ is an AF algebra. Now we define an isomorphism θ from $\mathfrak{A}(X \setminus \{m\})$ to $\mathfrak{B}(X \setminus \{m\})$ as the direct sum of the ψ^x's and β_{X_j}'s. We get that (from Lemmas 2 and 3)

$$\overline{\theta} \circ \eta_{\mathsf{e}^{\mathfrak{A}}} = \overline{\theta} \circ \left(\bigoplus_{j=1}^{n} \eta_{\mathsf{e}^{\mathfrak{A}} \cdot \pi_j} \right) = \bigoplus_{j=1}^{n} \overline{\theta_j} \circ \eta_{\mathsf{e}^{\mathfrak{A}} \cdot \pi_j}$$

$$= \bigoplus_{j=1}^{n} \eta_{\mathsf{e}^{\mathfrak{B}} \cdot \pi_j} \circ \beta_{\{m\}} = \left(\bigoplus_{j=1}^{n} \eta_{\mathsf{e}^{\mathfrak{B}} \cdot \pi_j} \right) \circ \beta_{\{m\}} = \eta_{\mathsf{e}^{\mathfrak{B}}} \circ \beta_{\{m\}},$$

where the θ_j's denote the corresponding ψ^x's and β_{X_j}'s. Hence, by Theorem 2.2 of [12], $\mathfrak{A} \cong \mathfrak{B}$.

If $X_{\blacksquare} = \emptyset$ the result is due to Elliott's classification result [20], and if $X_{\square} = \{m\}$ the theorem follows easily by making modifications to the above proof.

Remark 2. Let $\mathfrak{A}$ and $\mathfrak{B}$ be graph C^*-algebras that are C^*-algebras over X, so that $\mathfrak{A}(X_i)$ and $\mathfrak{B}(X_i)$ are tight C^*-algebras over X_i, for $i = 1, 2, \ldots, n$. Assume that

$$0 \to \mathfrak{A}(X_i)/\mathfrak{A}(X_i \setminus \{x_i\}) \to \mathfrak{A}(X_i \cup \{m\})/\mathfrak{A}(X_i \setminus \{x_i\}) \to \mathfrak{A}(X_i \cup \{m\})/\mathfrak{A}(X_i) \to 0$$

is essential whenever $\mathfrak{A}(X_i)$ is $\mathcal{O}_\infty$-absorbing, where x_i is the greatest element of X_i. Assume that there exists an isomorphism $\alpha : \mathsf{FK}_X^+(\mathfrak{A}) \to \mathsf{FK}_X^+(\mathfrak{B})$. Assume moreover, that $\mathfrak{A}(\{m\})$ is an AF algebra and that the set of $x \in X$ for which $\mathfrak{A}(\{x\})$ is an AF algebra is hereditary. Then $\mathfrak{A} \otimes \mathbb{K} \cong \mathfrak{B} \otimes \mathbb{K}$. This follows from the proof above.

The above extensions are essential, e.g., if $\mathfrak{A}(\{x_i\})$ is the least ideal of $\mathfrak{A}(\{x_i, m\})$, for all $i = 1, 2, \ldots, n$, and the remark applies to the cases[4]

1. 4.E.1, where we view the algebra $\mathfrak{A}$ that is tight over the space 4.E as a C^*-algebra over $a \leftarrow b \rightarrow c$ as indicated by the assignment $b \rightarrow a \leftarrow b \rightarrow c$.
2. 4.1E.1 and 4.1E.3, where we view the algebra $\mathfrak{A}$ that is tight over the space 4.1E as a C^*-algebra over $a \leftarrow b \rightarrow c$ as indicated by the assignment

$$\begin{array}{ccc} b & \rightarrow & a \\ \downarrow & & \uparrow \\ c & \leftarrow & b \end{array}$$

3. 4.3E.1, where we view the algebra $\mathfrak{A}$ that is tight over the space 4.3E as a C^*-algebra over $a \leftarrow b \rightarrow c$ as indicated by the assignment

$$\begin{array}{ccc} b & \rightarrow & b \\ & \swarrow & \downarrow \\ a & & c \end{array}$$

The following proposition follows from the results in [19].

Proposition 3. *Let $\mathfrak{A}$ be a graph C^*-algebra with exactly one nontrivial ideal $\mathfrak{I}$. If $\mathfrak{A}$ is not an AF algebra, then $0 \rightarrow \mathfrak{I} \otimes \mathbb{K} \rightarrow \mathfrak{A} \otimes \mathbb{K} \rightarrow \mathfrak{A}/\mathfrak{I} \otimes \mathbb{K} \rightarrow 0$ is a full extension.*

Using the UCT for accordion spaces (see [30] and [4]) and for many other four-point spaces under the added assumption of real rank zero as described in [1], the cases 3.6.2, 3.6.3, 4.38.8, 4.38.9, 4.38.B, can be classified using the following theorem.

Theorem 5. *Let $\mathfrak{A}$ and $\mathfrak{B}$ be graph C^*-algebras that are tight C^*-algebras over X, with X_i being a singleton, for each $i = 1, 2, \ldots, n$. Suppose there exists an isomorphism $\alpha : \mathsf{FK}_X^+(\mathfrak{A}) \rightarrow \mathsf{FK}_X^+(\mathfrak{B})$ which lifts to an invertible element in $KK(X; \mathfrak{A}, \mathfrak{B})$. Then $\mathfrak{A} \otimes \mathbb{K} \cong \mathfrak{B} \otimes \mathbb{K}$.*

Proof. If $\mathfrak{A}(\{m\})$ is an AF algebra, the result follows from Theorem 4. Suppose $\mathfrak{A}(\{m\})$ is an $\mathcal{O}_\infty$-absorbing simple C^*-algebra and that $\mathfrak{A}$ and $\mathfrak{B}$ are stable C^*-algebras. Then by Lemma 3 and Proposition 3, $\overline{\pi}_i \circ \eta_{\mathfrak{e}_\mathfrak{A}} : \mathfrak{A}(\{m\}) \rightarrow \mathcal{Q}(\mathfrak{A}(X_i))$ and $\overline{\pi}_i \circ \eta_{\mathfrak{e}_\mathfrak{B}} : \mathfrak{B}(\{m\}) \rightarrow \mathcal{Q}(\mathfrak{B}(X_i))$ are full extensions, for all $i = 1, 2, \ldots, n$. Hence, by Lemma 4, $\eta_{\mathfrak{e}_\mathfrak{A}}$ and $\eta_{\mathfrak{e}_\mathfrak{B}}$ are full extensions. The theorem now follows from the results of [15].

[4]Here we specify how we view the algebras as algebras over $a \leftarrow b \rightarrow c$ by providing a continuous map from the primitive ideal space to $\{a, b, c\}$

5.3.2 Primitive Ideal Space with n Minimal Elements

Assumption 2. For this subsection, let $n > 1$ be a fixed integer, and let $X_i = \mathsf{X}_{l_i}$ for $i = 1, 2, \ldots, n$, where $l_1, l_2, \ldots, l_n$ are fixed positive integers. Let, moreover,

$$X = \{M\} \sqcup X_1 \sqcup X_2 \sqcup \cdots \sqcup X_n$$

and define a partial order on X as follows. The element M is the greatest element of X, and for each $i = 1, 2, \ldots, n$, if $x, y \in X_i$ then $x \leq y$ in X if and only if $x \leq y$ in X_i. There are no other relations between the elements of X.

Lemma 5. *Let $\mathfrak{A}$ be a tight C^*-algebra over X and let $Y \in \mathbb{O}(X_\square \setminus \{M\})$ be given. Consider the extensions*

$$\mathfrak{e} : 0 \to \mathfrak{A}(\{M\}) \to \mathfrak{A} \to \mathfrak{A}(X \setminus \{M\}) \to 0$$

and

$$\iota_{\mathfrak{A},Y} \cdot \mathfrak{e} : 0 \to \mathfrak{A}(\{M\}) \to \mathfrak{A}(Y \cup \{M\}) \to \mathfrak{A}(Y) \to 0$$

where $\iota_{\mathfrak{A},Y} : \mathfrak{A}(Y) \to \mathfrak{A}(X \setminus \{M\})$ is the usual embedding. Then $\eta_{\iota_{\mathfrak{A},Y} \cdot \mathfrak{e}} = \eta_{\mathfrak{e}} \circ \iota_{\mathfrak{A},Y}$.

Proof. Note that the diagram

$$\begin{array}{ccccccccc}
0 & \longrightarrow & \mathfrak{A}(\{M\}) & \longrightarrow & \mathfrak{A}(Y \cup \{M\}) & \longrightarrow & \mathfrak{A}(Y) & \longrightarrow & 0 \\
 & & \| & & \downarrow & & \downarrow \iota_{\mathfrak{A},Y} & & \\
0 & \longrightarrow & \mathfrak{A}(\{M\}) & \longrightarrow & \mathfrak{A} & \longrightarrow & \mathfrak{A}(X \setminus \{M\}) & \longrightarrow & 0
\end{array}$$

commutes. Hence, by Theorem 2.2 of [12], $\eta_{\iota_{\mathfrak{A},Y} \cdot \mathfrak{e}} = \eta_{\mathfrak{e}} \circ \iota_{\mathfrak{A},Y}$.

Lemma 6. *Suppose the following diagram of C^*-algebras with short exact rows is commutative*

$$\begin{array}{ccccccccc}
0 & \longrightarrow & \mathfrak{B} & \xrightarrow{\iota_1} & \mathfrak{E}_1 & \xrightarrow{\pi_1} & \mathfrak{A}_1 & \longrightarrow & 0 \\
 & & \| & & \downarrow \varphi_1 & & \downarrow \varphi_2 & & \\
0 & \longrightarrow & \mathfrak{B} & \xrightarrow{\iota_2} & \mathfrak{E}_2 & \xrightarrow{\pi_2} & \mathfrak{A}_2 & \longrightarrow & 0.
\end{array}$$

1. *If $\varphi_2(\mathfrak{A}_1)$ is a hereditary sub-C^*-algebra of $\mathfrak{A}_2$, then $\varphi_1(\mathfrak{E}_1)$ is a hereditary sub-C^*-algebra of $\mathfrak{E}_2$.*
2. *If $\varphi_2(\mathfrak{A}_1)$ is full in $\mathfrak{A}_2$, then $\varphi_1(\mathfrak{E}_1)$ is full in $\mathfrak{E}_2$.*

Proof. We first prove 1. Let $x \in \mathfrak{E}_1$ and $y \in \mathfrak{E}_2$ such that $0 \leq y \leq \varphi_1(x)$. Since $\varphi_2(\mathfrak{A}_1)$ is a hereditary sub-C^*-algebra of $\mathfrak{A}_2$, we have that there exists $z \in \varphi_1(\mathfrak{E}_1)$ such that $\pi_2(y) = \pi_2(z)$. Thus, $y - z \in \mathfrak{B}$. Since the map on the ideals is the identity, we have that $y - z \in \varphi_1(\mathfrak{E}_1)$. Hence, $y \in \varphi_1(\mathfrak{E}_1)$. Therefore, $\varphi_1(\mathfrak{E}_1)$ is a hereditary sub-C^*-algebra of $\mathfrak{E}_2$.

We now prove 2. Let $x \in \mathfrak{E}_2$. Since $\varphi_2(\mathfrak{A}_1)$ is full in $\mathfrak{A}_2$, there exists y in the ideal of $\mathfrak{E}_2$ generated by $\varphi_1(\mathfrak{E}_1)$ such that $x - y \in \mathfrak{B}$. Since the map on the ideals is the identity, we have that $y - z \in \varphi_1(\mathfrak{E}_1)$. Hence, x is in the ideal of $\mathfrak{E}_2$ generated by $\varphi_1(\mathfrak{E}_1)$.

Lemma 7. *Let* $\mathfrak{e} : 0 \to \mathfrak{I} \to \mathfrak{A} \to \bigoplus_{k=1}^n \mathfrak{A}_k \to 0$ *be an extension and let* $\iota_k : \mathfrak{A}_k \to \bigoplus_{k=1}^n \mathfrak{A}_k$ *be the inclusion. Suppose* $\eta_\mathfrak{e} \circ \iota_k$ *is full for each* k*. Then* $\eta_\mathfrak{e}$ *is full.*

Proof. Let $(a_1, a_2, \ldots, a_n)$ be a nonzero positive element in $\bigoplus_{k=1}^n \mathfrak{A}_k$. Without loss of generality, we may assume that $a_1 \neq 0$. Note that ideal in $\mathcal{Q}(\mathfrak{I})$ generated by $\eta_\mathfrak{e}(a_1, \ldots, a_n)$ contains the ideal in $\mathcal{Q}(\mathfrak{I})$ generated by $\eta_\mathfrak{e} \circ \iota_1(a_1)$. Since $\eta_\mathfrak{e} \circ \iota_k$ is full, we have that the ideal in $\mathcal{Q}(\mathfrak{I})$ generated by $\eta_\mathfrak{e} \circ \iota_1(a_1)$ is $\mathcal{Q}(\mathfrak{I})$. Thus, the ideal in $\mathcal{Q}(\mathfrak{I})$ generated by $\eta_\mathfrak{e}(a_1, \ldots, a_n)$ is $\mathcal{Q}(\mathfrak{I})$.

The following result applies to the cases 3.3.1, 3.3.5, 4.F.6, 4.F.8, 4.F.E, 4.A.2, 4.A.F, 4.A.E.

Theorem 6. *Let* $\mathfrak{A}$ *and* $\mathfrak{B}$ *be graph* C^**-algebras that are tight* C^**-algebras over* X *such that each of* $\mathfrak{A}(X_i)$, $\mathfrak{B}(X_i)$ *are either* AF *algebras or* $\mathcal{O}_\infty$*-absorbing. Suppose there exists an isomorphism* $\alpha : \mathsf{FK}_X^+(\mathfrak{A}) \to \mathsf{FK}_X^+(\mathfrak{B})$ *and* $\mathfrak{A}(\{M\})$ *is an* AF *algebra. Then* $\mathfrak{A} \otimes \mathbb{K} \cong \mathfrak{B} \otimes \mathbb{K}$.

Proof. We may assume that $\mathfrak{A}$ and $\mathfrak{B}$ are stable C^*-algebras. Note that for each $x \in X$, $\mathfrak{A}(\{x\})$ is an AF algebra if and only if $\mathfrak{B}(\{x\})$ is an AF algebra, and $\mathfrak{A}(\{x\})$ is $\mathcal{O}_\infty$-absorbing if and only if $\mathfrak{B}(\{x\})$ is $\mathcal{O}_\infty$-absorbing (since there exists a positive isomorphism from $K_0(\mathfrak{A}(\{x\}))$ to $K_0(\mathfrak{B}(\{x\}))$. Specifically, $\mathfrak{B}(\{M\})$ is an AF algebra. First we assume that $X_\blacksquare \neq \emptyset$ and $X_\square \setminus \{M\} \neq \emptyset$.

Note that $\alpha_{X_\square} : K_0(\mathfrak{A}(X_\square)) \to K_0(\mathfrak{B}(X_\square))$ is a positive isomorphism and that $\mathfrak{A}(X_\square)$ and $\mathfrak{B}(X_\square)$ are AF algebras. Thus there exists an isomorphism $\beta : \mathfrak{A}(X_\square) \to \mathfrak{B}(X_\square)$ such that $K_0(\beta) = \alpha_{X_\square}$ (by Elliott's classification result [20]). Since $\mathfrak{A}(X_\square)$ and $\mathfrak{B}(X_\square)$ are AF algebras and β is an $X_\square$-equivariant isomorphism, we have that $K_0(\beta_Y) = \alpha_Y$ for all $Y \in \mathbb{LC}(X)$ such that $Y \subseteq X_\square$. In particular, $K_0(\beta_{\{x\}}) = \alpha_{\{x\}}$ for all $x \in X_\square$.

Let

$$\mathfrak{e}_\mathfrak{A} : 0 \to \mathfrak{A}(\{M\}) \to \mathfrak{A} \to \mathfrak{A}(X \setminus \{M\}) \to 0,$$

and

$$\mathfrak{e}_\mathfrak{B} : 0 \to \mathfrak{B}(\{M\}) \to \mathfrak{B} \to \mathfrak{B}(X \setminus \{M\}) \to 0.$$

Since β is an $X_\square$-equivariant isomorphism, by Lemma 5 above and Theorem 2.2 of [12], for $Y \in \mathbb{O}(X_\square \setminus \{M\})$

$$\overline{\beta}_{\{M\}} \circ \eta_{e_{\mathfrak{A}}} \circ \iota_{\mathfrak{A},Y} = \eta_{e_B} \circ \iota_{\mathfrak{B},Y} \circ \beta_Y$$

for all $Y \in \mathbb{O}(X_\square \setminus \{M\})$, where $\iota_{\mathfrak{A},Y} : \mathfrak{A}(Y) \to \mathfrak{A}(X \setminus \{M\})$ and $\iota_{\mathfrak{B},Y} : \mathfrak{B}(Y) \to \mathfrak{B}(X \setminus \{M\})$ are the canonical embeddings.

Since the given α induces an isomorphism from $\mathsf{FK}^+_{X_\blacksquare \cup \{M\}}(\mathfrak{A}(X_\blacksquare \cup \{M\}))$ to $\mathsf{FK}^+_{X_\blacksquare \cup \{M\}}(\mathfrak{B}(X_\blacksquare \cup \{M\}))$, by Lemma 5, Theorem 2.3 of [13], Theorem 4.14 of [30], Kirchberg [25], and Theorem 3.3 of [15], there exists an $X_\blacksquare$-equivariant isomorphism $\psi : \mathfrak{A}(X_\blacksquare) \to \mathfrak{B}(X_\blacksquare)$ such that $K_*(\psi) = \alpha_{X_\blacksquare}$ and

$$[\overline{\beta}_{\{M\}} \circ \eta_{e_{\mathfrak{A}}} \circ \iota_{\mathfrak{A},X_\blacksquare}] = [\eta_{e_{\mathfrak{B}}} \circ \iota_{\mathfrak{B},X_\blacksquare} \circ \psi]$$

in $KK^1(\mathfrak{A}(X_\blacksquare), \mathfrak{B}(\{M\}))$. By Corollary 5.6 of [15], $\eta_{e_{\mathfrak{A}}} \circ \iota_{\mathfrak{A},X_i}$ and $\eta_{e_{\mathfrak{B}}} \circ \iota_{\mathfrak{B},X_i}$ are full extensions for each $i = 1, 2, \ldots, n$ with X_i being $\mathcal{O}_\infty$-absorbing (i.e., $X_i \subseteq X_\blacksquare$). Thus, both $\eta_{e_{\mathfrak{A}}} \circ \iota_{\mathfrak{A},X_\blacksquare}$ and $\eta_{e_{\mathfrak{B}}} \circ \iota_{\mathfrak{B},X_\blacksquare}$ are full extensions since, respectively, $\mathfrak{A}(X_\blacksquare) = \bigoplus_{i \in \{1,2,\ldots,n\}, X_i \subseteq X_\blacksquare} \mathfrak{A}(X_i)$ and $\mathfrak{B}(X_\blacksquare) = \bigoplus_{i \in \{1,2,\ldots,n\}, X_i \subseteq X_\blacksquare} \mathfrak{B}(X_i)$. Hence, $\overline{\beta}_{\{M\}} \circ \eta_{e_{\mathfrak{A}}} \circ \iota_{\mathfrak{A},X_\blacksquare}$ and $\eta_{e_{\mathfrak{B}}} \circ \iota_{\mathfrak{B},X_\blacksquare} \circ \psi$ are full extensions.

Let $\pi_{\mathfrak{A},X_\square \setminus \{M\}} : \mathfrak{A}(X \setminus \{M\}) \to \mathfrak{A}(X_\square \setminus \{M\})$, $\pi_{\mathfrak{A},X_\blacksquare} : \mathfrak{A}(X \setminus \{M\}) \to \mathfrak{A}(X_\blacksquare)$, $\pi_{\mathfrak{B},X_\square \setminus \{M\}} : \mathfrak{B}(X \setminus \{M\}) \to \mathfrak{B}(X_\square \setminus \{M\})$, $\pi_{\mathfrak{B},X_\blacksquare} : \mathfrak{B}(X \setminus \{M\}) \to \mathfrak{B}(X_\blacksquare)$ be the canonical projections. Note that the range of $\eta_{e_{\mathfrak{A}}} \circ \iota_{\mathfrak{A},X_\square \setminus \{M\}}$ and the range of $\eta_{e_{\mathfrak{A}}} \circ \iota_{\mathfrak{A},X_\blacksquare}$ are orthogonal and the range of $\eta_{e_{\mathfrak{B}}} \circ \iota_{\mathfrak{B},X_\square \setminus \{M\}}$ and the range of $\eta_{e_{\mathfrak{B}}} \circ \iota_{\mathfrak{B},X_\blacksquare}$ are orthogonal. Moreover,

$$\eta_{e_{\mathfrak{A}}} = \eta_{e_{\mathfrak{A}}} \circ \iota_{\mathfrak{A},X_\square \setminus \{M\}} \circ \pi_{\mathfrak{A},X_\square \setminus \{M\}} + \eta_{e_{\mathfrak{A}}} \circ \iota_{\mathfrak{A},X_\blacksquare} \circ \pi_{\mathfrak{A},X_\blacksquare}$$

$$\eta_{e_{\mathfrak{B}}} = \eta_{e_{\mathfrak{B}}} \circ \iota_{\mathfrak{B},X_\square \setminus \{M\}} \circ \pi_{\mathfrak{B},X_\square \setminus \{M\}} + \eta_{e_{\mathfrak{B}}} \circ \iota_{\mathfrak{B},X_\blacksquare} \circ \pi_{\mathfrak{B},X_\blacksquare}.$$

We claim that there exist full hereditary sub-C^*-algebras $\mathscr{E}_1$ and $\mathscr{E}_2$ of $\mathfrak{A}$ and $\mathfrak{B}$, respectively, such that $\mathscr{E}_1 \cong \mathscr{E}_2$. Then by Theorem 2.8 of [7], $\mathfrak{A} \otimes \mathbb{K} \cong \mathfrak{B} \otimes \mathbb{K}$.

Choose full projections $p_1, q_1 \in \mathfrak{A}(X_\blacksquare)$ and $p_2, q_2 \in \mathfrak{A}(X_\square \setminus \{M\})$ such that $p_1 + p_2$ is orthogonal to $q_1 + q_2$ in $\mathfrak{A}(X \setminus \{M\})$ (to do this, we use stability, and that graph algebras with finitely many ideals satisfies Condition (K) and hence are of real rank zero). Therefore, $\eta_{e_{\mathfrak{A}}}(p_1 + p_2) \neq 1_{\mathscr{Q}(\mathfrak{A}(\{M\}))}$ since $\eta_{e_{\mathfrak{A}}}(p_1 + p_2)$ is orthogonal to $\eta_{e_{\mathfrak{A}}}(q_1 + q_2)$. Set $e_1 = \psi(p_1)$, $e_2 = \beta_{X_\square \setminus \{M\}}(p_2)$, $f_1 = \psi(q_1)$, and $f_2 = \beta_{X_\square \setminus \{M\}}(q_2)$. Then $e_1 + e_2$ and $f_1 + f_2$ are nonzero orthogonal projections. So, $\eta_{e_{\mathfrak{B}}}(e_1 + e_2) \neq 1_{\mathscr{Q}(\mathfrak{B}(\{M\}))}$.

Set $e = \overline{\beta}_{\{M\}} \circ \eta_{e_{\mathfrak{A}}} \circ \iota_{\mathfrak{A},X_\square \setminus \{M\}}(p_2) = \eta_{e_{\mathfrak{B}}} \circ \iota_{\mathfrak{B},X_\square \setminus \{M\}} \circ \beta_{X_\square \setminus \{M\}}(p_2)$ and set $f = (1_{\mathscr{Q}(\mathfrak{B}(\{M\}))} - e)$. Let $j_\blacksquare : p_1 \mathfrak{A}(X_\blacksquare) p_1 \to \mathfrak{A}(X_\blacksquare)$ and $j_\square : p_2 \mathfrak{A}(X_\square \setminus \{M\}) p_2 \to \mathfrak{A}(X_\square \setminus \{M\})$ be the usual embeddings. Note that

$$e \overline{\beta}_{\{M\}} \circ \eta_{e_{\mathfrak{A}}} \circ \iota_{\mathfrak{A},\blacksquare} \circ j_\blacksquare(x) = \overline{\beta}_{\{M\}} \circ \eta_{e_{\mathfrak{A}}} \circ \iota_{\mathfrak{A},\blacksquare} \circ j_\blacksquare(x) e = 0$$

and

$$e\left(\eta_{e_{\mathfrak{B}}} \circ \iota_{\mathfrak{B},X_{\blacksquare}} \circ \psi \circ j_{\blacksquare}(x)\right)$$
$$= \left(\eta_{e_{\mathfrak{B}}} \circ \iota_{\mathfrak{B},X_{\square}\setminus\{M\}} \circ \beta_{X_{\square}\setminus\{M\}}(p_2)\right) \cdot \left(\eta_{e_{\mathfrak{B}}} \circ \iota_{\mathfrak{B},X_{\blacksquare}} \circ \psi \circ j_{\blacksquare}(x)\right) = 0$$

as well as

$$\left(\eta_{e_{\mathfrak{B}}} \circ \iota_{\mathfrak{B},X_{\blacksquare}} \circ \psi \circ j_{\blacksquare}(x)\right) e$$
$$= \left(\eta_{e_{\mathfrak{B}}} \circ \iota_{\mathfrak{B},X_{\blacksquare}} \circ \psi \circ j_{\blacksquare}(x)\right) \cdot \left(\eta_{e_{\mathfrak{B}}} \circ \iota_{\mathfrak{B},X_{\square}\setminus\{M\}} \circ \beta_{X_{\square}\setminus\{M\}}(p_2)\right) = 0$$

for all $x \in p_1\mathfrak{A}(X_{\blacksquare})p_1$. Hence, we have injective homomorphisms $\overline{\beta}_{\{M\}} \circ \eta_{e_{\mathfrak{A}}} \circ \iota_{\mathfrak{A},\blacksquare} \circ j_{\blacksquare}$ and $\eta_{e_{\mathfrak{B}}} \circ \iota_{\mathfrak{B},X_{\blacksquare}} \circ \psi \circ j_{\blacksquare}$ from $p_1\mathfrak{A}(X_{\blacksquare})p_1$ to $f\mathscr{Q}(\mathfrak{B}(\{M\}))f$.

Since $\mathfrak{B}(\{M\})$ is an AF algebra, by Corollary 2.11 of [39] f lifts to a projection f' in $\mathscr{M}(\mathfrak{B}(\{M\}))$. Note that there exists an isomorphism γ from $f'\mathscr{M}(\mathfrak{B}(\{M\}))f'$ to $\mathscr{M}(f'\mathfrak{B}(\{M\})f')$ that is the identity on $f'\mathfrak{B}(\{M\})f'$ (see II.7.3.14, p. 147 of [5]). Thus, we have an isomorphism $\overline{\gamma}$ from $f\mathscr{Q}(\mathfrak{B}(\{M\}))f$ to $\mathscr{Q}(f'\mathfrak{B}(\{M\})f')$ such that the diagram

$$\begin{array}{ccccccccc} 0 & \longrightarrow & f'\mathfrak{B}(\{M\})f' & \longrightarrow & f'\mathscr{M}(\mathfrak{B}(\{M\}))f' & \longrightarrow & f\mathscr{Q}(\mathfrak{B}(\{M\}))f & \longrightarrow & 0 \\ & & \| & & \downarrow{\scriptstyle\gamma} & & \downarrow{\scriptstyle\overline{\gamma}} & & \\ 0 & \longrightarrow & f'\mathfrak{B}(\{M\})f' & \longrightarrow & \mathscr{M}(f'\mathfrak{B}(\{M\})f') & \longrightarrow & \mathscr{Q}(f'\mathfrak{B}(\{M\})f') & \longrightarrow & 0 \end{array}$$

is commutative. By Corollary 5.6 of [15], $\eta_{e_{\mathfrak{A}}} \circ \iota_{\mathfrak{A},X_i}$ and $\eta_{e_{\mathfrak{B}}} \circ \iota_{\mathfrak{B},X_i}$ are full extensions for each $i = 1, 2, \ldots, n$ with X_i being $\mathscr{O}_\infty$-absorbing (i.e., $X_i \subseteq X_{\blacksquare}$). Thus, by Lemma 7, $\eta_{e_{\mathfrak{A}}} \circ \iota_{\mathfrak{A},X_{\blacksquare}}$ and $\eta_{e_{\mathfrak{B}}} \circ \iota_{\mathfrak{B},X_{\blacksquare}}$ are full extensions since $\mathfrak{A}(X_{\blacksquare}) = \bigoplus_{i\in\{1,2,\ldots,n\}, X_i\subseteq X_{\blacksquare}} \mathfrak{A}(X_i)$ and $\mathfrak{B}(X_{\blacksquare}) = \bigoplus_{i\in\{1,2,\ldots,n\}, X_i\subseteq X_{\blacksquare}} \mathfrak{B}(X_i)$.

Hence, $\overline{\beta}_{\{M\}} \circ \eta_{e_{\mathfrak{A}}} \circ \iota_{\mathfrak{A},X_{\blacksquare}}$ and $\eta_{e_{\mathfrak{B}}} \circ \iota_{\mathfrak{B},X_{\blacksquare}} \circ \psi$ are full extensions. Thus, $\overline{\beta}_{\{M\}} \circ \eta_{e_{\mathfrak{A}}} \circ \iota_{\mathfrak{A},X_{\blacksquare}}(p_1)$ is a norm-full projection in $\mathscr{Q}(\mathfrak{B}(\{M\}))$. Since $\overline{\beta}_{\{M\}} \circ \eta_{e_{\mathfrak{A}}} \circ \iota_{\mathfrak{A},X_{\blacksquare}}(p_1) \leq f$, we have that f is a norm-full projection in $\mathscr{Q}(\mathfrak{B}(\{M\}))$. By Lemma 3.3 of [16], we have that f' is a norm-full projection in $\mathscr{M}(\mathfrak{B}(\{M\}))$ since $\mathfrak{B}(\{M\})$ has an approximate identity consisting of projections. Since $\mathfrak{B}(\{M\})$ is an AF algebra, by Lemma 3.10 of [13], $\mathfrak{B}(\{M\})$ has the corona factorization property. Thus, f' is Murray-von Neumann equivalent to $1_{\mathscr{M}(\mathfrak{B}(\{M\}))}$. Thus, $f'\mathfrak{B}(\{M\})f' \cong \mathfrak{B}(\{M\})$ which implies that $f'\mathfrak{B}(\{M\})f'$ is a stable C^*-algebra since $\mathfrak{B}(\{M\})$ is a stable C^*-algebra.

Let ι be the embedding of $f'\mathfrak{B}(\{M\})f'$ into $\mathfrak{B}(\{M\})$, $\tilde{\iota}$ be the embedding of $f'\mathscr{M}(\mathfrak{B}(\{M\}))f'$ into $\mathscr{M}(\mathfrak{B}(\{M\}))$, and $\overline{\iota}$ be the embedding of $f\mathscr{Q}(\mathfrak{B}(\{M\}))f$ into $\mathscr{Q}(\mathfrak{B}(\{M\}))$. Note that the following diagram

$$\begin{array}{ccccccccc}
0 & \longrightarrow & f'\mathfrak{B}(\{M\})f' & \longrightarrow & f'\mathcal{M}(\mathfrak{B}(\{M\}))f' & \longrightarrow & f\mathcal{Q}(\mathfrak{B}(\{M\}))f & \longrightarrow & 0 \\
 & & \downarrow{\scriptstyle \iota} & & \downarrow{\scriptstyle \tilde{\iota}} & & \downarrow{\scriptstyle \overline{\iota}} & & \\
0 & \longrightarrow & \mathfrak{B}(\{M\}) & \longrightarrow & \mathcal{M}(\mathfrak{B}(\{M\})) & \longrightarrow & \mathcal{Q}(\mathfrak{B}(\{M\})) & \longrightarrow & 0
\end{array}$$

is commutative. Note that the range of $\eta_{\mathfrak{e}_{\mathfrak{B}}} \circ \iota_{\mathfrak{B},X_{\blacksquare}} \circ \psi \circ j_{\blacksquare}$ and the range of $\overline{\beta}_{\{M\}} \circ \eta_{\mathfrak{e}_{\mathfrak{A}}} \circ \iota_{\mathfrak{A},X_{\blacksquare}} \circ j_{\blacksquare}$ are contained in $f\mathcal{Q}(\mathfrak{B}(\{M\}))f$. Let $\mathfrak{e}_1$ be the extension defined by $\overline{\gamma} \circ \overline{\iota}^{-1} \circ \overline{\beta}_{\{M\}} \circ \eta_{\mathfrak{e}_{\mathfrak{A}}} \circ \iota_{\mathfrak{A},X_{\blacksquare}} \circ j_{\blacksquare}$ and let $\mathfrak{e}_2$ be the extension defined by $\overline{\gamma} \circ \overline{\iota}^{-1} \circ \eta_{\mathfrak{e}_{\mathfrak{B}}} \circ \iota_{\mathfrak{B},X_{\blacksquare}} \circ \psi \circ j_{\blacksquare}$. Then

$$\overline{\iota} \circ \overline{\gamma}^{-1} \circ \eta_{\mathfrak{e}_1} = \overline{\beta}_{\{M\}} \circ \eta_{\mathfrak{e}_{\mathfrak{A}}} \circ \iota_{\mathfrak{A},X_{\blacksquare}} \circ j_{\blacksquare} \quad \text{and} \quad \overline{\iota} \circ \overline{\gamma}^{-1} \circ \eta_{\mathfrak{e}_2} = \eta_{\mathfrak{e}_{\mathfrak{B}}} \circ \iota_{\mathfrak{B},X_{\blacksquare}} \circ \psi \circ j_{\blacksquare}$$

Since $\eta_{\mathfrak{e}_{\mathfrak{A}}}(p_1 + p_2) \neq 1_{\mathcal{Q}(\mathfrak{A}(\{M\}))}$ and $\eta_{\mathfrak{e}_{\mathfrak{B}}}(e_1 + e_2) \neq 1_{\mathcal{Q}(\mathfrak{B}(\{M\}))}$ and since $\overline{\beta}_{\{M\}}$ and ψ are isomorphisms, we have that $\overline{\beta}_{\{M\}} \circ \eta_{\mathfrak{e}_{\mathfrak{A}}} \circ \iota_{\mathfrak{A},X_{\blacksquare}} \circ j_{\blacksquare}(p_1) \neq f$ and $\eta_{\mathfrak{e}_{\mathfrak{B}}} \circ \iota_{\mathfrak{B},X_{\blacksquare}} \circ \psi \circ j_{\blacksquare}(p_1) \neq f$. Thus, $\eta_{\mathfrak{e}_1}(p_1)$ and $\eta_{\mathfrak{e}_2}(p_1)$ are not equal to $1_{\mathcal{Q}(f'\mathfrak{B}(\{M\})f')}$. Therefore, $\mathfrak{e}_1$ and $\mathfrak{e}_2$ are non-unital full extensions. Since

$$[\overline{\beta}_{\{M\}} \circ \eta_{\mathfrak{e}_{\mathfrak{A}}} \circ \iota_{\mathfrak{A},X_{\blacksquare}}] = [\eta_{\mathfrak{e}_{\mathfrak{B}}} \circ \iota_{\mathfrak{B},X_{\blacksquare}} \circ \psi]$$

in $KK^1(\mathfrak{A}(X_{\blacksquare}), \mathfrak{B}(\{M\}))$, since ι induces an element in $KK(f'\mathfrak{B}(\{M\})f', \mathfrak{B}(\{M\}))$ which is invertible, and since $\overline{\gamma}$ is an isomorphism, we have that $[\eta_{\mathfrak{e}_1}] = [\eta_{\mathfrak{e}_2}]$ in $KK^1(p_1\mathfrak{A}(X_{\blacksquare})p_1, f'\mathfrak{B}(\{M\})f')$. Since $f'\mathfrak{B}(\{M\})f' \cong \mathfrak{B}(\{M\})$, we have that $f'\mathfrak{B}(\{M\})f'$ has the corona factorization property. Thus, there exists a unitary u' in $\mathcal{M}(f'\mathfrak{B}(\{M\})f')$ such that

$$\mathrm{Ad}(\overline{u'}) \circ \eta_{\mathfrak{e}_1} = \eta_{\mathfrak{e}_2},$$

where $\overline{u'}$ is the image of u' in $\mathcal{Q}(f'\mathfrak{B}(\{M\})f')$. Let $u = \tilde{\iota} \circ \gamma^{-1}(u')$. Then u is a partial isometry in $\mathcal{M}(\mathfrak{B}(\{M\}))$ such that $u^*u = f' = uu^*$ and

$$\mathrm{Ad}(\overline{u}) \circ \overline{\beta}_{\{M\}} \circ \eta_{\mathfrak{e}_{\mathfrak{A}}} \circ \iota_{\mathfrak{A},X_{\blacksquare}} \circ j_{\blacksquare} = \eta_{\mathfrak{e}_{\mathfrak{B}}} \circ \iota_{\mathfrak{B},X_{\blacksquare}} \circ \psi \circ j_{\blacksquare}$$

where $\overline{u}$ is the image of u in $\mathcal{Q}(\mathfrak{B}(\{M\}))$. Set $v = u + 1_{\mathcal{M}(\mathfrak{B}(\{M\}))} - f'$ and let $\overline{v}$ be the image of v in $\mathcal{Q}(\mathfrak{B}(\{M\}))$. Note that $\overline{v} = \overline{u} + e$ and

$$\mathrm{Ad}(\overline{v}) \circ \overline{\beta}_{\{M\}} \circ \eta_{\mathfrak{e}_{\mathfrak{A}}} \circ \iota_{\mathfrak{A},X_{\square}\setminus\{M\}} \circ j_{\square} = \overline{\beta}_{\{M\}} \circ \eta_{\mathfrak{e}_{\mathfrak{A}}} \circ \iota_{\mathfrak{A},X_{\square}\setminus\{M\}} \circ j_{\square}$$

$$\mathrm{Ad}(\overline{v}) \circ \overline{\beta}_{\{M\}} \circ \eta_{\mathfrak{e}_{\mathfrak{A}}} \circ \iota_{\mathfrak{A},X_{\blacksquare}} \circ j_{\blacksquare} = \eta_{\mathfrak{e}_{\mathfrak{B}}} \circ \iota_{\mathfrak{B},X_{\blacksquare}} \circ \psi \circ j_{\blacksquare}.$$

Let $a_1 \in p_1\mathfrak{A}(X_\blacksquare)p_1$ and $a_2 \in p_2\mathfrak{A}(X_\square \setminus \{M\})p_2$. Then

$$\begin{aligned}
&\overline{v}\left(\overline{\beta}_{\{M\}} \circ \eta_{e_\mathfrak{A}} \circ \iota_{\mathfrak{A},X_\blacksquare} \circ j_\blacksquare(a_1) + \overline{\beta}_{\{M\}} \circ \eta_{e_\mathfrak{A}} \circ \iota_{\mathfrak{A},X_\square\setminus\{M\}} \circ j_\square(a_2)\right)\overline{v}^* \\
&\quad = \eta_{e_\mathfrak{B}} \circ \iota_{\mathfrak{B},X_\blacksquare} \circ \psi \circ j_\blacksquare(a_1) + \overline{\beta}_{\{M\}} \circ \eta_{e_\mathfrak{A}} \circ \iota_{\mathfrak{A},X_\square\setminus\{M\}} \circ j_\square(a_2) \\
&\quad = \eta_{e_\mathfrak{B}} \circ \iota_{\mathfrak{B},X_\blacksquare} \circ \psi \circ j_\blacksquare(a_1) + \eta_{e_\mathfrak{B}} \circ \iota_{\mathfrak{B},X_\square\setminus\{M\}} \circ \beta_{X_\square\setminus\{M\}} \circ j_\square(a_2) \\
&\quad = \eta_{e_\mathfrak{B}} \circ (\psi \circ j_\blacksquare(a_1) + \beta_{X_\square\setminus\{M\}} \circ j_\square(a_2)).
\end{aligned}$$

Hence,

$$\mathrm{Ad}(\overline{v}) \circ \overline{\beta}_{\{M\}} \circ \eta_{e_\mathfrak{A}} \circ (\iota_{\mathfrak{A},X_\blacksquare} \circ j_\blacksquare + \iota_{\mathfrak{A},X_\square\setminus\{M\}} \circ j_\square) = \eta_{e_\mathfrak{B}} \circ (\psi \circ j_\blacksquare + \beta_{X_\square\setminus\{M\}} \circ j_\square). \tag{5.1}$$

Note that the Busby invariant of the extension

$$0 \to \mathfrak{A}(\{M\}) \to \mathscr{E}_1 \to (p_1 + p_2)\left(\mathfrak{A}(X_\blacksquare) \oplus \mathfrak{A}(X_\square \setminus \{M\})\right)(p_1 + p_2) \to 0$$

is given by $\eta_{e_\mathfrak{A}} \circ (\iota_{\mathfrak{A},X_\blacksquare} \circ j_\blacksquare + \iota_{\mathfrak{A},X_\square\setminus\{M\}} \circ j_\square)$ and the Busby invariant of the extension

$$0 \to \mathfrak{B}(\{M\}) \to \mathscr{E}_2 \to (e_1 + e_2)\left(\mathfrak{B}(X_\blacksquare) \oplus \mathfrak{B}(X_\square \setminus \{M\})\right)(e_1 + e_2) \to 0$$

is given by $\eta_{e_\mathfrak{B}} \circ (\kappa_\blacksquare + \kappa_\square)$, where $\kappa_\blacksquare : e_1\mathfrak{B}(X_\blacksquare)e_1 \to \mathfrak{B}(X_\blacksquare)$ and $\kappa_\square : e_2\mathfrak{B}(X_\square \setminus \{M\})e_2 \to \mathfrak{B}(X_\square \setminus \{M\})$ are the natural embeddings. Hence, by Eq. (5.1), Theorem 2.2 of [12], and the five lemma, $\mathscr{E}_1 \cong \mathscr{E}_2$. By Lemma 6, $\mathscr{E}_1$ is isomorphic to a full hereditary sub-C^*-algebra of $\mathfrak{A}$ and $\mathscr{E}_2$ is isomorphic to a full hereditary sub-C^*-algebra of $\mathfrak{B}$. We have just proved the claim.

If $X_\blacksquare = \emptyset$ the result is due to Elliott's classification result [20], and if $X_\square \setminus \{M\} = \emptyset$ the theorem follows easily by making modifications to the above proof.

Remark 3. Let $\mathfrak{A}$ and $\mathfrak{B}$ be graph C^*-algebras satisfying Condition (K) that are C^*-algebras over X such that each of $\mathfrak{A}(X_i), \mathfrak{B}(X_i)$ are either AF algebras or $\mathcal{O}_\infty$-absorbing and such that $\mathfrak{A}(X_i)$ and $\mathfrak{B}(X_i)$ are tight C^*-algebras over X_i, whenever $\mathfrak{A}(X_i)$ and $\mathfrak{B}(X_i)$ are $\mathcal{O}_\infty$-absorbing. Assume that there exists an isomorphism $\alpha : \mathsf{FK}_X^+(\mathfrak{A}) \to \mathsf{FK}_X^+(\mathfrak{B})$. Assume moreover, that $\mathfrak{A}(\{M\})$ is an AF algebra and that for every ideal $\mathfrak{I}$ of $\mathfrak{A}$, we have that $\mathfrak{I} \subseteq \mathfrak{A}(\{M\})$ or $\mathfrak{A}(\{M\}) \subseteq \mathfrak{I}$. Then $\mathfrak{A} \otimes \mathbb{K} \cong \mathfrak{B} \otimes \mathbb{K}$. This follows from the proof above together with Corollary 5.6 of [15] and applies to the cases[5]

1. $\mathsf{4.1E.4}$ and $\mathsf{4.1E.C}$, where we view the algebra $\mathfrak{A}$ that is tight over the space $\mathsf{4.1E}$ as a C^*-algebra over $a \to b \leftarrow c$ as indicated by the assignment

[5]Here we specify how we view the algebras as algebras over $a \to b \leftarrow c$ by providing a continuous map from the primitive ideal space to $\{a, b, c\}$

$$\begin{array}{ccc} a & \rightarrow & b \\ \downarrow & & \uparrow \\ b & \leftarrow & c \end{array}$$

2. 4.1F.4 and 4.1F.C, where we view the algebra $\mathfrak{A}$ that is tight over the space 4.1F as a C^*-algebra over $a \rightarrow b \leftarrow c$ as indicated by the assignment

$$\begin{array}{ccc} a & & c \\ \downarrow & \swarrow & \\ b & \rightarrow & b \end{array}$$

The following result resolves the cases 3.3.2, 3.3.3, 4.A.1, 4.A.3, 4.A.7.

Theorem 7. *Let $\mathfrak{A}$ and $\mathfrak{B}$ be graph C^*-algebras that are tight C^*-algebras over X, with X_i being a singleton, for each $i = 1, 2, \ldots, n$. Suppose there exists an isomorphism $\alpha : \mathsf{FK}^+_X(\mathfrak{A}) \rightarrow \mathsf{FK}^+_X(\mathfrak{B})$ such that α lifts to an invertible element in $KK(X; \mathfrak{A}, \mathfrak{B})$. Then $\mathfrak{A} \otimes \mathbb{K} \cong \mathfrak{B} \otimes \mathbb{K}$.*

Proof. Note that we may assume that $\mathfrak{A}$ and $\mathfrak{B}$ are stable C^*-algebras. If $\mathfrak{A}(\{M\})$ is an AF algebra, then the theorem follows from Theorem 6. Suppose $\mathfrak{A}(\{M\})$ is $\mathcal{O}_\infty$-absorbing. Then $\mathfrak{B}(\{M\})$ is $\mathcal{O}_\infty$-absorbing. Hence, by Proposition 3 and Lemma 7, the extensions

$$0 \rightarrow \mathfrak{A}(\{M\}) \rightarrow \mathfrak{A} \rightarrow \mathfrak{A}(X \setminus \{M\}) \rightarrow 0,$$
$$0 \rightarrow \mathfrak{B}(\{M\}) \rightarrow \mathfrak{B} \rightarrow \mathfrak{B}(X \setminus \{M\}) \rightarrow 0$$

are full extensions. The theorem now follows from the results of Theorem 4.6 of [15].

5.4 A Pullback Technique

The main idea of this section is to write the algebra as a pullback of extensions we can classify coherently. The problem is that classification usually does not give us unique isomorphisms on the algebra level. But when the quotient is an AF algebra we can in certain cases use that the KK-class of the isomorphism is unique. The main idea here is similar to the main idea of Sect. 5.3.

Lemma 8. *For each $i = 1, 2$, let there be given C^*-algebras $\mathfrak{A}_i$, $\mathfrak{B}_i$, and $\mathfrak{C}_i$ together with $*$-homomorphisms $\alpha_i : \mathfrak{A}_i \rightarrow \mathfrak{C}_i$ and $\beta_i : \mathfrak{B}_i \rightarrow \mathfrak{C}_i$. Let $\mathfrak{P}_i$ denote the pullback of $\mathfrak{A}_i$ and $\mathfrak{B}_i$ along α_i and β_i, for each $i = 1, 2$.*

Assume that there are isomorphisms $\varphi_{\mathfrak{A}} : \mathfrak{A}_1 \rightarrow \mathfrak{A}_2$, $\varphi_{\mathfrak{B}} : \mathfrak{B}_1 \rightarrow \mathfrak{B}_2$ and $\varphi_{\mathfrak{C}} : \mathfrak{C}_1 \rightarrow \mathfrak{C}_2$, such that the following diagram commutes:

$$\begin{array}{ccccc}
\mathfrak{A}_1 & \xrightarrow{\alpha_1} & \mathfrak{C}_1 & \xleftarrow{\beta_1} & \mathfrak{B}_1 \\
\downarrow{\scriptstyle \varphi_{\mathfrak{A}}} & & \downarrow{\scriptstyle \varphi_{\mathfrak{C}}} & & \downarrow{\scriptstyle \varphi_{\mathfrak{B}}} \\
\mathfrak{A}_2 & \xrightarrow{\alpha_2} & \mathfrak{C}_2 & \xleftarrow{\beta_2} & \mathfrak{B}_2 .
\end{array}$$

Then we get a canonically induced isomorphism from $\mathfrak{P}_1$ *to* $\mathfrak{P}_2$.

Proof. The existence of the $*$-homomorphism from $\mathfrak{P}_1$ to $\mathfrak{P}_2$ follows from the universal property of the pullback. That this $*$-homomorphism is an isomorphism also follows from the universal property.

Lemma 9. *Let* $\mathfrak{I}$ *and* $\mathfrak{J}$ *be ideals of a* C^**-algebra* $\mathfrak{A}$ *satisfying* $\mathfrak{I} \cap \mathfrak{J} = 0$. *Then* $\mathfrak{A}$ *is the pullback of* $\mathfrak{A}/\mathfrak{I}$ *and* $\mathfrak{A}/\mathfrak{J}$ *along the quotient maps* $\mathfrak{A}/\mathfrak{I} \to \mathfrak{A}/(\mathfrak{I}+\mathfrak{J})$ *and* $\mathfrak{A}/\mathfrak{J} \to \mathfrak{A}/(\mathfrak{I}+\mathfrak{J})$.

Proof. This follows from Proposition 3.1 of [32] by noting that we have a commuting diagram

$$\begin{array}{ccccc}
\mathfrak{I} & \longrightarrow & \mathfrak{A} & \longrightarrow & \mathfrak{A}/\mathfrak{I} \\
\| & & \downarrow & & \downarrow \\
\mathfrak{I} & \longrightarrow & \mathfrak{A}/\mathfrak{J} & \longrightarrow & \mathfrak{A}/(\mathfrak{I}+\mathfrak{J})
\end{array}$$

with short exact rows.

The signatures 4.E.4 and 4.E.5 are covered by the following theorem.

Theorem 8. *Let* $\mathfrak{A}$ *and* $\mathfrak{B}$ *be graph* C^**-algebras that are tight over* X, *where* X *is some finite* T_0 *space. Assume that there exists an isomorphism* $\alpha : \mathsf{FK}_X^+(\mathfrak{A}) \to \mathsf{FK}_X^+(\mathfrak{B})$. *Assume, moreover, that we have disjoint open subsets* O_0 *and* O_1 *of* X. *Let*

$$Y_0 = X \setminus O_1, \qquad Y_1 = X \setminus O_0, \qquad \textit{and} \qquad Z = X \setminus (O_0 \cup O_1).$$

Assume also $Z \neq \emptyset$ *and that* $\mathfrak{A}(Z)$ *is an* AF *algebra.*

For each $i = 0, 1$, *if* $\mathfrak{A}(O_i)$ *is* $\mathcal{O}_\infty$*-absorbing, then we assume that:*

1. *There exist two disjoint clopen subsets* Y_i^1 *and* Y_i^2 *of* Y_i *(with the subspace topology) such that* $Y_i = Y_i^1 \cup Y_i^2$ *and* $O_i \subseteq Y_i^1$.
2. *The ideal lattice of* $\mathfrak{A}(O_i)$ *is linear, i.e.,* $O_i \cong \mathsf{X}_j$ *for some* j.
3. $\mathfrak{A}(O_i)$ *is an essential ideal of* $\mathfrak{A}(Y_i^1)$.
4. $\mathfrak{A}(\{m_i\})$ *is essential in* $\mathfrak{A}(\{m_i\} \cup (Y_i^1 \setminus O_i))$, *where* m_i *is the least element of* O_i.

Then $\mathfrak{A} \otimes \mathbb{K} \cong \mathfrak{B} \otimes \mathbb{K}$.

Proof. We may assume that $\mathfrak{A}$ and $\mathfrak{B}$ are stable C^*-algebras. Note that for each locally closed subset Y of X, $\mathfrak{A}(Y)$ is an AF algebra if and only if $\mathfrak{B}(Y)$ is an AF algebra, and $\mathfrak{A}(Y)$ is $\mathcal{O}_\infty$-absorbing if and only if $\mathfrak{B}(Y)$ is $\mathcal{O}_\infty$-absorbing (since there exists a positive isomorphism from $K_0(\mathfrak{A}(Y))$ to $K_0(\mathfrak{B}(Y))$). Specifically $\mathfrak{B}(X \setminus (O_0 \cup O_1))$ is an AF algebra.

Note that the diagram

$$\begin{array}{ccccc} 0 & \longrightarrow & \mathfrak{A}(O_1) & = & \mathfrak{A}(O_1) \\ \downarrow & & \downarrow & & \downarrow \\ \mathfrak{A}(O_0) & \longrightarrow & \mathfrak{A} & \longrightarrow & \mathfrak{A}(Y_1) \\ \| & & \downarrow & & \downarrow \\ \mathfrak{A}(O_0) & \longrightarrow & \mathfrak{A}(Y_0) & \longrightarrow & \mathfrak{A}(Z) \end{array}$$

is commutative with short exact rows and columns, analogously for $\mathfrak{B}$.

If both $\mathfrak{A}(O_0)$ and $\mathfrak{A}(O_1)$ are AF algebras, then it follows from the permanence properties of AF algebras that $\mathfrak{A}$ is an AF algebra, and thus also $\mathfrak{B}$. In this case the theorem follows from Elliott's classification result [20].

Now assume that $\mathfrak{A}(O_0)$ is an AF algebra and that $\mathfrak{A}(O_1)$ is $\mathcal{O}_\infty$-absorbing. Let $Z_1^1 = Z \setminus Y_1^2$ and $Z_1^2 = Y_1^2$. Then Z_1^1 and Z_1^2 are locally closed subsets of X, and Z is the disjoint union of Z_1^1 and Z_1^2. Since $\mathfrak{A}(Y_0)$ and $\mathfrak{B}(Y_0)$ are extensions of AF algebras, these are themselves AF algebras. Since $\alpha_{Y_0} : K_0(\mathfrak{A}(Y_0)) \to K_0(\mathfrak{B}(Y_0))$ is a positive isomorphism, there exists an isomorphism $\beta : \mathfrak{A}(Y_0) \to \mathfrak{B}(Y_0)$ such that $K_0(\beta) = \alpha_{Y_0}$ (by Elliott's classification result [20]). Since $\mathfrak{A}(Y_0)$ and $\mathfrak{B}(Y_0)$ are AF algebras and β is an Y_0-equivariant isomorphism, we have that $K_0(\beta_Y) = \alpha_Y$ for all $Y \in \mathbb{LC}(X)$ such that $Y \subseteq Y_0$.

Let

$$\mathfrak{e}^{\mathfrak{A}} : 0 \to \mathfrak{A}(O_1) \to \mathfrak{A}(Y_1^1) \to \mathfrak{A}(Z^1) \to 0,$$

and

$$\mathfrak{e}^{\mathfrak{B}} : 0 \to \mathfrak{B}(O_1) \to \mathfrak{B}(Y_1^1) \to \mathfrak{B}(Z^1) \to 0.$$

Since $\alpha : \mathsf{FK}_X^+(\mathfrak{A}) \to \mathsf{FK}_X^+(\mathfrak{B})$ is an isomorphism, we also have an isomorphism $\alpha_{Y_1^1} : \mathsf{FK}_{Y_1^1}^+(\mathfrak{A}) \to \mathsf{FK}_{Y_1^1}^+(\mathfrak{B})$. So by Theorem 4.14 of [30], Kirchberg [25], and Theorem 3.3 of [15], there exists an isomorphism $\varphi : \mathfrak{A}(O_1) \to \mathfrak{B}(O_1)$ such that $K_*(\varphi) = \alpha_{O_1}$, and

$$[\eta_{\mathfrak{e}^{\mathfrak{B}}} \circ \beta_{O_{Z^1}}] = [\overline{\varphi} \circ \eta_{\mathfrak{e}^{\mathfrak{A}}}]$$

in $KK^1(\mathfrak{A}(Z^1), \mathfrak{B}(O_1))$, since $KK(\beta_{Z^1})$ is the unique lifting of α_{Z^1}.

As in the proof of Proposition 6.3 of [15], Corollary 5.3 of [15] implies that $\eta_{\mathfrak{e}^{\mathfrak{A}}}$ and $\eta_{\mathfrak{e}^{\mathfrak{B}}}$ are full extensions, and thus also the extensions with Busby maps $\eta_{\mathfrak{e}^{\mathfrak{B}}} \circ \beta_{Z^1}$ and $\overline{\varphi} \circ \eta_{\mathfrak{e}^{\mathfrak{A}}}$ are full. Since the extensions are non-unital and $\mathfrak{B}(O_1)$ satisfies the corona factorization property, there exists a unitary $u \in \mathcal{M}(\mathfrak{B}(O_1))$ such that

$$\eta_{\mathfrak{e}^{\mathfrak{B}}} \circ \beta_{Z^1} = \mathrm{Ad}(\overline{u}) \circ \overline{\varphi} \circ \eta_{\mathfrak{e}^{\mathfrak{A}}}$$

where $\overline{u}$ is the image of u in the corona algebra (this follows from [22] and [28]). Hence, by Theorem 2.2 of [12], there exists an isomorphism $\eta : \mathfrak{A}(Y_1^1) \to \mathfrak{B}(Y_1^1)$ such that $(\mathrm{Ad}(\overline{u}) \circ \varphi, \eta, \beta_{Z_1^1})$ is an isomorphism from $\mathfrak{e}^{\mathfrak{A}}$ to $\mathfrak{e}^{\mathfrak{B}}$.

Since the extension

$$0 \to \mathfrak{A}(O_1) \to \mathfrak{A}(Y_1) \to \mathfrak{A}(Z) \to 0$$

is the direct sum of the extensions

$$0 \to \mathfrak{A}(O_1) \to \mathfrak{A}(Y_1^1) \to \mathfrak{A}(Z^1) \to 0$$

and

$$0 \to 0 \to \mathfrak{A}(Z^2) \to \mathfrak{A}(Z^2) \to 0$$

and analogously for $\mathfrak{B}$, we get an isomorphism from $0 \to \mathfrak{A}(O_1) \to \mathfrak{A}(Y_1) \to \mathfrak{A}(Z) \to 0$ to $0 \to \mathfrak{B}(O_1) \to \mathfrak{B}(Y_1) \to \mathfrak{B}(Z) \to 0$, which is equal to β_Z on the quotient. Now the theorem follows from Lemmas 9 and 8.

Now assume instead that both $\mathfrak{I}$ and $\mathfrak{J}$ are $\mathcal{O}_\infty$-absorbing. The proof is similar to the case above. Instead of lifting $\alpha_{Y_0} : K_0(\mathfrak{A}(Y_0)) \to K_0(\mathfrak{B}(Y_0))$ to $\beta : \mathfrak{A}(Y_0) \to \mathfrak{B}(Y_0)$ we just lift $\alpha_Z : K_0(\mathfrak{A}(Z)) \to K_0(\mathfrak{B}(Z))$ to $\beta : \mathfrak{A}(Z) \to \mathfrak{B}(Z)$. Then we do as above first for the extensions corresponding to the relative open subset O_0 of Y_0 and then for the extensions corresponding to the relative open subset O_1 of Y_1. As above, the theorem then follows from Lemmas 9 and 8.

5.5 Ad Hoc Methods

In this section we present arguments that resolve the classification question for some examples of tempered ideal spaces which are not covered by the general results above. Most of the results are based on knowing strong classification for smaller ideal spaces, as explained below. Our results of this nature, presented in [17], are of a rather limited scope, and require restrictions on the K-theory, requiring the K-groups to be finitely generated, or even for the graph C^*-algebra to be unital. We will see this idea in use in a very clear form in the two open cases for three primitive ideals (cf. Sect. 5.5.1) and in more complicated four-point cases.

Our starting point is

Theorem 9. *Let $\mathfrak{A}_1$ and $\mathfrak{A}_2$ be graph C^*-algebras that are tight C^*-algebras over a finite T_0-space X and let $U \in \mathbb{O}(X)$ be non-empty. Let $\mathfrak{e}_i$ be the extension $0 \to \mathfrak{A}_i(U) \otimes \mathbb{K} \to \mathfrak{A}_i \otimes \mathbb{K} \to \mathfrak{A}_i(X \setminus U) \otimes \mathbb{K} \to 0$. Suppose*

1. *$\mathfrak{e}_i$ is a full extension;*
2. *There exists an invertible element $\alpha \in KK(X; \mathfrak{A}_1, \mathfrak{A}_2)$; and*
3. *The induced invertible element $\alpha_Y \in KK(\mathfrak{A}_1(Y) \otimes \mathbb{K}, \mathfrak{A}_2(Y) \otimes \mathbb{K})$ lifts to an isomorphism from $\mathfrak{A}_1(Y) \otimes \mathbb{K}$ to $\mathfrak{A}_2(Y) \otimes \mathbb{K}$ for $Y = U$ and $Y = X \setminus U$.*

Then $\mathfrak{A}_1 \otimes \mathbb{K} \cong \mathfrak{A}_2 \otimes \mathbb{K}$.

Proof. By 3, there exists an isomorphism $\varphi_Y : \mathfrak{A}_1(Y) \otimes \mathbb{K} \to \mathfrak{A}_2(Y) \otimes \mathbb{K}$ for $Y = U$ and $Y = X \setminus U$ such that $KK(\varphi_Y) = \alpha_Y$. It follows from 1 that $\mathfrak{e}_i$ are essential, so by Theorem 3.3 of [15], $\alpha_{X \setminus U} \times [\eta_{\mathfrak{e}_2}] = [\eta_{\mathfrak{e}_1}] \times \alpha_U$. Therefore, $KK(\varphi_{X \setminus U}) \times [\eta_{\mathfrak{e}_2}] = [\eta_{\mathfrak{e}_1}] \times KK(\varphi_U)$. Hence, by Proposition 6.1 and Lemma 4.5 of [15], we have that $\mathfrak{A}_1 \otimes \mathbb{K} \cong \mathfrak{A}_2 \otimes \mathbb{K}$.

Definition 7. For a T_0 topological space X, we will consider classes $\mathscr{C}_X$ of separable, nuclear C^*-algebras in the bootstrap category of Rosenberg and Schochet $\mathcal{N}$ such that

1. Any element in $\mathscr{C}_X$ is a C^*-algebra over X;
2. If $\mathfrak{A}$ and $\mathfrak{B}$ are in $\mathscr{C}_X$ and there exists an invertible element α in $KK(X; \mathfrak{A}, \mathfrak{B})$ which induces an isomorphism from $\mathsf{FK}_X^+(\mathfrak{A})$ to $\mathsf{FK}_X^+(\mathfrak{B})$, then there exists an isomorphism $\varphi : \mathfrak{A} \to \mathfrak{B}$ such that $KK(\varphi) = \alpha_X$, where α_X is the element in $KK(\mathfrak{A}, \mathfrak{B})$ induced by α.

Remark 4. Let X be a finite T_0-space, let U be an open subset of X, and let $\mathscr{C}_U$ and $\mathscr{C}_{X \setminus U}$ be classes of C^*-algebras satisfying the conditions of Definition 7. If $\mathfrak{A}_1$ and $\mathfrak{A}_2$ are separable C^*-algebras such that $\mathfrak{A}_1(U), \mathfrak{A}_2(U) \in \mathscr{C}_U$ and $\mathfrak{A}_1(X \setminus U)$, $\mathfrak{A}_2(X \setminus U) \in \mathscr{C}_{X \setminus U}$, then 3 in Theorem 9 holds.

Let $\mathscr{C}_X$ and $\mathscr{C}_Y$ be classes of C^*-algebras satisfy the conditions in Definition 7. Let $\mathscr{C}_{X \sqcup Y}$ be the classes of C^*-algebras consisting of elements $\mathfrak{A} \oplus \mathfrak{B}$ with $\mathfrak{A} \in \mathscr{C}_X$ and $\mathfrak{B} \in \mathscr{C}_Y$. Then $\mathscr{C}_{X \sqcup Y}$ satisfies the conditions in Definition 7.

Remark 5. Here we will provide some examples of classes satisfying the conditions in Definition 7.

1. By [25], the class all stable, nuclear, separable, $\mathcal{O}_\infty$-absorbing C^*-algebras that are tight over a finite T_0-space satisfy the conditions in Definition 7.

By Corollary 3.10 and Theorem 3.13 of [17] and by the results of [11], the following classes of C^*-algebras satisfies the conditions in Definition 7.

2. Let $\mathscr{C}_{\mathsf{X}_n}$ be the class of nuclear, separable, tight C^*-algebras $\mathfrak{A}$ over X_n such that $\mathfrak{A}$ is stable, $\mathfrak{A}(\{n\})$ is a Kirchberg algebra, $\mathfrak{A}([1, n-1])$ is an AF-algebra, and $K_i(\mathfrak{A}[Y])$ is finitely generated for all $Y \in \mathbb{LC}(\mathsf{X}_n)$.

3. Let $\mathscr{C}'_{\mathsf{X}_2}$ be the class of unital graph C^*-algebras with exactly one non-trivial ideal with the ideal being an AF algebra and the quotient $\mathcal{O}_\infty$-absorbing, simple C^*-algebras. Let $\mathscr{C}_{\mathsf{X}_2}$ be the class of C^*-algebras $\mathfrak{A}$ such that $\mathfrak{A} \cong \mathfrak{B} \otimes \mathbb{K}$ for some $\mathfrak{B} \in \mathscr{C}'_{\mathsf{X}_2}$.

   By [20], the following class of C^*-algebras satisfy the conditions in Definition 7.

4. Let $\mathscr{C}_X$ be the class of stable AF-algebras over X.

5.5.1 Linear Spaces

This case is solved in [17], and the reader is referred there for details. However, since this is the most basic case in which our approach via Theorem 9 is applied, we will explain the methods for the benefit of the reader.

Lemma 10. *Let $\mathfrak{A}$ be a graph C^*-algebra such that $\mathfrak{A}$ is a tight C^*-algebra over X_n.*

1. *If $\mathfrak{A}(\{n\})$ and $\mathfrak{A}(\{1\})$ are $\mathcal{O}_\infty$-absorbing and $\mathfrak{A}([2,n-1])$ is an AF-algebra, then*

$$\mathfrak{e} : 0 \to \mathfrak{A}([2,n]) \otimes \mathbb{K} \to \mathfrak{A} \otimes \mathbb{K} \to \mathfrak{A}(\{1\}) \otimes \mathbb{K} \to 0$$

   is a full extension.
2. *If $\mathfrak{A}([k,n])$ and $\mathfrak{A}([1,k-2])$ are AF-algebras and $\mathfrak{A}(\{k-1\})$ is $\mathcal{O}_\infty$-absorbing, then*

$$\mathfrak{e} : 0 \to \mathfrak{A}([k,n]) \otimes \mathbb{K} \to \mathfrak{A} \otimes \mathbb{K} \to \mathfrak{A}([1,k-1]) \otimes \mathbb{K} \to 0$$

   is a full extension.
3. *If $\mathfrak{A}([k,n])$ and $\mathfrak{A}([1,k-2])$ are AF-algebras and $\mathfrak{A}(\{k-1\})$ is $\mathcal{O}_\infty$-absorbing, then*

$$\mathfrak{e} : 0 \to \mathfrak{A}([k-1,n]) \otimes \mathbb{K} \to \mathfrak{A} \otimes \mathbb{K} \to \mathfrak{A}([1,k-2]) \otimes \mathbb{K} \to 0$$

   is a full extension.

Proof. In [17], we prove 1 and 2. We now prove 3. Note that

$$0 \to \mathfrak{A}(\{k-1\}) \otimes \mathbb{K} \to \mathfrak{A}([k-2,k-1]) \otimes \mathbb{K} \to \mathfrak{A}(\{k-2\}) \otimes \mathbb{K} \to 0$$

is full since this is an essential extension and $\mathfrak{A}(\{k-1\})$ is $\mathcal{O}_\infty$-absorbing. Since $\mathfrak{A}([k,n])$ is the largest AF-ideal of $\mathfrak{A}([k-1,n])$ and $\mathfrak{A}([k-1,n])/\mathfrak{A}([k,n]) = \mathfrak{A}(\{k-1\})$ is $\mathcal{O}_\infty$-absorbing, by Proposition 3.10 of [19] and Lemma 1.5 of [13], $0 \to \mathfrak{A}([k,n]) \otimes \mathbb{K} \to \mathfrak{A}([k-1,n]) \otimes \mathbb{K} \to \mathfrak{A}(\{k-1\}) \otimes \mathbb{K} \to 0$ is full.

By Proposition 3.2 of [14], $0 \to \mathfrak{A}([k-1,n]) \otimes \mathbb{K} \to \mathfrak{A}([k-2,n]) \otimes \mathbb{K} \to \mathfrak{A}(\{k-2\}) \otimes \mathbb{K} \to 0$ is full. Since $\mathfrak{A}(\{k-2\}) = \mathfrak{A}([k-2,n])/\mathfrak{A}([k-1,n])$ is an essential of $\mathfrak{A}/\mathfrak{A}([k-1,n])$, the extension in 3 is full by Proposition 5.4 of [15].

To solve the cases 3.7.5 and 4.3F.9, we now argue as follows:

Theorem 10. *Let $\mathfrak{A}_1$ and $\mathfrak{A}_2$ be graph C^*-algebras that are tight C^*-algebras over X_n. Suppose*

1. *$\mathfrak{A}_i(\{n\})$ and $\mathfrak{A}_i(\{1\})$ are $\mathcal{O}_\infty$-absorbing;*
2. *$\mathfrak{A}_i([2,n-1])$ is an AF-algebra; and*
3. *The K-groups of $\mathfrak{A}_i$ are finitely generated.*

Then $\mathfrak{A}_1 \otimes \mathbb{K} \cong \mathfrak{A}_2 \otimes \mathbb{K}$ if and only if $\mathsf{FK}^+_{\mathsf{X}_n}(\mathfrak{A}_1 \otimes \mathbb{K}) \cong \mathsf{FK}^+_{\mathsf{X}_n}(\mathfrak{A}_2 \otimes \mathbb{K})$.

Proof. Let $\mathfrak{e}_i$ be the extension

$$0 \to \mathfrak{A}_i([2,n]) \otimes \mathbb{K} \to \mathfrak{A}_i \otimes \mathbb{K} \to \mathfrak{A}_i(\{1\}) \otimes \mathbb{K} \to 0.$$

By Lemma 10, $\mathfrak{e}_i$ is a full extension. Thus, Assumption 1 of Theorem 9 holds. Suppose $\alpha : \mathsf{FK}^+_{\mathsf{X}_n}(\mathfrak{A}_1 \otimes \mathbb{K}) \to \mathsf{FK}^+_{\mathsf{X}_n}(\mathfrak{A}_2 \otimes \mathbb{K})$ is an isomorphism. Lift α to an invertible element $x \in KK(\mathsf{X}_n; \mathfrak{A}_1 \otimes \mathbb{K}, \mathfrak{A}_2 \otimes \mathbb{K})$, such a lifting exists by Theorem 4.14 of [30]. Therefore, Assumption 2 of Theorem 9 holds.

Note now that x induces invertible elements $r^{[2,n]}_{\mathsf{X}_n}(x)$ in $KK([2,n]; \mathfrak{A}_1([2,n]) \otimes \mathbb{K}, \mathfrak{A}_2([2,n]) \otimes \mathbb{K})$ and $r^{[1]}_{\mathsf{X}_n}(x)$ in $KK(\mathfrak{A}(\{1\}) \otimes \mathbb{K}, \mathfrak{A}_2(\{1\}) \otimes \mathbb{K})$. Note that $\mathfrak{A}_i([2,n])$ has a smallest ideal $\mathfrak{A}_i(\{n\})$ which is $\mathcal{O}_\infty$-absorbing and the quotient $\mathfrak{A}_i([2,n-1])$ is an AF algebra. By Theorem 3.9 of [17], there exists an isomorphism $\varphi : \mathfrak{A}_1([2,n]) \otimes \mathbb{K} \to \mathfrak{A}_2([2,n]) \otimes \mathbb{K}$ such that $KL(\varphi)$ is the (necessarily invertible) element in $KL(\mathfrak{A}_1([2,n]), \mathfrak{A}_2([2,n]))$ induced by x. Since the K-theory of $\mathfrak{A}_1$ is finitely generated, $KL(\mathfrak{A}_1([2,n]), \mathfrak{A}_2([2,n])) = KK(\mathfrak{A}_1([2,n]), \mathfrak{A}_2([2,n]))$. Thus, $KK(\varphi)$ is the invertible element in $KK(\mathfrak{A}_1([2,n]), \mathfrak{A}_2([2,n]))$ induced by x. By the Kirchberg-Phillips classification, there exists an isomorphism $\psi : \mathfrak{A}_1(\{1\}) \otimes \mathbb{K} \to \mathfrak{A}_2(\{1\}) \otimes \mathbb{K}$ lifting $r^{[1]}_{\mathsf{X}_n}(x)$. We have just shown that Assumption 3 of Theorem 9 holds.

By Theorem 9, we can conclude that $\mathfrak{A}_1 \otimes \mathbb{K} \cong \mathfrak{A}_2 \otimes \mathbb{K}$.

Similarly, one solves 3.7.2, 4.3F.2, and 4.3F.4 using

Theorem 11. *Let $\mathfrak{A}_1$ and $\mathfrak{A}_2$ be graph C^*-algebras that are tight C^*-algebras over X_n. Suppose*

1. *$\mathfrak{A}_i([k,n])$ and $\mathfrak{A}_i([1,k-2])$ are AF algebras;*
2. *$\mathfrak{A}_i(\{k-1\})$ is $\mathcal{O}_\infty$-absorbing; and*
3. *The K-groups of $\mathfrak{A}_i$ are finitely generated.*

Then $\mathfrak{A}_1 \otimes \mathbb{K} \cong \mathfrak{A}_2 \otimes \mathbb{K}$ if and only if $\mathsf{FK}^+_{\mathsf{X}_n}(\mathfrak{A}_1 \otimes \mathbb{K}) \cong \mathsf{FK}^+_{\mathsf{X}_n}(\mathfrak{A}_2 \otimes \mathbb{K})$.

A proof is given in [17].

5.5.2 Accordion Spaces

Lemma 11. *Let $\mathfrak{A}$ be a graph C^*-algebra with signature* 4.F.x*, and let $\mathfrak{I}$ be the smallest ideal of $\mathfrak{A}$.*

1. *When* x = 3, 5, 7, 9, A, B, D*, then the extension $0 \to \mathfrak{I} \otimes \mathbb{K} \to \mathfrak{A} \otimes \mathbb{K} \to \mathfrak{A}/\mathfrak{I} \otimes \mathbb{K} \to 0$ is full.*
2. *When* x = 2, 4, C*, then the extension $0 \to \mathfrak{I} \otimes \mathbb{K} \to \mathfrak{A} \otimes \mathbb{K} \to \mathfrak{A}/\mathfrak{I} \otimes \mathbb{K} \to 0$ is full provided that $\mathfrak{A}$ is unital.*

Proof. First note that the extension $0 \to \mathfrak{I} \otimes \mathbb{K} \to \mathfrak{A} \otimes \mathbb{K} \to \mathfrak{A}/\mathfrak{I} \otimes \mathbb{K} \to 0$ is essential. Hence, in the case 4.F.x for x = 3, 5, 7, 9, B, D the extension is full since $\mathfrak{I} \otimes \mathbb{K}$ is a simple, purely infinite, stable C^*-algebra, which implies that $\mathcal{Q}(\mathfrak{I} \otimes \mathbb{K})$ is simple. If A is unital and Y is the space 4.F.x for x = 2, 4, and C, then the extension is full since in this case $\mathfrak{I} \cong \mathbb{K}$ and $\mathcal{Q}(\mathbb{K})$ is simple. We are left with showing the extension is full for the case 4.F.A. This case follows from Proposition 5.4 and Corollary 5.6 of [15].

Lemma 12. *Let $\mathfrak{A}$ be a graph C^*-algebra with tempered signature* 4.3F.x *for* x = 5, 6, A, D*. Then the ideal lattice of $\mathfrak{A}$ is $0 \trianglelefteq \mathfrak{I}_1 \trianglelefteq \mathfrak{I}_2 \trianglelefteq \mathfrak{I}_3 \trianglelefteq \mathfrak{A}$ and the extension $0 \to \mathfrak{I}_2 \otimes \mathbb{K} \to \mathfrak{A} \otimes \mathbb{K} \to \mathfrak{A}/\mathfrak{I}_2 \otimes \mathbb{K} \to 0$ is full.*

Proof. We will show that $\mathfrak{e} : 0 \to \mathfrak{I}_2 \otimes \mathbb{K} \to \mathfrak{I}_3 \otimes \mathbb{K} \to \mathfrak{I}_3/\mathfrak{I}_2 \otimes \mathbb{K} \to 0$ is a full extension. By Lemma 10, $\mathfrak{e}$ is a full extension for x = 5, A, D. Consider the case x = 6. Note that $\mathfrak{I}_2$ and $\mathfrak{I}_3/\mathfrak{I}_1$ are isomorphic to non-*AF* graph C^*-algebras with exactly one nontrivial ideal. Therefore, by Proposition 3,

$$0 \to \mathfrak{I}_1 \otimes \mathbb{K} \to \mathfrak{I}_2 \otimes \mathbb{K} \to \mathfrak{I}_2/\mathfrak{I}_1 \otimes \mathbb{K} \to 0$$

$$0 \to \mathfrak{I}_2/\mathfrak{I}_1 \otimes \mathbb{K} \to \mathfrak{I}_3/\mathfrak{I}_1 \otimes \mathbb{K} \to \mathfrak{I}_3/\mathfrak{I}_2 \otimes \mathbb{K} \to 0$$

are full extensions. By Proposition 3.2 of [14], $\mathfrak{e}$ is a full extension. The lemma now follows from Proposition 5.4 of [15].

Lemma 13. *Let $\mathfrak{A}$ be a graph C^*-algebra with tempered signature* 4.39.x *for* x = 2, 6, 9, A, B, C, D*, or* E*. Let $\mathfrak{I}$ be the greatest proper ideal of $\mathfrak{A}$.*

1. *If $\mathfrak{A}$ is unital, then the extension $0 \to \mathfrak{I} \otimes \mathbb{K} \to \mathfrak{A} \otimes \mathbb{K} \to \mathfrak{A}/\mathfrak{I} \otimes \mathbb{K} \to 0$ is full.*
2. *When* x = 9, B, C, D*, the extension $0 \to \mathfrak{I} \otimes \mathbb{K} \to \mathfrak{A} \otimes \mathbb{K} \to \mathfrak{A}/\mathfrak{I} \otimes \mathbb{K} \to 0$ is full.*

Proof. Suppose $\mathfrak{A}$ is unital. Using the general theory of graph C^*-algebras with this specific ideal structure, we have that $\mathfrak{I}$ is stable. Since $\mathfrak{A}/\mathfrak{I}$ is simple and unital, the conclusion now follows from Lemma 1.5 and Proposition 1.6 of [13]. We now prove the extension $0 \to \mathfrak{I} \otimes \mathbb{K} \to \mathfrak{A} \otimes \mathbb{K} \to \mathfrak{A}/\mathfrak{I} \otimes \mathbb{K} \to 0$ is always full for the spaces 4.39.x with x = 9, B, C, D. Note that $\mathfrak{I} = \mathfrak{I}_1 \oplus \mathfrak{I}_2$ with $\mathfrak{I}_1$ simple and $\mathfrak{I}_2$ a tight C^*-algebra over X_2. By Lemma 4.5 of [17] and Corollaries 5.3 and 5.6 of [15], we have $0 \to \mathfrak{I}_2 \otimes \mathbb{K} \to \mathfrak{A}/\mathfrak{I}_1 \otimes \mathbb{K} \to (\mathfrak{A}/\mathfrak{I}) \otimes \mathbb{K} \to 0$ is full.

Since $\mathfrak{A}/\mathfrak{J}_2 \otimes \mathbb{K}$ is a non-*AF* graph C^*-algebra with exactly one nontrivial ideal, the extension $0 \to \mathfrak{J}_1 \otimes \mathbb{K} \to \mathfrak{A}/\mathfrak{J}_2 \otimes \mathbb{K} \to \mathfrak{A}/\mathfrak{J} \otimes \mathbb{K} \to 0$ is a full extension (cf. Proposition 3). Thus, by Lemma 4, $0 \to \mathfrak{J} \otimes \mathbb{K} \to \mathfrak{A} \otimes \mathbb{K} \to \mathfrak{A}/\mathfrak{J} \otimes \mathbb{K} \to 0$ is full.

Using the above lemmas and the Universal Coefficient Theorem of Bentmann and Köhler [4], we get the following cases:

Corollary 2. *Let $\mathfrak{A}$ and $\mathfrak{B}$ be graph C^*-algebras that are tight over a finite accordion space X. Assume that there exists an isomorphism from $\mathsf{FK}_X^+(\mathfrak{A})$ to $\mathsf{FK}_X^+(\mathfrak{B})$. If*

1. *$\mathfrak{A}$ and $\mathfrak{B}$ both have tempered signature 4.F.7, 4.F.9, 4.39.B, 4.39.C, or*
2. *$\mathfrak{A}$ and $\mathfrak{B}$ both have finitely generated K-theory and have tempered signature 4.F.3, 4.F.A, 4.F.B, 4.39.9, 4.39.D, 4.3F.5, 4.3F.D, or*
3. *$\mathfrak{A}$ and $\mathfrak{B}$ both are unital and have tempered signature 4.F.2, 4.F.4, 4.F.5, 4.F.C, 4.F.D, 4.39.2, 4.39.6, 4.39.A, 4.39.E, 4.3F.6, 4.3F.A,*

then $\mathfrak{A} \otimes \mathbb{K} \cong \mathfrak{B} \otimes \mathbb{K}$.

Proof. By the above lemmas, all the extensions are full. Note that the specified ideal and quotient for each space belongs to classes of C^*-algebras satisfying the conditions in Definition 7. Hence, the result now follows from Theorem 9 and the UCT for accordion spaces.

5.5.3 *Y-Shaped Spaces*

Lemma 14. *Let $\mathfrak{A}$ be a graph C^*-algebra with tempered signature 4.1F.x for $\mathsf{x} =$ 2, 5, 6, 7, or D, and let $\mathfrak{J}_1$ be the smallest ideal of $\mathfrak{A}$ and let $\mathfrak{J}_2$ be the ideal of $\mathfrak{A}$ containing $\mathfrak{J}_1$ such that $\mathfrak{J}_2/\mathfrak{J}_1$ is simple.*

1. *When $\mathsf{x} =$ 2, 6, 7, or D, the extension $0 \to \mathfrak{J}_2 \otimes \mathbb{K} \to \mathfrak{A} \otimes \mathbb{K} \to \mathfrak{A}/\mathfrak{J}_2 \otimes \mathbb{K} \to 0$ is full.*
2. *When $\mathsf{x} =$ 5, the extension $0 \to \mathfrak{J}_2 \otimes \mathbb{K} \to \mathfrak{A} \otimes \mathbb{K} \to \mathfrak{A}/\mathfrak{J}_2 \otimes \mathbb{K} \to 0$ is full if $\mathfrak{A}$ is unital.*

Proof. Let $\mathfrak{J}_1$ and $\mathfrak{J}_2$ be the maximal ideals of $\mathfrak{A}$ containing $\mathfrak{J}_2$. Suppose $\mathsf{x} = 2, 6, 7$, or D. Then, by Lemma 10, Proposition 3.2 of [14], and Corollaries 5.3 and 5.6 of [15], $0 \to \mathfrak{J}_2 \otimes \mathbb{K} \to \mathfrak{J}_\ell \otimes \mathbb{K} \to \mathfrak{J}_\ell/\mathfrak{J}_2 \otimes \mathbb{K} \to 0$ is full. Hence, by Lemma 7, $0 \to \mathfrak{J}_2 \otimes \mathbb{K} \to \mathfrak{A} \otimes \mathbb{K} \to \mathfrak{A}/\mathfrak{J}_2 \otimes \mathbb{K} \to 0$ is full.

Suppose that the signature is 4.1F.5 and $\mathfrak{A}$ is unital. Assume that $\mathfrak{J}_1/\mathfrak{J}_2$ is an AF-algebra and $\mathfrak{J}_2/\mathfrak{J}_2$ is purely infinite. By Lemma 10, $0 \to \mathfrak{J}_2 \otimes \mathbb{K} \to \mathfrak{J}_2 \otimes \mathbb{K} \to \mathfrak{J}_2/\mathfrak{J}_2 \otimes \mathbb{K} \to 0$ is full. Since $\mathfrak{A}$ is a unital graph C^*-algebra, we have that $\mathfrak{J}_2/\mathfrak{J}_1 \cong \mathbb{K}$. Therefore, $0 \to \mathfrak{J}_2/\mathfrak{J}_1 \otimes \mathbb{K} \to \mathfrak{J}_1/\mathfrak{J}_1 \otimes \mathbb{K} \to \mathfrak{J}_1/\mathfrak{J}_2 \otimes \mathbb{K} \to 0$ is full. Since $\mathfrak{J}_2$ is stably isomorphic to a non-AF graph C^*-algebra with exactly one nontrivial ideal, by Proposition 3, $0 \to \mathfrak{J}_1 \otimes \mathbb{K} \to \mathfrak{J}_2 \otimes \mathbb{K} \to \mathfrak{J}_2/\mathfrak{J}_1 \otimes \mathbb{K} \to 0$ is full. By Proposition 3.2 of [14],

$$0 \to \mathfrak{I}_2 \otimes \mathbb{K} \to \mathfrak{I}_1 \otimes \mathbb{K} \to \mathfrak{I}_1/\mathfrak{I}_2 \otimes \mathbb{K} \to 0$$

is full. Hence, by Lemma 7, $0 \to \mathfrak{I}_2 \otimes \mathbb{K} \to \mathfrak{A} \otimes \mathbb{K} \to \mathfrak{A}/\mathfrak{I}_2 \otimes \mathbb{K} \to 0$ is full.

Lemma 15. *Let $\mathfrak{A}$ be a graph C^*-algebra with tempered signature 4.3E.x for x = 3, 4, 5, 9, B, or D, and let $\mathfrak{I}_1$ and $\mathfrak{I}_2$ be the minimal ideals of $\mathfrak{A}$.*

1. *When x = 3, 4, 5, B, D, the extension $0 \to (\mathfrak{I}_1 \oplus \mathfrak{I}_2) \otimes \mathbb{K} \to \mathfrak{A} \otimes \mathbb{K} \to \mathfrak{A}/(\mathfrak{I}_1 \oplus \mathfrak{I}_2) \otimes \mathbb{K} \to 0$ is a full extension.*
2. *When x = 9, and $\mathfrak{A}$ is unital, then $0 \to (\mathfrak{I}_1 \oplus \mathfrak{I}_2) \otimes \mathbb{K} \to \mathfrak{A} \otimes \mathbb{K} \to \mathfrak{A}/(\mathfrak{I}_1 \oplus \mathfrak{I}_2) \otimes \mathbb{K} \to 0$ is a full extension.*

Proof. Suppose x = 4, 5, B, or D. Let $\mathfrak{I}$ be the ideal of $\mathfrak{A}$ containing $(\mathfrak{I}_1 \oplus \mathfrak{I}_2)$ such that $\mathfrak{I}/(\mathfrak{I}_1 \oplus \mathfrak{I}_2)$ is simple. Note that the push forward extension of the extension $0 \to (\mathfrak{I}_1 \oplus \mathfrak{I}_2) \otimes \mathbb{K} \to \mathfrak{I} \otimes \mathbb{K} \to \mathfrak{I}/(\mathfrak{I}_1 \oplus \mathfrak{I}_2) \otimes \mathbb{K} \to 0$ via the coordinate projection $(\mathfrak{I}_1 \oplus \mathfrak{I}_2) \to \mathfrak{I}_i$ is a full extension since it is isomorphic to a non-*AF* graph C^*-algebras with exactly one nontrivial ideal. Therefore, by Lemma 4, $0 \to (\mathfrak{I}_1 \oplus \mathfrak{I}_2) \otimes \mathbb{K} \to \mathfrak{I} \otimes \mathbb{K} \to \mathfrak{I}/(\mathfrak{I}_1 \oplus \mathfrak{I}_2) \otimes \mathbb{K} \to 0$ is a full extension. By Proposition 5.4 of [15], $0 \to (\mathfrak{I}_1 \oplus \mathfrak{I}_2) \otimes \mathbb{K} \to \mathfrak{A} \otimes \mathbb{K} \to \mathfrak{A}/(\mathfrak{I}_1 \oplus \mathfrak{I}_2) \otimes \mathbb{K} \to 0$ is a full extension since $\mathfrak{I}/(\mathfrak{I}_1 \oplus \mathfrak{I}_2) \otimes \mathbb{K}$ is an essential ideal of $\mathfrak{A}/(\mathfrak{I}_1 \oplus \mathfrak{I}_2) \otimes \mathbb{K}$.

We now prove the extension is full for the case x = 3. Note that in this case $\mathfrak{I}_1 \otimes \mathbb{K}$ and $\mathfrak{I}_2 \otimes \mathbb{K}$ are purely infinite, simple C^*-algebras. Let $\mathfrak{I}$ be the ideal of $\mathfrak{A}$ containing $(\mathfrak{I}_1 \oplus \mathfrak{I}_2)$ such that $\mathfrak{I}/(\mathfrak{I}_1 \oplus \mathfrak{I}_2)$ is simple. By Lemmas 3 and 4, $0 \to (\mathfrak{I}_1 \oplus \mathfrak{I}_2) \otimes \mathbb{K} \to \mathfrak{I} \otimes \mathbb{K} \to \mathfrak{I}/(\mathfrak{I}_1 \oplus \mathfrak{I}_2) \otimes \mathbb{K} \to 0$ is a full extension. The conclusion now follows from Proposition 5.4 of [15] since $\mathfrak{I}/(\mathfrak{I}_1 \oplus \mathfrak{I}_2) \otimes \mathbb{K}$ is an essential ideal of $\mathfrak{A}/(\mathfrak{I}_1 \oplus \mathfrak{I}_2) \otimes \mathbb{K}$.

Suppose x = 9 and $\mathfrak{A}$ is unital. Then $\mathfrak{I}_i$ is either $\mathbb{K}$ or a stable, purely infinite, simple C^*-algebra. Let $\mathfrak{I}$ be the ideal containing $\mathfrak{I}_1 \oplus \mathfrak{I}_2$ such that $\mathfrak{I}/(\mathfrak{I}_1 \oplus \mathfrak{I}_2)$ is simple. Note that the signature of $\mathfrak{I}$ is 3.6. By Lemma 3, the push forward extension of the extension $0 \to (\mathfrak{I}_1 \oplus \mathfrak{I}_2) \otimes \mathbb{K} \to \mathfrak{I} \otimes \mathbb{K} \to \mathfrak{I}/(\mathfrak{I}_1 \oplus \mathfrak{I}_2) \otimes \mathbb{K} \to 0$ via the coordinate projection $(\mathfrak{I}_1 \oplus \mathfrak{I}_2) \otimes \mathbb{K} \to \mathfrak{I}_i \otimes \mathbb{K}$ is essential, and hence full since $\mathcal{Q}(\mathfrak{I}_i \otimes \mathbb{K})$ is simple. Thus, by Lemma 4, $0 \to (\mathfrak{I}_1 \oplus \mathfrak{I}_2) \otimes \mathbb{K} \to \mathfrak{I} \otimes \mathbb{K} \to \mathfrak{I}/(\mathfrak{I}_1 \oplus \mathfrak{I}_2) \otimes \mathbb{K} \to 0$ is full. By Proposition 5.4 of [15], $0 \to (\mathfrak{I}_1 \oplus \mathfrak{I}_2) \otimes \mathbb{K} \to \mathfrak{A} \otimes \mathbb{K} \to \mathfrak{A}/(\mathfrak{I}_1 \oplus \mathfrak{I}_2) \otimes \mathbb{K} \to 0$ is a full extension since $\mathfrak{I}/(\mathfrak{I}_1 \oplus \mathfrak{I}_2)$ is an essential ideal of $\mathfrak{A}/(\mathfrak{I}_1 \oplus \mathfrak{I}_2)$.

Lemma 16. *Let $\mathfrak{A}$ be a graph C^*-algebra with tempered signature 4.3E.7. Let $\mathfrak{I}$ be the ideal of $\mathfrak{A}$ such that $\mathfrak{A}/\mathfrak{I}$ is simple. Then $0 \to \mathfrak{I} \otimes \mathbb{K} \to \mathfrak{A} \otimes \mathbb{K} \to \mathfrak{A}/\mathfrak{I} \otimes \mathbb{K} \to 0$ is a full extension.*

Proof. Let $\mathfrak{I}_1$ and $\mathfrak{I}_2$ be the minimal ideals of $\mathfrak{A}$ which is contained in $\mathfrak{I}$. Since $\mathfrak{I}/(\mathfrak{I}_1 + \mathfrak{I}_2)$ is a non-unital, purely infinite, simple C^*-algebra, we have that $0 \to \mathfrak{I}/(\mathfrak{I}_1 + \mathfrak{I}_2) \otimes \mathbb{K} \to \mathfrak{A}/(\mathfrak{I}_1 + \mathfrak{I}_2) \otimes \mathbb{K} \to \mathfrak{A}/\mathfrak{I} \otimes \mathbb{K} \to 0$ is a full extension. The conclusion of the lemma now follows from Corollary 5.3 of [15].

Lemma 17. *Let $\mathfrak{A}$ be a graph C^*-algebra with tempered signature 4.1F.E. Let $\mathfrak{I}$ be the smallest ideal of $\mathfrak{A}$. Then $0 \to \mathfrak{I} \otimes \mathbb{K} \to \mathfrak{A} \otimes \mathbb{K} \to \mathfrak{A}/\mathfrak{I} \otimes \mathbb{K} \to 0$ is a full extension.*

Proof. Let $\mathfrak{I}_1$ be the ideal of $\mathfrak{A}$ such that $\mathfrak{I}_1$ contains $\mathfrak{I}$ and $\mathfrak{I}_1/\mathfrak{I}$ is simple. Since $\mathfrak{I}_1$ is stably isomorphic to a non-*AF* graph C^*-algebra with exactly one nontrivial ideal, we have that $0 \to \mathfrak{I} \otimes \mathbb{K} \to \mathfrak{I}_1 \otimes \mathbb{K} \to \mathfrak{I}_1/\mathfrak{I} \otimes \mathbb{K} \to 0$ is full. Since $\mathfrak{I}_1/\mathfrak{I}$ is an essential ideal of $\mathfrak{A}/\mathfrak{I}$, the conclusion of the lemma follows from Proposition 5.4 of [15].

Using the above lemmas and the results of [1], we get the following:

Corollary 3. *Let $\mathfrak{A}$ and $\mathfrak{B}$ be graph C^*-algebras with signature either* 4.1F *or* 4.3E*, and assume that there exists an isomorphism from $\mathsf{FK}_X^+(\mathfrak{A})$ to $\mathsf{FK}_X^+(\mathfrak{B})$. If*

1. *$\mathfrak{A}$ and $\mathfrak{B}$ both have tempered signature* 4.1F.7*,* 4.1F.E*,* 4.3E.3*,* 4.3E.7*, or* 4.3E.D*, or*
2. *$\mathfrak{A}$ and $\mathfrak{B}$ both have finitely generated K-theory and have tempered signature* 4.1F.D*,* 4.3E.4 *or* 4.3E.5*, or*
3. *$\mathfrak{A}$ and $\mathfrak{B}$ both are unital and have tempered signature* 4.1F.2*,* 4.1F.5*,* 4.1F.6*,* 4.3E.9 *or* 4.3E.B*,*

then $\mathfrak{A} \otimes \mathbb{K} \cong \mathfrak{B} \otimes \mathbb{K}$.

Proof. By the above lemmas, all the extensions are full. Note that the specified ideal and quotient for each space belongs to classes of C^*-algebras satisfying the conditions in Definition 7. Hence, the result now follows from Theorem 9.

5.5.4 O-Shaped Spaces

Lemma 18. *Let $\mathfrak{A}$ be a graph C^*-algebra that is a tight C^*-algebra over the O-shaped space* 4.3B.7*. Let $\mathfrak{I}$ be the smallest ideal of $\mathfrak{A}$ and let $\mathfrak{I}_1$ and $\mathfrak{I}_2$ be the ideals of $\mathfrak{A}$ which contain $\mathfrak{I}$ and $\mathfrak{I}_k/\mathfrak{I}$ is simple. Then $0 \to (\mathfrak{I}_1 + \mathfrak{I}_2) \otimes \mathbb{K} \to \mathfrak{A} \otimes \mathbb{K} \to \mathfrak{A}/(\mathfrak{I}_1 + \mathfrak{I}_2) \otimes \mathbb{K} \to 0$ is a full extension.*

Proof. Note that $\mathfrak{A}/\mathfrak{I}$ is a tight C^*-algebra over the space 3.6.5. Then by Lemma 4, $0 \to (\mathfrak{I}_1 + \mathfrak{I}_2)/\mathfrak{I} \otimes \mathbb{K} \to \mathfrak{A}/\mathfrak{I} \otimes \mathbb{K} \to \mathfrak{A}/(\mathfrak{I}_1 + \mathfrak{I}_2) \otimes \mathbb{K} \to 0$ is a full extension since $\mathfrak{I}_1/\mathfrak{I}$ and $\mathfrak{I}_2/\mathfrak{I}$ are purely infinite, simple C^*-algebras. Also, since $\mathfrak{I}$ is an essential ideal of $\mathfrak{I}_1 + \mathfrak{I}_2$ and since $\mathfrak{I}$ is a purely infinite, simple C^*-algebra, we have that $0 \to \mathfrak{I} \otimes \mathbb{K} \to (\mathfrak{I}_1 + \mathfrak{I}_2) \otimes \mathbb{K} \to (\mathfrak{I}_1 + \mathfrak{I}_2)/\mathfrak{I} \otimes \mathbb{K} \to 0$ is a full extension. The conclusion of the lemma now follows from Proposition 3.2 of [14] since $\mathfrak{A}/(\mathfrak{I}_1 + \mathfrak{I}_2)$ is simple.

Lemma 19. *Let $\mathfrak{A}$ be a graph C^*-algebra that is a tight C^*-algebra over the O-shaped space* 4.3B.E*. Let $\mathfrak{I}$ be the smallest ideal of $\mathfrak{A}$. Then $0 \to \mathfrak{I} \otimes \mathbb{K} \to \mathfrak{A} \otimes \mathbb{K} \to \mathfrak{A}/\mathfrak{I} \otimes \mathbb{K} \to 0$ is a full extension.*

Proof. Let $\mathfrak{I}_1$ and $\mathfrak{I}_2$ be the ideals of $\mathfrak{A}$ which contain $\mathfrak{I}$ and $\mathfrak{I}_k/\mathfrak{I}$ is simple. Since $\mathfrak{I}_k \otimes \mathbb{K}$ is isomorphic to a graph C^*-algebra with exactly one non-trivial ideal and $\mathfrak{I}_k \otimes \mathbb{K}$ is not an *AF* algebra, by Proposition 3, we have that $0 \to \mathfrak{I} \otimes \mathbb{K} \to \mathfrak{I}_k \otimes \mathbb{K} \to \mathfrak{I}_k/\mathfrak{I} \otimes \mathbb{K} \to 0$ is a full extension. By Lemma 7, $0 \to \mathfrak{I} \otimes \mathbb{K} \to (\mathfrak{I}_1 + \mathfrak{I}_2) \otimes \mathbb{K} \to (\mathfrak{I}_1 + \mathfrak{I}_2)/\mathfrak{I} \otimes \mathbb{K} \to 0$ is a full extension.

The conclusion of the lemma now follows from Proposition 5.4 of [13] since $(\mathfrak{I}_1 + \mathfrak{I}_2)/\mathfrak{I} \otimes \mathbb{K}$ is an essential ideal of $\mathfrak{A}/\mathfrak{I}$.

Using the above lemmas and the results of [2], we get the following cases:

Corollary 4. *Let $\mathfrak{A}$ and $\mathfrak{B}$ be graph C^*-algebras that are tight over a O-shaped space X. Assume that there exists an isomorphism from $\mathsf{FK}_X^+(\mathfrak{A})$ to $\mathsf{FK}_X^+(\mathfrak{B})$. If $\mathfrak{A}$ and $\mathfrak{B}$ both have tempered signature* 4.3B.7 *or* 4.3B.E*, then $\mathfrak{A} \otimes \mathbb{K} \cong \mathfrak{B} \otimes \mathbb{K}$.*

Proof. By the above lemmas, all the extensions are full. Note that the specified ideal and quotient for each space belongs to classes of C^*-algebras satisfying the conditions in Definition 7. Hence, the result now follows from Theorem 9.

5.6 Summary of Results

In this final section, we index our results. Cases that are open are indicated by "?". Cases that are solved in general are marked by "$\surd$", and if we need to impose conditions of finitely generated K-theory or unitality, this is indicated by "$\surd_{f.g.}$" or "$\surd_1$", respectively.

5.6.1 *One Point Spaces*

Having nothing new to add, we include the simple case only for completeness.

1.0.x			
0	□	√	Theorem 2, p. 96
1	■	√	Theorem 3, p. 97

5.6.2 *Two Point Spaces*

This case was solved in [19], so again we include it only for completeness.

2.1.x			
0	□ ⟶ □	√	Theorem 2, p. 96
1	□ ⟶ ■	√	Proposition 1, p. 97
2	■ ⟶ □	√	Proposition 2, p. 98
3	■ ⟶ ■	√	Theorem 3, p. 97

5.6.3 Three Point Spaces

We resolve the case of three primitive ideal spaces here, up to a condition of finite generation which must be imposed in the cases of signature 3.7.2 and 3.7.5. We do not know if this condition is necessary.

3.3.x			
0	□ ⟶ □ ⟵ □	√	Theorem 2, p. 96
1	□ ⟶ □ ⟵ ■	√	Theorem 6, p. 105
2	□ ⟶ ■ ⟵ □	√	Theorem 7, p. 110
3	□ ⟶ ■ ⟵ ■	√	Theorem 7, p. 110
5	■ ⟶ □ ⟵ ■	√	Theorem 6, p. 105
7	■ ⟶ ■ ⟵ ■	√	Theorem 3, p. 97

3.6.x			
0	□ ⟵ □ ⟶ □	√	Theorem 2, p. 96
1	□ ⟵ □ ⟶ ■	√	Theorem 4, p. 100
2	□ ⟵ ■ ⟶ □	√	Theorem 5, p. 103
3	□ ⟵ ■ ⟶ ■	√	Theorem 5, p. 103
5	■ ⟵ □ ⟶ ■	√	Theorem 4, p. 100
7	■ ⟵ ■ ⟶ ■	√	Theorem 3, p. 97

3.7.x			
0	□ ⟶ □ ⟶ □	√	Theorem 2, p. 96
1	□ ⟶ □ ⟶ ■	√	Proposition 1, p. 97
2	□ ⟶ ■ ⟶ □	$\surd_{f.g.}$	Theorem 11, p. 116
3	□ ⟶ ■ ⟶ ■	√	Proposition 1, p. 97
4	■ ⟶ □ ⟶ □	√	Proposition 2, p. 98
5	■ ⟶ □ ⟶ ■	$\surd_{f.g.}$	Theorem 10, p. 116
6	■ ⟶ ■ ⟶ □	√	Proposition 2, p. 98
7	■ ⟶ ■ ⟶ ■	√	Theorem 3, p. 97

5.6.4 Four Point Spaces

In this section, we present our results for the case of four primitive ideals. As will be obvious below, the strength of our results varies dramatically with the nature of the spaces. In general, we can say quite a lot about all spaces apart from 4.E, 4.1E, and 4.3B. It may be interesting to note what makes these spaces difficult to handle; indeed the case 4.E is an accordion space in which a general UCT is known to hold, but it differs from the other accordion spaces by having poor separation properties when it comes to establishing fullness. The O-shaped spaces are also hard to separate fully, but have the added difficulty that no general UCT is known for them.

5.6.4.1 Accordion Spaces

4.E.x			
0	$\square \longrightarrow \square \longleftarrow \square \longrightarrow \square$	$\surd$	Theorem 2, p. 96
1	$\square \longrightarrow \blacksquare \longleftarrow \square \longrightarrow \square$	$\surd$	Remark 2, p. 102
2	$\square \longrightarrow \square \longleftarrow \blacksquare \longrightarrow \square$	?	
3	$\square \longrightarrow \blacksquare \longleftarrow \blacksquare \longrightarrow \square$	?	
4	$\square \longrightarrow \square \longleftarrow \square \longrightarrow \blacksquare$	$\surd$	Theorem 8, p. 111
5	$\square \longrightarrow \blacksquare \longleftarrow \square \longrightarrow \blacksquare$	$\surd$	Theorem 8, p. 111
6	$\square \longrightarrow \square \longleftarrow \blacksquare \longrightarrow \blacksquare$	?	
7	$\square \longrightarrow \blacksquare \longleftarrow \blacksquare \longrightarrow \blacksquare$	?	
8	$\blacksquare \longrightarrow \square \longleftarrow \square \longrightarrow \square$	?	
9	$\blacksquare \longrightarrow \blacksquare \longleftarrow \square \longrightarrow \square$	?	
A	$\blacksquare \longrightarrow \square \longleftarrow \blacksquare \longrightarrow \square$	?	
B	$\blacksquare \longrightarrow \blacksquare \longleftarrow \blacksquare \longrightarrow \square$	?	
C	$\blacksquare \longrightarrow \square \longleftarrow \square \longrightarrow \blacksquare$	?	
D	$\blacksquare \longrightarrow \blacksquare \longleftarrow \square \longrightarrow \blacksquare$	?	
E	$\blacksquare \longrightarrow \square \longleftarrow \blacksquare \longrightarrow \blacksquare$	?	
F	$\blacksquare \longrightarrow \blacksquare \longleftarrow \blacksquare \longrightarrow \blacksquare$	$\surd$	Theorem 3, p. 97

4.F.x			
0	$\square \longrightarrow \square \longleftarrow \square \longleftarrow \square$	$\surd$	Theorem 2, p. 96
1	$\square \longrightarrow \blacksquare \longleftarrow \square \longleftarrow \square$	$\surd$	Proposition 1, p. 97
2	$\square \longrightarrow \square \longleftarrow \blacksquare \longleftarrow \square$	$\surd_1$	Corollary 2, p. 118
3	$\square \longrightarrow \blacksquare \longleftarrow \blacksquare \longleftarrow \square$	$\surd_{f.g.}$	Corollary 2, p. 118
4	$\square \longrightarrow \square \longleftarrow \square \longleftarrow \blacksquare$	$\surd_1$	Corollary 2, p. 118
5	$\square \longrightarrow \blacksquare \longleftarrow \square \longleftarrow \blacksquare$	$\surd_1$	Corollary 2, p. 118
6	$\square \longrightarrow \square \longleftarrow \blacksquare \longleftarrow \blacksquare$	$\surd$	Theorem 6, p. 105
7	$\square \longrightarrow \blacksquare \longleftarrow \blacksquare \longleftarrow \blacksquare$	$\surd$	Corollary 2, p. 118
8	$\blacksquare \longrightarrow \square \longleftarrow \square \longleftarrow \square$	$\surd$	Theorem 6, p. 105
9	$\blacksquare \longrightarrow \blacksquare \longleftarrow \square \longleftarrow \square$	$\surd$	Corollary 2, p. 118
A	$\blacksquare \longrightarrow \square \longleftarrow \blacksquare \longleftarrow \square$	$\surd_{f.g.}$	Corollary 2, p. 118
B	$\blacksquare \longrightarrow \blacksquare \longleftarrow \blacksquare \longleftarrow \square$	$\surd_{f.g.}$	Corollary 2, p. 118
C	$\blacksquare \longrightarrow \square \longleftarrow \square \longleftarrow \blacksquare$	$\surd_1$	Corollary 2, p. 118
D	$\blacksquare \longrightarrow \blacksquare \longleftarrow \square \longleftarrow \blacksquare$	$\surd_1$	Corollary 2, p. 118
E	$\blacksquare \longrightarrow \square \longleftarrow \blacksquare \longleftarrow \blacksquare$	$\surd$	Theorem 6, p. 105
F	$\blacksquare \longrightarrow \blacksquare \longleftarrow \blacksquare \longleftarrow \blacksquare$	$\surd$	Theorem 3, p. 97

4.39.x				4.3F.x			
0	$\square \longleftarrow \square \longrightarrow \square \longrightarrow \square$	$\surd$	Theorem 2, p. 96	0	$\square \longrightarrow \square \longrightarrow \square \longrightarrow \square$	$\surd$	Theorem 2, p. 96
1	$\square \longleftarrow \square \longrightarrow \square \longrightarrow \blacksquare$	$\surd$	Theorem 4, p. 100	1	$\square \longrightarrow \square \longrightarrow \square \longrightarrow \blacksquare$	$\surd$	Proposition 1, p. 97
2	$\square \longleftarrow \square \longrightarrow \blacksquare \longrightarrow \square$	$\surd_1$	Corollary 2, p. 118	2	$\square \longrightarrow \square \longrightarrow \blacksquare \longrightarrow \square$	$\surd_{f.g.}$	Theorem 11, p. 116
3	$\square \longleftarrow \square \longrightarrow \blacksquare \longrightarrow \blacksquare$	$\surd$	Theorem 4, p. 100	3	$\square \longrightarrow \square \longrightarrow \blacksquare \longrightarrow \blacksquare$	$\surd$	Proposition 1, p. 97
4	$\blacksquare \longleftarrow \square \longrightarrow \square \longrightarrow \square$	$\surd$	Theorem 4, p. 100	4	$\square \longrightarrow \blacksquare \longrightarrow \square \longrightarrow \square$	$\surd_{f.g.}$	Theorem 11, p. 116
5	$\blacksquare \longleftarrow \square \longrightarrow \square \longrightarrow \blacksquare$	$\surd$	Theorem 4, p. 100	5	$\square \longrightarrow \blacksquare \longrightarrow \square \longrightarrow \blacksquare$	$\surd_{f.g.}$	Corollary 2, p. 118
6	$\blacksquare \longleftarrow \square \longrightarrow \blacksquare \longrightarrow \square$	$\surd_1$	Corollary 2, p. 118	6	$\square \longrightarrow \blacksquare \longrightarrow \blacksquare \longrightarrow \square$	$\surd_1$	Corollary 2, p. 118
7	$\blacksquare \longleftarrow \square \longrightarrow \blacksquare \longrightarrow \blacksquare$	$\surd$	Theorem 4, p. 100	7	$\square \longrightarrow \blacksquare \longrightarrow \blacksquare \longrightarrow \blacksquare$	$\surd$	Proposition 1, p. 97
8	$\square \longleftarrow \blacksquare \longrightarrow \square \longrightarrow \square$	$\surd$	Proposition 2, p. 98	8	$\blacksquare \longrightarrow \square \longrightarrow \square \longrightarrow \square$	$\surd$	Proposition 2, p. 98
9	$\square \longleftarrow \blacksquare \longrightarrow \square \longrightarrow \blacksquare$	$\surd_{f.g.}$	Corollary 2, p. 118	9	$\blacksquare \longrightarrow \square \longrightarrow \square \longrightarrow \blacksquare$	$\surd_{f.g.}$	Theorem 10, p. 116
A	$\square \longleftarrow \blacksquare \longrightarrow \blacksquare \longrightarrow \square$	$\surd_1$	Corollary 2, p. 118	A	$\blacksquare \longrightarrow \square \longrightarrow \blacksquare \longrightarrow \square$	$\surd_1$	Corollary 2, p. 118
B	$\square \longleftarrow \blacksquare \longrightarrow \blacksquare \longrightarrow \blacksquare$	$\surd$	Corollary 2, p. 118	B	$\blacksquare \longrightarrow \square \longrightarrow \blacksquare \longrightarrow \blacksquare$	?	
C	$\blacksquare \longleftarrow \blacksquare \longrightarrow \square \longrightarrow \square$	$\surd$	Corollary 2, p. 118	C	$\blacksquare \longrightarrow \blacksquare \longrightarrow \square \longrightarrow \square$	$\surd$	Proposition 2, p. 98
D	$\blacksquare \longleftarrow \blacksquare \longrightarrow \square \longrightarrow \blacksquare$	$\surd_{f.g.}$	Corollary 2, p. 118	D	$\blacksquare \longrightarrow \blacksquare \longrightarrow \square \longrightarrow \blacksquare$	$\surd_{f.g.}$	Corollary 2, p. 118
E	$\blacksquare \longleftarrow \blacksquare \longrightarrow \blacksquare \longrightarrow \square$	$\surd_1$	Corollary 2, p. 118	E	$\blacksquare \longrightarrow \blacksquare \longrightarrow \blacksquare \longrightarrow \square$	$\surd$	Proposition 2, p. 98
F	$\blacksquare \longleftarrow \blacksquare \longrightarrow \blacksquare \longrightarrow \blacksquare$	$\surd$	Theorem 3, p. 97	F	$\blacksquare \longrightarrow \blacksquare \longrightarrow \blacksquare \longrightarrow \blacksquare$	$\surd$	Theorem 3, p. 97

5.6.4.2 Fan Spaces

4.A.x			
0	□ □ □ ↘ ↓ ↙ □	√	Theorem 2, p. 96
1	□ □ □ ↘ ↓ ↙ ■	√	Theorem 7, p. 110
2	□ □ ■ ↘ ↓ ↙ □	√	Theorem 6, p. 105
3	□ □ ■ ↘ ↓ ↙ ■	√	Theorem 7, p. 110
6	□ ■ ■ ↘ ↓ ↙ □	√	Theorem 6, p. 105
7	□ ■ ■ ↘ ↓ ↙ ■	√	Theorem 7, p. 110
E	■ ■ ■ ↘ ↓ ↙ □	√	Theorem 6, p. 105
F	■ ■ ■ ↘ ↓ ↙ ■	√	Theorem 3, p. 97

4.38.x			
0	□ ↙ ↓ ↘ □ □ □	√	Theorem 2, p. 96
1	□ ↙ ↓ ↘ □ □ ■	√	Theorem 4, p. 100
3	□ ↙ ↓ ↘ □ ■ ■	√	Theorem 4, p. 100
7	□ ↙ ↓ ↘ ■ ■ ■	√	Theorem 4, p. 100
8	■ ↙ ↓ ↘ □ □ □	√	Theorem 5, p. 103
9	■ ↙ ↓ ↘ □ □ ■	√	Theorem 5, p. 103
B	■ ↙ ↓ ↘ □ ■ ■	√	Theorem 5, p. 103
F	■ ↙ ↓ ↘ ■ ■ ■	√	Theorem 3, p. 97

5.6.4.3 *Y*-Shaped Spaces

4.1F.x			
0	□ □ ↓ ↙ □ → □	$\surd$	Theorem 2, p. 96
1	□ □ ↓ ↙ □ → ■	$\surd$	Proposition 1, p. 97
2	□ □ ↓ ↙ ■ → □	$\surd_1$	Corollary 3, p. 120
3	□ □ ↓ ↙ ■ → ■	$\surd$	Proposition 1, p. 97
4	□ ■ ↓ ↙ □ → □	$\surd$	Remark 3, p. 109
5	□ ■ ↓ ↙ □ → ■	$\surd_1$	Corollary 3, p. 120
6	□ ■ ↓ ↙ ■ → □	$\surd_1$	Corollary 3, p. 120
7	□ ■ ↓ ↙ ■ → ■	$\surd$	Corollary 3, p. 120
C	■ ■ ↓ ↙ □ → □	$\surd$	Remark 3, p. 109
D	■ ■ ↓ ↙ □ → ■	$\surd_{f.g.}$	Corollary 3, p. 120
E	■ ■ ↓ ↙ ■ → □	$\surd$	Corollary 3, p. 120
F	■ ■ ↓ ↙ ■ → ■	$\surd$	Theorem 3, p. 97

4.3E.x			
0	□ → □ ↙ ↓ □ □	$\surd$	Theorem 2, p. 96
1	□ → □ ↙ ↓ □ ■	$\surd$	Remark 2, p. 102
3	□ → □ ↙ ↓ ■ ■	$\surd$	Corollary 3, p. 120
4	□ → ■ ↙ ↓ □ □	$\surd_{f.g.}$	Corollary 3, p. 120
5	□ → ■ ↙ ↓ □ ■	$\surd_{f.g.}$	Corollary 3, p. 120
7	□ → ■ ↙ ↓ ■ ■	$\surd$	Corollary 3, p. 120
8	■ → □ ↙ ↓ □ □	$\surd$	Proposition 2, p. 98
9	■ → □ ↙ ↓ □ ■	$\surd_1$	Corollary 3, p. 120
B	■ → □ ↙ ↓ ■ ■	$\surd_1$	Corollary 3, p. 120
C	■ → ■ ↙ ↓ □ □	$\surd$	Proposition 2, p. 98
D	■ → ■ ↙ ↓ □ ■	$\surd$	Corollary 3, p. 120
F	■ → ■ ↙ ↓ ■ ■	$\surd$	Theorem 3, p. 97

5.6.4.4 O-Shaped Spaces

4.1E.x			
0	□→□ ↓ ↑ □←□	√	Theorem 2, p. 96
1	□→□ ↓ ↑ ■←□	√	Remark 2, p. 102
3	□→■ ↓ ↑ ■←□	√	Remark 2, p. 102
4	□→□ ↓ ↑ □←■	√	Remark 3, p. 109
5	□→□ ↓ ↑ ■←■	?	
7	□→■ ↓ ↑ ■←■	?	
C	■→□ ↓ ↑ □←■	√	Remark 3, p. 109
D	■→□ ↓ ↑ ■←■	?	
F	■→■ ↓ ↑ ■←■	?	

4.3B.x			
0	□→□ ↓ ↓ □→□	√	Theorem 2, p. 96
1	□→□ ↓ ↓ □→■	√	Proposition 1, p. 97
2	□→□ ↓ ↓ ■→□	?	
3	□→□ ↓ ↓ ■→■	?	
6	□→■ ↓ ↓ ■→□	?	
7	□→■ ↓ ↓ ■→■	√	Corollary 4, p. 121
8	■→□ ↓ ↓ □→□	√	Proposition 2, p. 98
9	■→□ ↓ ↓ □→■	?	
A	■→□ ↓ ↓ ■→□	?	
B	■→□ ↓ ↓ ■→■	?	
E	■→■ ↓ ↓ ■→□	√	Corollary 4, p. 121
F	■→■ ↓ ↓ ■→■	√	Theorem 3, p. 97

Acknowledgements We gratefully acknowledge support from the NordForsk Research Network "Operator Algebra and Dynamics" (grant #11580) and the Danish National Research Foundation through the Centre for Symmetry and Deformation (DNRF92).

References

1. Arklint, S., Restorff, G., Ruiz, E.: Filtrated K-theory for real rank zero C^*-algebras. Internat. J. Math. **23**(8), 1250,078, 19 (2012). DOI 10.1142/S0129167X12500784. URL http://dx.doi.org/10.1142/S0129167X12500784
2. Arklint, S.E., Bentmann, R., Katsura, T.: Reduction of filtered K-theory and a characterization of Cuntz-Krieger algebras (2013), arXiv:1301.7223v1
3. Bates, T., Hong, J.H., Raeburn, I., Szymański, W.: The ideal structure of the C^*-algebras of infinite graphs. Illinois J. Math. **46**(4), 1159–1176 (2002). URL http://projecteuclid.org/getRecord?id=euclid.ijm/1258138472
4. Bentmann, R., Köhler, M.: Universal coefficient theorems for C^*-algebras over finite topological spaces (2011), arXiv:1101.5702
5. Blackadar, B.: Operator algebras, *Encyclopaedia of Mathematical Sciences*, vol. 122. Springer-Verlag, Berlin (2006). Theory of C^*-algebras and von Neumann algebras, Operator Algebras and Non-commutative Geometry, III
6. Brinkmann, G., McKay, B.D.: Posets on up to 16 points. Order **19**(2), 147–179 (2002). DOI 10.1023/A:1016543307592. URL http://dx.doi.org/10.1023/A:1016543307592
7. Brown, L.G.: Stable isomorphism of hereditary subalgebras of C^*-algebras. Pacific J. Math. **71**(2), 335–348 (1977)
8. Carlsen, T.M., Eilers, S., Tomforde, M.: Index maps in the K-theory of graph algebras. J. K-Theory **9**(2), 385–406 (2012). DOI 10.1017/is011004017jkt156. URL http://dx.doi.org/10.1017/is011004017jkt156
9. Drinen, D., Tomforde, M.: Computing K-theory and Ext for graph C^*-algebras. Illinois J. Math. **46**(1), 81–91 (2002). URL http://projecteuclid.org/getRecord?id=euclid.ijm/1258136141
10. Drinen, D., Tomforde, M.: The C^*-algebras of arbitrary graphs. Rocky Mountain J. Math. **35**(1), 105–135 (2005). DOI 10.1216/rmjm/1181069770. URL http://dx.doi.org/10.1216/rmjm/1181069770
11. Eilers, S., Katsura, T.: Semiprojectivity and properly infinite projections in graph C^*-algebra. In preparation
12. Eilers, S., Loring, T.A., Pedersen, G.K.: Morphisms of extensions of C^*-algebras: pushing forward the Busby invariant. Adv. Math. **147**(1), 74–109 (1999). DOI 10.1006/aima.1999.1834. URL http://dx.doi.org/10.1006/aima.1999.1834
13. Eilers, S., Restorff, G., Ruiz, E.: Classification of extensions of classifiable C^*-algebras. Adv. Math. **222**(6), 2153–2172 (2009). DOI 10.1016/j.aim.2009.07.014. URL http://dx.doi.org/10.1016/j.aim.2009.07.014
14. Eilers, S., Restorff, G., Ruiz, E.: On graph C^*-algebras with a linear ideal lattice. Bull. Malays. Math. Sci. Soc. (2) **33**(2), 233–241 (2010)
15. Eilers, S., Restorff, G., Ruiz, E.: Classifying C^*-algebras with both finite and infinite subquotients. J. Funct. Anal. (2013). DOI 10.1016/j.jfa.2013.05.006. In press
16. Eilers, S., Restorff, G., Ruiz, E.: The ordered K-theory of a full extension. Canad. J. Math. (2013). DOI 10.4153/CJM-2013-015-7. In press
17. Eilers, S., Restorff, G., Ruiz, E.: Strong classification of extensions of classifiable C^*-algebras (2013), arXiv:1301.7695v1
18. Eilers, S., Ruiz, E., Sørensen, A.P.W.: Amplified graph C^*-algebras. Münster J. Math. **5**, 121–150 (2012)

19. Eilers, S., Tomforde, M.: On the classification of nonsimple graph C^*-algebras. Math. Ann. **346**(2), 393–418 (2010). DOI 10.1007/s00208-009-0403-z. URL http://dx.doi.org/10.1007/s00208-009-0403-z
20. Elliott, G.A.: On the classification of inductive limits of sequences of semisimple finite-dimensional algebras. J. Algebra **38**(1), 29–44 (1976)
21. Elliott, G.A.: Towards a theory of classification. Adv. Math. **223**(1), 30–48 (2010). DOI 10.1016/j.aim.2009.07.018. URL http://dx.doi.org/10.1016/j.aim.2009.07.018
22. Elliott, G.A., Kucerovsky, D.: An abstract Voiculescu-Brown-Douglas-Fillmore absorption theorem. Pacific J. Math. **198**(2), 385–409 (2001). DOI 10.2140/pjm.2001.198.385. URL http://dx.doi.org/10.2140/pjm.2001.198.385
23. Jeong, J.A.: Real rank of C^*-algebras associated with graphs. J. Aust. Math. Soc. **77**(1), 141–147 (2004). DOI 10.1017/S1446788700010211. URL http://dx.doi.org/10.1017/S1446788700010211
24. Kirchberg, E.: The classification of purely infinite C^*-algebras using Kasparov's theory (1994). Preprint, third draft
25. Kirchberg, E.: Das nicht-kommutative Michael-Auswahlprinzip und die Klassifikation nicht-einfacher Algebren. In: C^*-algebras (Münster, 1999), pp. 92–141. Springer, Berlin (2000)
26. Kirchberg, E., Rørdam, M.: Non-simple purely infinite C^*-algebras. Amer. J. Math. **122**(3), 637–666 (2000). URL http://muse.jhu.edu/journals/american_journal_of_mathematics/v122/122.3kirchberg.pdf
27. Kirchberg, E., Rørdam, M.: Infinite non-simple C^*-algebras: absorbing the Cuntz algebras $\mathcal{O}_\infty$. Adv. Math. **167**(2), 195–264 (2002). DOI 10.1006/aima.2001.2041. URL http://dx.doi.org/10.1006/aima.2001.2041
28. Kucerovsky, D., Ng, P.W.: The corona factorization property and approximate unitary equivalence. Houston J. Math. **32**(2), 531–550 (electronic) (2006)
29. Meyer, R., Nest, R.: C^*-algebras over topological spaces: the bootstrap class. Münster J. Math. **2**, 215–252 (2009)
30. Meyer, R., Nest, R.: C*-algebras over topological spaces: filtrated K-theory. Canad. J. Math. **64**(2), 368–408 (2012). DOI 10.4153/CJM-2011-061-x. URL http://dx.doi.org/10.4153/CJM-2011-061-x
31. Murphy, G.J.: C^*-algebras and operator theory. Academic Press Inc., Boston, MA (1990)
32. Pedersen, G.K.: Pullback and pushout constructions in C^*-algebra theory. J. Funct. Anal. **167**(2), 243–344 (1999). DOI 10.1006/jfan.1999.3456. URL http://dx.doi.org/10.1006/jfan.1999.3456
33. Phillips, N.C.: A classification theorem for nuclear purely infinite simple C^*-algebras. Doc. Math. **5**, 49–114 (electronic) (2000)
34. Raeburn, I.: Graph algebras, *CBMS Regional Conference Series in Mathematics*, vol. 103. Published for the Conference Board of the Mathematical Sciences, Washington, DC (2005)
35. Restorff, G.: Classification of non-simple C^*-algebras. Ph.D. thesis, University of Copenhagen (2008). URL http://www.math.ku.dk/~restorff/papers/thesis.pdf
36. Rørdam, M.: Classification of extensions of certain C^*-algebras by their six term exact sequences in K-theory. Math. Ann. **308**(1), 93–117 (1997). DOI 10.1007/s002080050067. URL http://dx.doi.org/10.1007/s002080050067
37. Tomforde, M.: Uniqueness theorems and ideal structure for Leavitt path algebras. J. Algebra **318**(1), 270–299 (2007). DOI 10.1016/j.jalgebra.2007.01.031. URL http://dx.doi.org/10.1016/j.jalgebra.2007.01.031
38. Toms, A.S., Winter, W.: Strongly self-absorbing C^*-algebras. Trans. Amer. Math. Soc. **359**(8), 3999–4029 (2007). DOI 10.1090/S0002-9947-07-04173-6. URL http://dx.doi.org/10.1090/S0002-9947-07-04173-6
39. Zhang, S.: K_1-groups, quasidiagonality, and interpolation by multiplier projections. Trans. Amer. Math. Soc. **325**(2), 793–818 (1991). DOI 10.2307/2001648. URL http://dx.doi.org/10.2307/2001648

Chapter 6
Remarks on the Pimsner-Voiculescu Embedding

George A. Elliott and Zhuang Niu

Abstract Irrational extended rotation algebras are shown to be C*-alloys in the sense of Exel (C R Math Acad Sci Soc R Can (2012), arXiv:1204.0486).

Keywords C^*-alloys • C^*-blends • Irrational extended rotation algebras

Mathematics Subject Classification (2010): 46L05, 46L55.

6.1 Introduction

In [16], Pimsner and Voiculescu showed that the irrational rotation C*-algebra A_θ can be embedded in an AF C*-algebra. This construction has been studied (and generalized) extensively; see for instance [13], [17], and [14].

In [9], the authors described a more canonical form of the Pimsner-Voiculescu embedding (avoiding the infinitely many choices implicit in the original construction), at least for a generic set of $\theta \in \mathbb{R} \setminus \mathbb{Q}$. In any case, i.e., for any $\theta \in \mathbb{R} \setminus \mathbb{Q}$, the C*-algebra A_θ was shown in [9] to embed naturally in a C*-algebra generated by two commutative AF algebras, which in the present note we shall show forms what Exel has called an alloy of these two algebras.

It might be remarked, incidentally, that the fact proved in [7] that A_θ is an A$\mathbb{T}$ algebra can also be used to construct an embedding of A_θ in an AF algebra, using

G.A. Elliott (✉)
Department of Mathematics, University of Toronto, Toronto, ON, Canada M5S 2E4,
e-mail: elliott@math.toronto.edu

Z. Niu
Department of Mathematics, University of Wyoming, Laramie, WY 82071, USA
e-mail: zniu@uwyo.edu

T.M. Carlsen et al. (eds.), *Operator Algebra and Dynamics*, Springer Proceedings in Mathematics & Statistics 58, DOI 10.1007/978-3-642-39459-1_6,

classification theory for simple $A\mathbb{T}$ algebras (which implies that any simple $A\mathbb{T}$ algebra can be embedded in an AF algebra with the same ordered K_0-group—see [6]).

Another application of the fact that A_θ is $A\mathbb{T}$ is to re-derive and generalize the uniqueness result for approximate homomorphisms from A_θ into a finite-dimensional C*-algebra used in [16] (and studied and extended in [14], [17], and [8]). (This uniqueness result was somewhat special in that it considered only approximate homomorphisms defined by (direct sums of) what might be called Voiculescu pairs of unitary matrices—powers of the matrices introduced by Voiculescu in [19].)

Namely, one has uniqueness for arbitrary approximate homomorphisms from A_θ into a finite-dimensional C*-algebra: any two such maps—defined (let us say) as pairs of unitary elements of the codomain algebra with approximately the same commutation relation as the two canonical unitary generators of A_θ—, which in the natural sense agree K-theoretically on the approximate Rieffel projection (cf. [11, 13]) must be approximately unitarily equivalent. (Presumably, this holds for more general codomain algebras, with a more inclusive invariant, as it does for exact homomorphisms; see e.g. [12].) To see this, it is sufficient to consider exact homomorphisms into an asymptotic sequence algebra $\prod B_n / \bigoplus B_n$, with the B's finite-dimensional, which agree exactly on K_0 (here, we are using the result of [15] that the K_0-group of A_θ is generated by the classes of the unit and the Rieffel projection). Then, using just that the domain is a real rank zero $A\mathbb{T}$ algebra, and that a circle algebra (continuous matrix-valued functions on the circle) is weakly semiprojective (i.e., has stable relations), we see that the problem is reduced to showing that, if A is the real rank zero inductive limit of a sequence of (direct sums of) circle algebras, $A_1 \to A_2 \to \cdots$, in which (because of real rank zero—cf. [5]) we may suppose that for each i the compositions of the map $A_i \to A_{i+1}$ with any two irreducible representations of A_{i+1} in the same connected component are approximately unitarily equivalent, then any two (exact) homomorphisms of A_{i+1} into a finite-dimensional C*-algebra which agree on K_0 are approximately unitarily equivalent on A_i—which is immediate.

Neither of the techniques that we have just described (using [7]) would yield as elementary an embedding construction as in [16], as both use facts concerning the ordered K_0-group of A_θ (which were not known until after the work of Rieffel in [18] and Pimsner and Voiculescu in [15] and [16])—for instance that it is the ordered K_0-group of an AF algebra (because it is totally ordered—see [2], [3], [4], and [1]). Let us just remark that, using just the original Pimsner-Voiculescu uniqueness theorem, interpreted as a special case of the more general uniqueness theorem established in the preceding paragraph, with the comparison between the two maps stated in terms of the Rieffel–Loring K_0-class, and with purely qualitative estimates (no specific relation between epsilon and delta—the required summability obtained simply by passing to a subsequence), a one-sided intertwining argument embedding A_θ into an AF algebra with K_0-group $\mathbb{Z}+\mathbb{Z}\theta$ can be constructed (taking the unit into a projection with class $1 \in \mathbb{Z}+\mathbb{Z}\theta$, and the Rieffel projection into a projection with class θ), knowing only that such an AF algebra exists.

6.2 C^*-Blends and C^*-Alloys

C*-blends and C*-alloys were introduced by Ruy Exel in [10] to describe a situation in which two C*-algebras may be said to act on each other. Let us recall:

Definition 1 (3.1 of [10]). Consider a quintuple $\chi = (A, B, i, j, X)$, where A, B and X are C*-algebras, and

$$i : A \to \mathrm{M}(X) \quad \text{and} \quad j : B \to \mathrm{M}(X)$$

are *-homomorphisms of A and B into the multiplier algebra of X. Also consider the linear maps

$$i \circledast j : a \otimes b \in A \dot{\otimes} B \mapsto i(a)j(b) \in \mathrm{M}(X),$$

and

$$j \circledast i : b \otimes a \in B \dot{\otimes} A \mapsto j(b)i(a) \in \mathrm{M}(X),$$

where $\dot{\otimes}$ denotes the algebraic tensor product.

The system χ was said in [10] to be

1. A C*-blend if the ranges of $i \circledast j$ and $j \circledast i$ are contained in X and are dense, and
2. A C*-alloy if, in addition to (1), the maps $i \circledast j$ and $j \circledast i$ are injective.

Besides crossed-product C*-algebras, several examples of C*-blends were studied in [10]. In this note, we provide additional examples. We shall show that the irrational extended rotation algebras introduced in [9] are C*-blends, and that they are in fact C*-alloys.

6.3 Irrational Extended Rotation Algebras are C^*-Alloys

Consider the C*-algebra $\mathrm{C}(\mathbb{T})$ as the canonical sub-C*-algebra of $\mathrm{L}^\infty(\mathbb{T})$, and denote by σ the automorphism of $\mathrm{L}^\infty(\mathbb{T})$ induced by translation by $e^{2\pi i\theta}$:

$$f(z) \mapsto f(e^{2\pi i\theta}z).$$

Note that $\mathrm{C}(\mathbb{T})$ is invariant under the action of σ.

Let $\{p_i\}_{i\in\Lambda_1}$ and $\{q_j\}_{i\in\Lambda_2}$ be two collections of subintervals of $\mathbb{T}$, and denote again by p_i and q_j the spectral projections of the canonical unitary $f(z) = z$ in $\mathrm{L}^\infty(\mathbb{T})$ corresponding to the subintervals p_i and q_j.

Let us consider the following two commutative C*-algebras:

$$\mathrm{C}(\Omega_u) := \mathrm{C}^*(\mathrm{C}(\mathbb{T}) \cup \{\sigma^{-k}(p_i);\ i \in \Lambda_1, k \in \mathbb{Z}\}) \subseteq \mathrm{L}^\infty(\mathbb{T})$$

and

$$C(\Omega_v) := C^*(C(\mathbb{T}) \cup \{\sigma^k(q_j);\ j \in \Lambda_2, k \in \mathbb{Z}\}) \subseteq L^\infty(\mathbb{T})$$

where Ω_u and Ω_v denote the spectra of these algebras. Denote by u and v the canonical generators of $C(\mathbb{T})$ inside $C(\Omega_u)$ and $C(\Omega_v)$, respectively.

Definition 2. For an irrational number θ, and two given collections of subintervals $\{p_i\}_{i\in\Lambda_1}$ and $\{q_j\}_{j\in\Lambda_2}$ of the unit circle $\mathbb{T}$, we shall refer to the universal C*-algebra generated by $C(\Omega_u)$ and $C(\Omega_v)$ with respect to the relations

1. $uv = e^{2\pi i\theta}vu$,
2. $u\sigma^k(q_j)u^* = \sigma^{k+1}(q_j)$ for any $j \in \Lambda_2$ and $k \in \mathbb{Z}$, and
3. $v\sigma^{-k}(p_i)v^* = \sigma^{-k-1}(p_i)$ for any $i \in \Lambda_1$ and $k \in \mathbb{Z}$

as the (irrational) extended rotation algebra, and denote it by $\mathscr{B}_\theta = \mathscr{B}_\theta(\{p_i\}, \{q_j\})$.

It is clear from the definition that the commutative C*-algebras $C(\Omega_u)$ and $C(\Omega_v)$ can be alternatively described as

$$C(\Omega_u) := C^*\{u, v^{-k}p_iv^k;\ i \in \Lambda_1, k \in \mathbb{Z}\} \subseteq \mathscr{B}_\theta$$

and

$$C(\Omega_v) := C^*\{v, u^kq_ju^{-k};\ j \in \Lambda_2, k \in \mathbb{Z}\} \subseteq \mathscr{B}_\theta,$$

respectively. In what follows, we shall call any finite product of $\{v^{-k}p_iv^k;\ i \in \Lambda_1, k \in \mathbb{Z}\}$ a spectral projection in $C(\Omega_u)$ and any finite product of $\{u^kq_ju^{-k};\ j \in \Lambda_2, k \in \mathbb{Z}\}$ a spectral projection in $C(\Omega_v)$.

Lemma 1. *For any continuous function f with norm one on the spectrum of u, and for any $\varepsilon > 0$, there is $\delta > 0$ such that if g is a continuous function with norm one on the spectrum of v with* Supp(g) *contained in an open interval of length at most δ, then*

$$\|f(u)g(v)\|^2 < \varepsilon + \int |f|^2,$$

where the integral is over the circle with normalized Lebesgue measure.

Proof. Choose a polynomial $P = c_{-n}z^{-n} + \cdots + c_{-1}z^{-1} + c_0 + c_1z + c_2z^2 + \cdots c_nz^n$ such that

$$\left\| P(u) - |f|^2(u) \right\| < \varepsilon/2.$$

Note that

$$c_0 = \int P$$

and thus

$$\left\| c_0 - \int |f|^2 \right\| < \varepsilon/2.$$

Since θ is irrational, there exists $\delta > 0$ such that

$$(u^{-i}g(v)u^i)g(v) = 0, \quad -n \leq i \leq n,\ i \neq 0,$$

for any continuous function g on the spectrum of v with $\mathrm{Supp}(g)$ contained in an open interval of length at most δ. Then

$$g(v)u^i g(v) = 0, \quad -n \leq i \leq n,\ i \neq 0,$$

and hence

$$g(v)P(u)g(v) = c_0 g^2(v).$$

Therefore, if $\|g\| = 1$, then

$$\begin{aligned}
\|f(u)g(v)\|^2 &= \left\| g(v)\,|f|^2\,(u)g(v) \right\| \\
&\leq \|g(v)P(u)g(v)\| + \varepsilon/2 \\
&= \left\| c_0 g^2(v) \right\| + \varepsilon/2 \\
&< (\int |f|^2) \left\| g^2(v) \right\| + \varepsilon \\
&\leq \varepsilon + \int |f|^2,
\end{aligned}$$

as desired.

Lemma 2. *For any spectral projections $p \in \mathrm{C}(\Omega_u)$ and $q \in \mathrm{C}(\Omega_v)$, one has that*

$$pq \in \overline{\mathrm{C}(\Omega_v) \cdot \mathrm{C}(\Omega_u)}.$$

Proof. Fix an arbitrary $\varepsilon > 0$ for the time being.

One has decompositions

$$p = f(u) + f_- + f_+$$

and

$$q = g(v) + g_- + g_+$$

with f and g positive and continuous, and f_- and f_+ positive and with only one point of discontinuity on the spectrum of u, and g_- and g_+ positive and with only one point of discontinuity on the spectrum of v.

Moreover, one may assume that each of $\mathrm{Supp}(f_-)$ and $\mathrm{Supp}(f_+)$ is inside an open interval with length at most $\varepsilon/2$.

Choose continuous functions $\tilde{f}_-$ and $\tilde{f}_+$ such that

$$f_- \leq \tilde{f}_- \leq 1 \quad \text{and} \quad f_+ \leq \tilde{f}_+ \leq 1,$$

and each of $\mathrm{Supp}(\tilde{f}_-)$ and $\mathrm{Supp}(\tilde{f}_+)$ is contained inside an open interval with length at most ε.

With δ chosen as in Lemma 1 with respect to both $\tilde{f}_-$ and $\tilde{f}_+$, and the fixed ε, choose g_- and g_+ such that each of $\mathrm{Supp}(g_-)$ and $\mathrm{Supp}(g_+)$ is in an open interval with length at most $\delta/2$.

Note that

$$\begin{aligned} pq &= (f + f_- + f_+)(g + g_- + g_+) \\ &= f(g + g_- + g_+) + (f_- + f_+)g + f_-g_- + f_-g_+ + f_+g_- + f_+g_+. \end{aligned}$$

Consider the element f_-g_-. Choose a positive continuous function $\tilde{g}_-$ such that

$$g_- \leq \tilde{g}_- \leq 1,$$

and $\mathrm{Supp}(\tilde{g}_-)$ is contained in an open interval with length at most δ. Then,

$$\begin{aligned} &\|f_-g_-\|^2 = \|f_-g_-^2f_-\| \leq \|f_-(\tilde{g}_-)^2f_-\| = \\ &\|\tilde{g}_-(f_-)^2\tilde{g}_-\| \leq \left\|\tilde{g}_-(\tilde{f}_-)^2\tilde{g}_-\right\| = \left\|\tilde{f}_-\tilde{g}_-\right\|^2. \end{aligned}$$

By Lemma 1, one has

$$\begin{aligned} \left\|\tilde{f}_-\tilde{g}_-\right\|^2 &\leq \varepsilon + \int (\tilde{f}_-)^2 \\ &\leq \varepsilon + \varepsilon = 2\varepsilon, \end{aligned}$$

and hence

$$\|f_-g_-\| \leq \sqrt{2\varepsilon}.$$

Similarly,

$$f_-g_+ \leq \sqrt{2\varepsilon}, \quad f_+g_- \leq \sqrt{2\varepsilon}, \quad \text{and} \quad f_+g_+ \leq \sqrt{2\varepsilon},$$

and so

$$\|pq - (f(g + g_- + g_+) + (f_- + f_+)g)\| \leq 4\sqrt{2\varepsilon}.$$

Since f and g are continuous, one has (as by Proposition 3.4 of [10], a crossed product is a C*-blend)

$$f(g + g_- + g_+) \in \overline{\mathrm{C}(\Omega_v) \cdot \mathrm{C}(\Omega_u)} \quad \text{and} \quad (f_- + f_+)g \in \overline{\mathrm{C}(\Omega_v) \cdot \mathrm{C}(\Omega_u)}.$$

Hence,

$$\mathrm{dist}(pq, \overline{\mathrm{C}(\Omega_v) \cdot \mathrm{C}(\Omega_u)}) \leq 4\sqrt{2\varepsilon}.$$

Since ε is arbitrary, one has $pq \in \overline{\mathrm{C}(\Omega_v) \cdot \mathrm{C}(\Omega_u)}$, as desired.

Theorem 1. *The irrational extended rotation algebra $\mathscr{B}_\theta$ is a C*-blend of* $\mathrm{C}(\Omega_u)$ *and* $\mathrm{C}(\Omega_v)$*; that is,*

$$\overline{\mathrm{C}(\Omega_v) \cdot \mathrm{C}(\Omega_u)} = \mathscr{B}_\theta = \overline{\mathrm{C}(\Omega_u) \cdot \mathrm{C}(\Omega_v)}.$$

Proof. It is sufficient to prove $\overline{\mathrm{C}(\Omega_v) \cdot \mathrm{C}(\Omega_u)} = \mathscr{B}_\theta$, as the second equality follows on taking adjoints. Fix an arbitrary word

$$w' p'_1 q'_1 p'_2 q'_2 \cdots p'_n q'_n \in \mathscr{B}_\theta,$$

where w' is a word in $\{u, u^*, v, v^*\}$, p'_i is a spectral projection of u, and q'_i is a spectral projection of v.

By Lemma 2, $p'_1 q'_1 \in \overline{\mathrm{C}(\Omega_v) \cdot \mathrm{C}(\Omega_u)}$, and hence $w' p'_1 q'_1 p'_2 q'_2 \cdots p'_n q'_n$ is in the closure of the vector space spanned by the words

$$w'' q''_1 p''_2 q''_2 \cdots p''_n q''_n$$

with w'' a word in $\{u, u^*, v, v^*\}$, p''_i a spectral projection of u, and q''_i a spectral projection of v (but still with the same n).

Repeating this procedure, one obtains eventually

$$w' p'_1 q'_1 p'_2 q'_2 \cdots p'_n q'_n \in \overline{\mathrm{C}(\Omega_v) \cdot \mathrm{C}(\Omega_u)}.$$

This shows that

$$\mathscr{B}_\theta = \overline{\mathrm{C}(\Omega_v) \cdot \mathrm{C}(\Omega_u)},$$

as desired.

Now, let us show further that $\mathscr{B}_\theta$ is a C*-alloy of $C(\Omega_u)$ and $C(\Omega_v)$ if either $\{p_i\}_{i\in\Lambda_1}$ or $\{q_j\}_{j\in\Lambda_1}$ is a collection of half-open intervals with the same orientation. First, let us prove the corresponding statement for crossed-product C*-algebras.

Theorem 2. *Let A be a unital C*-algebra and let σ be an automorphism of A. Then*

$$(A, C(\mathbb{T}), A \rtimes_\sigma \mathbb{Z})$$

is a C-alloy.*

Proof. Recall that by Proposition 3.4 of [10], any crossed product is a C*-blend.

Let

$$a = \sum_{i=1}^{n} b_i \otimes c_i \in A \dot{\otimes} C(\mathbb{T}).$$

Denote by

$$\mathbb{E} : A \rtimes_\sigma \mathbb{Z} \to A$$

the canonical conditional expectation. Note that

$$\mathbb{E}(d_1 e d_2) = d_1 \mathbb{E}(e) d_2 \quad \text{and} \quad \mathbb{E}(e) \in \mathbb{C}1$$

for any $d_1, d_2 \in A$ and $e \in C(\mathbb{T})$.

Suppose that

$$\sum_{i=1}^{n} b_i c_i = 0 \in A \rtimes_\sigma \mathbb{Z}.$$

Without loss of generality, one may assume that $\{b_1, b_2, \ldots, b_n\}$ are linearly independent. Noting that

$$\sum_{i=1}^{n} b_i c_i u^k = 0 \in A \rtimes_\sigma \mathbb{Z}, \quad k \in \mathbb{Z},$$

where u is the canonical unitary, and applying the conditional expectation to both sides, one has

$$\sum_{i=1}^{n} b_i \mathbb{E}(c_i u^k) = 0, \quad k \in \mathbb{Z}.$$

Since $\mathbb{E}(c_i u^k) \in \mathbb{C}1$ and $\{b_1, b_2, \ldots, b_n\}$ are linearly independent, one has

$$\mathbb{E}(c_i u^k) = 0, \quad k \in \mathbb{Z},\ 1 \le i \le n,$$

and hence (by Fourier theory for $C(\mathbb{T})$) $c_1 = c_2 = \cdots = c_n = 0$. So $a = 0$.

The same argument shows that if $\sum_{i=1}^{n} c_i b_i = 0$, then $a = 0$. This shows that

$$(A, \mathrm{C}(\mathbb{T}), A \rtimes_\sigma \mathbb{Z})$$

is a C*-alloy.

Theorem 3. *Let θ be an irrational number, and assume that $\{p_i\}_{i\in\Lambda_1}$ or $\{q_j\}_{j\in\Lambda_1}$ is a collection of half-open intervals with the same orientation. Then*

$$(\mathrm{C}(\Omega_u), \mathrm{C}(\Omega_v), \mathscr{B}_\theta)$$

is a C-alloy.*

Proof. By Theorem 1, the extended rotation algebra $\mathscr{B}_\theta$ is a C*-blend of $\mathrm{C}(\Omega_u)$ and $\mathrm{C}(\Omega_v)$. We only have to show that the multiplication maps are one-to-one. The argument is similar to that of Theorem 2. (We do not actually use Theorem 2.)

Let us prove the theorem in the case that $\{q_j, j \in \Lambda_1\}$ are half-open intervals with the same orientation. The other case can be proved in the same way. Note that in this case, an element of $\mathrm{C}(\Omega_v)$ is zero if and only if all its Fourier coefficients are zero.

Set

$$c = \sum_{i=1}^{n} a_i \otimes b_i \in \mathrm{C}(\Omega_u)\dot{\otimes}\mathrm{C}(\Omega_v).$$

By Proposition 3.4 of [9], there is a canonical conditional expectation

$$\mathbb{E}_u : \mathscr{B}_\theta \to \mathrm{C}(\Omega_u)$$

such that

$$\mathbb{E}_u(d_1 e d_2) = d_1 \mathbb{E}_u(e) d_2 \quad \text{and} \quad \mathbb{E}(e) \in \mathbb{C}1$$

for any $d_1, d_2 \in \mathrm{C}(\Omega_u)$ and $e \in \mathrm{C}(\Omega_v)$.

Suppose that $\sum_{i=1}^{n} a_i b_i = 0$. Without loss of generality, one may assume that $\{a_1, a_2, \ldots, a_n\}$ are linearly independent. Then the same argument as that of Theorem 2 shows that

$$\mathbb{E}_u(b_i v^k) = 0, \quad k \in \mathbb{Z},\ 1 \leq i \leq n.$$

Hence (by Fourier theory for $L^2(\mathbb{T})$), $b_1 = b_2 = \cdots = b_n = 0$, and so $c = 0$.

The same argument also shows that if $\sum_{i=1}^{n} b_i a_i = 0$, then one also has that $c = 0$. So

$$(\mathrm{C}(\Omega_u), \mathrm{C}(\Omega_v), \mathscr{B}_\theta)$$

is a C*-alloy.

Remark 1. By Corollary 5.8 of [9], if there are nonzero minimal projections in each of $C(\Omega_u)$ and $C(\Omega_v)$, then they are orthogonal. Hence, in this case, the C*-algebra $(C(\Omega_u), C(\Omega_v), \mathscr{B}_\theta)$ is not a C*-alloy.

References

1. Effros, E.G., Shen, C.L.: The geometry of finite rank dimension groups. Illinois J. Math. **25**(1), 27–38 (1981). URL http://projecteuclid.org/getRecord?id=euclid.ijm/1256047361
2. Elliott, G.A.: On the classification of inductive limits of sequences of semisimple finite-dimensional algebras. J. Algebra **38**(1), 29–44 (1976)
3. Elliott, G.A.: A property of totally ordered abelian groups. C. R. Math. Rep. Acad. Sci. Canada **1**(2), 63–66 (1978/79)
4. Elliott, G.A.: On totally ordered groups, and K_0. In: Ring theory (Proc. Conf., Univ. Waterloo, Waterloo, 1978), *Lecture Notes in Math.*, vol. 734, pp. 1–49. Springer, Berlin (1979)
5. Elliott, G.A.: On the classification of C^*-algebras of real rank zero. J. Reine Angew. Math. **443**, 179–219 (1993). DOI 10.1515/crll.1993.443.179. URL http://dx.doi.org/10.1515/crll.1993.443.179
6. Elliott, G.A.: A classification of certain simple C^*-algebras. II. J. Ramanujan Math. Soc. **12**(1), 97–134 (1997)
7. Elliott, G.A., Evans, D.E.: The structure of the irrational rotation C^*-algebra. Ann. of Math. (2) **138**(3), 477–501 (1993). DOI 10.2307/2946553. URL http://dx.doi.org/10.2307/2946553
8. Elliott, G.A., Loring, T.A.: AF embeddings of $C(\mathbf{T}^2)$ with a prescribed K-theory. J. Funct. Anal. **103**(1), 1–25 (1992). DOI 10.1016/0022-1236(92)90130-B. URL http://dx.doi.org/10.1016/0022-1236(92)90130-B
9. Elliott, G.A., Niu, Z.: The extended rotation algebra: Adjoining spectral projections to rotation algebras. J. Reine Angew. Math. **665**, 1–71 (2012). DOI 10.1515/CRELLE.2011.112
10. Exel, R.: Blends and alloys. C. R. Math. Acad. Sci. Soc. R. Can. **35**(3), 77–114 (2013)
11. Exel, R., Loring, T.A.: Invariants of almost commuting unitaries. J. Funct. Anal. **95**(2), 364–376 (1991). DOI 10.1016/0022-1236(91)90034-3. URL http://dx.doi.org/10.1016/0022-1236(91)90034-3
12. Lin, H.: Homomorphisms from AH-algebras (2012), arXiv:1202.4631
13. Loring, T.A.: K-theory and asymptotically commuting matrices. Canad. J. Math. **40**(1),197–216 (1988). DOI 10.4153/CJM-1988-008-9. URL http://dx.doi.org/10.4153/CJM-1988-008-9
14. Loring, T.A.: Berg's technique for pseudo-actions with applications to AF embeddings. Canad. J. Math. **43**(1), 119–157 (1991). DOI 10.4153/CJM-1991-008-5. URL http://dx.doi.org/10.4153/CJM-1991-008-5
15. Pimsner, M., Voiculescu, D.: Exact sequences for K-groups and Ext-groups of certain cross-product C^*-algebras. J. Operator Theory **4**(1), 93–118 (1980)
16. Pimsner, M., Voiculescu, D.: Imbedding the irrational rotation C^*-algebra into an AF-algebra. J. Operator Theory **4**(2), 201–210 (1980)
17. Pimsner, M.V.: Embedding some transformation group C^*-algebras into AF-algebras. Ergodic Theory Dynam. Systems **3**(4), 613–626 (1983). DOI 10.1017/S0143385700002182. URL http://dx.doi.org/10.1017/S0143385700002182
18. Rieffel, M.A.: C^*-algebras associated with irrational rotations. Pacific J. Math. **93**(2), 415–429 (1981). URL http://projecteuclid.org/getRecord?id=euclid.pjm/1102736269
19. Voiculescu, D.: Asymptotically commuting finite rank unitary operators without commuting approximants. Acta Sci. Math. (Szeged) **45**(1–4), 429–431 (1983)

Chapter 7
Graph C^*-Algebras with a T_1 Primitive Ideal Space

James Gabe

Abstract We give necessary and sufficient conditions which a graph should satisfy in order for its associated C^*-algebra to have a T_1 primitive ideal space. We give a description of which one-point sets in such a primitive ideal space are open, and use this to prove that any purely infinite graph C^*-algebra with a T_1 (in particular Hausdorff) primitive ideal space, is a c_0-direct sum of Kirchberg algebras. Moreover, we show that graph C^*-algebras with a T_1 primitive ideal space canonically may be given the structure of a $C(\tilde{\mathbb{N}})$-algebra, and that isomorphisms of their $\tilde{\mathbb{N}}$-filtered K-theory (without coefficients) lift to $E(\tilde{\mathbb{N}})$-equivalences, as defined by Dadarlat and Meyer.

Keywords Graph C^*-algebras • Primitive ideal space • Filtered K-theory

Mathematics Subject Classification (2010): 46L55, 46L35, 46M15, 19K35.

7.1 Introduction

When classifying non-simple C^*-algebras a lot of focus has been on C^*-algebras with finitely many ideals. However, Dadarlat and Meyer recently proved in [2] a Universal Multicoefficient Theorem in equivariant E-theory for separable C^*-algebras over second countable, zero-dimensional, compact Hausdorff spaces. In particular, together with the strong classification result of Kirchberg [7], this shows that any separable, nuclear, $\mathcal{O}_\infty$-absorbing C^*-algebra with a zero-dimensional,

J. Gabe (✉)
Department of Mathematical Sciences, University of Copenhagen Ø, Universitetsparken 5, DK-2100 Copenhagen Ø, Denmark
e-mail: gabe@math.ku.dk

T.M. Carlsen et al. (eds.), *Operator Algebra and Dynamics*, Springer Proceedings in Mathematics & Statistics 58, DOI 10.1007/978-3-642-39459-1_7,

compact Hausdorff primitive ideal space, for which all simple subquotients are in the classical bootstrap class, is strongly classified by its filtered total K-theory. This suggests and motivates the study of C^*-algebras with infinitely many ideals, in the eyes of classification.

In this paper we consider graph C^*-algebras with a T_1 primitive ideal space, i.e. a primitive ideal space in which every one-point set is closed. Clearly our main interest are such graph C^*-algebras with infinitely many ideals, since any finite T_1 space is discrete. In Sect. 7.2 we recall the definition of graph C^*-algebras and many of the related basic concepts. In particular, we give a complete description of the primitive ideal space of a graph C^*-algebra. In Sect. 7.3 we find necessary and sufficient condition which a graph should satisfy in order for the induced C^*-algebra to have a T_1 primitive ideal space. In Sect. 7.4 we prove that a lot of subsets of such primitive ideal spaces are both closed and open. In particular, we give a complete description of when one-point sets are open. We use this to show that any purely infinite graph C^*-algebra with a T_1 primitive ideal space is a c_0-direct sum of Kirchberg algebras. Moreover, we show that any graph C^*-algebra with a T_1 primitive ideal space may be given a canonical structure of a (not necessarily continuous) $C(\tilde{\mathbb{N}})$-algebra, where $\tilde{\mathbb{N}}$ is the one-point compactification of $\mathbb{N}$. As an ending remark, we prove that $\tilde{\mathbb{N}}$-filtered K-theory classifies these $C(\tilde{\mathbb{N}})$-algebras up to $E(\tilde{\mathbb{N}})$-equivalence, as defined by Dadarlat and Meyer in [2].

7.2 Preliminaries

We recall the definition of a graph C^*-algebra and many related definitions and properties. Let $E = (E^0, E^1, r, s)$ be a countable directed graph, i.e. a graph with countably many vertices E^0, countably many edges E^1 and a range and source map $r, s \colon E^1 \to E^0$ respectively. A vertex $v \in E^0$ is called a sink if $s^{-1}(v) = \emptyset$ and an infinite emitter if $|s^{-1}(v)| = \infty$. A graph with no infinite emitters is called row-finite.

We define the graph C^*-algebra of E, $C^*(E)$, to be the universal C^*-algebra generated by a family of mutually orthogonal projections $\{p_v : v \in E^0\}$ and partial isometries with mutually orthogonal ranges $\{s_e : e \in E^1\}$, subject to the following Cuntz-Krieger relations

1. $s_e^* s_e = p_{r(e)}$ for $e \in E^1$,
2. $s_e s_e^* \leq p_{s(e)}$ for $e \in E^1$,
3. $p_v = \sum_{e \in s^{-1}(v)} s_e s_e^*$ for $v \in E^0$ such that $0 < |s^{-1}(v)| < \infty$.

By universality there is a gauge action $\gamma \colon \mathbb{T} \to \mathrm{Aut}(C^*(E))$ such that $\gamma_z(p_v) = p_v$ and $\gamma_z(s_e) = z s_e$ for $v \in E^0, e \in E^1$ and $z \in \mathbb{T}$. An ideal in $C^*(E)$ is said to be gauge-invariant if is invariant under γ. All ideals are assumed to be two-sided and closed.

If $\alpha_1, \ldots, \alpha_n$ are edges such that $r(\alpha_i) = s(\alpha_{i+1})$ for $i = 1, \ldots, n-1$, then we say that $\alpha = (\alpha_1, \ldots, \alpha_n)$ is a path, with source $s(\alpha) = s(\alpha_1)$ and range

$r(\alpha) = r(\alpha_n)$. A loop is a path of positive length such that the source and range coincide, and this vertex is called the base of the loop. A loop α is said to have an exit, if there exist $e \in E^1$ and $i = 1, \dots, n$ such that $s(e) = s(\alpha_i)$ but $e \neq \alpha_i$. A loop α is called simple if $s(\alpha_i) \neq s(\alpha_j)$ for $i \neq j$. A graph E is said to have condition (K) if each vertex $v \in E^0$ is the base of no (simple) loop or is the base of at least two simple loops. It turns out that a graph E has condition (K) if and only if every ideal in $C^*(E)$ is gauge-invariant if and only if $C^*(E)$ has real rank zero.

For $v, w \in E^0$ we write $v \geq w$ if there is a path α with $s(\alpha) = v$ and $r(\alpha) = w$. A subset H of E^0 is called hereditary if $v \geq w$ and $v \in H$ implies that $w \in H$. A subset H of E^0 is called saturated if whenever $v \in E^0$ satisfies $0 < |s^{-1}(v)| < \infty$ and $r(s^{-1}(v)) \subseteq H$ then $v \in H$. If X is a subset of E^0 then we let $\Sigma H(X)$ denote the smallest hereditary and saturated set containing X. If H is hereditary and saturated we define

$$H_\infty^{\text{fin}} = \{v \in E^0 \backslash H : |s^{-1}(v)| = \infty \text{ and } 0 < |s^{-1}(v) \cap r^{-1}(E^0 \backslash H)| < \infty\},$$
$$H_\infty^{\emptyset} = \{v \in E^0 \backslash H : |s^{-1}(v)| = \infty \text{ and } s^{-1}(v) \cap r^{-1}(E^0 \backslash H) = \emptyset\}.$$

By [1, Theorem 3.6] there is a one-to-one correspondence between pairs (H, B), where $H \subseteq E^0$ is hereditary and saturated and $B \subseteq H_\infty^{\text{fin}}$, and the gauge-invariant ideals of $C^*(E)$. In fact, this is a lattice isomorphism when the different sets are given certain lattice structures. The ideal corresponding to (H, B) is denoted $J_{H,B}$ and if $B = \emptyset$ we denote it by J_H.

A non-empty subset $M \subseteq E^0$ is called a maximal tail if the following three conditions are satisfied.

1. If $v \in E^0, w \in M$ and $v \geq w$ then $v \in M$.
2. If $v \in M$ and $0 < |s^{-1}(v)| < \infty$ then there exists $e \in E^1$ such that $s(e) = v$ and $r(e) \in M$.
3. For every $v, w \in M$ there exists $y \in M$ such that $v \geq y$ and $w \geq y$.

Note that $E^0 \backslash M$ is hereditary by 1 and saturated by 2. Moreover, by 3 it follows that $(E^0 \backslash M)_\infty^{\emptyset}$ is either empty or consists of exactly one vertex. We let $\mathscr{M}(E)$ denote the set of all maximal tails in E, and let $\mathscr{M}_\gamma(E)$ denote the set of all maximal tails M in E such that each loop in M has an exit in M. We let $\mathscr{M}_\tau(E) = \mathscr{M}(E) \backslash \mathscr{M}_\gamma(E)$.

If $X \subseteq E^0$ then define

$$\Omega(X) = \{w \in E^0 \backslash X : w \not\geq v \text{ for all } v \in X\}.$$

Note that if M is a maximal tail, then $\Omega(M) = E^0 \backslash M$. For a vertex $v \in E^0$, $E^0 \backslash \Omega(v)$ is a maximal tail if and only if v is a sink, an infinite emitter or if v is the base of a loop.

We define the set of breaking vertices to be

$$BV(E) = \{v \in E^0 : |s^{-1}(v)| = \infty \text{ and } 0 < |s^{-1}(v) \backslash r^{-1}(\Omega(v))| < \infty\}.$$

Hence an infinite emitter v is a breaking vertex if and only if $v \in \Omega(v)_\infty^{\text{fin}}$.

We should warn the reader that it is customary to say that the elements of H_∞^{fin} are called breaking vertices of H, where H is hereditary and saturated. In these terms, a vertex $v \in E^0$ is a breaking vertex, i.e. $v \in BV(E)$, if and only if v is a breaking vertex of $\Omega(v)$. But beware, a vertex which is a breaking vertex of some hereditary and saturated set need not be a breaking vertex in general. In order to avoid confusion we will only use the term breaking vertex as a name for the elements in $BV(E)$.

In [6] they define for each $N \in \mathscr{M}_\tau(E)$ and $t \in \mathbb{T}$ a (primitive) ideal $R_{N,t}$, and prove that there is a bijection

$$\mathscr{M}_\gamma(E) \sqcup BV(E) \sqcup (\mathscr{M}_\tau(E) \times \mathbb{T}) \to \text{Prim}C^*(E)$$

given by

$$\begin{aligned} \mathscr{M}_\gamma(E) \ni M &\mapsto J_{\Omega(M),\Omega(M)_\infty^{\text{fin}}} \\ BV(E) \ni v &\mapsto J_{\Omega(v),\Omega(v)_\infty^{\text{fin}}\setminus\{v\}} \\ \mathscr{M}_\tau(E) \times \mathbb{T} \ni (N,t) &\mapsto R_{N,t}. \end{aligned}$$

In [6], Hong and Szymański give a complete description of the hull-kernel topology on $\text{Prim}C^*(E)$ in terms of the maximal tails and breaking vertices. In order to describe this we use the following notation. Whenever $M \in \mathscr{M}_\tau(E)$ there is a unique (up to cyclic permutation) simple loop $L = (\alpha_1, \ldots, \alpha_n)$ in M such that $M = \{v \in E^0 : v \geq s(\alpha_i) \text{ for some } i\}$, and we denote by L_M^0 the set of vertices $\{s(\alpha_1), \ldots, s(\alpha_n)\}$. If $Y \subseteq \mathscr{M}_\tau(E)$ we let

$$\begin{aligned} Y_{\min} &:= \{U \in Y : \text{for all } U' \in Y, U' \neq U \text{ there is no path from } L_U^0 \text{ to } L_{U'}^0\}, \\ Y_\infty &:= \{U \in Y : \text{for all } V \in Y_{\min} \text{ there is no path from } L_U^0 \text{ to } L_V^0\}. \end{aligned}$$

Due to a minor mistake in [6] the description of the topology is however not entirely correct. We will give a correct description below and explain what goes wrong in the original proof in Remark 1.

Theorem 1 (Hong-Szymański). *Let E be a countable directed graph. Let $X \subseteq \mathscr{M}_\gamma(E), W \subseteq BV(E), Y \subseteq \mathscr{M}_\tau(E)$, and let $D(U) \subseteq \mathbb{T}$ for each $U \in Y$. If $M \in \mathscr{M}_\gamma(E), v \in BV(E), N \in \mathscr{M}_\tau(E)$, and $z \in \mathbb{T}$, then the following hold.*

1. *$M \in \overline{X}$ if and only if one of the following three conditions holds.*
   a. *$M \in X$,*
   b. *$M \subseteq \bigcup X$ and $\Omega(M)_\infty^{\emptyset} = \emptyset$,*
   c. *$M \subseteq \bigcup X$ and $|s^{-1}(\Omega(M)_\infty^{\emptyset}) \cap r^{-1}(\bigcup X)| = \infty$.*
2. *$v \in \overline{X}$ if and only if $v \in \bigcup X$ and $|s^{-1}(v) \cap r^{-1}(\bigcup X)| = \infty$.*
3. *$(N, z) \in \overline{X}$ if and only if $N \subseteq \bigcup X$.*

4. *$M \in \overline{W}$ if and only if either*

   a. *$M \subseteq E^0 \setminus \bigcap_{w \in W} \Omega(w)$ and $\Omega(M)^{\emptyset}_{\infty} = \emptyset$, or*
   b. *$M \subseteq E^0 \setminus \bigcap_{w \in W} \Omega(w)$ and $|s^{-1}(\Omega(M)^{\emptyset}_{\infty}) \cap r^{-1}(E^0 \setminus \bigcap_{w \in W} \Omega(w))| = \infty$.*

5. *$v \in \overline{W}$ if and only if either*

   a. *$v \in W$, or*
   b. *$v \in E^0 \setminus \bigcap_{w \in W} \Omega(w)$ and $|s^{-1}(v) \cap r^{-1}(E^0 \setminus \bigcap_{w \in W} \Omega(w))| = \infty$.*

6. *$(N, z) \in \overline{W}$ if and only if $N \subseteq E^0 \setminus \bigcap_{w \in W} \Omega(w)$.*
7. *M is in the closure of $\{(U, t) : U \in Y, t \in D(U)\}$ if and only if one of the following four conditions holds.*

   a. *$M \subseteq \bigcup Y_{\infty}$ and $\Omega(M)^{\emptyset}_{\infty} = \emptyset$,*
   b. *$M \subseteq \bigcup Y_{\infty}$ and $|s^{-1}(\Omega(M)^{\emptyset}_{\infty}) \cap r^{-1}(\bigcup Y_{\infty})| = \infty$,*
   c. *$M \subseteq \bigcup Y_{\min}$ and $\Omega(M)^{\emptyset}_{\infty} = \emptyset$,*
   d. *$M \subseteq \bigcup Y_{\min}$ and $|s^{-1}(\Omega(M)^{\emptyset}_{\infty}) \cap r^{-1}(\bigcup Y_{\min})| = \infty$.*

8. *v is in the closure of $\{(U, t) : U \in Y, t \in D(U)\}$ if and only if either*

   a. *$v \in \bigcup Y_{\infty}$ and $|s^{-1}(v) \cap r^{-1}(\bigcup Y_{\infty})| = \infty$, or*
   b. *$v \in \bigcup Y_{\min}$ and $|s^{-1}(v) \cap r^{-1}(\bigcup Y_{\min})| = \infty$.*

9. *(N, z) is in the closure of $\{(U, t) : U \in Y, t \in D(U)\}$ if and only one of the following three conditions holds.*

   a. *$N \subseteq \bigcup Y_{\infty}$,*
   b. *$N \notin Y_{\min}$ and $N \subseteq \bigcup Y_{\min}$,*
   c. *$N \in Y_{\min}$ and $z \in \overline{D(N)}$.*

Before explaining what goes wrong in the original proof we give an example that illustrates the mistake in the original theorem.

Example 1. Let E denote the graph

$$v \Longleftarrow u \Longrightarrow w,$$

i.e. $E^0 = \{u, v, w\}$ and u emits infinitely to both v and w. This has the maximal tails $\{u\}, \{u, v\}, \{u, w\}$. Note that

$$J_{\Omega(\{u,v\}), \Omega(\{u,v\})^{\text{fin}}_{\infty}} = J_{\{w\}} \quad \text{and} \quad J_{\Omega(\{u\}), \Omega(\{u\})^{\text{fin}}_{\infty}} = J_{\{v,w\}}.$$

Since $J_{\{w\}} \subseteq J_{\{v,w\}}$ it follows that $\{u\} \in \overline{\{\{u, v\}\}}$.

The original theorem [6, Theorem 3.4] states that $\{u\} \in \overline{\{\{u, v\}\}}$ if and only if

$$\{u\} \subseteq \{u, v\} \text{ and } s^{-1}(\Omega(\{u\})^{\emptyset}_{\infty}) \cap r^{-1}(\Omega(\{u, v\})) \text{ is finite.}$$

However,

$$s^{-1}(\Omega(\{u\})_\infty^\emptyset) \cap r^{-1}(\Omega(\{u, v\})) = s^{-1}(\{u\}) \cap r^{-1}(\{w\})$$

is infinite, which is a contradiction.

In the proof of the original theorem they prove, correctly, what in the above example corresponds to the statement $\{u\} \in \{\{u, v\}\}$ if and only if $u \notin \{w\}_\infty^{\mathrm{fin}}$. It is the latter statement which they reformulate incorrectly, as is described in the remark below.

Remark 1. The minor mistake in the original proof of Theorem 1 is an error which occurs in the proofs of Lemma 3.3 and Theorem 3.4 of [6]. We will explain what goes wrong. Suppose that M is a maximal tail, K is a hereditary and saturated set such that $K \subseteq \Omega(M)$, and that $B \subseteq K_\infty^{\mathrm{fin}}$. Note that $B \backslash \Omega(M)_\infty^\emptyset \subseteq \Omega(M) \cup \Omega(M)_\infty^{\mathrm{fin}}$. Hence if $w \in \Omega(M)_\infty^\emptyset$ then $J_{K,B} \subseteq J_{\Omega(M),\Omega(M)_\infty^{\mathrm{fin}}}$ if and only if $w \notin B$, since $w \notin \Omega(M) \cup \Omega(M)_\infty^{\mathrm{fin}}$. In the cases we consider we have that $w \in B$ if and only if $w \in K_\infty^{\mathrm{fin}}$. Now it is claimed that $w \notin K_\infty^{\mathrm{fin}}$ if and only if $s^{-1}(w) \cap r^{-1}(K)$ is finite. However, this is not the case. If both $s^{-1}(w) \cap r^{-1}(K)$ and $s^{-1}(w) \cap r^{-1}(E^0 \backslash K)$ are infinite then $w \notin K_\infty^{\mathrm{fin}}$. The correct statement would be that $w \notin K_\infty^{\mathrm{fin}}$ if and only if $|s^{-1}(w) \cap r^{-1}(E^0 \backslash K)| = \infty$.

A similar thing occurs in the case where $v \in BV(E)$. Here we have, in the cases we consider, that $J_{K,B} \subseteq J_{\Omega(v),\Omega(v)_\infty^{\mathrm{fin}} \backslash \{v\}}$ if and only if $v \notin K_\infty^{\mathrm{fin}}$. Again, the correct statement becomes $v \notin K_\infty^{\mathrm{fin}}$ if and only if $|s^{-1}(v) \cap r^{-1}(E^0 \backslash K)| = \infty$.

After changing these minor mistakes, one obtains Theorem 1 above.

7.3 T_1 Primitive Ideal Space

Recall that a topological space is said to satisfy the separation axiom T_1 if every one-point set is closed. In particular, every Hausdorff space is a T_1 space. For a C^*-algebra A the primitive ideal space Prim$A$ is T_1 exactly if every primitive ideal is a maximal ideal. All of our ideals are assumed to be two-sided and closed.

As shown in [1], every gauge-invariant primitive ideal of a graph C^*-algebra may be represented by a maximal tail or by a breaking vertex. The following lemma shows that we only need to consider maximal tails.

Lemma 1. *Let E be a graph such that* Prim$(C^*(E))$ *is T_1. Then E has no breaking vertices.*

Proof. Suppose E has a breaking vertex v. Then

$$J_{\Omega(v),\Omega(v)_\infty^{\mathrm{fin}} \backslash \{v\}} \text{ and } J_{\Omega(v),\Omega(v)_\infty^{\mathrm{fin}}}$$

are primitive ideals of $C^*(E)$, the former being a proper ideal of the latter by [1, Corollary 3.10]. Hence

$$J_{\Omega(v),\Omega(v)_\infty^{\mathrm{fin}}} \in \overline{\{J_{\Omega(v),\Omega(v)_\infty^{\mathrm{fin}}\setminus\{v\}}\}}$$

and thus $C^*(E)$ can not have a T_1 primitive ideal space. □

It turns out that it might be helpful to consider gauge-invariant ideals which are maximal in the following sense.

Definition 1. Let E be a countable directed graph and let J be a proper ideal of $C^*(E)$. We say that J is a *maximal gauge-invariant ideal* if J is gauge-invariant and if J and $C^*(E)$ are the only gauge-invariant ideals containing J.

The following theorem gives a complete description of the graphs whose induced C^*-algebras have a T_1 primitive ideal space.

Theorem 2. *Let E be a countable directed graph. The following are equivalent.*

1. *$C^*(E)$ has a T_1 primitive ideal space,*
2. *E has no breaking vertices, and whenever M and N are maximal tails such that M is a proper subset of N, then $\Omega(M)_\infty^\emptyset$ is non-empty, and*

$$|s^{-1}(\Omega(M)_\infty^\emptyset) \cap r^{-1}(N)| < \infty,$$

3. *E has no breaking vertices, and $J_{\Omega(M),\Omega(M)_\infty^{\mathrm{fin}}}$ is a maximal gauge-invariant ideal in $C^*(E)$ for any maximal tail M,*
4. *E has no breaking vertices, and the map $M \mapsto J_{\Omega(M),\Omega(M)_\infty^{\mathrm{fin}}}$ is a bijective map from the set of maximal tails of E onto the set of all maximal gauge-invariant ideals of $C^*(E)$.*

The last condition in 2 of the theorem may look complicated but it is easy to describe. It says, that if $M \subsetneq N$ are maximal tails then M must contain an infinite emitter v which only emits edges out of M, and only emits finitely many edges to N. Note that this is equivalent to $v \in \Omega(N)_\infty^{\mathrm{fin}}$.

Proof. We start by proving 1 ⇔ 2. By Lemma 1 we may restrict to the case where E has no breaking vertices. The proof is just a translation of Theorem 1 into our setting. We have four cases.

Case 1: Let $M, N \in \mathcal{M}_\gamma(E)$. By Theorem 1 we have $M \in \overline{\{N\}}$ if and only if one of the following three holds: (*i*) $M = N$, (*ii*) $M \subsetneq N$ and $\Omega(M)_\infty^\emptyset = \emptyset$, (*iii*) $M \subsetneq N$, $\Omega(M)_\infty^\emptyset \neq \emptyset$ and

$$|s^{-1}(\Omega(M)_\infty^\emptyset) \cap r^{-1}(N)| = \infty.$$

We eliminate the possibilities (*ii*) and (*iii*) exactly by imposing the conditions in 2.

Case 2: Let $(M,z) \in \mathcal{M}_\tau(E)\times\mathbb{T}$ and $N \in \mathcal{M}_\gamma(E)$. By Theorem 1, $(M,z) \in \overline{\{N\}}$ if and only if $M \subseteq N$. Since $M \in \mathcal{M}_\tau(E)$ it follows that $\Omega(M)_\infty^\emptyset = \emptyset$ and thus the conditions in 2 says $M \nsubseteq N$.

Case 3: Let $(N,t) \in \mathscr{M}_\tau(E) \times \mathbb{T}$ and $M \in \mathscr{M}_\gamma(E)$. Note that $\{N\}_{\min} = \{N\}$ and $\{N\}_\infty = \emptyset$. By Theorem 1 we have $M \in \overline{\{(N,t)\}}$ if and only if one of the following two holds: (i) $M \subseteq N$ and $\Omega(M)^\emptyset_\infty = \emptyset$, (ii) $M \subseteq N$, $\Omega(M)^\emptyset_\infty \neq \emptyset$ and

$$|s^{-1}(\Omega(M)^\emptyset_\infty) \cap r^{-1}(N)| = \infty.$$

Conditions (i) and (ii) do not hold exactly when assuming the conditions of 2.

Case 4: Let $(M,z), (N,t) \in \mathscr{M}_\tau(E) \times \mathbb{T}$. By Theorem 1 we have $(M,z) \in \overline{\{(N,t)\}}$ if and only if either $M \subsetneq N$, or $M = N$ and $z = t$. Note that condition 9a of the theorem can never be satisfied. Since the maximal tail M satisfies $\Omega(M)^\emptyset_\infty = \emptyset$ the conditions of 2 say $M \subseteq N$ if and only if $M = N$ thus finishing $1 \Leftrightarrow 2$.

We will prove $1 \Rightarrow 3$. In order to simplify matters, we replace E with its desingularisation F (see [4]) thus obtaining a row-finite graph without sinks. Since E has no breaking vertices by Lemma 1, there is a canonical one-to-one correspondence between $\mathscr{M}(E)$ and $\mathscr{M}(F)$ and a lattice isomorphism between the ideal lattices of $C^*(E)$ and $C^*(F)$ such that $M' \mapsto M$ implies $J_{\Omega(M'),\Omega(M')^{\mathrm{fin}}_\infty} \mapsto J_{\Omega(M)}$. In this case $J_{\Omega(M'),\Omega(M')^{\mathrm{fin}}_\infty}$ is a maximal gauge-invariant ideal if and only if $J_{\Omega(M)}$ is a maximal gauge-invariant ideal and thus it suffices to prove that $J_{\Omega(M)}$ is a maximal gauge-invariant ideal in $C^*(F)$ for $M \in \mathscr{M}(F)$.

Suppose $J_{\Omega(M)} \subseteq J_H$ for some hereditary and saturated set $H \neq F^0$. Since F is row-finite without sinks we may find an infinite path α in $F \backslash H$. Let

$$N = \{v \in F : v \geq s(\alpha_j) \text{ for some } j\}$$

which is a maximal tail such that $N \subseteq F^0 \backslash H$. Hence $\Omega(M) \subseteq H \subseteq \Omega(N)$ which implies $N \subseteq M$. Since F is row-finite, $\Omega(N)^\emptyset_\infty$ is empty, and thus since $1 \Leftrightarrow 2$, $M = N$. Hence $H = \Omega(M)$ and thus $1 \Rightarrow 3$.

We will prove $3 \Rightarrow 4$. Again, we let F be the desingularisation of E and note that 4 holds for F if and only if it holds for E. Note that 3 implies that the map in 4 is well-defined, and this is clearly injective. Let H be a hereditary and saturated set in F such that J_H is a maximal gauge-invariant ideal in $C^*(F)$. As above, we may find a maximal tail M such that $H \subseteq \Omega(M)$ which implies $J_H \subseteq J_{\Omega(M)}$. Since J_H is a maximal gauge-invariant ideal, $H = \Omega(M)$ which proves surjectivity of the map and finishes $3 \Rightarrow 4$.

For $4 \Rightarrow 1$ we may again replace E by its desingularisation F. Since $1 \Leftrightarrow 2$ and F is row-finite, 1 is equivalent to the following: if $M \subseteq N$ are maximal tails then $M = N$, since $\Omega(M)^\emptyset_\infty = \emptyset$ for every maximal tail M. Let $M \subseteq N$ be maximal tails in F. Then $J_{\Omega(N)} \subseteq J_{\Omega(M)}$ are maximal gauge-invariant ideals and thus $N = M$, which finishes the proof. □

Definition 2. Let E be a countable directed graph. If E satisfies one (and hence all) of the conditions in Theorem 2, then we say that E is a T_1 *graph*.

For row-finite graphs the above theorem simplifies significantly.

Corollary 1. *Let E be a row-finite graph. The following are equivalent.*

1. *E is a T_1 graph,*
2. *If $M \subseteq N$ are maximal tails, then $M = N$,*
3. *$J_{\Omega(M)}$ is a maximal gauge-invariant ideal in $C^*(E)$ for any maximal tail M,*
4. *The map $M \mapsto J_{\Omega(M)}$ is a bijective map from the set of maximal tails of E onto the set of all maximal gauge-invariant ideals of $C^*(E)$,*

Proof. Since E is row-finite it has no breaking vertices and $\Omega(M)^{\emptyset}_{\infty}$ is empty for any maximal tail M. Hence it follows from Theorem 2. □

We will end this section by constructing a class of graph C^*-algebras, all of which have a non-discrete T_1 primitive ideal space.

Example 2. Let B be a simple AF-algebra and let F be a Bratteli diagram of B as in [3], such that the vertex set F^0 is partitioned into vertex sets $F^0_n = \{w^1_n, \dots, w^{k_n}_n\}$ and every edge with a source in F^0_n has range in F^0_{n+1}. Let $G_1, G_2, \dots$ be a sequence of graphs all of which have no non-trivial hereditary and saturated sets. Construct a graph E as follows:

$$E^0 = F^0 \cup \bigcup_{n=1}^{\infty} G^0_n,$$

$$E^1 = F^1 \cup \bigcup_{n=1}^{\infty} G^1_n \cup \bigcup_{n=1}^{\infty} \{e^1_n, \dots e^{k_n}_n\}$$

where the range and source maps do not change on $F^1 \cup \bigcup_{n=1}^{\infty} G^1_n$ and where $s(e^j_n) = w^j_n$ and $r(e^j_n) \in G^0_n$.

Using that F and each G_n have no non-trivial hereditary and saturated sets we get that the maximal tails of E are

$$M_n = \bigcup_{k=1}^{n} F^0_k \cup G^0_n,$$

$$M_\infty = \bigcup_{k=1}^{\infty} F^0_k = F^0.$$

Hence no maximal tail is contained in another and thus the primitive ideal space of $C^*(E)$ is T_1. For any of these maximal tails M, each vertex in M emits only finitely many edges to $\Omega(M)$ and thus $\Omega(M)^{\text{fin}}_{\infty}$ is empty. The quotients $C^*(E)/J_{\Omega(M_n)}$ are Morita equivalent $C^*(G_n)$ and $C^*(E)/J_{\Omega(M_\infty)} = C^*(F)$ which is Morita equivalent to B.

If, in addition, each G_n has condition (K) then one can verify that Prim$C^*(E)$ is homeomorphic to $\tilde{\mathbb{N}} = \mathbb{N} \cup \{\infty\}$, the one-point compactification of $\mathbb{N}$. Such a homeomorphism may be given by

$$\tilde{\mathbb{N}} \ni n \mapsto J_{\Omega(M_n)} \in \mathrm{Prim} C^*(E).$$

7.4 Clopen Maximal Gauge-Invariant Ideals

Whenever a subset of a topological space is both closed and open, then we say that the set is *clopen*. In this section we give a description of which one-point sets in the primitive ideal space of a T_1 graph are clopen. In fact, we describe which maximal gauge-invariant ideals in the primitive ideal space correspond to clopen sets. We use this description to show that every purely infinite graph C^*-algebra with a T_1 primitive ideal space is a c_0-direct sum of Kirchberg algebras. Moreover, we prove that graph C^*-algebras with a T_1 primitive ideal space are canonically $C(\tilde{\mathbb{N}})$-algebras, which are classified up to $E(\tilde{\mathbb{N}})$-equivalence by their $\tilde{\mathbb{N}}$-filtered K-theory.

In order to describe the clopen maximal gauge-invariant ideals, we need a notion of when a maximal tail distinguishes itself from all other maximal ideals in a certain way.

Definition 3. Let E be a T_1 graph and let M be a maximal tail in E. We say that M is *isolated* if either

1. M contains a vertex which is not contained in any other maximal tail, or
2. $\Omega(M)^{\emptyset}_{\infty}$ is non-empty and

$$|s^{-1}(\Omega(M)^{\emptyset}_{\infty}) \cap r^{-1}\left(\bigcup_{N \in \mathscr{M}_M(E)} N\right)| < \infty.$$

where $\mathscr{M}_M(E)$ denotes the set of all maximal tails N such that $M \subseteq N$.

This definition may look strange but it turns out that a maximal tail corresponds to a clopen maximal gauge-invariant ideal if and only if it is isolated, see Theorem 3.

Remark 2. For a row-finite T_1 graph E the above definition simplifies, since $\Omega(M)^{\emptyset}_{\infty}$ is empty for any maximal tail M. Hence, in this case, a maximal tail is isolated if and only if it contains a vertex which is not contained in any other maximal tail.

Example 3. Consider the two graphs

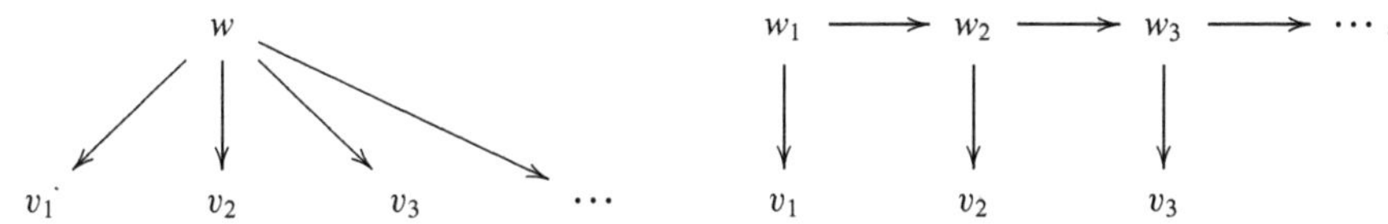

The latter graph is the desingularisation of the former but without changing sinks to tails. The maximal tails of the former graph are given by $N_n = \{w, v_n\}$ and $N_\infty = \{w\}$. The maximal tails of the latter graph are

$$\begin{aligned} M_n &= \{w_1, \ldots, w_n, v_n\}, \\ M_\infty &= \{w_1, w_2, \ldots\}. \end{aligned}$$

Hence both graphs are easily seen to be T_1 graphs. All the maximal tails N_n and M_n for $n \in \mathbb{N}$ are easily seen to be isolated, and by Remark 2, M_∞ is not isolated. Since $\Omega(N_\infty) = \{w\}$ and

$$|s^{-1}(w) \cap r^{-1}\left(\bigcup_{N \in \mathscr{M}_{N_\infty}(E)} N\right)| = \infty$$

we note that N_∞ is not isolated. In fact, by Corollary 4 below, N_∞ would be isolated if and only if M_∞ was isolated.

The latter graph is an example of a graph in Example 2, with $B = \mathbb{C}$ and each G_n consisting of one vertex and no edges. Since the graph has condition (K), the primitive ideal space is homeomorphic to $\tilde{\mathbb{N}} = \mathbb{N} \cup \{\infty\}$, the one-point compactification of $\mathbb{N}$, by the map

$$\tilde{\mathbb{N}} \ni n \mapsto J_{\Omega(M_n)}.$$

It turns out that many maximal tails are isolated, as can be seen in the following lemma.

Lemma 2. *Let E be a T_1 graph and let M be a maximal tail which contains a sink or a loop. Then M is isolated.*

Proof. Let $v \in M$ be the sink or the base of a loop in M, and note that $\Omega(v)^{\emptyset}_{\infty} = \emptyset$. If N is a maximal tail such that $v \in N$ then $E^0 \backslash \Omega(v) \subseteq N$ and since $\Omega(v)^{\emptyset}_{\infty}$ is empty, $N = E^0 \backslash \Omega(v)$ by Theorem 2. Hence v is not contained in any other maximal tail than M and thus M is isolated. □

The following is the main theorem of this section, mainly due to all the corollaries following it.

Theorem 3. *Let E be a countable directed graph for which the primitive ideal space of $C^*(E)$ is T_1, and let M be a maximal tail in E. Then*

$$\{\mathfrak{p} \in \mathrm{Prim}C^*(E) : J_{\Omega(M),\Omega(M)^{\mathrm{fin}}_\infty} \subseteq \mathfrak{p}\} \subseteq \mathrm{Prim}C^*(E)$$

is a clopen set if and only if M is isolated.

In particular, if $M \in \mathscr{M}_\gamma(E)$, then the one-point set

$$\{J_{\Omega(M),\Omega(M)^{\mathrm{fin}}_\infty}\} \subseteq \mathrm{Prim}C^*(E)$$

is clopen if and only if M is isolated, and if $M \in \mathscr{M}_\tau(E)$ then

$$\{R_{M,t} : t \in \mathbb{T}\} \subseteq \mathrm{Prim}C^*(E)$$

is a clopen set homeomorphic to the circle S^1.

Proof. To ease notation define

$$U_M := \{\mathfrak{p} \in \mathrm{Prim}C^*(E) : J_{\Omega(M),\Omega(M)^{\mathrm{fin}}_\infty} \subseteq \mathfrak{p}\}.$$

By definition U_M is closed. By [6, Lemma 2.6] it follows that if J is a gauge-invariant ideal, $M \in \mathscr{M}_\tau(E)$ and $t \in \mathbb{T}$, then $J \subseteq R_{M,t}$ if and only if $J \subseteq J_{\Omega(M),\Omega(M)^{\mathrm{fin}}_\infty}$. We will use this fact several times throughout the proof, without mentioning it.

Suppose U_M is clopen. If $M \in \mathscr{M}_\tau(E)$ then M contains a loop and is thus isolated by Lemma 2. Hence we may suppose that $M \in \mathscr{M}_\gamma(E)$ for which it follows that $U_M = \{J_{\Omega(M),\Omega(M)^{\mathrm{fin}}_\infty}\}$. Since U_M is open there is a unique ideal J such that

$$\{J_{\Omega(M),\Omega(M)^{\mathrm{fin}}_\infty}\} = \{\mathfrak{p} \in \mathrm{Prim}C^*(E) : J \not\subseteq \mathfrak{p}\}.$$

Suppose J is not gauge-invariant. Then we can find a $z \in \mathbb{T}$ such that $\gamma_z(J) \neq J$. Note that $\gamma_z(J) \not\subseteq \gamma_z(J_{\Omega(M),\Omega(M)^{\mathrm{fin}}_\infty}) = J_{\Omega(M),\Omega(M)^{\mathrm{fin}}_\infty}$. Further, for an arbitrary $\mathfrak{p} \in \mathrm{Prim}C^*(E)\backslash\{J_{\Omega(M),\Omega(M)^{\mathrm{fin}}_\infty}\}$, we have $\gamma_z(J) \subseteq \gamma_z(\mathfrak{p})$, since $J \subseteq \mathfrak{p}$. Since γ_z fixes $J_{\Omega(M),\Omega(M)^{\mathrm{fin}}_\infty}$ it induces a bijection from $\mathrm{Prim}C^*(E)\backslash\{J_{\Omega(M),\Omega(M)^{\mathrm{fin}}_\infty}\}$ to itself and thus $\gamma_z(J) \subseteq \mathfrak{p}$ for any primitive ideal $\mathfrak{p} \neq J_{\Omega(M),\Omega(M)^{\mathrm{fin}}_\infty}$. However, this contradicts the uniqueness of J, and thus J must be gauge-invariant.

Since J is gauge-invariant, $J = J_{H,B}$ for a hereditary and saturated set H and $B \subseteq H^{\mathrm{fin}}_\infty$. If $H \not\subseteq \Omega(M)$ then any vertex $v \in H$ such that $v \in M$ is not contained in any other maximal tail, since $J_{H,B} \subseteq J_{\Omega(N),\Omega(N)^{\mathrm{fin}}_\infty}$ for any maximal tail $N \neq M$. Hence we may restrict to the case where $H \subseteq \Omega(M)$. Since $J_{H,B} \not\subseteq J_{\Omega(M),\Omega(M)^{\mathrm{fin}}_\infty}$, $B \not\subseteq \Omega(M) \cup \Omega(M)^{\mathrm{fin}}_\infty$. It is easily observed that $B\backslash\Omega(M)^{\emptyset}_\infty \subseteq \Omega(M) \cup \Omega(M)^{\mathrm{fin}}_\infty$ and hence it follows that $\Omega(M)^{\emptyset}_\infty = \{w\}$ for some vertex w and that $w \in B$. Recall that $\mathscr{M}_M(E) = \{N \in \mathscr{M}(E) : M \subseteq N\}$. Since $w \in H^{\mathrm{fin}}_\infty$ we have

$$|s^{-1}(w) \cap r^{-1}(E^0\backslash H)| < \infty$$

and since $\bigcup_{N \in \mathscr{M}_M(E)} N \subseteq E^0\backslash H$ it follows that

$$|s^{-1}(w) \cap r^{-1}(\bigcup_{N \in \mathscr{M}_M(E)} N)| < \infty.$$

Thus M is isolated.

Now suppose that M is an isolated maximal tail. If M contains a vertex v which is not contained in any other maximal tail, then $J_{\Sigma H(v)} \not\subseteq J_{\Omega(M),\Omega(M)^{\text{fin}}_\infty}$ and $J_{\Sigma H(v)} \subseteq J_{\Omega(N),\Omega(N)^{\text{fin}}_\infty}$ for any maximal tail $N \neq M$. Hence

$$U_M = \{\mathfrak{p} \in \text{Prim}C^*(E) : J_{\Sigma H(v)} \not\subseteq \mathfrak{p}\}$$

and thus U_M is clopen. Now suppose that every vertex of M is contained in some other maximal tail. Let $H = \bigcap_{N \in \mathscr{M}_M(E)} \Omega(N)$ which is hereditary and saturated. Since M is isolated, $\Omega(M)^{\emptyset}_\infty = \{w\}$ for some vertex w. Moreover, since M is isolated and $E^0 \backslash H = \bigcup_{N \in \mathscr{M}_M(E)} N$ it follows that $w \in H^{\text{fin}}_\infty$. Hence $J_{H,\{w\}} \not\subseteq J_{\Omega(M),\Omega(M)^{\text{fin}}_\infty}$ and $J_{H,\{w\}} \subseteq J_{\Omega(N),\Omega(N)^{\text{fin}}_\infty}$ for any $N \in \mathscr{M}_M(E)$ by Theorem 2. Now, as above, $J_{\Sigma H(w)} \subseteq J_{\Omega(N),\Omega(N)^{\text{fin}}_\infty}$ for any $N \notin \mathscr{M}_M(E)$ and $J_{\Sigma H(w)} \not\subseteq J_{\Omega(N),\Omega(N)^{\text{fin}}_\infty}$ for $N \in \mathscr{M}_M(E)$. Hence

$$U_M = \{\mathfrak{p} : J_{H,\{w\}} \not\subseteq \mathfrak{p}\} \cap \{\mathfrak{p} : J_{\Sigma H(w)} \not\subseteq \mathfrak{p}\}$$

is the intersection of two open sets, and is thus clopen.

For the ‘in particular’ part note that if $M \in \mathscr{M}_\gamma(E)$ then $U_M = \{J_{\Omega(M),\Omega(M)^{\text{fin}}_\infty}\}$. If $M \in \mathscr{M}_\tau(E)$ then M contains a loop and is thus isolated by Lemma 2. Hence

$$U_M = \{R_{M,t} : t \in \mathbb{T}\}$$

is clopen. By Theorem 1 it follows that this set is homeomorphic to the circle S^1.

□

Corollary 2. *Let E be a T_1 graph and $\mathfrak{p} \in \text{Prim}C^*(E)$ be a primitive ideal. Then $\{\mathfrak{p}\}$ is clopen if and only if $\mathfrak{p} = J_{\Omega(M),\Omega(M)^{\text{fin}}_\infty}$ for an isolated maximal tail $M \in \mathscr{M}_\gamma(E)$.*

Corollary 3. *Let E be a T_1 graph and suppose that every maximal tail in E is isolated. Then*

$$\text{Prim}C^*(E) \cong \bigsqcup_{M \in \mathscr{M}_\gamma} \star \sqcup \bigsqcup_{M \in \mathscr{M}_\tau} S^1$$

is a disjoint union, where $\star$ is a one-point topological space and S^1 is the circle.

In particular, if E in addition has condition (K) then Prim$C^*(E)$ *is discrete.*

If two graphs E and F have Morita equivalent C^*-algebras, then the corresponding ideal lattices are canonically isomorphic. Hence, if E and F have no breaking vertices, there is an induced one-to-one correspondence between the maximal tails in E and F. The following corollary is immediate from Theorem 3.

Corollary 4. *Let E and F be T_1 graphs such that $C^*(E)$ and $C^*(F)$ are Morita equivalent. Then a maximal tail in E is isolated if and only if the corresponding maximal tail in F is isolated.*

Our main application of the above theorem is the following corollary.

Corollary 5. *Any purely infinite graph C^*-algebra with a T_1 (in particular Hausdorff) primitive ideal space is isomorphic to a c_0-direct sum of Kirchberg algebras.*

Proof. Let E be a T_1 graph such that $C^*(E)$ is purely infinite. By [5, Theorem 2.3] E has condition (K) and every maximal tail in E contains a loop, and is thus isolated by Lemma 2. By Corollary 3 the primitive ideal space $\mathrm{Prim} C^*(E)$ is discrete. Hence $C^*(E)$ is the c_0-direct sum of all its simple ideals, which are Kirchberg algebras. □

We also have another application of the above theorem.

Corollary 6. *Let A be a graph C^*-algebra for which the primitive ideal space is T_1. Let J be the ideal generated by all the direct summands in A corresponding to $A/J_{\Omega(M),\Omega(M)^{\mathrm{fin}}_\infty}$ where M is an isolated maximal tail. Then A/J is an AF-algebra.*

Proof. Note that the ideal is well-defined by Theorem 3, since $J_{\Omega(M),\Omega(M)^{\mathrm{fin}}_\infty}$ is a direct summand in A for every isolated maximal tail M. By Corollary 4 it suffices to prove this up to Morita equivalence. Hence we may assume that there is a row-finite graph E such that $C^*(E) = A$. Let V denote the set of all vertices which are contained in exactly one maximal tail. For any isolated maximal tail M, the direct summand in A which corresponds to $A/J_{\Omega(M)}$ is $J_{\Sigma H(v)}$ where v is any vertex in M which is not contained in any other maximal tail. Hence $J = J_{\Sigma H(V)}$ since this is the smallest ideal containing all $J_{\Sigma H(v)}$ for $v \in V$. By Lemma 2 any vertex which is the base of a loop, is in V. Hence the graph $E \setminus \Sigma H(V)$ contains no loops and thus $A/J = C^*(E \setminus \Sigma H(V))$ is an AF-algebra. □

Remark 3. By an analogous argument as given in the proof of Corollary 6, we get the following result. Let A be a real rank zero graph C^*-algebra for which the primitive ideal space is T_1. Then A contains a (unique) purely infinite ideal J such that A/J is an AF-algebra.

In fact, we could define V in the proof of Corollary 6 to be the set of all vertices which are the base of some loop. Then $J = J_{\Sigma H(V)}$ would be the direct sum of all simple purely infinite ideals in A, and A/J would again be an AF-algebra. Note that this ideal, in general, is not the same as the one defined in Corollary 6.

Remark 4. Let $\tilde{\mathbb{N}} = \mathbb{N} \cup \{\infty\}$ be the one-point compactification of $\mathbb{N}$. We may give any graph C^*-algebra A with a T_1 primitive ideal space a canonical structure of a $C(\tilde{\mathbb{N}})$-algebra. In fact, list all of the direct summands in A corresponding to $A/J_{\Omega(M),\Omega(M)^{\mathrm{fin}}_\infty}$ for M an isolated maximal tail, as $J_1, J_2, \ldots$. By letting

$$A(\{n\}) = J_n, \text{ and } A(\{n, n+1, \ldots, \infty\}) = A/\bigoplus_{k=1}^{n-1} J_k,$$

then A gets the structure of a C^*-algebra over $\tilde{\mathbb{N}}$ which is the same as a (not necessarily continuous) $C(\tilde{\mathbb{N}})$-algebra (see e.g. [8]). This structure is unique up to an automorphism functor σ_* on $\mathfrak{C}^*\mathfrak{alg}(\tilde{\mathbb{N}})$, the category of $C(\tilde{\mathbb{N}})$-algebras, where $\sigma : \tilde{\mathbb{N}} \to \tilde{\mathbb{N}}$ is a homeomorphism. Moreover, by Corollary 6, the fibre A_∞ is an AF-algebra.

Using the structure of a $C(\tilde{\mathbb{N}})$-algebra we may construct an $\tilde{\mathbb{N}}$-filtered K-theory functor as in [2]. In fact, let $C(\tilde{\mathbb{N}}, \mathbb{Z})$ be the ring of locally constant maps $\tilde{\mathbb{N}} \to \mathbb{Z}$. If A is a $C(\tilde{\mathbb{N}})$-algebra then the K-theory $K_*(A)$ has the natural structure as a $\mathbb{Z}/2$-graded $C(\tilde{\mathbb{N}}, \mathbb{Z})$-module. Similarly, let Λ be the ring of Böckstein operations, and let $C(\tilde{\mathbb{N}}, \Lambda)$ be the ring of locally constant maps $\tilde{\mathbb{N}} \to \Lambda$. If A is a $C(\tilde{\mathbb{N}})$-algebra then the total K-theory $\underline{K}(A)$ has the natural structure as a $C(\tilde{\mathbb{N}}, \Lambda)$-module. It is this latter invariant, that Dadarlat and Meyer proved a UMCT for. We refer the reader to [2] for a more detailed definition.

We end this paper by showing that for T_1 graph C^*-algebras given $C(\tilde{\mathbb{N}})$-algebra structures as in Remark 4, an isomorphism of $\tilde{\mathbb{N}}$-filtered K-theory (without coefficients) lifts to an $E(\tilde{\mathbb{N}})$-equivalence. Note that this is not true in general by [2, Example 6.14].

Proposition 1. *Let A and B be graph C^*-algebras with T_1 primitive ideal spaces, and suppose that these have the structure of $C(\tilde{\mathbb{N}})$-algebras as in Remark 4. Then $K_*(A) \cong K_*(B)$ as $\mathbb{Z}/2$-graded $C(\tilde{\mathbb{N}}, \mathbb{Z})$-modules if and only if A and B are $E(\tilde{\mathbb{N}})$-equivalent.*

In addition, if A and B are continuous $C(\tilde{\mathbb{N}})$-algebras, then $K_*(A) \cong K_*(B)$ as $\mathbb{Z}/2$-graded $C(\tilde{\mathbb{N}}, \mathbb{Z})$-modules if and only if A and B are $KK(\tilde{\mathbb{N}})$-equivalent.

Proof. Clearly an $E(\tilde{\mathbb{N}})$-equivalence induces an isomorphism of $\tilde{\mathbb{N}}$-filtered K-theory. Suppose that $\phi = (\phi_0, \phi_1)\colon K_*(A) \to K_*(B)$ is an isomorphism of $\mathbb{Z}/2$-graded $C(\tilde{\mathbb{N}}, \mathbb{Z})$-modules. By the UMCT of Dadarlat and Meyer [2, Theorem 6.11], it suffices to lift ϕ to an isomorphism of $\tilde{\mathbb{N}}$-filtered total K-theory. Since the K_1-groups are free, $K_0(D; \mathbb{Z}/n) = K_0(D) \otimes \mathbb{Z}/n$ for $D \in \{A, B\}$. Hence define

$$\phi_0^n = \phi_0 \otimes id_{\mathbb{Z}/n}\colon K_0(A; \mathbb{Z}/n) \to K_0(B; \mathbb{Z}/n)$$

which are isomorphisms for each $n \in \mathbb{N}$. Since the fibres A_∞ and B_∞ are AF-algebras by Corollary 6, $K_1(A_\infty; \mathbb{Z}/n) = K_1(B_\infty; \mathbb{Z}/n) = 0$ for each $n \in \mathbb{N}$. Since the map $K_0(D; \mathbb{Z}/n) \to K_0(D_\infty; \mathbb{Z}/n)$ is clearly surjective, and $K_1(D_\infty; \mathbb{Z}/n) = 0$, it follows by six-term exactness that

$$K_1(D; \mathbb{Z}/n) \cong K_1(D(\mathbb{N}); \mathbb{Z}/n) \cong \bigoplus_{k \in \mathbb{N}} K_1(D_k; \mathbb{Z}/n)$$

for $D \in \{A, B\}$ and $n \in \mathbb{N}$. Since $\phi_*\colon K_*(A) \to K_*(B)$ is an isomorphism of $\mathbb{Z}/2$-graded $C(\tilde{\mathbb{N}}, \mathbb{Z})$-modules, ϕ_* restricts to an isomorphism $\phi_{*,k}\colon K_*(A_k) \to K_*(B_k)$ for each $k \in \mathbb{N}$. By the UCT of Rosenberg and Schochet [9] we may lift these isomorphisms to invertible KK-elements, and in particular also to isomorphisms of the total K-theory $\phi_{*,k}\colon \underline{K}(A_k) \to \underline{K}(B_k)$. Here we used that the fibres are graph C^*-algebras and thus satisfy the UCT (see e.g. [10, Remark A.11.13]). Now define the group isomorphisms $\underline{\phi}_0\colon \underline{K}_0(A) \to \underline{K}_0(B)$ to be the isomorphism induced by ϕ_0 and each ϕ_0^n, and $\underline{\phi}_1\colon \underline{K}_1(A) \to \underline{K}_1(B)$ to be the composition

$$\underline{K}_1(A) \cong \bigoplus_{k\in\mathbb{N}} \underline{K}_1(A_k) \xrightarrow{\bigoplus_k \underline{\phi}_{1,k}} \bigoplus_{k\in\mathbb{N}} \underline{K}_1(B_k) \cong \underline{K}_1(B),$$

where $\underline{K}_i(D) = K_i(D) \oplus \bigoplus_{n\in\mathbb{N}} K_i(D;\mathbb{Z}/n)$. It is straight forward to check that $\underline{\phi} = (\underline{\phi}_0, \underline{\phi}_1)\colon \underline{K}(A) \to \underline{K}(B)$ is an isomorphism of $C(\tilde{\mathbb{N}}, \Lambda)$-modules.

If A and B are continuous $C(\tilde{\mathbb{N}})$-algebras then $E(\tilde{\mathbb{N}})$- and $KK(\tilde{\mathbb{N}})$-theory agree by [2, Theorem 5.4]. □

Acknowledgements The author would like to thank the referee for useful suggestions. The author would also like to thank his PhD-advisers Søren Eilers and Ryszard Nest for valuable discussions. This research was supported by the Danish National Research Foundation through the Centre for Symmetry and Deformation (DNFR92).

References

1. Bates, T., Hong, J.H., Raeburn, I., Szymański, W.: The ideal structure of the C^*-algebras of infinite graphs. Illinois J. Math. **46**(4), 1159–1176 (2002). URL http://projecteuclid.org/getRecord?id=euclid.ijm/1258138472
2. Dadarlat, M., Meyer, R.: E-theory for C*-algebras over topological spaces. J. Funct. Anal. **263**(1), 216–247 (2012). DOI 10.1016/j.jfa.2012.03.022. URL http://dx.doi.org/10.1016/j.jfa.2012.03.022
3. Drinen, D.: Viewing AF-algebras as graph algebras. Proc. Amer. Math. Soc. **128**(7), 1991–2000 (2000). DOI 10.1090/S0002-9939-99-05286-7. URL http://dx.doi.org/10.1090/S0002-9939-99-05286-7
4. Drinen, D., Tomforde, M.: The C^*-algebras of arbitrary graphs. Rocky Mountain J. Math. **35**(1), 105–135 (2005). DOI 10.1216/rmjm/1181069770. URL http://dx.doi.org/10.1216/rmjm/1181069770
5. Hong, J.H., Szymański, W.: Purely infinite Cuntz-Krieger algebras of directed graphs. Bull. London Math. Soc. **35**(5), 689–696 (2003). DOI 10.1112/S0024609303002364. URL http://dx.doi.org/10.1112/S0024609303002364
6. Hong, J.H., Szymański, W.: The primitive ideal space of the C^*-algebras of infinite graphs. J. Math. Soc. Japan **56**(1), 45–64 (2004). DOI 10.2969/jmsj/1191418695. URL http://dx.doi.org/10.2969/jmsj/1191418695
7. Kirchberg, E.: Das nicht-kommutative Michael-Auswahlprinzip und die Klassifikation nicht-einfacher Algebren. In: C^*-algebras (Münster, 1999), pp. 92–141. Springer, Berlin (2000)
8. Meyer, R., Nest, R.: C^*-algebras over topological spaces: the bootstrap class. Münster J. Math. **2**, 215–252 (2009)
9. Rosenberg, J., Schochet, C.: The Künneth theorem and the universal coefficient theorem for Kasparov's generalized K-functor. Duke Math. J. **55**(2), 431–474 (1987). DOI 10.1215/S0012-7094-87-05524-4. URL http://dx.doi.org/10.1215/S0012-7094-87-05524-4
10. Tomforde, M.: Extensions of graph C^*-algebras. Ph.D. thesis, Dartmouth College (2002)

Chapter 8
The Law of Large Numbers for the Free Multiplicative Convolution

Uffe Haagerup and Sören Möller

Abstract In classical probability the law of large numbers for the multiplicative convolution follows directly from the law for the additive convolution. In free probability this is not the case. The free additive law was proved by D. Voiculescu in 1986 for probability measures with bounded support and extended to all probability measures with first moment by J.M. Lindsay and V. Pata in 1997, while the free multiplicative law was proved only recently by G. Tucci in 2010. In this paper we extend Tucci's result to measures with unbounded support while at the same time giving a more elementary proof for the case of bounded support. In contrast to the classical multiplicative convolution case, the limit measure for the free multiplicative law of large numbers is not a Dirac measure, unless the original measure is a Dirac measure. We also show that the mean value of $\ln x$ is additive with respect to the free multiplicative convolution while the variance of $\ln x$ is not in general additive. Furthermore we study the two parameter family $(\mu_{\alpha,\beta})_{\alpha,\beta\geq 0}$ of measures on $(0,\infty)$ for which the S-transform is given by $S_{\mu_{\alpha,\beta}}(z) = (-z)^{\beta}(1+z)^{-\alpha}, 0 < z < 1$.

Keywords Free probability • Free multiplicative law • Law of large numbers • Free convolution

Mathematics Subject Classification (2010): 46L54, 60F05.

U. Haagerup (✉)
Department of Mathematical Sciences, University of Copenhagen, Universitetsparken 5, DK-2100 Copenhagen Ø, Denmark
e-mail: haagerup@math.ku.dk

S. Möller
Department of Mathematics and Computer Science, University of Southern Denmark, Campusvej 55, 5230 Odense M, Denmark
e-mail: moeller@imada.sdu.dk

T.M. Carlsen et al. (eds.), *Operator Algebra and Dynamics*, Springer Proceedings in Mathematics & Statistics 58, DOI 10.1007/978-3-642-39459-1_8,

8.1 Introduction

In classical probability the weak law of large numbers is well known (see for instance [14, Corollary 5.4.11]), both for additive and multiplicative convolution of Borel measures on $\mathbb{R}$, respectively, $[0, \infty)$.

Going from classical probability to free probability, one could ask if similar results exist for the additive and multiplicative free convolutions $\boxplus$ and $\boxtimes$ as defined by D. Voiculescu in [16] and [17] and extended to unbounded probability measures by H. Bercovici and D. Voiculescu in [4]. The law of large numbers for the free additive convolution of measures with bounded support is an immediate consequence of D. Voiculescu's work in [16] and J. M. Lindsay and V. Pata proved it for measures with first moment in [11, Corollary 5.2].

Theorem 1 ([11, Corollary 5.2]). *Let μ be a probability measure on $\mathbb{R}$ with existing mean value α, and let $\psi_n: \mathbb{R} \to \mathbb{R}$ be the map $\psi_n(x) = \frac{1}{n}x$. Then*

$$\dot{\psi}_n(\underbrace{\mu \boxplus \cdots \boxplus \mu}_{n \text{ times}}) \to \delta_\alpha$$

where convergence is weak and δ_x denotes the Dirac measure at $x \in \mathbb{R}$.

Here $\dot{\phi}(\mu)$ denotes the image measure of μ under ϕ for a Borel measurable function $\phi: \mathbb{R} \to \mathbb{R}$, respectively, $[0, \infty) \to [0, \infty)$.

In classical probability the multiplicative law follows directly from the additive law. This is not the case in free probability, here a multiplicative law requires a separate proof. This has been proved by G.H. Tucci in [15, Theorem 3.2] for measures with bounded support using results on operator algebras from [6] and [8]. In this paper we give an elementary proof of Tucci's theorem which also shows that the theorem holds for measures with unbounded support.

Theorem 2. *Let μ be a probability measure on $[0, \infty)$ and let $\phi_n: [0, \infty) \to [0, \infty)$ be the map $\phi_n(x) = x^{\frac{1}{n}}$. Set $\delta = \mu(\{0\})$. If we denote*

$$\nu_n = \dot{\phi}_n(\mu_n) = \dot{\phi}_n(\underbrace{\mu \boxtimes \cdots \boxtimes \mu}_{n \text{ times}})$$

then ν_n converges weakly to a probability measure ν on $[0, \infty)$. If μ is a Dirac measure on $[0, \infty)$ then $\nu = \mu$. Otherwise ν is the unique measure on $[0, \infty)$ characterised by $\nu\left(\left[0, \frac{1}{S_\mu(t-1)}\right]\right) = t$ for all $t \in (\delta, 1)$ and $\nu(\{0\}) = \delta$. The support of the measure ν is the closure of the interval

$$(a, b) = \left(\left(\int_0^\infty x^{-1} \mathrm{d}\mu(x)\right)^{-1}, \int_0^\infty x \mathrm{d}\mu(x)\right),$$

where $0 \leq a < b \leq \infty$.

Note that unlike the additive case, the multiplicative limit distribution is only a Dirac measure if μ is a Dirac measure. Furthermore S_μ and hence (by [17, Theorem 2.6]) μ can be reconstructed from the limit measure.

We start by recalling some definitions and proving some preliminary results in Sect. 8.2, which then in Sect. 8.3 are used to prove Theorem 2. In Sect. 8.4 we prove some further formulas in connection with the limit law, which we in Sect. 8.5 apply to the two parameter family $(\mu_{\alpha,\beta})_{\alpha,\beta\geq 0}$ of measures on $(0,\infty)$ for which the S-transform is given by $S_{\mu_{\alpha,\beta}}(z) = \frac{(-z)^\beta}{(1+z)^\alpha}$, $0 < z < 1$.

8.2 Preliminaries

We start with recalling some results we will use and proving some technical tools necessary for the proof of Theorem 2. At first we recall the definition and some properties of Voiculescu's S-transform for measures on $[0,\infty)$ with unbounded support as defined by H. Bercovici and D. Voiculescu in [4].

Definition 1 ([4, Sect. 6]). Let μ be a probability measure on $[0,\infty)$ and assume that $\delta = \mu(\{0\}) < 1$. We define $\psi_\mu(u) = \int_0^\infty \frac{tu}{1-tu} d\mu(t)$ and denote its inverse in a neighbourhood of $(\delta - 1, 0)$ by χ_μ. Now we define the *S-transform* of μ by $S_\mu(z) = \frac{z+1}{z}\chi_\mu(z)$ for $z \in (\delta - 1, 0)$.

Lemma 1 ([4, Proposition 6.8]). *Let μ be a probability measure on $[0,\infty)$ with $\delta = \mu(\{0\}) < 1$ then S_μ is decreasing on $(\delta - 1, 0)$ and positive. Moreover, if $\delta > 0$ we have $S_\mu(z) \to \infty$ if $z \to \delta - 1$.*

Lemma 2. *Let μ be a probability measure on $[0,\infty)$ with $\delta = \mu(\{0\}) < 1$. Assume that μ is not a Dirac measure, then $S'_\mu(z) < 0$ for $z \in (\delta - 1, 0)$. In particular S_μ is strictly decreasing on $(\delta - 1, 0)$.*

Proof. For $u \in (-\infty, 0)$,

$$\psi'_\mu(u) = \int_0^\infty \frac{t}{(1-ut)^2} d\mu(t) > 0. \tag{8.1}$$

Moreover $\lim_{u\to 0-} \psi_\mu(u) = 0$ and $\lim_{u\to -\infty} \psi_\mu(u) = \delta - 1$. Hence ψ_μ is a strictly increasing homeomorphism of $(-\infty, 0)$ onto $(\delta - 1, 0)$. For $u \in (-\infty, 0)$, we have

$$S_\mu(\psi_\mu(u)) = \frac{\psi_\mu(u) + 1}{\psi_\mu(u)} \cdot u.$$

Hence

$$\frac{d}{du}\left(\ln S_\mu(\psi_\mu(u))\right) = -\frac{\psi'_\mu(u)}{\psi_\mu(u)(\psi_\mu(u)+1)} + \frac{1}{u} = \frac{\psi_\mu(u)(\psi_\mu(u)+1) - u\psi'_\mu(u)}{u\psi_\mu(u)(\psi_\mu(u)+1)} \tag{8.2}$$

where the denominator is positive and the nominator is equal to

$$\begin{aligned}&\left(\int_0^\infty \frac{ut}{1-ut}\mathrm{d}\mu(t)\right)\cdot\left(\int_0^\infty \frac{1}{1-ut}\mathrm{d}\mu(t)\right)-\int_0^\infty \frac{ut}{(1-ut)^2}\mathrm{d}\mu(t)\\ &=\frac{u}{2}\int_0^\infty\int_0^\infty \frac{s+t}{(1-us)(1-ut)}\mathrm{d}\mu(s)\mathrm{d}\mu(t)\\ &\quad-\frac{u}{2}\int_0^\infty\int_0^\infty\left(\frac{s}{(1-us)^2}+\frac{t}{(1-ut)^2}\right)\mathrm{d}\mu(s)\mathrm{d}\mu(t)\\ &=-\frac{u^2}{2}\int_0^\infty\int_0^\infty \frac{(s-t)^2}{(1-us)^2(1-ut)^2}\mathrm{d}\mu(s)\mathrm{d}\mu(t)\end{aligned}$$

where we have used that

$$(s+t)(1-us)(1-ut)-s(1-ut)^2-t(1-us)^2=-u(s-t)^2.$$

Since μ is not a Dirac measure,

$$(\mu\times\mu)\left(\{(s,t)\in[0,\infty)^2 : s\neq t\}\right)>0$$

and thus

$$\int_0^\infty\int_0^\infty \frac{(s-t)^2}{(1-us)^2(1-ut)^2}\mathrm{d}\mu(s)\mathrm{d}\mu(t)>0$$

which shows that the right hand side of (8.2) is strictly positive. Hence

$$\frac{\mathrm{d}}{\mathrm{d}z}\left(\ln S_\mu(z)\right)<0$$

for $z\in(\delta-1,0)$, which proves the lemma. □

Remark 1. Furthermore, by [4, Proposition 6.1] and [4, Proposition 6.3] ψ_μ and χ_μ are analytic in a neighbourhood of $(-\infty,0)$, respectively, $(-1,0)$, hence S_μ is analytic in a neighbourhood of $(\delta-1,0)$.

Lemma 3 ([4, Corollary 6.6]). *Let μ and ν be probability measures on $[0,\infty)$, none of them being δ_0, then we have $S_{\mu\boxtimes\nu}=S_\mu S_\nu$.*

Next we have to determine the image of S_μ. Here we closely follow the argument given for measures with compact support by F. Larsen and the first author in [6, Theorem 4.4].

Lemma 4. *Let μ be a probability measure on $[0,\infty)$ not being a Dirac measure, then $S_\mu((\delta-1,0))=(b^{-1},a^{-1})$, where a, b and δ are defined as in Theorem 2.*

Proof. First assume $\delta = 0$. Observe that for $u \to \infty$ we have

$$\int_0^\infty \frac{u}{1+ut} \mathrm{d}\mu(t) \to \int_0^\infty \frac{1}{t} \mathrm{d}\mu(t) = a^{-1} \quad \text{and} \quad \int_0^\infty \frac{ut}{1+ut} \mathrm{d}\mu(t) \to 1.$$

Hence

$$\frac{-\psi_\mu(-u)}{u(\psi_\mu(-u)+1)} = \left(\int_0^\infty \frac{ut}{1+ut} \mathrm{d}\mu(t)\right) \left(\int_0^\infty \frac{u}{1+ut} \mathrm{d}\mu(t)\right)^{-1} \to a \quad \text{for } u \to \infty.$$

Similarly, for $u \to 0$ we have

$$\int_0^\infty \frac{t}{1+ut} \mathrm{d}\mu(t) \to \int_0^\infty t \mathrm{d}\mu(t) = b \quad \text{and} \quad \int_0^\infty \frac{1}{1+ut} \mathrm{d}\mu(t) \to 1.$$

Hence

$$\frac{-\psi_\mu(-u)}{u(\psi_\mu(-u)+1)} = \frac{\int_0^\infty \frac{t}{1+ut} \mathrm{d}\mu(t)}{\int_0^\infty \frac{1}{1+ut} \mathrm{d}\mu(t)} \to b \quad \text{for } u \to 0.$$

As χ_μ is the inverse of ψ_μ we have

$$S_\mu(\psi_\mu(-u)) = \frac{\psi_\mu(-u)+1}{\psi_\mu(-u)} \chi_\mu(\psi_\mu(-u)) = \frac{u(\psi_\mu(-u)+1)}{-\psi_\mu(-u)}.$$

By (8.1) and Lemma 2 ψ_μ is strictly increasing and continuous and S_μ is strictly decreasing and continuous so $S_\mu(\psi_\mu((-\infty, 0))) = S_\mu((-1, 0)) = (b^{-1}, a^{-1})$.

If now $\delta > 0$ we have by Lemma 1 that $S_\mu(z) \to \infty$ for $z \to \delta - 1$, so in this case continuity gives us $S_\mu((\delta - 1, 0)) = (b^{-1}, \infty)$, which is as desired as $a = 0$ in this case. □

8.3 Proof of the Main Result

Let μ be a probability measure on $[0, \infty)$ and let ν be as defined in Theorem 2. If μ is a Dirac measure, then $\nu_n = \mu$ for all n and hence $\nu_n \to \nu = \mu$ weakly, so the theorem holds in this case. In the following we can therefore assume that μ is not a Dirac measure. We start by assuming further that $\mu(\{0\}) = 0$, and will deal with the case $\mu(\{0\}) > 0$ in Remark 2.

Lemma 5. *For all $t \in (0, 1)$ and all $n \geq 1$ we have*

$$\int_0^\infty \left(1 + \frac{1-t}{t} S_\mu(t-1)^n x^n\right)^{-1} \mathrm{d}\nu_n(x) = t.$$

Proof. Let $t \in (0,1)$ and set $z = t-1$. By Definition 1 we have

$$\begin{aligned}
z+1 &= \psi_{\mu_n}(\chi_{\mu_n}(z))+1 \\
&= \int_0^\infty \frac{\chi_{\mu_n}(z)x}{1-\chi_{\mu_n}(z)x}\mathrm{d}\mu_n(x)+1 \\
&= \int_0^\infty \frac{1}{1-\chi_{\mu_n}(z)x}\mathrm{d}\mu_n(x) \\
&= \int_0^\infty \left(1-\frac{z}{z+1}S_{\mu_n}(z)x\right)^{-1}\mathrm{d}\mu_n(x) \\
&= \int_0^\infty \left(1-\frac{z}{z+1}S_{\mu}(z)^n x\right)^{-1}\mathrm{d}\mu_n(x).
\end{aligned}$$

In the last equality we use multiplicativity of the S-transform from Lemma 3.

Now substitute $t = z+1$ and afterwards $y^n = x$ and use the definition of ν_n to get

$$\begin{aligned}
t &= \int_0^\infty \left(1+\frac{1-t}{t}S_\mu(t-1)^n x\right)^{-1}\mathrm{d}\mu_n(x) \\
&= \int_0^\infty \left(1+\frac{1-t}{t}S_\mu(t-1)^n y^n\right)^{-1}\mathrm{d}\nu_n(y).
\end{aligned}$$
□

Now, using this lemma, we can prove the following characterisation of the weak limit of ν_n.

Lemma 6. *For all $t \in (0,1)$ we have $t = \lim_{n\to\infty} \nu_n\left(\left[0, \frac{1}{S_\mu(t-1)}\right]\right)$.*

Proof. Fix $t \in (0,1)$ and let $t' \in (0,t)$. Then

$$\begin{aligned}
t' &= \int_0^\infty \left(1+\frac{1-t'}{t'}S_\mu(t'-1)^n x^n\right)^{-1}\mathrm{d}\nu_n(x) \\
&\le \int_0^\infty \left(1+\frac{1-t}{t}S_\mu(t'-1)^n x^n\right)^{-1}\mathrm{d}\nu_n(x) \\
&\le \int_0^{\frac{1}{S_\mu(t-1)}} 1\mathrm{d}\nu_n(x) + \int_{\frac{1}{S_\mu(t-1)}}^\infty \left(1+\frac{1-t}{t}S_\mu(t'-1)^n x^n\right)^{-1}\mathrm{d}\nu_n(x) \\
&\le \int_0^{\frac{1}{S_\mu(t-1)}} 1\mathrm{d}\nu_n(x) + \int_{\frac{1}{S_\mu(t-1)}}^\infty \left(1+\frac{1-t}{t}\left(\frac{S_\mu(t'-1)}{S_\mu(t-1)}\right)^n\right)^{-1}\mathrm{d}\nu_n(x) \\
&\le \nu_n\left(\left[0,\frac{1}{S_\mu(t-1)}\right]\right) + \left(1+\frac{1-t}{t}\left(\frac{S_\mu(t'-1)}{S_\mu(t-1)}\right)^n\right)^{-1}.
\end{aligned}$$

Here the first inequality holds as $t' \leq t$ while $S_\mu(t'-1)^n x^n > 0$, the second holds as $1 + \frac{1-t}{t} S_\mu(t'-1)^n x^n \geq 0$, and the last because ν_n is a probability measure.

By Lemma 2, $S_\mu(t-1)$ is strictly decreasing, and hence $\frac{S_\mu(t'-1)}{S_\mu(t-1)} > 1$. This implies

$$\lim_{n\to\infty} \left(1 + \frac{1-t}{t}\left(\frac{S_\mu(t'-1)}{S_\mu(t-1)}\right)^n\right)^{-1} = 0.$$

And hence

$$t' \leq \liminf_{n\to\infty} \nu_n\left(\left[0, \frac{1}{S_\mu(t-1)}\right]\right).$$

As this holds for all $t' \in (0, t)$ we have

$$t \leq \liminf_{n\to\infty} \nu_n\left(\left[0, \frac{1}{S_\mu(t-1)}\right]\right). \tag{8.3}$$

On the other hand if $t'' \in (t, 1)$ we get

$$\begin{aligned} t'' &= \int_0^\infty \left(1 + \frac{1-t''}{t''} S_\mu(t''-1)^n x^n\right)^{-1} d\nu_n(x) \\ &\geq \int_0^\infty \left(1 + \frac{1-t}{t} S_\mu(t''-1)^n x^n\right)^{-1} d\nu_n(x) \\ &\geq \int_0^{\frac{1}{S(t-1)}} \left(1 + \frac{1-t}{t} S_\mu(t''-1)^n x^n\right)^{-1} d\nu_n(x) \\ &\geq \int_0^{\frac{1}{S(t-1)}} \left(1 + \frac{1-t}{t} \frac{S_\mu(t''-1)^n}{S_\mu(t-1)^n}\right)^{-1} d\nu_n(x) \\ &\geq \nu_n\left(\left[0, \frac{1}{S_\mu(t-1)}\right]\right) \cdot \left(1 + \frac{1-t}{t}\left(\frac{S_\mu(t''-1)}{S_\mu(t-1)}\right)^n\right)^{-1}. \end{aligned}$$

Here the first inequality holds as $t'' > t$ while $S_\mu(t''-1)x^n \geq 0$, and the second to last inequality holds as $S_\mu(t-1)$ is decreasing.

Again as $S_\mu(t-1)$ is strictly decreasing we have $\frac{S_\mu(t''-1)}{S_\mu(t-1)} < 1$, hence

$$\lim_{n\to\infty} \left(1 + \frac{1-t}{t}\left(\frac{S_\mu(t''-1)}{S_\mu(t-1)}\right)^n\right)^{-1} = 1.$$

This implies

$$t'' \geq \limsup_{n\to\infty} \nu_n\left(\left[0, \frac{1}{S_\mu(t-1)}\right]\right).$$

As this holds for all $t'' \in (t, 1)$ we have

$$t \geq \limsup_{n\to\infty} \nu_n\left(\left[0, \frac{1}{S_\mu(t-1)}\right]\right). \tag{8.4}$$

Combining (8.3) and (8.4) we get

$$t = \lim_{n\to\infty} \nu_n\left(\left[0, \frac{1}{S_\mu(t-1)}\right]\right)$$

as desired. □

For proving weak convergence of ν_n to ν it remains to show that ν_n vanishes in limit outside of the support of ν.

Lemma 7. *For all $x \leq a$ and $y \geq b$ we have $\nu_n([0, x]) \to 0$, respectively, $\nu_n([0, y]) \to 1$.*

Proof. To prove the first convergence, let $t \leq a$ and $s \in (0, 1)$. Now we have that $t \leq \frac{1}{S_\mu(s-1)}$ from Lemma 4 and hence

$$\limsup_{n\to\infty} \nu_n([0, t]) \leq \limsup_{n\to\infty} \nu_n\left(\left[0, \frac{1}{S_\mu(s-1)}\right]\right) = s.$$

Here the inequality holds because ν_n is a positive measure and the equality comes from Lemma 6. As this holds for all $s \in (0, 1)$ we have $\limsup_{n\to\infty} \nu_n([0, t]) \leq 0$ and hence $\limsup_{n\to\infty} \nu_n([0, t]) = 0$ by positivity of the measure.

For the second convergence we proceed in the same manner, by letting $t \geq b$ and $s \in (0, 1)$. Now we have that $t \geq \frac{1}{S_\mu(s-1)}$ from Lemma 4 and hence

$$\liminf_{n\to\infty} \nu_n([0, t]) \geq \liminf_{n\to\infty} \nu_n\left(\left[0, \frac{1}{S_\mu(s-1)}\right]\right) = s.$$

Again the inequality holds because ν_n is a positive measure and the equality comes from Lemma 6. As this holds for all $s \in (0, 1)$ we have $\limsup_{n\to\infty} \nu_n([0, t]) \geq 1$ and hence $\limsup_{n\to\infty} \nu_n([0, t]) = 1$ as ν_n is a probability measure. □

Lemmas 6 and 7 now prove Theorem 2 without any assumptions on bounded support as weak convergence of measures is equivalent to point-wise convergence of distribution functions for all but countably many $x \in [0, \infty)$.

Remark 2. In the case $\delta = \mu(\{0\}) > 0$, S_μ is only defined on $(\delta - 1, 0)$ and $S_\mu(z) \to \infty$ when $z \to \delta - 1$. This implies that Lemma 5 only holds for $t \in (\delta, 1)$, with a similar proof. Similarly, Lemma 6 only holds for $t \in (\delta, 1)$, and in the proof we have to assume $t' \in (\delta, t)$. Similarly, in the proof of Lemma 7 we have to assume

$s \in (\delta, 1)$. Moreover, in Lemma 7 the statement, $0 \le x \le a$ implies $\nu_n([0,x]) \to 0$ for $n \to \infty$, should be changed to $a = 0$ and $\nu_n(\{0\}) = \delta = \nu(\{0\})$ for all $n \in \mathbb{N}$.

Using our result we can prove the following corollary, generalizing a theorem ([8, Theorem 2.2]) by H. Schultz and the first author.

Let $(\mathcal{M}, \tau)$ be a finite von Neumann algebra $\mathcal{M}$ with a normal faithful tracial state τ. In [7, Proposition 3.9] the definition of Brown's spectral distribution measure μ_T was extended to all operators $T \in \mathcal{M}^\Delta$, where $\mathcal{M}^\Delta$ is the set of unbounded operators affiliated with $\mathcal{M}$ for which $\tau(\ln^+(|T|)) < \infty$.

Corollary 1. *If T is an R-diagonal in $\mathcal{M}^\Delta$ then $\dot{\phi}(\mu_{(T^*)^n T^n}) \to \dot{\psi}(\mu_T)$ weakly, where $\psi(z) = |z|^2$, $z \in \mathbb{C}$, and $\phi_n(x) = x^{1/n}$ for $x \ge 0$.*

Proof. By [7, Proposition 3.9] we have $\mu_{T^*T}^{\boxtimes n} = \mu_{(T^*)^n T^n}$ and by Theorem 2 we have $\dot{\phi}(\mu_{T^*T}^{\boxtimes n}) \to \nu$ weakly. On the other hand observe that $\nu = \dot{\psi}(\mu_T)$ by [7, Theorem 4.17] which gives the result. □

Remark 3. In [8, Theorem 1.5] it was shown that $\dot{\phi}_n(\mu_{(T^*)^n T^n}) \to \dot{\psi}(\mu_T)$ weakly for all bounded operators $T \in \mathcal{M}$. It would be interesting to know, whether this limit law can be extended to all $T \in \mathcal{M}^\Delta$.

8.4 Further Formulas for the S-Transform

In this section we present some further formulas for the S-transform of measures on $[0, \infty)$, obtained by similar means as in the preceding sections and use those to investigate the difference between the laws of large numbers for classical and free probability. From now on we assume $\mu(\{0\}) = 0$. Therefore μ can be considered as a probability measure on $(0, \infty)$.

We start with a technical lemma which will be useful later.

Lemma 8. *We have the following identities*

$$\int_0^1 \ln^2\left(\frac{t}{1-t}\right) dt = \frac{\pi^2}{3}$$

$$\int_0^1 \ln^2 t \, dt = 2$$

$$\int_0^1 \ln^2(1-t) dt = 2$$

$$\int_0^1 \ln t \ln(1-t) dt = 2 - \frac{\pi^2}{6}.$$

Proof. For the first identity we start with the substitution $x = \frac{t}{1-t}$ which gives us $t = \frac{x}{1+x}$ and $\mathrm{d}t = \frac{\mathrm{d}x}{(1+x)^2}$ and hence

$$\begin{aligned}\int_0^1 \ln^2\left(\frac{t}{1-t}\right)\mathrm{d}t &= \int_0^\infty \frac{\ln^2 x}{(1+x)^2}\mathrm{d}x \\ &= \frac{\mathrm{d}^2}{\mathrm{d}\alpha^2}\int_0^\infty \frac{x^\alpha}{(1+x)^2}\mathrm{d}x\bigg|_{\alpha=0} \\ &= \frac{\mathrm{d}^2}{\mathrm{d}\alpha^2}B(1+\alpha,1-\alpha)\bigg|_{\alpha=0} \\ &= \frac{\mathrm{d}^2}{\mathrm{d}\alpha^2}\frac{\pi\alpha}{\sin(\pi\alpha)}\bigg|_{\alpha=0} \\ &= \frac{\mathrm{d}^2}{\mathrm{d}\alpha^2}\left(1-\frac{(\pi\alpha)^2}{3!}+\cdots\right)^{-1}\bigg|_{\alpha=0} \\ &= \frac{\mathrm{d}^2}{\mathrm{d}\alpha^2}\left(1+\frac{\pi^2}{6}\alpha^2+\cdots\right)\bigg|_{\alpha=0} = \frac{\pi^2}{3}\end{aligned}$$

where $B(\cdot,\cdot)$ denotes the Beta function. The second and the third identity follow from the substitution $t \mapsto \exp(-x)$, respectively, $1-t \mapsto \exp(-x)$.

Finally, the last identity follows by observing

$$\begin{aligned}\frac{\pi^2}{3} &= \int_0^1 \ln^2\left(\frac{t}{1-t}\right)\mathrm{d}t \\ &= \int_0^1 \ln^2 t + \ln^2(1-t) - 2\ln t\ln(1-t)\mathrm{d}t \\ &= 4 - 2\int_0^1 \ln t\ln(1-t)\mathrm{d}t\end{aligned}$$

which gives the desired result. □

Now we prove two propositions calculating the expectations of $\ln x$ and $\ln^2 x$ both for μ and ν expressed by the S-transform of μ.

Proposition 1. *Let* μ *be a probability measure on* $(0,\infty)$ *and let* ν *be as defined in Theorem 2. Then* $\int_0^\infty |\ln x|\,\mathrm{d}\mu(x) < \infty$ *if and only if* $\int_0^1 \left|\ln S_\mu(t-1)\right|\mathrm{d}t < \infty$ *and if and only if* $\int_0^\infty |\ln x|\,\mathrm{d}\nu(x) < \infty$. *If these integrals are finite, then*

$$\int_0^\infty \ln x\mathrm{d}\mu(x) = -\int_0^1 \ln S_\mu(t-1)\mathrm{d}t = \int_0^\infty \ln x\mathrm{d}\nu(x).$$

Proof. For $x > 0$, put $\ln^+ x = \max(\ln x, 0)$ and $\ln^- x = \max(-\ln x, 0)$. Then one easily checks that

$$\ln^+ x \le \ln(x+1) \le \ln^+ x + \ln 2$$

and by replacing x by $\frac{1}{x}$ it follows that

$$\ln^- x \le \ln\left(\frac{x+1}{x}\right) \le \ln^- x + \ln 2.$$

Hence

$$\int_0^\infty \ln^+ x \mathrm{d}\mu(x) < \infty \Leftrightarrow \int_0^\infty \ln(x+1)\mathrm{d}\mu(x) < \infty$$

and

$$\int_0^\infty \ln^- x \mathrm{d}\mu(x) < \infty \Leftrightarrow \int_0^\infty \ln\left(\frac{x+1}{x}\right)\mathrm{d}\mu(x) < \infty.$$

We prove next that

$$\int_0^\infty \ln(x+1)\mathrm{d}\mu(x) = \int_0^\infty \ln^- u\psi'_\mu(-u)\mathrm{d}u \tag{8.5}$$

and

$$\int_0^\infty \ln\left(\frac{x+1}{x}\right)\mathrm{d}\mu(x) = \int_0^\infty \ln^+ u\psi'_\mu(-u)\mathrm{d}u. \tag{8.6}$$

Recall from (8.1), that

$$\psi'_\mu(-u) = \int_0^\infty \frac{t}{(1+ut)^2}\mathrm{d}\mu(t), \quad u > 0.$$

Hence by Tonelli's theorem

$$\int_0^\infty \ln^+ u\psi'_\mu(-u)\mathrm{d}u = \int_1^\infty \ln u\psi'_\mu(-u)\mathrm{d}u = \int_0^\infty \int_1^\infty \frac{x}{(1+ux)^2}\ln u\mathrm{d}u\mathrm{d}\mu(x)$$

and similarly,

$$\int_0^\infty \ln^- u\psi'_\mu(-u)\mathrm{d}u = \int_0^\infty \int_0^1 \frac{x}{(1+ux)^2}\ln\left(\frac{1}{u}\right)\mathrm{d}u\mathrm{d}\mu(x).$$

By partial integration, we have

$$\int_1^\infty \frac{x}{(1+ux)^2}\ln u\mathrm{d}u = \left[-\frac{\ln u}{1+ux}+\ln\left(\frac{u}{1+ux}\right)\right]_{u=1}^{u=\infty} = \ln\left(\frac{x+1}{x}\right)$$

and similarly,

$$\begin{aligned}\int_0^1 \frac{x}{(1+ux)^2}\ln\left(\frac{1}{u}\right)\mathrm{d}u &= \left[\frac{\ln u}{1+ux}-\ln\left(\frac{u}{1+ux}\right)\right]_{u=0}^{u=1}\\ &= \left[\frac{ux}{1+ux}\ln u+\ln(1+ux)\right]_{u=0}^{u=1} = \ln(x+1)\end{aligned}$$

which proves (8.5) and (8.6). Therefore

$$\int_0^\infty |\ln x|\,\mathrm{d}\mu(x) < \infty \Leftrightarrow \int_0^\infty |\ln u|\,\psi'_\mu(-u)\mathrm{d}u < \infty$$

and substituting $x = \psi_\mu(-u) + 1$ we get

$$\int_0^\infty |\ln u|\,\psi'_\mu(-u)\mathrm{d}u = \int_0^1 \left|\ln\left(-\chi_\mu(t-1)\right)\right|\mathrm{d}t = \int_0^1 \left|\ln\left(\frac{t}{1-t}\right)+\ln S_\mu(t-1)\right|\mathrm{d}t.$$

Since $\int_0^1 \left|\ln\left(\frac{t}{1-t}\right)\right|\mathrm{d}t < \infty$ it follows that

$$\int_0^\infty |\ln u|\,\psi'_\mu(-u)\mathrm{d}u < \infty \Leftrightarrow \int_0^1 \left|\ln S_\mu(t-1)\right|\mathrm{d}t < \infty.$$

If μ is not a Dirac measure, the substitution $x = S_\mu(t-1)^{-1}, 0 < t < 1$ gives $t = \nu((0,x])$ for $a < x < b$, where as before $a = \left(\int_0^\infty x^{-1}\mathrm{d}\mu(x)\right)^{-1}$ and $b = \int_0^\infty x\mathrm{d}\mu(x)$. The measure ν is concentrated on the interval (a,b). Hence

$$\int_0^\infty |\ln x|\,\mathrm{d}\nu(x) = \int_a^b |\ln x|\,\mathrm{d}\nu(x) = \int_0^1 \left|\ln\left(\frac{1}{S_\mu(t-1)}\right)\right|\mathrm{d}t = \int_0^1 \left|\ln S_\mu(t-1)\right|\mathrm{d}t.$$

This proves the first statement in Proposition 1. If all three integrals in that statement are finite, we get

$$\begin{aligned}\int_0^\infty \ln x\mathrm{d}\mu(x) &= \int_0^\infty \ln(x+1)\mathrm{d}\mu(x) - \int_0^\infty \ln\left(\frac{x+1}{x}\right)\mathrm{d}\mu(x)\\ &= \int_0^\infty \left(\ln^- u - \ln^+ u\right)\psi'_\mu(-u)\mathrm{d}u = -\int_0^\infty \ln u\psi'_\mu(-u)\mathrm{d}u.\end{aligned}$$

By the substitution $t = \psi_\mu(-u) + 1$ we get

$$\int_0^1 \ln\left(-\chi_\mu(t-1)\right) dt = \int_0^1 \left(\ln\left(\frac{1-t}{t}\right) + \ln S_\mu(t-1)\right) dt = \int_0^1 \ln S_\mu(t-1)dt.$$

Hence $\int_0^\infty \ln x d\mu(x) = -\int_0^1 \ln S_\mu(t-1)dt$. Moreover, by the substitution $x = S_\mu(t-1)^{-1}, 0 < t < 1$ we get

$$\int_0^\infty \ln x d\mu(x) = \int_0^1 \ln\left(\frac{1}{S_\mu(t-1)}\right) dt = \int_0^\infty \ln x d\nu(x).$$

Finally, if $\mu = \delta_x$, $x \in (0, \infty)$, this identity holds trivially, because $\nu = \delta_x$ and $S_\nu(z) = \frac{1}{x}, 0 < z < 1$. □

Corollary 2. *Let μ_1 and μ_2 be probability measures on $(0, \infty)$. If $\mathbb{E}_{\mu_1}(\ln x)$ and $\mathbb{E}_{\mu_2}(\ln x)$ exist then $\mathbb{E}_{\mu_1 \boxtimes \mu_2}(\ln x)$ also exists and*

$$\mathbb{E}_{\mu_1 \boxtimes \mu_2}(\ln x) = \mathbb{E}_{\mu_1}(\ln x) + \mathbb{E}_{\mu_2}(\ln x)$$

where $\mathbb{E}_\mu(f) = \int_0^\infty f(x)d\mu(x)$.

Proof. The statement follows directly from Proposition 1 and multiplicativity of the S-transform. □

For further use, we define the map ρ for a probability measure μ on $(0, \infty)$ by

$$\rho(\mu) = \int_0^1 \ln\left(\frac{1-t}{t}\right) \ln S_\mu(t-1)dt.$$

Note that $\rho(\mu)$ is well-defined and non-negative for all probability measures on $(0, \infty)$ because

$$\ln\left(\tfrac{1-t}{t}\right) \ln S_\mu(t-1) = \ln\left(\tfrac{1-t}{t}\right) \ln\left(\frac{S_\mu(t-1)}{S_\mu(-\frac{1}{2})}\right) + \ln\left(\tfrac{1-t}{t}\right) S_\mu\left(-\tfrac{1}{2}\right), \tag{8.7}$$

where the first term on the right hand side is non-negative for all $t \in (0, 1)$ and the second term is integrable with integral 0.

Lemma 9. *Let μ be a probability measure on $(0, \infty)$, then*

$$0 \le \rho(\mu) \le \frac{\pi}{\sqrt{3}} \left(\int_0^1 \ln^2 S_\mu(t-1)dt\right)^{1/2}.$$

Furthermore, $\rho(\mu) = 0$ if and only if μ is a Dirac measure. Moreover, equality holds in the right inequality if and only if $S_\mu(z) = \left(\frac{z}{1+z}\right)^\gamma$ for some $\gamma > 0$ and in *this case* $\rho(\mu) = \gamma\frac{\pi^2}{3}$. *Additionally, if* μ_1, μ_2 *are probability measures on* $(0,\infty)$ *we have* $\rho(\mu_1 \boxtimes \mu_2) = \rho(\mu_1) + \rho(\mu_2)$.

Proof. We already have observed $\rho \geq 0$. For the second inequality observe that

$$\rho(\mu)^2 \leq \left(\int_0^1 \ln^2\left(\frac{1-t}{t}\right)\mathrm{d}t\right)\left(\int_0^1 \ln^2 S_\mu(t-1)\mathrm{d}t\right)$$

by the Cauchy-Schwarz-inequality, where the first term equals $\frac{\pi^2}{3}$ by Lemma 8.

If $\mu = \delta_a$ for some $a > 0$ we have $S_\mu(z) = \frac{1}{a}$, hence $\ln S_\mu(t-1)$ is constant so the oddity of $\ln(\frac{1-t}{t})$ gives us $\rho(\mu) = 0$. On the other hand, if $\rho(\mu) = 0$, the first term in (8.7) has to integrate to 0, but by symmetry of $\ln\left(\frac{1-t}{t}\right)$ and the fact that S_μ is decreasing, this implies that S_μ must be constant, hence μ is a Dirac measure.

Equality in the second inequality, by the Cauchy-Schwarz inequality happens precisely if $\ln S_\mu(t-1) = \gamma \ln(\frac{1-t}{t})$ for some $\gamma > 0$ which is the case if and only if $S_\mu(t-1) = \left(\frac{1-t}{t}\right)^\gamma$, and in this case $\rho(\mu) = \gamma\frac{\pi^2}{3}$ by Lemma 8.

For the last formula we use multiplicity of the S-transform to get

$$\begin{aligned}\rho(\mu_1 \boxtimes \mu_2) &= \int_0^1 \ln\left(\frac{1-t}{t}\right)\ln S_{\mu_1\boxtimes\mu_2}(t-1)\mathrm{d}t\\ &= \int_0^1 \ln\left(\frac{1-t}{t}\right)(\ln S_{\mu_1}(t-1) + \ln S_{\mu_2}(t-1))\,\mathrm{d}t\\ &= \rho(\mu_1) + \rho(\mu_2).\end{aligned}$$

□

Proposition 2. *Let* μ *be a probability measure on* $(0,\infty)$*, and let* ν *be defined as in Theorem 2. Then*

$$\int_0^\infty \ln^2 x\mathrm{d}\mu(x) = \int_0^1 \ln^2 S_\mu(t-1)\mathrm{d}t + 2\rho(\mu)$$

$$\int_0^\infty \ln^2 x\mathrm{d}\nu(x) = \int_0^1 \ln^2 S_\mu(t-1)\mathrm{d}t$$

$$\mathbb{V}_\mu(\ln x) = \mathbb{V}_\nu(\ln x) + 2\rho(\mu)$$

as equalities of numbers in $[0,\infty]$*, where* $\mathbb{V}_\sigma(\ln x)$ *denotes the variance of* $\ln x$ *with respect to a probability measure* σ *on* $(0,\infty)$*. Moreover,*

$$0 \leq \rho(\mu) \leq \frac{\pi}{\sqrt{3}}\mathbb{V}_\nu(\ln x)^{\frac{1}{2}}.$$

Proof. We first prove the following identity

$$\int_0^\infty \ln^2 u\psi'_\mu(-u)\mathrm{d}u = \int_0^\infty \ln^2 x\mathrm{d}\mu(x) + \frac{\pi^2}{3}. \tag{8.8}$$

Since $\psi'(-u) = \int_0^\infty \frac{x}{(1+ux)^2}\mathrm{d}x$, we get by Tonelli's theorem, that

$$\int_0^\infty \ln^2 u\psi_\mu'(-u)\mathrm{d}u = \int_0^\infty \left(\int_0^\infty \ln^2 u\frac{x}{(1+ux)^2}\mathrm{d}u\right)\mathrm{d}\mu(x)$$
$$= \int_0^\infty \left(\int_0^\infty \ln^2\left(\frac{v}{x}\right)\frac{\mathrm{d}v}{(1+v)^2}\right)\mathrm{d}\mu(x).$$

Note next that

$$\int_0^\infty \ln^2\left(\frac{v}{x}\right)\frac{\mathrm{d}v}{(1+v)^2} = c_0 + c_1\ln x + c_2\ln^2 x$$

where $c_0 = \int_0^\infty \frac{\ln^2 v}{(1+v)^2}\mathrm{d}v$, $c_1 = -2\int_0^\infty \frac{\ln v}{(1+v)^2}\mathrm{d}v$, and $c_2 = \int_0^\infty \frac{1}{(1+v)^2}\mathrm{d}v = 1$. Moreover, by the substitution $v = \frac{1}{w}$ one gets $c_1 = -c_1$ and hence $c_1 = 0$. Finally, by the substitution $v = \frac{t}{1-t}, 0 < t < 1$ and Lemma 8,

$$c_0 = \int_0^1 \ln^2\left(\frac{t}{1-t}\right)\mathrm{d}t = \frac{\pi^2}{3}.$$

Hence

$$\int_0^\infty \ln^2 u\psi_\mu(-u)\mathrm{d}u = \int_0^\infty \left(\ln^2 x + \frac{\pi^2}{3}\right)\mathrm{d}\mu(x)$$

which proves (8.8). Next by the substitution $t = \psi_\mu(-u) + 1$, we have

$$\int_0^\infty \ln^2 u\psi_\mu'(-u)\mathrm{d}u = \int_0^1 \ln^2\left(-\chi_\mu(t-1)\right)\mathrm{d}t = \tag{8.9}$$
$$\int_0^1 \left(\ln\tfrac{1-t}{t} + \ln S_\mu(t-1)\right)^2\mathrm{d}t.$$

Since $t \mapsto \ln\left(\frac{1-t}{t}\right)$ is square integrable on $(0,1)$ the right hand side of (8.9) is finite if and only if

$$\int_0^1 \ln\left(S_\mu(t-1)\right)^2\mathrm{d}t < \infty.$$

Hence by (8.8) and (8.9) this condition is equivalent to

$$\int_0^\infty \ln^2 x\mathrm{d}\mu(x) < \infty,$$

so to prove the first equation in Proposition 2 is suffices to consider the case, where the two above integrals are finite. In that case $\rho(\mu) < \infty$ by Lemma 9. Thus by Lemma 8 and the definition of $\rho(\mu)$,

$$\int_0^1 \left(\ln\left(\frac{1-t}{t}\right) + \ln S_\mu(t-1)\right)^2 \mathrm{d}t = \int_0^1 \ln^2\left(S_\mu(t-1)\right) \mathrm{d}t + 2\rho(\mu) + \frac{\pi^2}{3}.$$

Hence by (8.8) and (8.9)

$$\int_0^\infty \ln^2 x \mathrm{d}\mu(x) = \int_0^1 \ln^2\left(S_\mu(t-1)\right) \mathrm{d}t + 2\rho(\mu).$$

The second equality in Proposition 2

$$\int_0^\infty \ln^2 x \mathrm{d}\nu(x) = \int_0^1 \ln^2 S_\mu(t-1)\mathrm{d}t$$

follows from the substitution $x = S_\mu(t-1)^{-1}$ in case μ is not a Dirac measure, and it is trivially true for Dirac measures. By the first two equalities in Proposition 2, we have

$$\int_0^\infty \ln^2 x \mathrm{d}\mu(x) = \int_0^\infty \ln^2 x \mathrm{d}\nu(x) + 2\rho(\mu). \tag{8.10}$$

If both sides of this equality are finite, then by Proposition 1,

$$\int_0^\infty \ln x \mathrm{d}\mu(x) = \int_0^\infty \ln x \mathrm{d}\nu(x)$$

where both integrals are well-defined. Combined with (8.10) we get

$$\mathbb{V}_\mu(\ln x) = \mathbb{V}_\nu(\ln x) + 2\rho(\mu) \tag{8.11}$$

and if $\int_0^\infty \ln^2 x \mathrm{d}\mu(x) = +\infty$, both sides of (8.11) must be infinite by (8.10).

As the S-transform behaves linearly when scaling the probability distribution in the sense that the image measure μ_c of μ under $x \mapsto cx$ for $c > 0$ gives us $S_{\mu_c}(z) = c^{-1} S_\mu(z)$ we have for ρ that

$$\begin{aligned}\rho(\mu_c) &= \int_0^1 \ln\left(\frac{1-t}{t}\right) \ln(c^{-1} S_\mu(t-1))\mathrm{d}t \\ &= \int_0^1 \ln\left(\frac{1-t}{t}\right) \ln S_\mu(t-1)\mathrm{d}t + \int_0^1 \ln\left(\frac{1-t}{t}\right) c^{-1}\mathrm{d}t = \rho(\mu) + 0\end{aligned}$$

by anti-symmetry of the second term around $t = \frac{1}{2}$. Using this for $c = \exp(\mathbb{E}_\nu(\ln x))$, we get

$$\begin{aligned}
\rho(\mu) = \rho(\mu_c) &\le \frac{\pi}{\sqrt{3}} \left(\int_0^1 \left(\ln S_\mu(t-1) - \mathbb{E}_\nu\left(\ln x\right)\right)^2 \mathrm{d}t \right)^{\frac{1}{2}} \\
&= \frac{\pi}{\sqrt{3}} \left(\int_0^1 \left(\ln S_\mu(t-1)^2 - 2\mathbb{E}_\nu\left(\ln x\right)^2 + \mathbb{E}_\nu\left(\ln x\right)^2\right) \mathrm{d}t \right)^{\frac{1}{2}} \\
&= \frac{\pi}{\sqrt{3}} \left(\mathbb{V}_\nu(\ln x)\right)^{\frac{1}{2}}. \qquad \square
\end{aligned}$$

Now we can use the preceding lemmas to investigate the different behavior of the multiplicative law of large numbers in classical and free probability. Note that in classical probability for a family of identically distributed independent random variables $(X_i)_{i=1}^\infty$ we have the identity $\mathbb{V}(\ln(\prod_{i=1}^n X_i)) = n\mathbb{V}(\ln X_1)$. In free probability by Propositions 1 and 2 we have instead

$$\begin{aligned}
&\mathbb{V}_{\mu^{\boxtimes n}}(\ln t) \\
&= \int_0^\infty \ln^2 t \mathrm{d}(\mu^{\boxtimes n})(t) - \left(\int_0^\infty \ln t \mathrm{d}(\mu^{\boxtimes n})(t) \right)^2 \\
&= \int_0^1 \ln^2 S_{\mu^{\boxtimes n}}(t-1)\mathrm{d}z + 2\rho(\mu^{\boxtimes n}) - \left(- \int_{-1}^0 \ln S_{\mu^{\boxtimes n}}(z)\mathrm{d}z \right)^2 \\
&= n^2 \int_0^1 \ln^2 S_\mu(t-1)\mathrm{d}z + 2n\rho(\mu) - n^2 \left(\int_{-1}^0 \ln S_\mu(z)\mathrm{d}z \right)^2 \\
&= n^2\mathbb{V}_\nu(\ln x) + 2n\rho(\mu).
\end{aligned}$$

Hence $\mathbb{V}_{\mu^{\boxtimes n}}(\ln t) = n\mathbb{V}_\mu(\ln t) + n(n-1)\mathbb{V}_\nu(\ln t) > n\mathbb{V}_\mu(\ln t)$ for $n \ge 2$ if μ is not a Dirac measure and $\mathbb{V}_\nu(\ln t) < \infty$, which shows that the variance of $\ln t$ is not in general additive.

Lemma 10. *Let μ be a probability measure on $(0, \infty)$ and let ν be defined as in Theorem 2. Then*

$$\int_0^\infty x^\gamma \mathrm{d}\mu(x) = \frac{\sin(\pi\gamma)}{\pi\gamma} \int_0^1 \left(\frac{1-t}{t} S_\mu(t-1) \right)^{-\gamma} \mathrm{d}t$$

for $-1 < \gamma < 1$ and

$$\int_0^\infty x^\gamma \mathrm{d}\nu(x) = \int_0^1 S_\mu(t-1)^{-\gamma} \mathrm{d}t$$

for $\gamma \in \mathbb{R}$ as equalities of numbers in $[0, \infty]$.

Proof. By Tonelli's theorem followed by the substitution $u = yx$ we get

$$\begin{aligned}\int_0^\infty y^{-\gamma}\psi'_\mu(-y)\mathrm{d}y &= \int_0^\infty \int_0^\infty \frac{y^{-\gamma}x}{(1+yx)^2}\mathrm{d}y\mathrm{d}\mu(x)\\ &= \int_0^\infty x^\gamma \int_0^\infty \frac{u^{-\gamma}}{(1+u)^2}\mathrm{d}u\mathrm{d}\mu(x)\\ &= B(1-\gamma,1+\gamma)\int_0^\infty x^\gamma \mathrm{d}\mu(x),\end{aligned}$$

where $B(s,t) = \int_0^\infty \frac{u^{s-1}}{(1+u)^{s+t}}\mathrm{d}u$ is the Beta function. But $B(1-\gamma,1+\gamma) = \frac{\sin(\pi\gamma)}{\pi\gamma}$ by well-known properties of B. Substitute now $x = -\chi_\mu(-z)$ and $z = 1-t$ to get

$$\int_0^\infty x^{-\gamma}\psi'_\mu(-x)\mathrm{d}x = \int_0^1 \left(-\chi_\mu(-z)\right)^{-\gamma}\mathrm{d}z = \int_0^1 \left(\frac{1-t}{t}S_\mu(t-1)\right)^{-\gamma}\mathrm{d}t,$$

which gives the first identity. The second identity follows from the substitution $x = S_\mu(t-1)^{-1}$ and the properties of ν from Theorem 2. □

8.5 Examples

In this section we will investigate a two parameter family of distributions for which there can be made explicit calculations.

Proposition 3. *Let $\alpha, \beta \geq 0$. There exists a probability measure $\mu_{\alpha,\beta}$ on $(0,\infty)$ which S-transform is given by*

$$S_{\mu_{\alpha,\beta}}(z) = \frac{(-z)^\beta}{(1+z)^\alpha}.$$

Furthermore, these measures form a two-parameter semigroup, multiplicative under $\boxtimes$ induced by multiplication of $(\alpha,\beta) \in [0,\infty) \times [0,\infty)$.

Proof. Note first that $\alpha = \beta = 0$ gives $S_{\mu_{0,0}} = 1$, which by uniqueness of the S-transform results in $\mu_{0,0} = \delta_1$, hence we can in the following assume $(\alpha,\beta) \neq (0,0)$.

Define the function $v_{\alpha,\beta}\colon \mathbb{C}\setminus[0,1] \to \mathbb{C}$ by

$$v_{\alpha,\beta}(z) = \beta\ln(-z) - \alpha\ln(1+z)$$

for all $z \in \mathbb{C}\setminus[0,1]$.

In the following we for $z \in \mathbb{C}$ denote by $\arg z \in [-\pi,\pi]$ its argument. Assume $z = x + \mathrm{i}y$ and $y > 0$ then

$$\ln(-z) = \frac{1}{2}\ln\left(x^2+y^2\right) + \mathrm{i}\arg(-x-\mathrm{i}y)$$

where $\arg(-x-\mathrm{i}y) < 0$, which implies that $\ln(\mathbb{C}^+) \subseteq \mathbb{C}^-$. Similarly, if we assume $z = x+\mathrm{i}y$ and $y > 0$ then

$$\ln(1+z) = \frac{1}{2}\ln\left((x+1)^2+y^2\right) + \mathrm{i}\arg((x+1)+\mathrm{i}y)$$

where $\arg((x+1)+\mathrm{i}y) > 0$, which implies that $-\ln(1+\mathbb{C}^+) \subseteq \mathbb{C}^-$ and hence $v_{\alpha,\beta}(\mathbb{C}^+) \subseteq \mathbb{C}^-$. Furthermore, we observe that for all $z \in \mathbb{C}$, $v_{\alpha,\beta}(\overline{z}) = \overline{v_{\alpha,\beta}(z)}$. By [4, Theorem 6.13 (ii)] these results imply that there exists a unique $\boxtimes$-infinitely divisible measure $\mu_{\alpha,\beta}$ with the S-transform

$$S_{\mu_{\alpha,\beta}}(z) = \exp(v(z)) = \exp(\beta\ln(-z) - \alpha\ln(1+z)) = \frac{(-z)^\beta}{(1+z)^\alpha}.$$

The semigroup property follows from multiplicativity of the S-transform. □

The existence of $\mu_{\alpha,0}$ was previously proven by T. Banica, S.T. Belinschi, M. Capitaine and B. Collins in [2] as a special case of free Bessel laws. The case $\mu_{\alpha,\alpha}$ is known as a Boolean stable law from O. Arizmendi and T. Hasebe [1].

Furthermore, there is a clear relationship between the measures $\mu_{\alpha,\beta}$ and $\mu_{\beta,\alpha}$.

Lemma 11. *Let $\alpha,\beta \geq 0$, $(\alpha,\beta) \neq (0,0)$ and let $\zeta\colon (0,\infty) \to (0,\infty)$ be the map $\zeta(t) = t^{-1}$. Then we have $\mu_{\beta,\alpha} = \dot{\zeta}(\mu_{\alpha,\beta})$, where $\dot{\zeta}$ denotes the image measure under the map ζ.*

Proof. Put $\sigma = \dot{\zeta}(\mu_{\alpha,\beta})$. Then by the proof of [7, Proposition 3.13],

$$S_\sigma(z) = \frac{1}{S_{\mu_{\alpha,\beta}}(-1-z)} = \frac{(-z)^\alpha}{(1+z)^\beta} = S_{\mu_{\beta,\alpha}}$$

for $0 < z < 1$. Hence $\sigma = \mu_{\beta,\alpha}$. □

Lemma 12. *Let $(\alpha,\beta) \neq (0,0)$. Denote the limit measure corresponding to $\mu_{\alpha,\beta}$ by $\nu_{\alpha,\beta}$. Then $\nu_{\alpha,\beta}$ is uniquely determined by the formula*

$$F_{\alpha,\beta}\left(\frac{t^\alpha}{(1-t)^\beta}\right) = t$$

for $0 < t < 1$, where $F_{\alpha,\beta}(x) = \nu_{\alpha,\beta}((0,x])$ is the distribution function of $\nu_{\alpha,\beta}$.

Proof. The lemma follows directly from Lemma 3 and Theorem 2. □

For $\beta = 0$ and $\alpha > 0$,

$$F_{\alpha,0}(x) = \begin{cases} x^{\frac{1}{\alpha}}, & 0 < x < 1 \\ 1, & x \geq 1. \end{cases}$$

Similarly, for $\alpha = 0$ and $\beta > 0$

$$F_{0,\beta}(x) = \begin{cases} 0, & 0 < x < 1 \\ (1-x)^{-\frac{1}{\beta}}, & x \geq 1. \end{cases}$$

Hence $\nu_{0,\beta}$ is the Pareto distribution with scale parameter 1 and shape parameter $\frac{1}{\beta}$.

Moreover, if $\alpha = \beta > 0$ we get $F_{\alpha,\alpha}(x) = (1 + x^{-1/\alpha})^{-1}$ for $x \in (0, \infty)$, which we recognize as the image measure of the Burr distribution with parameters $(1, \alpha^{-1})$ (or equivalently the Fisk or log-logistic distribution (cf. [9, p. 54]) with scale parameter 1 and shape parameter α^{-1}) under the map $x \mapsto x^{-1}$.

On the other hand, we can make some observations about the distribution $\mu_{\alpha,\beta}$, too. For the cases $(\alpha, \beta) = (1, 0)$ and $(\alpha, \beta) = (0, 1)$ we can recognize the measures $\mu_{1,0}$ and $\mu_{0,1}$ from their S-transform, as $S_{\mu_{1,0}}(z) = (1+z)^{-1}$ is the S-transform of the free Poisson distributions with shape parameter 1 (cf. [18, p. 34]), which is given by

$$\mu_{1,0} = \frac{1}{2\pi}\sqrt{\frac{4-x}{x}}1_{(0,4)}(x)\mathrm{d}x,$$

while $S_{\mu_{0,1}}(z) = -z$ according to Lemma 11 is the S-transform of the image of the above free Poisson distribution under the map $t \mapsto t^{-1}$,

$$\mu_{0,1} = \frac{1}{2\pi}\frac{\sqrt{4x-1}}{x^2}1_{(\frac{1}{4},\infty)}(x)\mathrm{d}x,$$

which is the same as the free stable distribution with parameters $\alpha = 1/2$ and $\rho = 1$ as described by H. Bercovici, V. Pata and P. Biane in [3, Appendix A1]. More generally, $\mu_{0,\beta}$ is the same as the free stable distribution $\upsilon_{\alpha,\rho}$ with $\alpha = \frac{1}{\beta+1}$ and $\rho = 1$, because by [3, Appendix A4] $\upsilon_{\alpha,1}$ is characterized by $\Sigma_{\upsilon_{\alpha,1}}(y) = \left(\frac{-y}{1-y}\right)^{\frac{1}{\alpha}-1}$, $y \in (-\infty, 0)$, and it is easy to check that

$$S_{\upsilon_{\alpha,0}}(z) = \Sigma_{\upsilon_{\alpha,0}}\left(\frac{z}{1+z}\right) = (-z)^{\frac{1}{\alpha}-1} = S_{\mu_{0,\frac{1}{\alpha}-1}}(z), \quad 0 < z < 1, 0 < \alpha < 1.$$

From the above observations, we now can describe a construction of the measures $\mu_{m,n}$.

Proposition 4. *Let m, n be nonnegative integers. Then the measure $\mu_{m,n}$ is given by*

$$\mu_{m,n} = \mu_{1,0}^{\boxtimes m} \boxtimes \mu_{0,1}^{\boxtimes n}.$$

Proof. By multiplicativity of the S-transform we have that

$$S_{\mu_{1,0}^{\boxtimes m}\boxtimes\mu_{0,1}^{\boxtimes n}}(z) = S_{\mu_{1,0}}(z)^m S_{\mu_{0,1}}(z)^n = \frac{(-z)^n}{(1+z)^m} = S_{\mu_{m,n}}(z),$$

which by uniqueness of the S-transform gives the desired result. □

Proposition 5. *For all* $\alpha, \beta \geq 0$

$$\mathbb{E}_{\mu_{\alpha,\beta}}(\ln x) = \beta - \alpha$$

$$\rho(\mu_{\alpha,\beta}) = \frac{\pi^2}{6}(\alpha+\beta)$$

$$\mathbb{V}_{\mu_{\alpha,\beta}}(\ln x) = (\alpha-\beta)^2 + \frac{\pi^2}{3}(\alpha\beta+\alpha+\beta).$$

Proof. These formulas follow easily from Propositions 1 and 2 and Lemma 8. □

Furthermore, we also can calculate explicitly all fractional moments of $\mu_{\alpha,\beta}$ by the following theorem.

Theorem 3. *Let* $\alpha, \beta > 0$ *and* $\gamma \in \mathbb{R}$ *then we have*

$$\int_0^\infty x^\gamma \mathrm{d}\mu_{\alpha,\beta}(x) = \begin{cases} \frac{\sin(\pi\gamma)}{\pi\gamma}\frac{\Gamma(1+\gamma+\gamma\alpha)\Gamma(1-\gamma-\gamma\beta)}{\Gamma(2+\gamma\alpha-\gamma\beta)} & -\frac{1}{1+\alpha} < \gamma < \frac{1}{1+\beta} \\ \infty & \textit{otherwise} \end{cases} \tag{8.12}$$

$$\int_0^\infty x^\gamma \mathrm{d}\mu_{\alpha,0}(x) = \begin{cases} \frac{\Gamma(1+\gamma+\gamma\alpha)}{\Gamma(1+\gamma)\Gamma(2+\gamma\alpha)} & \gamma > -\frac{1}{1+\alpha} \\ \infty & \textit{otherwise} \end{cases} \tag{8.13}$$

$$\int_0^\infty x^\gamma \mathrm{d}\mu_{0,\beta}(x) = \begin{cases} \frac{\Gamma(1-\gamma-\gamma\beta)}{\Gamma(1-\gamma)\Gamma(2-\gamma\beta)} & \gamma < \frac{1}{1+\beta} \\ \infty & \textit{otherwise.} \end{cases} \tag{8.14}$$

Proof. Let first $-1 < \gamma < 1$. Then (8.12)–(8.14) follow from Lemma 10 together with the formula $\Gamma(1+\gamma)\Gamma(1-\gamma) = \frac{\pi\gamma}{\sin(\pi\gamma)}$. Since $S_{\mu_{\alpha,0}}(z) = \frac{1}{(z+1)^\alpha}$ is analytic in a neighborhood of 0, $\mu_{\alpha,0}$ has finite moments of all orders. Therefore the functions

$$s \mapsto \int_0^\infty x^s \mathrm{d}\mu_{\alpha,0}(x)$$

$$s \mapsto \frac{\Gamma(1+s+s\alpha)}{\Gamma(1+s)\Gamma(2+s\alpha)}$$

are both analytic in the half-plane $\Re s > 0$ and they coincide for $s \in (0,1)$. Hence they are equal for all $s \in \mathbb{C}$ with $\Re s > 0$ which proves (8.13). By Lemma 11 (8.14) follows from (8.13). □

Remark 4. By Theorem 3 (8.12) we have

1. If $\beta > 0$, then $\int_0^\infty x \mathrm{d}\mu_{\alpha,\beta}(x) = \infty$. Hence $\sup(\mathrm{supp}(\mu_{\alpha,\beta})) = \infty$. Similarly, if $\alpha > 0$ then $\int_0^\infty x^{-1} \mathrm{d}\mu_{\alpha,\beta}(x) = \infty$. Hence $\inf(\mathrm{supp}(\mu_{\alpha,\beta})) = 0$.
2. If $\beta = 0$, then by Stirling's formula

$$\sup(\mathrm{supp}(\mu_{\alpha,0})) = \lim_{0\to\infty} \left(\int_0^\infty t^n \mathrm{d}\mu_{\alpha,0}(t) \right)^{\frac{1}{n}} = \frac{(\alpha+1)^{\alpha+1}}{\alpha^\alpha}.$$

Hence by Lemma 11, we have for $\alpha = 0$

$$\inf(\mathrm{supp}(\mu_{0,\beta})) = \frac{\beta^\beta}{(\beta+1)^{\beta+1}}.$$

Note that $\sup(\mathrm{supp}(\mu_{n,0})) = \frac{(n+1)^{n+1}}{n^n}, n \in \mathbb{N}$ was already proven by F. Larsen in [10, Proposition 4.1] and it was proven by T. Banica, S. T. Belinschi, M. Capitane and B. Collins in [2] that $\mathrm{supp}(\mu_{\alpha,0}) = \left[0, \frac{(\alpha+1)^{\alpha+1}}{\alpha^\alpha}\right]$. Note that this also follows from our Corollary 3.

If $\alpha = \beta$ it is also possible to calculate explicitly the density of $\mu_{\alpha,\alpha}$. To do this we require an additional lemma.

Lemma 13. *For $-1 < \gamma < 1$ and $-\pi < \theta < \pi$ we have*

$$\frac{\sin\theta}{\pi} \int_0^\infty \frac{t^\gamma}{t^2 + 2\cos(\theta)t + 1} \mathrm{d}t = \frac{\sin(\theta\gamma)}{\sin(\pi\gamma)}.$$

Proof. Note first that by the substitution $t = e^x$ we have

$$\int_0^\infty \frac{t^\gamma}{t^2 + 2\cos(\theta)t + 1} \mathrm{d}t = \frac{1}{2} \int_{-\infty}^\infty \frac{e^{\gamma x}}{\cosh x + \cos\theta} \mathrm{d}x.$$

The function

$$z \mapsto \frac{e^{\gamma x}}{\cosh x + \cos\theta}$$

is meromorphic with simple poles in $x = \pm \mathrm{i}(\pi - \theta) + p2\pi$, $p \in \mathbb{Z}$. Apply now the residue integral formula to this function on the boundary of

$$\{z \in \mathbb{C} : -R \leq \Re z \leq R, 0 \leq \Im z \leq 2\pi\}$$

and let $R \to \infty$. The result follows. □

The density of $\mu_{\alpha,\alpha}$ was computed by P. Biane [5, Sect. 5.4]. For completeness we include a different proof based on Theorem 3 and Lemma 13.

Theorem 4 ([5]). *Let $\alpha > 0$ then $\mu_{\alpha,\alpha}$ has the density $f_{\alpha,\alpha}(t)\mathrm{d}t$, where*

$$f_{\alpha,\alpha}(t) = \frac{\sin\left(\frac{\pi}{\alpha+1}\right)}{\pi t\left(t^{\frac{1}{\alpha+1}} + 2\cos\left(\frac{\pi}{\alpha+1}\right) + t^{-\frac{1}{\alpha+1}}\right)}$$

for $t \in (0,\infty)$. In particular $\mu_{1,1}$ has the density $(\pi\sqrt{t}(1+t))^{-1}\mathrm{d}t$ and $\mu_{2,2}$ has the density

$$\frac{\sqrt{3}}{2\pi(1 + t^{\frac{2}{3}} + t^{\frac{4}{3}})}\mathrm{d}t.$$

Proof. To prove this note that for $|\gamma| < \frac{1}{1+\alpha}$

$$\begin{aligned}\int_0^\infty x^\gamma f_{\alpha,\alpha}(x)\mathrm{d}x &= \int_0^\infty \frac{\sin\left(\frac{\pi}{\alpha+1}\right)(\alpha+1)y^{\gamma(\alpha+1)}}{\pi\left(y + 2\cos\left(\frac{\pi}{\alpha+1}\right) + y^{-1}\right)}\frac{\mathrm{d}y}{y} \\ &= \frac{(\alpha+1)\sin\left(\frac{\pi}{\alpha+1}\right)}{\pi}\int_0^\infty \frac{y^{\gamma(\alpha+1)}}{y^2 + 2\cos\left(\frac{\pi}{\alpha+1}\right)y + 1}\mathrm{d}y\end{aligned}$$

using the substitution $y = x^{\frac{1}{\alpha+1}}$. Now by Lemma 13 and Theorem 3 (8.12) we have

$$\int_0^\infty x^\gamma f_{\alpha,\alpha}(x)\mathrm{d}x = \int_0^\infty x^\gamma \mathrm{d}\mu_{\alpha,\alpha}(x) < \infty.$$

This implies by unique analytic continuation that the same formula holds for all $\gamma \in \mathbb{C}$ with $|\Re\gamma| < \frac{1}{\alpha+1}$. In particular

$$\int_0^\infty x^{\mathrm{i}s} f_{\alpha,\alpha}(x)\mathrm{d}x = \int_0^\infty x^{\mathrm{i}s}\mathrm{d}\mu_{\alpha,\alpha}(x)$$

for all $s \in \mathbb{R}$, which shows that the image measures under $x \mapsto \ln x$ of $f_{\alpha,\alpha}(x)\mathrm{d}x$ and $\mu_{\alpha,\alpha}$ have the same characteristic function. Hence $\mu_{\alpha,\alpha} = f_{\alpha,\alpha}(x)\mathrm{d}x$. □

Proposition 6. *For all $\alpha,\beta \geq 0$, $(\alpha,\beta) \neq (0,0)$, the measure $\mu_{\alpha,\beta}$ has a continuous density $f_{\alpha,\beta}(x)$, $(x > 0)$, with respect to the Lebesgue measure on $\mathbb{R}$ and*

$$\lim_{x\to 0+} x f_{\alpha,\beta}(x) = \lim_{x\to\infty} x f_{\alpha,\beta}(x) = 0. \tag{8.15}$$

Proof. By the method of proof of Theorem 4, the integral

$$h_{\alpha,\beta}(s) = \int_0^\infty x^{\mathrm{i}s}\mathrm{d}\mu_{\alpha,\beta}(x), \quad s \in \mathbb{R}$$

can be obtained by replacing γ by is in the formulas (8.12)–(8.14). Moreover,

$$h_{\alpha,\beta}(s) = \int_0^\infty \exp(\mathrm{i}st)\mathrm{d}\sigma_{\alpha,\beta}(t)$$

where $\sigma_{\alpha,\beta}$ is the image measure of $\mu_{\alpha,\beta}$ by the map $x \mapsto \log x$, $(x > 0)$. Hence by standard Fourier analysis, we know that if $h_{\alpha,\beta} \in L^1(\mathbb{R})$ then $\sigma_{\alpha,\beta}$ has a density $g_{\alpha,\beta} \in C_0(\mathbb{R})$ with respect to the Lebesgue measure on $\mathbb{R}$ and hence $\mu_{\alpha,\beta}$ has density $f_{\alpha,\beta}(x) = \frac{1}{x} g_{\alpha,\beta}(\log x)$ for $x > 0$, which satisfies the condition (8.15). To prove that $h_{\alpha,\beta} \in L^1(\mathbb{R})$ for all $\alpha, \beta \geq 0$, $(\alpha, \beta) \neq (0, 0)$, we observe first that

$$\Gamma(1-z)\Gamma(1+z) = \frac{\pi z}{\sin \pi z}, \quad z \in \mathbb{C} \setminus \mathbb{Z}$$

and hence by the functional equation of Γ

$$\Gamma(2-z)\Gamma(2+z) = \frac{\pi z(1-z^2)}{\sin \pi z}, \quad z \in \mathbb{C} \setminus \mathbb{Z}.$$

In particular, we have

$$|\Gamma(1+\mathrm{i}s)|^2 = \frac{\pi s}{\sinh \pi s}, \quad s \in \mathbb{R}$$

$$|\Gamma(2+\mathrm{i}s)|^2 = \frac{\pi s(1+s^2)}{\sinh \pi s}, \quad s \in \mathbb{R}.$$

Applying these formulas to (8.12)–(8.14) with γ replaced by is, we get

$$h_{\alpha,\beta}(s) = O\left(|s|^{-3/2}\right), \quad \text{for } s \to \pm\infty$$

for all choices of $\alpha, \beta \geq 0$, $(\alpha, \beta) \neq (0, 0)$. Thus by the continuity of $h_{\alpha,\beta}$ it follows that $h_{\alpha,\beta} \in L^1(\mathbb{R})$, which proves the proposition. □

Note that by Remark 4 it follows that $f_{\alpha,0}(x)$ can only be non-zero if $x \in \left(0, \frac{(\alpha+1)^{\alpha+1}}{\alpha^\alpha}\right)$ and $f_{0,\beta}(x)$ can only be non-zero if $x \in \left(\frac{\beta^\beta}{(\beta+1)^{\beta+1}}, \infty\right)$. Since we have seen, that $\mu_{0,\beta}$ coincides with the stable distribution $\upsilon_{\alpha,\rho}$ with $\alpha = \frac{1}{\beta+1}$ and $\rho = 1$ we have from [3, Appendix 4] that

Theorem 5 ([3]). *The map*

$$\phi \mapsto \frac{\sin \phi \sin^\beta(\beta\phi)}{\sin^{\beta+1}((\beta+1)\phi)}, \quad 0 < \phi < \frac{\pi}{\beta+1}$$

is a bijection of the interval $\left(0, \frac{\pi}{\beta+1}\right)$ *onto* $\left(\frac{\beta^\beta}{(\beta+1)^{\beta+1}}, \infty\right)$ *and*

$$f_{\mu_{0,\beta}}\left(\frac{\sin\phi\sin^{\beta}(\beta\phi)}{\sin^{\beta+1}((\beta+1)\phi)}\right)=\frac{\sin^{\beta+2}((\beta+1)\phi)}{\pi\sin^{\beta+1}(\beta\phi)},\quad 0<\phi<\frac{\pi}{\beta+1}. \tag{8.16}$$

Proof. We know that $\mu_{0,\beta}=\upsilon_{\frac{1}{\beta+1},1}$, the stable distribution with parameters $\alpha=\frac{1}{\beta+1}$ and $\rho=1$. Moreover, we have from [3, Proposition A1.4], that $\upsilon_{\alpha,1}$ has density $\psi_{\alpha,1}$ on the interval $\left(\alpha(1-\alpha)^{1/\alpha-1},\infty\right)$ given by

$$\psi_{\alpha,1}(x)=\frac{1}{\pi}\sin^{1+\frac{1}{\alpha}}\theta\sin^{-\frac{1}{\alpha}}((1-\alpha)\theta),$$

where $\theta\in(0,\pi)$ is the only solution to the equation

$$x=\sin^{-\frac{1}{\alpha}}\theta\sin^{\frac{1}{\alpha}-1}((1-\alpha)\theta)\sin\alpha\theta.$$

It is now easy to check that $f_{0,\beta}(x)=\psi_{\frac{1}{\beta+1},1}(x)$ has the form (8.16) by using the substitution $\phi=\frac{\theta}{\beta+1}$. □

Corollary 3. *The map*

$$\phi\mapsto\frac{\sin^{\alpha+1}((\alpha+1)\phi)}{\sin\phi\sin^{\alpha}(\alpha\phi)},\quad 0<\phi<\frac{\pi}{\alpha+1}$$

is a bijection of the interval $\left(0,\frac{\pi}{\alpha+1}\right)$ *onto* $\left(0,\frac{(\alpha+1)^{\alpha+1}}{\alpha^{\alpha}}\right)$ *and*

$$f_{\mu_{\alpha,0}}\left(\frac{\sin^{\alpha+1}((\alpha+1)\phi)}{\sin\phi\sin^{\alpha}(\alpha\phi)}\right)=\frac{\sin^{2}\phi\sin^{\alpha-1}(\alpha\phi)}{\pi\sin^{\alpha}((\alpha+1)\phi)},\quad 0<\phi<\frac{\pi}{\alpha+1}.$$

Proof. Since $\mu_{\alpha,0}$ is the image measure of $\mu_{0,\alpha}$ by the map $t\mapsto\frac{1}{t}$, $(t>0)$, we have

$$f_{\alpha,0}(x)=\frac{1}{x^{2}}f_{0,\alpha}\left(\frac{1}{x}\right),\quad x>0.$$

The corollary now follows from Theorem 5 by elementary calculations. □

We next use Biane's method to compute the density $f_{\alpha,\beta}$ for all $\alpha,\beta>0$.

Theorem 6. *Let* $\alpha,\beta>0$. *Then for each* $x>0$ *there are unique real numbers* $\phi_1,\phi_2>0$ *for which*

$$\pi=(\alpha+1)\phi_1+(\beta+1)\phi_2 \tag{8.17}$$

$$x=\frac{\sin^{\alpha+1}\phi_2}{\sin^{\beta+1}\phi_1}\sin^{\beta-\alpha}(\phi_1+\phi_2). \tag{8.18}$$

Moreover

$$f_{\mu_{\alpha,\beta}}(x) = \frac{\sin^{\beta+2}\phi_1}{\pi \sin^{\alpha}\phi_2} \sin^{\alpha-\beta-1}(\phi_1+\phi_2). \tag{8.19}$$

Proof. As $\mu_{\alpha,\beta}$ has the S-transform $S_{\mu_{\alpha,\beta}}(z) = \frac{(-z)^{\beta}}{(1+z)^{\alpha}}$ we by Definition 1 observe that

$$\chi_{\mu_{\alpha,\beta}}(z) = \frac{-(-z)^{\beta+1}}{(1+z)^{\alpha+1}} \quad \text{whence} \quad \psi_{\mu_{\alpha,\beta}}\left(-\frac{(-z)^{\beta+1}}{(1+z)^{\alpha+1}}\right) = z$$

for z in some complex neighborhood of $(-1,0)$. Now it is known that

$$G_{\mu}\left(\frac{1}{t}\right) = t\,(1+\psi_{\mu}(t))$$

for every probability measure on $(0,\infty)$. Hence

$$G_{\mu_{\alpha,\beta}}\left(-\frac{(1+z)^{\alpha+1}}{(-z)^{\beta+1}}\right) = -\frac{(-z)^{\beta+1}}{(1+z)^{\alpha}} \tag{8.20}$$

for z in a complex neighborhood of $(-1,0)$.

Let H denote the upper half plane in $\mathbb{C}$:

$$H = \{z \in \mathbb{C} : \Im z > 0\}.$$

For $z \in H$, put

$$\phi_1 = \phi_1(z) = \arg(1+z) \in (0,\pi)$$
$$\phi_2 = \phi_2(z) = \pi - \arg(z) \in (0,\pi).$$

Basic trigonometry applied to the triangle with vertices -1, 0 and z, shows that $\phi_1+\phi_2 < \pi$ and

$$\frac{\sin\phi_1}{|z|} = \frac{\sin\phi_2}{|1+z|} = \frac{\sin(\pi-\phi_1-\phi_2)}{1}.$$

Hence

$$|z| = \frac{\sin\phi_1}{\sin(\phi_1+\phi_2)} \quad \text{and} \quad |1+z| = \frac{\sin\phi_2}{\sin(\phi_1+\phi_2)}$$

from which

$$z = -\frac{\sin\phi_1}{\sin(\phi_1+\phi_2)}e^{i\phi_2} \quad \text{and} \quad \Im z = \frac{\sin\phi_1 \sin\phi_2}{\sin(\phi_1+\phi_2)}.$$

It follows that $\Phi: z \mapsto (\phi_1(z), \phi_2(z))$ is a diffeomorphism of H onto the triangle $T = \{(\phi_1, \phi_2) \in \mathbb{R}^2 : \phi_1, \phi_2 > 0, \phi_1 + \phi_2 < \pi\}$ with inverse

$$\Phi^{-1}(\phi_1, \phi_2) = -\frac{\sin\phi_1}{\sin(\phi_1+\phi_2)}e^{-i\phi_2}, \quad (\phi_1, \phi_2) \in T.$$

Put $H_{\alpha,\beta} = \{z \in H : (\alpha+1)\phi_1(z)+(\beta+1)\phi_2(z) < \pi\}$. Then $H_{\alpha,\beta} = \Phi^{-1}(T_{\alpha,\beta})$ where $T_{\alpha,\beta} = \{(\phi_1, \phi_2) \in T : (\alpha+1)\phi_1 + (\beta+1)\phi_2 < \pi\}$.

In particular $H_{\alpha,\beta}$ is an open connected subset of H. Put

$$F(z) = -\frac{(1+z)^{\alpha+1}}{(-z)^{\beta+1}}, \quad \Im z > 0.$$

Then

$$F(z) = \frac{|1+z|^{\alpha+1}}{|z|^{\beta+1}} e^{i((\alpha+1)\phi_1(z)+(\beta+1)\phi_2(z)-\pi)} \tag{8.21}$$

so for $z \in H_{\alpha,\beta}$, $\Im F(z) < 0$. Therefore $G_{\mu_{\alpha,\beta}}(F(z))$ is a well-defined analytic function on $H_{\alpha,\beta}$, and since $(-1, 0)$ is contained in the closure of $H_{\alpha,\beta}$ it follows from (8.20)

$$G_{\mu_{\alpha,\beta}}(F(z)) = \frac{1+z}{F(z)} \tag{8.22}$$

for z in some open subset of $H_{\alpha,\beta}$ and thus by analyticity it holds for all $z \in H_{\alpha,\beta}$.

Let $x > 0$ and assume that $\phi_1, \phi_2 > 0$ satisfy (8.17) and (8.18). Put

$$z = \Phi^{-1}(\phi_1, \phi_2) = -\frac{\sin\phi_1}{\sin(\phi_1+\phi_2)}e^{-i\phi_2}.$$

Then by (8.21)

$$F(z) = \frac{|1+z|^{\alpha+1}}{|z|^{\beta+1}} = \left(\frac{\sin\phi_2}{\sin(\phi_1+\phi_2)}\right)^{\alpha+1} \left(\frac{\sin(\phi_1+\phi_2)}{\sin\phi_1}\right)^{\beta+1} = x.$$

Since $\mu_{\alpha,\beta}$ has a continuous density $f_{\alpha,\beta}$ on $(0, \infty)$ by Proposition 6, the inverse Stieltjes transform gives

$$f_{\alpha,\beta}(x) = -\frac{1}{\pi} \lim_{w\to x, \Im w>0} \Im G_{\mu_{\alpha,\beta}}(w) = \frac{1}{\pi} \lim_{w\to x, \Im w<0} \Im G_{\mu_{\alpha,\beta}}(w).$$

For $0 < t < 1$, put $z_t = \Phi^{-1}(t\phi_1, t\phi_2)$. Then

$$z_t \in \Phi^{-1}\left(T_{\alpha,\beta}\right) = H_{\alpha,\beta}.$$

Thus $\Im F(z_t) < 0$. Moreover, $z_t \to z$ and $F(z_t) \to F(z) = x$ for $t \to 1^-$. Hence by (8.22),

$$f_{\alpha,\beta}(x) = \frac{1}{\pi} \lim_{t\to 1^-} \Im G_{\mu_{\alpha,\beta}}(F(z_t)) = \frac{1}{\pi} \lim_{t\to 1^-} \Im\left(\frac{z_t + 1}{F(z_t)}\right) = \frac{\Im z}{\pi x} = \frac{\sin\phi_1 \sin\phi_2}{\pi x \sin(\phi_1 + \phi_2)}$$

which proves (8.19). To complete the proof of Theorem 6, we only need to prove the existence and uniqueness of $\phi_1, \phi_2 > 0$. Assume that ϕ_1, ϕ_2 satisfy (8.17) then

$$\phi_1 = \frac{\pi - \theta}{\alpha + 1} \quad \text{and} \quad \phi_2 = \frac{\theta}{\beta + 1}$$

for a unique $\theta \in (0, \pi)$. Moreover,

$$\frac{d\phi_1}{d\theta} = -\frac{1}{\alpha + 1} \quad \text{and} \quad \frac{d\phi_2}{d\theta} = \frac{1}{\beta + 1}.$$

Hence, expressing $u = \frac{\sin^{\alpha+1}\phi_2}{\sin^{\beta+1}\phi_1} \sin^{\beta-\alpha}(\phi_1 + \phi_2)$ as a function $u(\theta)$ of θ, we get

$$\begin{aligned}(\alpha + 1)(\beta + 1)\frac{du(\theta)}{d\theta} &= (\beta + 1)^2 \cot\phi_1 + (\alpha + 1)^2 \cot\phi_2 - 2(\alpha - \beta)^2 \cot(\phi_1 + \phi_2) \\ &= \frac{A(\phi_1, \phi_2)}{\sin\phi_1 \sin\phi_2 \sin(\phi_1 + \phi_2)}\end{aligned}$$

where

$$A(\phi_1, \phi_2) = ((\alpha + 1)\sin\phi_1 \cos\phi_2 + (\beta + 1)\cos\phi_1 \sin\phi_2)^2 + (\alpha - \beta)^2 \sin^2\phi_1 \sin^2\phi_2.$$

For $\alpha \neq \beta$ $A(\phi_1, \phi_2) \geq (\alpha - \beta)^2 \sin^2\phi_1 \sin^2\phi_2 > 0$ and for $\alpha = \beta$ $A(\phi_1, \phi_2) = (\alpha + 1)^2 \sin(\phi_1 + \phi_2) > 0$. Hence $u(\theta)$ is a differentiable, strictly increasing function of θ, and it is easy to check that

$$\lim_{\theta\to 0^+} u(\theta) = 0 \quad \text{and} \quad \lim_{\theta\to \pi^-} u(\theta) = \infty.$$

Hence $u(\theta)$ is a bijection of $(0, \pi)$ onto $(0, \infty)$, which completes the proof of Theorem 6. □

Remark 5. It is much more complicated to express the densities $f_{\alpha,\beta}(x)$ directly as functions of x. This has been done for $\beta = 0$, $\alpha \in \mathbb{N}$ by K. Penson and

K. Życzkowski in [13] and extended to the case $\alpha \in \mathbb{Q}^+$ by W. Młotkowski, K. Penson and K. Życzkowski in [12, Theorem 3.1].

Acknowledgements The first author is supported by ERC Advanced Grant No. OAFPG 27731 and the Danish National Research Foundation through the Center for Symmetry and Deformation (DNRF92).

References

1. Arizmendi, O., Hasebe, T.: Classical and free infinite divisibility for Boolean stable laws (2012), arXiv:1205.1575
2. Banica, T., Belinschi, S.T., Capitaine, M., Collins, B.: Free Bessel laws. Canad. J. Math. **63**(1), 3–37 (2011). DOI 10.4153/CJM-2010-060-6. URL http://dx.doi.org/10.4153/CJM-2010-060-6
3. Bercovici, H., Pata, V.: Stable laws and domains of attraction in free probability theory. Ann. of Math. (2) **149**(3), 1023–1060 (1999). DOI 10.2307/121080. URL http://dx.doi.org/10.2307/121080. With an appendix by Philippe Biane
4. Bercovici, H., Voiculescu, D.: Free convolution of measures with unbounded support. Indiana Univ. Math. J. **42**(3), 733–773 (1993). DOI 10.1512/iumj.1993.42.42033. URL http://dx.doi.org/10.1512/iumj.1993.42.42033
5. Biane, P.: Processes with free increments. Math. Z. **227**(1), 143–174 (1998). DOI 10.1007/PL00004363. URL http://dx.doi.org/10.1007/PL00004363
6. Haagerup, U., Larsen, F.: Brown's spectral distribution measure for R-diagonal elements in finite von Neumann algebras. J. Funct. Anal. **176**(2), 331–367 (2000). DOI 10.1006/jfan.2000.3610. URL http://dx.doi.org/10.1006/jfan.2000.3610
7. Haagerup, U., Schultz, H.: Brown measures of unbounded operators affiliated with a finite von Neumann algebra. Math. Scand. **100**(2), 209–263 (2007)
8. Haagerup, U., Schultz, H.: Invariant subspaces for operators in a general II_1-factor. Publ. Math. Inst. Hautes Études Sci. (109), 19–111 (2009). DOI 10.1007/s10240-009-0018-7. URL http://dx.doi.org/10.1007/s10240-009-0018-7
9. Johnson, N.L., Kotz, S., Balakrishnan, N.: Continuous univariate distributions. Vol. 1, second edn. Wiley Series in Probability and Mathematical Statistics: Applied Probability and Statistics. John Wiley & Sons Inc., New York (1994). A Wiley-Interscience Publication
10. Larsen, F.: Powers of R-diagonal elements. J. Operator Theory **47**(1), 197–212 (2002)
11. Lindsay, J.M., Pata, V.: Some weak laws of large numbers in noncommutative probability. Math. Z. **226**(4), 533–543 (1997). DOI 10.1007/PL00004356. URL http://dx.doi.org/10.1007/PL00004356
12. Mlotkowski, W., Penson, K.A., Zyczkowski, K.: Densities of the Raney distributions (2012), arXiv:1211.7259
13. Penson, K.A., Zyczkowski, K.: Product of Ginibre matrices: Fuss-Catalan and Raney distributions. Phys. Rev. E **83**, 061,118 (2011), arXiv:1103.3453
14. Rosenthal, J.S.: A first look at rigorous probability theory, second edn. World Scientific Publishing Co. Pte. Ltd., Hackensack, NJ (2006)
15. Tucci, G.H.: Limits laws for geometric means of free random variables. Indiana Univ. Math. J. **59**(1), 1–13 (2010). DOI 10.1512/iumj.2010.59.3775. URL http://dx.doi.org/10.1512/iumj.2010.59.3775
16. Voiculescu, D.: Addition of certain noncommuting random variables. J. Funct. Anal. **66**(3), 323–346 (1986). DOI 10.1016/0022-1236(86)90062-5. URL http://dx.doi.org/10.1016/0022-1236(86)90062-5

17. Voiculescu, D.: Multiplication of certain noncommuting random variables. J. Operator Theory **18**(2), 223–235 (1987)
18. Voiculescu, D.V., Dykema, K.J., Nica, A.: Free random variables, *CRM Monograph Series*, vol. 1. American Mathematical Society, Providence, RI (1992). A noncommutative probability approach to free products with applications to random matrices, operator algebras and harmonic analysis on free groups

Chapter 9
Is Every Irreducible Shift of Finite Type Flow Equivalent to a Renewal System?

Rune Johansen

Abstract Is every irreducible shift of finite type flow equivalent to a renewal system? For the first time, this variation of a classic problem formulated by Adler is investigated, and several partial results are obtained in an attempt to find the range of the Bowen–Franks invariant over the set of renewal systems of finite type. In particular, it is shown that the Bowen–Franks group is cyclic for every member of a class of renewal systems known to attain all entropies realised by shifts of finite type, and several classes of renewal systems with non-trivial values of the invariant are constructed.

Keywords Reneweal systems • Symbolic dynamics • Shift spaces • Subshifts • Sofic shifts • Bowen–Franks group • Flow equivalence • Fischer cover

Mathematics Subject Classification (2010): 37B10.

9.1 Introduction

Here, a short introduction to the basic definitions and properties of shift spaces is given to make the present paper self-contained. For a thorough treatment of shift spaces see [12]. Let $\mathscr{A}$ be a finite set with the discrete topology. The *full shift* over $\mathscr{A}$ consists of the space $\mathscr{A}^{\mathbb{Z}}$ endowed with the product topology and the *shift map* $\sigma\colon \mathscr{A}^{\mathbb{Z}} \to \mathscr{A}^{\mathbb{Z}}$ defined by $\sigma(x)_i = x_{i+1}$ for all $i \in \mathbb{Z}$. Let $\mathscr{A}^*$ be the collection of finite words (also known as blocks) over $\mathscr{A}$. For $w \in \mathscr{A}^*$, $|w|$ will denote the

R. Johansen (✉)
Department of Mathematical Sciences, University of Copenhagen, Universitetsparken 5, DK-2100 Copenhagen Ø, Denmark
e-mail: rune@math.ku.dk

T.M. Carlsen et al. (eds.), *Operator Algebra and Dynamics*, Springer Proceedings in Mathematics & Statistics 58, DOI 10.1007/978-3-642-39459-1_9,

length of w. A subset $X \subseteq \mathscr{A}^{\mathbb{Z}}$ is called a *shift space* if it is invariant under the shift map and closed. For each $\mathscr{F} \subseteq \mathscr{A}^*$, define $\mathsf{X}_{\mathscr{F}}$ to be the set of bi-infinite sequences in $\mathscr{A}^{\mathbb{Z}}$ which do not contain any of the *forbidden words* from $\mathscr{F}$. A subset $X \subseteq \mathscr{A}^{\mathbb{Z}}$ is a shift space if and only if there exists $\mathscr{F} \subseteq \mathscr{A}^*$ such that $X = \mathsf{X}_{\mathscr{F}}$ (cf. [12, Proposition 1.3.4]). X is said to be a *shift of finite type* (SFT) if this is possible for a finite set $\mathscr{F}$.

The *language* of a shift space X is denoted $\mathscr{B}(X)$ and it is defined to be the set of all words which occur in at least one $x \in X$. The shift space X is said to be *irreducible* if there for every $u, w \in \mathscr{B}(X)$ exists $v \in \mathscr{B}(X)$ such that $uvw \in \mathscr{B}(X)$. For each $x \in X$, define the *left-ray* of x to be $x^- = \cdots x_{-2}x_{-1}$ and define the *right-ray* of x to be $x^+ = x_0x_1x_2\cdots$. The sets of all left-rays and all right-rays are, respectively, denoted X^- and X^+. Given a word or ray x, $\mathrm{rl}(x)$ and $\mathrm{ll}(x)$ will denote respectively the right-most and the left-most letter of x.

A *directed graph* is a quadruple $E = (E^0, E^1, r, s)$ consisting of countable sets E^0 and E^1, and maps $r, s\colon E^1 \to E^0$. A *path* $\lambda = e_1 \cdots e_n$ is a sequence of edges such that $r(e_i) = s(e_{i+1})$ for all $i \in \{1, \ldots n-1\}$. The vertices in E^0 are considered to be paths of length 0. For each $n \in \mathbb{N}_0$, the set of paths of length n is denoted E^n, and the set of all finite paths is denoted E^*. Extend the maps r and s to E^* by defining $s(e_1 \cdots e_n) = s(e_1)$ and $r(e_1 \cdots e_n) = r(e_n)$. A directed graph E is said to be *irreducible* (or transitive) if there for each pair of vertices $u, v \in E^0$ exists a path $\lambda \in E^*$ with $s(\lambda) = u$ and $r(\lambda) = v$. For a directed graph E, the *edge shift* (X_E, σ_E) is defined by $\mathsf{X}_E = \{x \in (E^1)^{\mathbb{Z}} \mid r(x_i) = s(x_{i+1}) \text{ for all } i \in \mathbb{Z}\}$.

A bijective, continuous and shift commuting map between two shift spaces is called a *conjugacy*, and when such a map exists, the two shift spaces are said to be *conjugate*. *Flow equivalence* is a weaker equivalence relation generated by conjugacy and *symbol expansion* [13]. Let A be the adjacency matrix of a directed graph E, then $\mathrm{BF}(A) = \mathbb{Z}^n/\mathbb{Z}^n(\mathrm{Id} - A)$ is called the *Bowen–Franks group* of A and it is an invariant of conjugacy of edge shifts. Let E and F be finite directed graphs for which the edge shifts X_E and X_F are irreducible and not flow equivalent to the trivial shift with one element, and let A_E and A_F be the corresponding adjacency matrices. Then X_E and X_F are flow equivalent if and only $\mathrm{BF}(A_E) \simeq \mathrm{BF}(A_F)$ and the signs $\operatorname{sgn}\det A_E$ and $\operatorname{sgn}\det A_F$ are equal [3]. Every SFT is conjugate to an edge shift, so this gives a complete flow equivalence invariant of irreducible SFTs. The pair consisting of the Bowen–Franks group and the sign of the determinant is called the *signed Bowen–Franks group*, and it is denoted BF_+. This invariant is easy to compute and easy to compare which makes it appealing to consider flow equivalence rather than conjugacy.

A *labelled graph* $(E, \mathscr{L})$ over an alphabet $\mathscr{A}$ consists of a directed graph E and a surjective labelling map $\mathscr{L}\colon E^1 \to \mathscr{A}$. Given a labelled graph $(E, \mathscr{L})$, define the shift space $(\mathsf{X}_{(E,\mathscr{L})}, \sigma)$ by setting $\mathsf{X}_{(E,\mathscr{L})} = \{(\mathscr{L}(x_i))_i \in \mathscr{A}^{\mathbb{Z}} \mid x \in \mathsf{X}_E\}$. The labelled graph $(E, \mathscr{L})$ is said to be a *presentation* of the shift space $\mathsf{X}_{(E,\mathscr{L})}$, and a *representative* of a word $w \in \mathscr{B}(\mathsf{X}_{(E,\mathscr{L})})$ is a path $\lambda \in E^*$ such that $\mathscr{L}(\lambda) = w$ with the natural extension of $\mathscr{L}$. Representatives of rays are defined analogously. Let $(E, \mathscr{L})$ be a labelled graph presenting X. For each $v \in E^0$, define the *predecessor* *set* of v to be the set of left-rays in X which have a presentation terminating at v. This is denoted $P_\infty^E(v)$, or just $P_\infty(v)$ when $(E, \mathscr{L})$ is understood from the context. The presentation $(E, \mathscr{L})$ is said to be *predecessor-separated* if $P_\infty^E(u) \neq P_\infty^E(v)$ when $u, v \in E^0$ and $u \neq v$.

A function $\pi: X_1 \to X_2$ between shift spaces X_1 and X_2 is said to be a *factor map* if it is continuous, surjective, and shift commuting. A shift space is called *sofic* [16] if it is the image of an SFT under a factor map. Every SFT is sofic, and a sofic shift which is not an SFT is called *strictly sofic*. Fischer proved that a shift space is sofic if and only if it can be presented by a finite labelled graph [2]. A sofic shift space is irreducible if and only if it can be presented by an irreducible labelled graph (see [12, Sect. 3.1]).

Let $(E, \mathscr{L})$ be a finite labelled graph which presents the sofic shift space $\mathsf{X}_{(E,\mathscr{L})}$, and let $\pi_\mathscr{L}: \mathsf{X}_E \to \mathsf{X}_{(E,\mathscr{L})}$ be the factor map induced by the labelling map $\mathscr{L}: E^1 \to \mathscr{A}$, then the SFT X_E is called a *cover* of the sofic shift $\mathsf{X}_{(E,\mathscr{L})}$, and $\pi_\mathscr{L}$ is called the covering map.

Let X be a shift space over an alphabet $\mathscr{A}$. A presentation $(E, \mathscr{L})$ of X is said to be *left-resolving* if no vertex in E^0 receives two edges with the same label. Fischer proved [2] that up to labelled graph isomorphism every irreducible sofic shift has a unique left-resolving presentation with fewer vertices than any other left-resolving presentation. This is called the *left Fischer cover* of X, and it is denoted $(F, \mathscr{L}_F)$.

For $x^+ \in X^+$, define the *predecessor set* of x^+ to be the set of left-rays which may precede x^+ in X, that is $P_\infty(x^+) = \{y^- \in X^- \mid y^- x^+ \in X\}$ (see [10, Sects. I and III] and [12, Exercise 3.2.8] for details). The *follower set* of a left-ray $x^- \in X^-$ is defined analogously. The *left Krieger cover* of the sofic shift space X is the labelled graph $(K, \mathscr{L}_K)$ where $K^0 = \{P_\infty(x^+) \mid x^+ \in X^+\}$, and where there is an edge labelled $a \in \mathscr{A}$ from $P \in K^0$ to $P' \in K^0$ if and only if there exists $x^+ \in X^+$ such that $P = P_\infty(ax^+)$ and $P' = P_\infty(x^+)$. A word $v \in \mathscr{B}(X)$ is said to be *intrinsically synchronising* if $uvw \in \mathscr{B}(X)$ whenever u and w are words such that $uv, vw \in \mathscr{B}(X)$. A ray is said to be *intrinsically synchronising* if it contains an intrinsically synchronising word as a factor. If a right-ray x^+ is intrinsically synchronising, then there is precisely one vertex in the left Fischer cover where a presentation of x^+ can start, and this vertex can be identified with the predecessor set $P_\infty(x^+)$ as a vertex in the Krieger cover. In this way, the left Fischer cover can be identified with the irreducible component of the left Krieger cover generated by the vertices that are predecessor sets of intrinsically synchronising right-rays [11, Lemma 2.7], [12, Exercise 3.3.4]. The interplay between the structure of the Fischer and Krieger covers is examined in detail in [8].

Let $\mathscr{A}$ be an alphabet, let $L \subseteq \mathscr{A}^*$ be a finite list of words over $\mathscr{A}$, and define $\mathscr{B}(L)$ to be the set of factors of elements of L^*. Then $\mathscr{B}(L)$ is the language of a shift space $\mathsf{X}(L)$ which is said to be the *renewal system* generated by L. L is said to be the *generating list* of $\mathsf{X}(L)$. A renewal system is an irreducible sofic shift since it can be presented by the labelled graph obtained by writing the generating words on loops starting and ending at a common vertex. This graph is called the *standard loop graph presentation* of $\mathsf{X}(L)$, and because of this presentation, renewal systems are called *loop systems* or *flower automata* in automata theory (e.g. [1]).

Simple examples show that not every sofic shift—or every SFT—is a renewal system [12, pp. 433], and these results naturally raise the following question, which was first asked by Adler: Is every irreducible shift of finite type conjugate to a renewal system? This question has been the motivation of most of the work done on renewal systems [4–6, 9, 14, 15, 17]. The analogous question for sofic shifts has a negative answer [17]. The aim of the present work has been to answer another natural variation of Adler's question: Is every irreducible SFT *flow equivalent* to a renewal system? To answer this question, it is sufficient to find the range of the Bowen–Franks invariant over the set of SFT renewal systems and check whether it is equal to the range over the set of irreducible SFTs. It is easy to check that a group G is the Bowen–Franks group of an irreducible SFT if and only if it is a finitely generated abelian group and that any combination of sign and Bowen–Franks group can be achieved by the Bowen–Franks invariant. Hence, the overall strategy of the investigation of the flow equivalence question has been to attempt to construct all these combinations of groups and signs. However, it is difficult to construct renewal systems attaining many of the values of the invariant. In fact, it is non-trivial to construct an SFT renewal system that is not flow equivalent to a full shift [7].

Section 9.2 concerns the left Fischer covers of renewal systems and gives conditions under which the Fischer covers of complicated renewal systems can be constructed from simpler building blocks with known presentations. Section 9.3 gives a flow classification of a class of renewal systems introduced in [6], while Sect. 9.4 uses the results of the previous two sections to construct classes of renewal systems with interesting values of the Bowen–Franks invariant.

9.2 Fischer Covers of Renewal Systems

In the attempt to find the range of the Bowen–Franks invariant over the set of SFT renewal systems, it is useful to be able to construct complicated renewal systems from simpler building blocks, but in general, it is non-trivial to study the structure of the renewal system $\mathsf{X}(L_1 \cup L_2)$ even if the renewal systems $\mathsf{X}(L_1)$ and $\mathsf{X}(L_2)$ are well understood. The goal of this section is to describe the structure of the left Fischer covers of renewal systems in order to give conditions under which the Fischer cover of $\mathsf{X}(L_1 \cup L_2)$ can be constructed when the Fischer covers of $\mathsf{X}(L_1)$ and $\mathsf{X}(L_2)$ are known.

Let L be a generating list and define $P_0(L) = \{\dots w_{-2}w_{-1}w_0 \mid w_i \in L\} \subseteq \mathsf{X}(L)^-$. $P_0(L)$ is the predecessor set of the central vertex in the standard loop graph of $\mathsf{X}(L)$, but it is not necessarily the predecessor set of a right-ray in $\mathsf{X}(L)^+$, so it does not necessarily correspond to a vertex in the left Fischer cover of $\mathsf{X}(L)$. If $p \in \mathscr{B}(\mathsf{X}(L))$ is a prefix of some word in L, define $P_0(L)p = \{\dots w_{-2}w_{-1}w_0 p \mid w_i \in L\} \subseteq \mathsf{X}(L)^-$.

Let L be a generating list. A triple (n_b, g, l) where $n_b, l \in \mathbb{N}$ and g is an ordered list of words $g_1, \dots, g_k \in L$ with $\sum_{i=1}^{k} |g_i| \geq n_b + l - 1$ is said to be a *partitioning* of the factor $v_{[n_b, n_b+l-1]} \in \mathscr{B}(\mathsf{X}(L))$ of $v = g_1 \cdots g_k$. The *beginning* of the partitioning is the word $v_{[1,n_b-1]}$, and the *end* is the word $v_{[n_b+l,|v|]}$. A partitioning of a right-ray $x^+ \in \mathsf{X}(L)^+$ is a pair $p = (n_b, (g_i)_{i \in \mathbb{N}})$ where $n_b \in \mathbb{N}$ and $g_i \in L$ such that $wx^+ = g_1 g_2 \cdots$ when w is the *beginning* consisting of the $n_b - 1$ first letters of the concatenation $g_1 g_2 \cdots$. Partitionings of left-rays are defined analogously.

Let $L \subseteq \mathscr{A}^*$ be a finite list, and let $w \in \mathscr{B}(\mathsf{X}(L)) \cup \mathsf{X}(L)^+$ be an allowed word or right-ray. Then w is said to be *left-bordering* if there exists a partitioning of w with empty beginning, and *strongly left-bordering* if every partitioning of w has empty beginning. Right-bordering words and left-rays are defined analogously.

Definition 1. Let $L \subseteq \mathscr{A}^*$ be finite, and let $(F, \mathscr{L}_F)$ be the left Fischer cover of $\mathsf{X}(L)$. A vertex $P \in F^0$ is said to be a *(universal) border point* for L if there exists a (strongly) left-bordering $x^+ \in X^+$ such that $P = P_\infty(x^+)$. An intrinsically synchronising word $w \in L^*$ is said to be a *generator* of the border point $P_\infty(w) = P_\infty(w^\infty)$, and it is said to be a *minimal generator* of P if no prefix of w is a generator of P.

The border points add information to the Fischer cover about the structure of the generating lists, and this information will be useful for studying $\mathsf{X}(L_1 \cup L_2)$ when the Fischer covers of $\mathsf{X}(L_1)$ and $\mathsf{X}(L_2)$ are known. If P is a (universal) border point of L and there is no ambiguity about which list is generating $X = \mathsf{X}(L)$, then the terminology will be abused slightly by saying that P is a (universal) border point of X or simply of the left Fischer cover.

Lemma 1. *Let L be a finite list generating a renewal system with left Fischer cover $(F, \mathscr{L}_F)$.*

1. *If $P \in F^0$ is a border point, then $P_0(L) \subseteq P$, and if P is universal then $P = P_0(L)$.*
2. *If $P_1, P_2 \in F^0$ are border points and if $w_1 \in L^*$ is a generator of P_1, then there exists a path with label w_1 from P_1 to P_2.*
3. *If $P_1 \in F^0$ is a border point and $w \in L^*$, then there exists a unique border point $P_2 \in F^0$ with a path labelled w from P_2 to P_1.*
4. *If $\mathsf{X}(L)$ is an SFT, then every border point of L has a generator.*
5. *If L has a strongly right-bordering word w, then $x^+ \in \mathsf{X}(L)^+$ is left-bordering if and only if $P_\infty(x^+)$ is a border point.*

Proof. (1) Choose a left-bordering $x^+ \in \mathsf{X}(L)^+$ such that $P = P_\infty(x^+)$ and note that $y^- x^+ \in \mathsf{X}(L)$ for each $y^- \in P_0(L)$. (2) Choose a left-bordering $x^+ \in \mathsf{X}(L)^+$ such that $P_2 = P_\infty(x^+)$. Then $P_\infty(w_1 x^+) = P_1$ since $w_1 x^+ \in \mathsf{X}(L)^+$ and w_1 is intrinsically synchronising, so there is a path labelled w_1 from P_1 to P_2. (3) Choose a left-bordering $x^+ \in \mathsf{X}(L)^+$ such that $P = P_\infty(x^+)$. Since $w \in L^*$, the right-ray wx^+ is also left-bordering. (4) Let $P = P_\infty(x^+)$ for some left-bordering $x^+ \in \mathsf{X}(L)^+$, and choose an intrinsically synchronising prefix $w \in L^*$ of x^+. Then $P_\infty(x^+) = P_\infty(w)$, so w is a generator of P. (5) If $P_\infty(x^+)$ is a border point, then $wx^+ \in \mathsf{X}(L)^+$, so x^+ must be left-bordering. The other implication holds by definition.

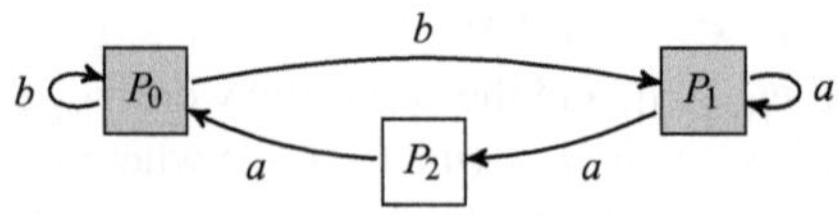

Fig. 9.1 Left Fischer cover of the SFT renewal system $\mathsf{X}(L)$ generated by $L = \{aa, aaa, b\}$ discussed in Example 1. The border points are coloured *grey*

In particular, the universal border point is unique when it exists. A predecessor set $P_\infty(x^+)$ can be a border point even though x^+ is not left-bordering.

Example 1. Consider the list $L = \{aa, aaa, b\}$ and the renewal system $\mathsf{X}(L)$. It is straightforward to check that $\mathsf{X}(L) = \mathsf{X}_{\mathscr{F}}$ for the set of forbidden words $\mathscr{F} = \{bab\}$, so this is an SFT. For this shift, there are three distinct predecessor sets:

$$P_0 = P_\infty(b\cdots) = \{\cdots x_{-1}x_0 \in \mathsf{X}(L)^- \mid x_0 = b \text{ or } x_{-1}x_0 = aa\},$$

$$P_1 = P_\infty(a^n b\cdots) = P_\infty(a^\infty) = \mathsf{X}(L)^-, \qquad n \geq 2,$$

$$P_2 = P_\infty(ab\cdots) = \{\cdots x_{-1}x_0 \in \mathsf{X}(L)^- \mid x_0 = a\}.$$

The information contained in these equations is sufficient to draw the left Krieger cover, and each set is the predecessor set of an intrinsically synchronising right-ray, so the left Fischer cover can be identified with the left Krieger cover. This graph is shown in Fig. 9.1. Here, P_0 is a universal border point because any right-ray starting with a b is strongly left bordering. The generating word b is a minimal generator of P_0. The vertex P_1 is a border point because $a^n b\cdots$ is left bordering for all $n \geq 2$. The word aa is a minimal generator of P_1, and aab is a non-minimal generator. The vertex P_2 is not a border point since there is no infinite concatenation x^+ of words from L such that $x^+ = ab\cdots$. Another way to see this is to note that every path terminating at P_2 has a as a suffix, so that P_0 is not a subset of P_2 which together with Lemma 1 implies that P_2 is not a border point. Note also that Lemma 1 means that there must be paths labelled b from P_0 to the two border points, and similarly, paths labelled aa and aab from P_1 to the two border points.

Consider two renewal systems $\mathsf{X}(L_1)$ and $\mathsf{X}(L_2)$. The *sum* $\mathsf{X}(L_1) + \mathsf{X}(L_2)$ is the renewal system $\mathsf{X}(L_1 \cup L_2)$. Generally, it is non-trivial to construct the Fischer cover of such a sum even if the Fischer covers of the summands are known.

Definition 2. Let L be a generating list with universal border point P_0 and let $(F, \mathscr{L}_F)$ be the left Fischer cover of $\mathsf{X}(L)$. L is said to be *left-modular* if for all $\lambda \in F^*$ with $r(\lambda) = P_0$, $\mathscr{L}_F(\lambda) \in L^*$ if and only if $s(\lambda)$ is a border point. *Right-modular* generating lists are defined analogously.

It is straightforward to check that the list considered in Example 1 is left-modular. When L is left-modular and there is no doubt about which generating list is used, the renewal system $\mathsf{X}(L)$ will also be said to be *left-modular*.

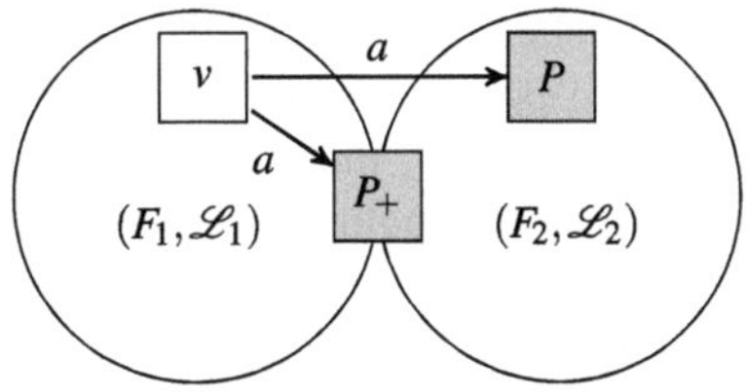

Fig. 9.2 The labelled graph $(F_+, \mathscr{L}_+)$. In $(F_1, \mathscr{L}_1)$, v emits an edge labelled a to P_1, so in $(F_+, \mathscr{L}_+)$, the corresponding vertex emits edges labelled a to every vertex corresponding to a border point $P \in F_2^0$

Lemma 2. *If L is a generating list with a strongly left-bordering word w_l and a strongly right-bordering word w_r, then it is both left- and right-modular.*

Proof. Let $(F, \mathscr{L}_F)$ be the left Fischer cover of $\mathsf{X}(L)$, let $P \in F^0$ be a border point, and choose $x^+ \in \mathsf{X}(L)^+$ such that $w_l x^+ \in \mathsf{X}(L)^+$. Assume that there is a path from P to $P_0(L) = P_\infty(w_l x^+)$ with label w. The word w_r has a partitioning with empty end, so there is a path labelled w_r terminating at P. It follows that $w_r w w_l x^+ \in \mathsf{X}(L)^+$, so $w \in L^*$. By symmetry, L is also right-modular.

For $i \in \{1,2\}$, let L_i be a left-modular generating list and let $X_i = \mathsf{X}(L_i)$ have alphabet $\mathscr{A}_i$ and left Fischer cover $(F_i, \mathscr{L}_i)$. Let $P_i \in F_i^0$ be the universal border point of L_i. Assume that $\mathscr{A}_1 \cap \mathscr{A}_2 = \emptyset$. The left Fischer cover of $X_1 + X_2$ will turn out to be the labelled graph $(F_+, \mathscr{L}_+)$ obtained by taking the union of $(F_1, \mathscr{L}_1)$ and $(F_2, \mathscr{L}_2)$, identifying the two universal border points P_1 and P_2, and adding certain connecting edges. To do this formally, introduce a new vertex P_+ and define $F_+^0 = (F_1^0 \cup F_2^0 \cup \{P_+\}) \setminus \{P_1, P_2\}$. Define maps $f_i \colon F_i^0 \to F_+^0$ such that for $v \in F_i^0 \setminus \{P_i\}$, $f_i(v)$ is the vertex in F_+^0 corresponding to v and such that $f_i(P_i) = P_+$. For each $e \in F_i^1$, define an edge $e' \in F_+^1$ such that $s(e') = f_i(s(e))$, $r(e') = f_i(r(e))$, and $\mathscr{L}_+(e') = \mathscr{L}_i(e)$. For each $e \in F_1^1$ with $r(e) = P_1$ and each non-universal border point $P \in F_2^0$, draw an additional edge $e' \in F_+^1$ with $s(e') = f_1(s(e))$, $r(e') = f_2(P)$, and $\mathscr{L}_+(e') = \mathscr{L}_1(e)$. Draw analogous edges for each $e \in F_2^1$ with $r(e) = P_2$ and every non-universal border point $P \in F_1^0$. This construction is illustrated in Fig. 9.2.

Proposition 1. *If L_1 and L_2 are left-modular generating lists with disjoint alphabets, then $L_1 \cup L_2$ is left-modular, the left Fischer cover of $\mathsf{X}(L_1 \cup L_2)$ is the graph $(F_+, \mathscr{L}_+)$ constructed above, and the vertex $P_+ \in F_+^0$ is the universal border point of $L_1 \cup L_2$.*

Proof. By construction, the labelled graph $(F_+, \mathcal{L}_+)$ is irreducible, left-resolving, and predecessor-separated, so it is the left Fischer cover of some sofic shift X_+ [12, Corollary 3.3.19]. Given $w \in L_1^*$, there is a path with label w in the left Fischer cover of X_1 from some border point $P \in F_1^0$ to the universal border point P_1 by Lemma 1. Hence, there is also a path labelled w in $(F_+, \mathcal{L}_+)$ from the vertex corresponding to P to the vertex P_+. This means that for every border point $Q \in F_2^0$, $(F_+, \mathcal{L}_+)$ contains a path labelled w from the vertex corresponding to P to the vertex corresponding to Q. By symmetry, it follows that every element of $(L_X \cup L_Y)^*$ has a presentation in $(F_+, \mathcal{L}_+)$. Hence, $\mathsf{X}(L_1 \cup L_2) \subset X_+$.

Assume that $awb \in \mathcal{B}(X_+)$ with $a, b \in \mathcal{A}_1$ and $w \in \mathcal{A}_2^*$. Then there must be a path labelled w in $(F_+, \mathcal{L}_+)$ from a vertex corresponding to a border point P of L_2 to P_+. By construction, this is only possible if there is also a path labelled w from P to P_2 in $(F_2, \mathcal{L}_2)$, but L_2 is left-modular, so this means that $w \in L_2^*$. By symmetry, $\mathsf{X}(L_1 \cup L_2) = X_+$, and P_+ is the universal border point by construction.

Let X be a shift space over the alphabet $\mathcal{A}$. Given $a \in \mathcal{A}$, $k \in \mathbb{N}$, and new symbols $a_1, \dots, a_k \notin \mathcal{A}$ consider the map $f_{a,k} \colon (\mathcal{A} \setminus \{a\}) \cup \{a_1, \dots, a_k\} \to \mathcal{A}$ defined by $f_{a,k}(a_i) = a$ for each $1 \le i \le k$ and $f_{a,k}(b) = b$ when $b \in \mathcal{A} \setminus \{a\}$. Let $F_{a,k} \colon ((\mathcal{A} \setminus \{a\}) \cup \{a_1, \dots, a_k\})^* \to \mathcal{A}^*$ be the natural extension of $f_{a,k}$. If $w \in \mathcal{A}^*$ contains l copies of the symbol a, then the preimage $F_{a,k}^{-1}(\{w\})$ is the set consisting of the k^l words that can be obtained by replacing the as by the symbols $a_1, \dots, a_k$.

Definition 3. Let $X = \mathsf{X}_{\mathcal{F}}$ be a shift space over the alphabet $\mathcal{A}$, let $a \in \mathcal{A}$, let $a_1, \dots, a_k \notin \mathcal{A}$, and let $F_{a,k}$ be defined as above. Then the shift space $X_{a,k} = \mathsf{X}_{F_{a,k}^{-1}(\mathcal{F})}$ is said to be the shift obtained from X by *fragmenting* a into $a_1, \dots, a_k$.

Note that this construction does not depend on the choice of $\mathcal{F}$ representing X, in particular, $\mathcal{B}(X_{a,k}) = F_{a,k}^{-1}(\mathcal{B}(X))$. Furthermore, $X_{a,k}$ is an SFT if and only if X is an SFT. If X is an irreducible sofic shift, then the left and right Fischer and Krieger covers of $X_{a,k}$ are obtained by replacing each edge labelled a in the corresponding cover of X by k edges labelled $a_1, \dots, a_k$. Note that X and $X_{a,k}$ are not generally conjugate or even flow equivalent. If $X = \mathsf{X}(L)$ is a renewal system, then $X_{a,k}$ is the renewal system generated by the list $L_{a,k} = F_{a,k}^{-1}(L)$.

Remark 1. Let A be the symbolic adjacency matrix of the left Fischer cover of an SFT renewal system $\mathsf{X}(L)$ with alphabet $\mathcal{A}$. Given $a \in \mathcal{A}$ and $k \in \mathbb{N}$, define $f \colon \mathcal{A} \to \mathbb{N}$ by $f(a) = k$ and $f(b) = 1$ for $b \neq a$. Extend f to the set of finite formal sums over $\mathcal{A}$ in the natural way and consider the integer matrix $f(A)$. Then $f(A)$ is the adjacency matrix of the underlying graph of the left Fischer cover of $\mathsf{X}(L_{a,k})$. For lists over disjoint alphabets, it follows immediately from the definitions that fragmentation and addition commute.

9.3 Entropy and Flow Equivalence

Hong and Shin [6] have constructed a class H of lists generating SFT renewal systems such that $\log\lambda$ is the entropy of an SFT if and only if there exists $L \in H$ with $h(\mathsf{X}(L)) = \log\lambda$, and this is arguably the most powerful general result known about the invariants of SFT renewal systems. In the following, the renewal systems generated by lists from H will be classified up to flow equivalence. As demonstrated in [7], it is difficult to construct renewal systems with non-cyclic Bowen–Franks groups and/or positive determinants directly, and this classification will yield hitherto unseen values of the invariant.

The construction of the class H of generating lists considered in [6] will be modified slightly since some of the details of the original construction are invisible up to flow equivalence. In particular, several words from the generating lists can be replaced by single symbols by using symbol reduction. Additionally, there are extra conditions on some of the variables in [6] which will be omitted here since the larger class can be classified without extra work.

Let $r \geq 2$ and let $n_1, \dots, n_r, c_1, \dots, c_r, d, N \in \mathbb{N}$, and let W be the set consisting of the following words:

- $\alpha_i = \alpha_{i,1} \cdots \alpha_{i,n_1}$ for $1 \leq i \leq c_1$
- $\tilde{\alpha}_i = \tilde{\alpha}_{i,1} \cdots \tilde{\alpha}_{i,n_1}$ for $1 \leq i \leq c_1$
- $\gamma_{k,i_k} = \gamma_{k,i_k,1} \cdots \gamma_{k,i_k,n_k}$ for $2 \leq k \leq r$ and $1 \leq i_k \leq c_k$
- $\alpha_{i_1}\gamma_{2,i_2} \cdots \gamma_{r,i_r}\beta_l^N$ for $1 \leq i_j \leq c_j$ and $1 \leq l \leq d$
- $\beta_l^N \tilde{\alpha}_{i_1}\gamma_{2,i_2} \cdots \gamma_{r,i_r}$ for $1 \leq i_j \leq c_j$ and $1 \leq l \leq d$.

The set of generating lists of this form will be denoted B.

Remark 2. Symbol reduction can be used to reduce the words α_i, $\tilde{\alpha}_i$, γ_{k,i_k}, and β_l^N to single letters [7, Lemmas 2.15 and 2.23], so up to flow equivalence, the list $W \in B$ considered above can be replaced by the list W' consisting of the one-letter words α_i, $\tilde{\alpha}_i$, and $\gamma_{k,i}$ as well as the words

- $\alpha_{i_1}\gamma_{2,i_2} \cdots \gamma_{r,i_r}\beta_l$ for $1 \leq i_j \leq c_j$ and $1 \leq l \leq d$
- $\beta_l \tilde{\alpha}_{i_1}\gamma_{2,i_2} \cdots \gamma_{r,i_r}$ for $1 \leq i_j \leq c_j$ and $1 \leq l \leq d$.

Furthermore, if

$$L = \{\alpha, \tilde{\alpha}, \alpha\gamma_2 \cdots \gamma_r\beta, \beta\tilde{\alpha}\gamma_2 \cdots \gamma_r\} \cup \{\gamma_k \mid 2 \leq k \leq r\}, \tag{9.1}$$

then $\mathsf{X}(W')$ can be obtained from $\mathsf{X}(L)$ by fragmenting α to $\alpha_1, \dots, \alpha_{c_1}$, β to $\beta_1, \dots, \beta_l$ and so on. Let R be the set of generating lists of the form given in (9.1).

Next consider generating lists $W_1, \dots, W_m \in B$ with disjoint alphabets, and let $W = \bigcup_{j=1}^m W_j$. Let $\tilde{W}$ be a finite set of words that do not share any letters with each other or with the words from W, and consider the generating list $W \cup \tilde{W}$. Let $\tilde{H}$ be the set of generating lists that can be constructed in this manner. Let μ be a Perron number. Then there exists $\tilde{L} \in \tilde{H}$ such that $\mathsf{X}(\tilde{L})$ is an SFT and $h(\mathsf{X}(\tilde{L})) = \log\mu$ [6].

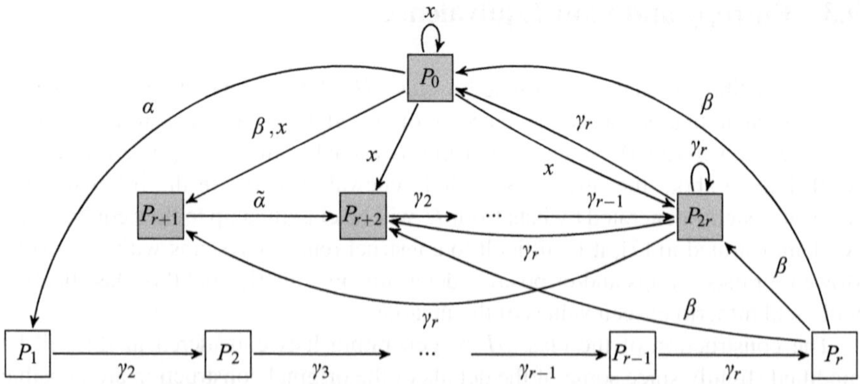

Fig. 9.3 Left Fischer cover of $\mathsf{X}(L)$ for L defined in (9.2). An edge labelled x from a vertex P to a vertex Q represents a collection of edges from P to Q such that Q receives an edge with each label from the set $\bigcup_{2 \le j \le r}\{\gamma_j\} \cup \{\alpha, \tilde{\alpha}\}$, i.e. the collection fills the gaps left by the edges which are labelled explicitly. The border points are coloured *grey*

Remark 3. If $W \cup \tilde{W} \in \tilde{H}$ as above, then symbol reduction can be used to show that $\mathsf{X}(W \cup \tilde{W})$ is flow equivalent to the renewal system generated by the union of W and $|\tilde{W}|$ new letters [7, Lemma 2.23], i.e. $\mathsf{X}(W \cup \tilde{W})$ is flow equivalent to a fragmentation of $\mathsf{X}(W \cup \{a\})$ when $a \notin \mathscr{A}(\mathsf{X}(W))$.

Consider a generating list $\tilde{L} \in \tilde{H}$ and $p \in \mathbb{N}$. For each letter $a \in \mathscr{A}(\mathsf{X}(\tilde{L}))$, introduce new letters $a_1, \dots, a_p \notin \mathscr{A}(\mathsf{X}(\tilde{L}))$, and let L denote the generating list obtained by replacing each occurrence of a in $\tilde{L}$ by the word $a_1 \cdots a_p$. Let H denote the set of generating lists that can be obtained from $\tilde{H}$ in this manner. Let λ be a weak Perron number. Then there exists $L \in H$ such that $\mathsf{X}(L)$ is an SFT and $h(\mathsf{X}(L)) = \log \lambda$ [6].

Remark 4. If L is obtained from $\tilde{L} \in \tilde{H}$ as above, then $\mathsf{X}(L) \sim_{\mathrm{FE}} \mathsf{X}(\tilde{L})$ since the modification can be achieved using symbol expansion of each $a \in \mathscr{A}(\mathsf{X}(\tilde{L}))$.

The next step is to prove that the building blocks in the class R introduced in Remark 2 are left-modular, and to construct the Fischer covers of the corresponding renewal systems. As the following lemmas show, this will allow a classification of the renewal systems generated by lists from H via addition and fragmentation. The first result follows immediately from Remarks 1 to 4.

Lemma 3. *For each $L \in H$, there exist $L_1, \dots, L_m \in R$ such that $\mathsf{X}(L)$ is flow equivalent to a fragmentation of $\mathsf{X}(\bigcup_{j=0}^{m} L_j)$, where $L_0 = \{a\}$ for some a that does not occur in $L_1, \dots, L_m$.*

Lemma 4. *If $L \in R$, then L is left-modular, $\mathsf{X}(L)$ is an SFT, and the left Fischer cover of $\mathsf{X}(L)$ is the labelled graph shown in Fig. 9.3.*

Proof. Let

$$L = \{\alpha, \tilde{\alpha}, \alpha\gamma_2\cdots\gamma_r\beta, \beta\tilde{\alpha}\gamma_2\cdots\gamma_r\} \cup \{\gamma_k \mid 2 \le k \le r\} \in R\,. \tag{9.2}$$

The word $\alpha\gamma_2\cdots\gamma_r\beta\beta\tilde{\alpha}\gamma_2\cdots\gamma_r$ is strongly left- and right-bordering, so L is left- and right-modular by Lemma 2. Let $P_0 = P_0(L)$. If $x^+ \in \mathsf{X}(L)^+$ does not have a suffix of a product of the generating words $\alpha\gamma_2\cdots\gamma_r\beta$ and $\beta\tilde{\alpha}\gamma_2\cdots\gamma_r$ as a prefix, then x^+ is strongly left-bordering, so $P_\infty(x^+) = P_0$. Hence, to determine the rest of the predecessor sets and thereby the vertices of the left Fischer cover, it is sufficient to consider right-rays that do have such a prefix.

Consider first $x^+ \in \mathsf{X}(L)^+$ such that $\beta x^+ \in \mathsf{X}(L)^+$. The letter β must come from either $\alpha\gamma_2\cdots\gamma_r\beta$ or $\beta\tilde{\alpha}\gamma_2\cdots\gamma_r$, so the beginning of a partitioning of βx^+ must be either empty or equal to $\alpha\gamma_2\cdots\gamma_r$. Assume first that every partitioning of βx^+ has beginning $\alpha\gamma_2\cdots\gamma_r$ (i.e. that $\tilde{\alpha}\gamma_2\cdots\gamma_r$ is not a prefix of x^+). In this case, βx^+ must be preceded by $\alpha\gamma_2\cdots\gamma_r$, and the corresponding predecessor sets are:

$$\begin{aligned}
P_\infty(\alpha\gamma_2\cdots\gamma_r\beta x^+) &= P_0\\
P_\infty(\gamma_2\cdots\gamma_r\beta x^+) &= P_0\alpha = P_1\\
&\ \,\vdots\\
P_\infty(\gamma_r\beta x^+) &= P_0\alpha\gamma_2\cdots\gamma_{r-1} = P_{r-1}\\
P_\infty(\beta x^+) &= P_0\alpha\gamma_2\cdots\gamma_{r-1}\gamma_r = P_r\,.
\end{aligned} \tag{9.3}$$

Assume now that there exists a partitioning of βx^+ with empty beginning (e.g. $x^+ = \beta\tilde{\alpha}\gamma_2\cdots\gamma_r^\infty$). The first word used in such a partitioning must be $\beta\tilde{\alpha}\gamma_2\cdots\gamma_r$. Replacing this word by the concatenation of the generating words $\alpha\gamma_2\cdots\gamma_r\beta$, $\tilde{\alpha}, \gamma_2, \ldots, \gamma_r$ creates a partitioning of βx^+ with beginning $\alpha\gamma_2\cdots\gamma_r$, so in this case:

$$\begin{aligned}
P_\infty(\alpha\gamma_2\cdots\gamma_r\beta x^+) &= P_0\\
P_\infty(\gamma_2\cdots\gamma_r\beta x^+) &= P_0 \cup P_0\alpha = P_0\\
&\ \,\vdots\\
P_\infty(\gamma_r\beta x^+) &= P_0 \cup P_0\alpha\gamma_2\cdots\gamma_{r-1} = P_0\\
P_\infty(\beta x^+) &= P_0 \cup P_0\alpha\gamma_2\cdots\gamma_{r-1}\gamma_r = P_0\,.
\end{aligned}$$

The argument above proves that there are no right-rays such that every partitioning of βx^+ has empty beginning.

It only remains to investigate right-rays that have a suffix of $\beta\tilde{\alpha}\gamma_2\cdots\gamma_r$ as a prefix. A partitioning of a right-ray $\gamma_r x^+$ may have empty beginning (e.g. $x^+ = \gamma_r^\infty$), beginning $\alpha\gamma_2\cdots\gamma_{r-1}$ (e.g. $x^+ = \beta\beta\tilde{\alpha}\gamma_2\cdots\gamma_r\cdots$ or $x^+ = \beta\tilde{\alpha}\gamma_2\cdots\gamma_r^\infty$), or beginning $\beta\tilde{\alpha}\gamma_2\cdots\gamma_{r-1}$ (e.g. $x^+ = \gamma_r^\infty$). Note that there is a partitioning with empty beginning if and only if there is a partitioning with beginning $\beta\tilde{\alpha}\gamma_2\cdots\gamma_{r-1}$. If there exists a partitioning of $\gamma_r x^+$ with beginning $\alpha\gamma_2\cdots\gamma_{r-1}$, then β must be a prefix of x^+, so the right-ray $\gamma_r x^+$ has already been considered above. Hence, it suffices to consider the case where there exists a partitioning of $\gamma_r x^+$ with empty beginning and a partitioning with beginning $\beta\tilde{\alpha}\gamma_2\cdots\gamma_{r-1}$ but no partitioning with beginning $\alpha\gamma_2\cdots\gamma_{r-1}$. In this case, the predecessor sets are

$$\begin{aligned}
P_\infty(\gamma_r x^+) &= P_0 \cup P_0\beta\tilde{\alpha}\gamma_2\cdots\gamma_{r-1} = P_{2r} \\
&\vdots \\
P_\infty(\gamma_2\cdots\gamma_r x^+) &= P_0 \cup P_0\beta\tilde{\alpha} = P_{r+2} \\
P_\infty(\tilde{\alpha}\gamma_2\cdots\gamma_r x^+) &= P_0 \cup P_0\beta = P_{r+1} \\
P_\infty(\beta\tilde{\alpha}\gamma_2\cdots\gamma_r x^+) &= P_0 \cup P_0\alpha\gamma_2\cdots\gamma_r = P_0 \,.
\end{aligned}$$

Now all right-rays have been investigated, so there are exactly $2r+1$ vertices in the left Krieger cover of $\mathsf{X}(L)$. The vertex P_0 is the universal border point, and the vertices $P_{r+1},\ldots,P_{2r}$ are border points, while none of the vertices $P_1,\ldots,P_r$ are border points. This gives the information needed to draw the left Fischer cover.

In [6] it is proved that all renewal systems in the class B are SFTs. That proof will also work for the related class R considered here, but the result also follows easily from the structure of the left Fischer cover constructed above [7, Lemma 5.46].

Lemma 5. *Let $L \in R$ and let X_f be a renewal system obtained from $\mathsf{X}(L)$ by fragmentation. Then the Bowen–Franks group of X_f is cyclic, and the determinant is given by (9.4).*

Proof. Let $L \in R$ be defined by (9.2). The symbolic adjacency matrix of the left Fischer cover of $\mathsf{X}(L)$ (shown in Fig. 9.3) is

$$A = \left(\begin{array}{c|ccccc|cccccc}
\gamma & \alpha & 0 & \cdots & 0 & 0 & \gamma+\beta & \tilde{\alpha}' & \gamma_2' & \cdots & \gamma_{r-2}' & \gamma_{r-1}' \\
\hline
0 & 0 & \gamma_2 & \cdots & 0 & 0 & & & & & & \\
0 & 0 & 0 & & 0 & 0 & & & & & & \\
\vdots & \vdots & & \ddots & & \vdots & & & 0 & & & \\
0 & 0 & 0 & & 0 & \gamma_r & & & & & & \\
\beta & 0 & 0 & \cdots & 0 & 0 & 0 & \beta & \beta & \cdots & \beta & \beta \\
\hline
0 & & & & & & 0 & \tilde{\alpha} & 0 & \cdots & 0 & 0 \\
0 & & & & & & 0 & 0 & \gamma_2 & & 0 & 0 \\
0 & & & & & & 0 & 0 & 0 & & 0 & 0 \\
\vdots & & & 0 & & & \vdots & & & \ddots & & \vdots \\
0 & & & & & & 0 & 0 & 0 & & 0 & \gamma_{r-1} \\
\gamma_r & & & & & & \gamma_r & \gamma_r & \gamma_r & \cdots & \gamma_r & \gamma_r
\end{array}\right),$$

where $\gamma = \alpha + \tilde{\alpha} + \sum_{k=2}^{r-1} \gamma_k$, $\tilde{\alpha}' = \gamma - \tilde{\alpha}$, and $\gamma_k' = \gamma - \gamma_k$. Index the rows and columns of A by $0, \dots, 2r$ in correspondence with the names used for the vertices above, and note that the column sums of the columns $0, r+1, \dots, 2r$ are all equal to $\alpha + \tilde{\alpha} + \beta + \sum_{k=2}^{r} \gamma_k$.

If X_f is a fragmentation of $\mathsf{X}(L)$, then the (non-symbolic) adjacency matrix A_f of the underlying graph of the left Fischer cover of X_f is obtained from A by replacing $\alpha, \tilde{\alpha}, \beta, \gamma_2, \dots, \gamma_r$ by positive integers (see Remark 1). To put $\mathrm{Id} - A_f$ into Smith normal form, begin by adding each row from number $r+1$ to $2r-1$ to the first row, and subtract the first column from column $r+1, \dots, 2r$ to obtain

$$
\mathrm{Id} - A_f \rightsquigarrow \left(\begin{array}{c|ccccc|ccccc}
1-\gamma & -\alpha & 0 & \cdots & 0 & 0 & -\beta & 0 & \cdots & 0 & -1 \\ \hline
0 & 1 & -\gamma_2 & \cdots & 0 & 0 & & & & & \\
0 & 0 & 1 & & 0 & 0 & & & & & \\
\vdots & \vdots & & \ddots & & \vdots & & & 0 & & \\
0 & 0 & 0 & & 1 & -\gamma_r & & & & & \\
-\beta & 0 & 0 & \cdots & 0 & 1 & \beta & 0 & \cdots & 0 & 0 \\ \hline
0 & & & & & & 1 & -\tilde{\alpha} & \cdots & 0 & 0 \\
0 & & & & & & 0 & 1 & & 0 & 0 \\
\vdots & & & 0 & & & \vdots & & \ddots & & \vdots \\
0 & & & & & & 0 & 0 & & 1 & -\gamma_{r-1} \\
-\gamma_r & & & & & & 0 & 0 & \cdots & 0 & 1
\end{array}\right) .
$$

Using row and column addition, this matrix can be further reduced to

$$
\rightsquigarrow \left(\begin{array}{c|ccc|ccc}
1-\gamma-b & 0 & \cdots & 0 & 0 & \cdots & t \\ \hline
0 & 1 & \cdots & 0 & & & \\
\vdots & \vdots & \ddots & \vdots & & 0 & \\
0 & 0 & \cdots & 1 & & & \\ \hline
0 & & & & 1 & \cdots & 0 \\
\vdots & & 0 & & \vdots & \ddots & \vdots \\
-\gamma_r & & & & 0 & \cdots & 1
\end{array}\right)
\qquad
\begin{array}{l}
b = \alpha\beta\gamma_2 \cdots \gamma_r \\
t = \tilde{\alpha}\gamma_2 \cdots \gamma_{r-1}(b - \beta) - 1 .
\end{array}
$$

Hence, the Bowen–Franks group of X_f is cyclic, and the determinant is

$$
\det(\mathrm{Id} - A) = 1 - \alpha - \tilde{\alpha} - \sum_{k=2}^{r} \gamma_k - (\alpha + \tilde{\alpha})\beta\gamma_2 \cdots \gamma_r + \alpha\tilde{\alpha}\beta(\gamma_2 \cdots \gamma_r)^2 . \tag{9.4}
$$

Theorem 1. *For each $L \in H$, the renewal system $\mathsf{X}(L)$ has cyclic Bowen–Franks group and determinant given by (9.5).*

Proof. By Lemma 3, there exist $L_1, \ldots, L_m \in R$, $L_0 = \{a\}$ for some letter a that does not appear in any of the lists, and a fragmentation Y_f of $Y = \mathsf{X}(\bigcup_{j=0}^{m} L_j)$ such that $Y_f \sim_{\mathrm{FE}} \mathsf{X}(L)$. For $1 \leq j \leq m$, let $L_j = \{\alpha_j, \tilde{\alpha}_j, \gamma_{j,k}, \alpha_j\gamma_{j,2}\cdots\gamma_{j,r_j}\beta_j, \beta_j\tilde{\alpha}_j\gamma_{j,2}\cdots\gamma_{j,r_j} \mid 2 \leq k \leq r_j\}, r_j \in \mathbb{N}$. Each L_j is left-modular by Lemma 4, so Y is an SFT, and the left Fischer cover of Y can be constructed using the technique from Sect. 9.2: Identify the universal border points in the left Fischer covers of $\mathsf{X}(L_0), \ldots, \mathsf{X}(L_m)$, and draw additional edges to the border points corresponding to the edges terminating at the universal border points in the individual left Fischer covers. Hence, the symbolic adjacency matrix A of the left Fischer cover of Y is

$$A = \left(\begin{array}{c|c|cccc|cccc|c|c}
\gamma & & \alpha_j & 0 & \cdots & 0 & \gamma+\beta_j & \tilde{\alpha}'_j & \cdots & \gamma'_{j,r_j-1} & \cdots & \gamma'_{i,k} \\ \hline
 & \ddots & & & & & & & & & & \\ \hline
0 & & 0 & \gamma_{j,2} & \cdots & 0 & & & & & & \\
0 & & 0 & 0 & & 0 & & & & & & \\
\vdots & & \vdots & & \ddots & \vdots & & & 0 & & & \\
0 & & 0 & 0 & & \gamma_{j,r} & & & & & & \\
\beta_j & & 0 & 0 & \cdots & 0 & 0 & \beta_j & \cdots & \beta_j & & \beta_j \\ \hline
0 & & & & & & 0 & \tilde{\alpha}_j & \cdots & 0 & & \\
0 & & & & & & 0 & 0 & & 0 & & \\
\vdots & & & 0 & & & \vdots & & \ddots & \vdots & & \\
0 & & & & & & 0 & 0 & & \gamma_{j,r_j-1} & & \\
\gamma_{j,r_j} & & & & & & \gamma_{j,r_j} & \gamma_{j,r_j} & \cdots & \gamma_{j,r_j} & & \gamma_{j,r_j} \\ \hline
 & & & & & & & & & & \ddots & \\ \hline
 & & & & & & & & & & & \ddots
\end{array}\right).$$

where $1 \leq j \leq m$, $\gamma = a + \sum_{j=1}^{m}\left(\alpha_j + \tilde{\alpha}_j + \sum_{k=2}^{r_j-1}\gamma_{j,k}\right)$, $\tilde{\alpha}'_j = \gamma - \tilde{\alpha}_j$, and $\gamma'_{j,k} = \gamma - \gamma_{j,k}$. This matrix has blocks of the same form as in the $m = 1$ case considered in Lemma 4. The jth block is shown together with the first row and column of the matrix—which contain the connections between the jth block and the universal border point P_0—and together with an extra column representing an arbitrary border point in a different block. Such a border point in another block will receive edges from the jth block with the same sources and labels as the edges that start in the jth block and terminate at the universal border point P_0.

Let Y_f be a fragmentation of Y. Then the (non-symbolic) adjacency matrix A_f of the underlying graph of the left Fischer cover of Y_f is obtained by replacing the entries of A by positive integers as described in Remark 1. In order to put $\mathrm{Id} - A_f$ into Smith normal form, first add rows $r_j + 1$ to $2r_j - 1$ in the jth block to the first row for each j, and then subtract the first column from every column corresponding to a border point in any block. In this way, $\mathrm{Id} - A_f$ is transformed into:

$$\left(\begin{array}{c|c|cccc|cccc|c}
1-\gamma & & -\alpha_j & 0 & \cdots & 0 & -\beta_j & 0 & \cdots & -1 & \\ \hline
 & \ddots & & & & & & & & & \\ \hline
0 & & 1 & -\gamma_{j,2} & \cdots & 0 & & & & & \\
0 & & 0 & 1 & & 0 & & & & & \\
\vdots & & \vdots & & \ddots & \vdots & & & 0 & & \\
0 & & 0 & 0 & & -\gamma_{j,r_j} & & & & & \\
-\beta_j & & 0 & 0 & \cdots & 1 & \beta_j & 0 & \cdots & 0 & \\ \hline
0 & & & & & & 1 & -\tilde{\alpha}_j & \cdots & 0 & \\
0 & & & & & & 0 & 1 & & 0 & \\
\vdots & & & & 0 & & \vdots & & \ddots & \vdots & \\
0 & & & & & & 0 & 0 & & -\gamma_{j,r_j-1} & \\
-\gamma_{j,r_j} & & & & & & 0 & 0 & \cdots & 1 & \\ \hline
 & & & & & & & & & & \ddots
\end{array}\right).$$

By using row and column addition, and by disregarding rows and columns where the only non-zero entry is a diagonal 1, $\mathrm{Id} - A$ can be further reduced to

$$\begin{pmatrix}
S & t_1 & t_2 & \cdots & t_m \\
-\gamma_{1,r_1} & 1 & 0 & & 0 \\
-\gamma_{2,r_2} & 0 & 1 & & 0 \\
\vdots & & & \ddots & \vdots \\
-\gamma_{m,r_m} & 0 & 0 & \cdots & 1
\end{pmatrix}
\qquad
\begin{array}{l}
b_j = \alpha_j \beta_j \gamma_{j,2} \cdots \gamma_{j,r_j} \\[1em]
t_j = \tilde{\alpha}_j \gamma_{j,2} \cdots \gamma_{j,r-1}(b_j - \beta_j) - 1 \ . \\[1em]
S = 1 - \gamma - \sum_{j=1}^{m} b_j
\end{array}$$

Hence, the Bowen–Franks group is cyclic and the determinant is

$$\det(\mathrm{Id} - A_f) = 1 - \gamma + \sum_{j=1}^{m} \left(\gamma_{j,r_j} t_j - b_j\right) \ . \tag{9.5}$$

With the results of [6], this gives the following result.

Corollary 1. *When* $\log \lambda$ *is the entropy of an SFT, there exists an SFT renewal system* $X(L)$ *with cyclic Bowen–Franks group such that* $h(\mathsf{X}(L)) = \log \lambda$.

9.4 Towards the Range of the Bowen–Franks Invariant

In the following, it will be proved that the range of the Bowen–Franks invariant over the class of SFT renewal systems contains a large class of pairs of signs and finitely generated abelian groups. First, the following special case will be used to show that every integer is the determinant of an SFT renewal system.

Example 2. Consider the generating list

$$L = \{a, \alpha, \tilde{\alpha}, \gamma, \alpha\gamma\beta, \beta\tilde{\alpha}\gamma\} . \tag{9.6}$$

By Lemma 4, L is left-modular, $\mathsf{X}(L)$ is an SFT, and the symbolic adjacency matrix of the left Fischer cover of $\mathsf{X}(L)$ is

$$A = \left(\begin{array}{c|cc|cc} a+\alpha+\tilde{\alpha} & \alpha & 0 & a+\alpha+\tilde{\alpha}+\beta & a+\alpha \\ \hline 0 & 0 & \gamma & 0 & 0 \\ \beta & 0 & 0 & 0 & \beta \\ \hline 0 & 0 & 0 & 0 & \tilde{\alpha} \\ \gamma & 0 & 0 & \gamma & \gamma \end{array} \right) . \tag{9.7}$$

By fragmenting $\mathsf{X}(L)$, it is possible to construct an SFT renewal system for which the (non-symbolic) adjacency matrix of the underlying graph of the left Fischer cover has this form with $a, \alpha, \tilde{\alpha}, \beta, \gamma \in \mathbb{N}$ as described in Remark 1. Let A_f be such a matrix. This is a special case of the shift spaces considered in Theorem 1, so the Bowen–Franks group is cyclic and the determinant is $\det(\mathrm{Id} - A_f) = \beta\alpha\tilde{\alpha}\gamma^2 - \alpha\beta\gamma - \tilde{\alpha}\beta\gamma - \alpha - \tilde{\alpha} - \gamma - a + 1$.

Theorem 2. *Any $k \in \mathbb{Z}$ is the determinant of an SFT renewal system with cyclic Bowen–Franks group.*

Proof. Consider the renewal system from Example 2 in the case $\alpha = \tilde{\alpha} = \beta = 1$, where the determinant is $\det(\mathrm{Id} - A_f) = \gamma^2 - 3\gamma - a - 1$, and note that the range of this polynomial is $\mathbb{Z}$.

All renewal systems considered until now have had cyclic Bowen–Franks groups, so the next goal is to construct a class of renewal systems exhibiting non-cyclic groups. Let $k \geq 2$, $\mathscr{A} = \{a_1, \dots, a_k\}$, and let $n_1, \dots, n_k \geq 2$ with $\max_i\{n_i\} > 2$. The goal is to define a generating list, L, for which $\mathsf{X}(L) = \mathsf{X}_{\mathscr{F}}$ with $\mathscr{F} = \{a_i^{n_i}\}$. For each $1 \leq i \leq k$, define

$$L_i = \{a_j a_i^l \mid j \neq i \text{ and } 0 < l < n_i - 1\} \\ \cup \{a_m a_j a_i^l \mid m \neq j \neq i \text{ and } 0 < l < n_i - 1\} . \tag{9.8}$$

Define $L = \bigcup_{i=1}^k L_i \neq \emptyset$, and $\mathsf{X}_{\mathrm{diag}(n_1,\dots,n_k)} = \mathsf{X}(L)$.

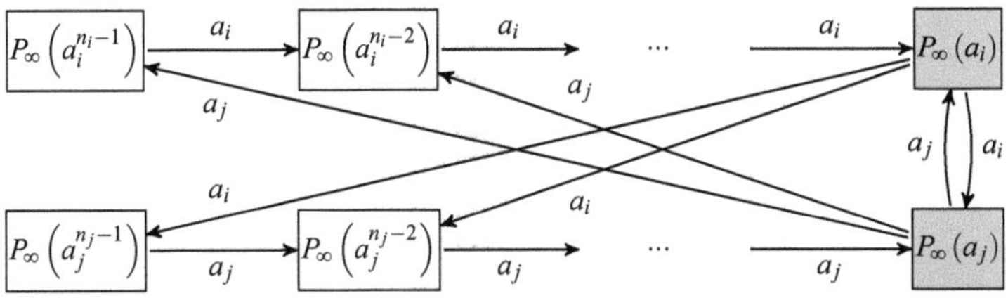

Fig. 9.4 Part of the left Fischer cover of $\mathsf{X}_{\text{diag}(n_1,\ldots,n_k)}$. The entire graph can be found by varying i and j. The border points are coloured *grey*

Lemma 6. *Define the renewal system* $\mathsf{X}_{diag(n_1,\ldots,n_k)}$ *as above. Then* $\mathsf{X}_{diag(n_1,\ldots,n_k)} = \mathsf{X}_{\mathscr{F}}$ *with* $\mathscr{F} = \{a_i^{n_i}\}$*, so* $\mathsf{X}_{diag(n_1,\ldots,n_k)}$ *is an SFT. The symbolic adjacency matrix of the left Fischer cover of* $\mathsf{X}_{diag(n_1,\ldots,n_k)}$ *is the matrix in (9.10).*

Proof. Note that for each i, $a_i^{n_i} \notin \mathscr{B}(\mathsf{X}_{\text{diag}(n_1,\ldots,n_k)})$ by construction. For $1 < l < n_i - 1$ and $j \neq i$ the word $a_j a_i^l$ has a partitioning in $\mathsf{X}_{\text{diag}(n_1,\ldots,n_k)}$ with empty beginning and end. Hence, $a_{i_1} a_{i_2}^{l_2} a_{i_3}^{l_3} \cdots a_{i_m}^{l_m}$ has a partitioning with empty beginning and end whenever $i_j \neq i_{j+1}$, $1 < l_j < n_{i_j}$ for all $1 < j < m$, and $0 < l_m < n_{i_m} - 1$. Given $i_1, \ldots, i_m \in \{1, \ldots, k\}$ with $i_j \neq i_{j+1}$ and $m \geq 2$, the word $a_{i_1} a_{i_2} \cdots a_{i_m}$ has a partitioning with empty beginning and end. Hence, every word that does not contain one of the words $a_i^{n_i}$ has a partitioning, so $\mathsf{X}_{\text{diag}(n_1,\ldots,n_k)} = \mathsf{X}_{\mathscr{F}}$ for $\mathscr{F} = \{a_i^{n_i}\}$.

To find the left Fischer cover of $\mathsf{X}_{\text{diag}(n_1,\ldots,n_k)}$, it is first necessary to determine the predecessor sets. Given $1 \leq i \leq k$ and $j \neq i$

$$\begin{aligned} P_\infty(a_i a_j \cdots) &= \{x^- \in \mathsf{X}^-_{\text{diag}(n_1,\ldots,n_k)} | x_{-n_i+1} \cdots x_0 \neq a_i^{n_i-1}\} \\ P_\infty(a_i^2 a_j \cdots) &= \{x^- \in \mathsf{X}^-_{\text{diag}(n_1,\ldots,n_k)} | x_{-n_i+2} \cdots x_0 \neq a_i^{n_i-2}\} \\ &\vdots \\ P_\infty(a_i^{n_i-1} a_j \cdots) &= \{x^- \in \mathsf{X}^-_{\text{diag}(n_1,\ldots,n_k)} | x_0 \neq a_i\}. \end{aligned} \tag{9.9}$$

Only the first of these predecessor sets is a border point. Equation 9.9 gives all the information necessary to draw the left Fischer cover of $\mathsf{X}_{\text{diag}(n_1,\ldots,n_k)}$. A part of the left Fischer cover is shown in Fig. 9.4, and the corresponding symbolic adjacency matrix is:

$$
\left(\begin{array}{c|c c|c}
\overbrace{\begin{matrix}0&\cdots&0&0\\ a_1&\cdots&0&0\\ &\ddots&&\\ 0&\cdots&a_1&0\end{matrix}}^{n_1-1} & \overbrace{\begin{matrix}a_1&\cdots&a_1&a_1\\ 0&\cdots&0&0\\ &\ddots&&\\ 0&\cdots&0&0\end{matrix}}^{n_2-1} & \cdots & \overbrace{\begin{matrix}a_1&\cdots&a_1&a_1\\ 0&\cdots&0&0\\ &\ddots&&\\ 0&\cdots&0&0\end{matrix}}^{n_k-1} \\ \hline
\begin{matrix}a_2&\cdots&a_2&a_2\\ 0&\cdots&0&0\\ &\ddots&&\\ 0&\cdots&0&0\end{matrix} & \begin{matrix}0&\cdots&0&0\\ a_2&\cdots&0&0\\ &\ddots&&\\ 0&\cdots&a_2&0\end{matrix} & \cdots & \begin{matrix}a_2&\cdots&a_2&a_2\\ 0&\cdots&0&0\\ &\ddots&&\\ 0&\cdots&0&0\end{matrix} \\ \hline
\vdots & \vdots & \ddots & \vdots \\ \hline
\begin{matrix}a_k&\cdots&a_k&a_k\\ 0&\cdots&0&0\\ &\ddots&&\\ 0&\cdots&0&0\end{matrix} & \begin{matrix}a_k&\cdots&a_k&a_k\\ 0&\cdots&0&0\\ &\ddots&&\\ 0&\cdots&0&0\end{matrix} & \cdots & \begin{matrix}0&\cdots&0&0\\ a_k&\cdots&0&0\\ &\ddots&&\\ 0&\cdots&a_k&0\end{matrix}
\end{array}\right). \tag{9.10}
$$

Let A be the (non-symbolic) adjacency matrix of the underlying graph of the left Fischer cover of $\mathsf{X}_{\mathrm{diag}(n_1,\dots,n_k)}$ constructed above. Then it is possible to do the following transformation by row and column addition

$$
\mathrm{Id}-A \rightsquigarrow \begin{pmatrix} 1 & 1-n_2 & 1-n_3 & \cdots & 1-n_k \\ 1-n_1 & 1 & 1-n_3 & \cdots & 1-n_k \\ 1-n_1 & 1-n_2 & 1 & \cdots & 1-n_k \\ \vdots & \vdots & \vdots & \ddots & \vdots \\ 1-n_1 & 1-n_2 & 1-n_3 & \cdots & 1 \end{pmatrix} \rightsquigarrow \begin{pmatrix} x & 1 & 1 & \cdots & 1 \\ -n_1 & n_2 & 0 & \cdots & 0 \\ -n_1 & 0 & n_3 & \cdots & 0 \\ \vdots & \vdots & \vdots & \ddots & \vdots \\ -n_1 & 0 & 0 & \cdots & n_k \end{pmatrix},
$$

where $x = 1-(k-1)n_1$. The determinant of this matrix is

$$
\det(\mathrm{Id}-A) = n_2\cdots n_k\left(x+\sum_{i=2}^{k}\frac{n_1}{n_i}\right) = -n_1n_2\cdots n_k\left(k-1-\sum_{i=1}^{k}\frac{1}{n_i}\right) < 0 .
$$

The inequality is strict since $k-1-\sum_{i=1}^{k}\frac{1}{n_i} > \frac{k}{2}-1 \geq 0$. Given concrete $n_1,\dots,n_k$, it is straightforward to compute the Bowen–Franks group of $\mathsf{X}_{\mathrm{diag}(n_1,\dots,n_k)}$, but it has not been possible to derive a general closed form for this group.

Proposition 2. *Let $n_1,\dots,n_k \geq 2$ with $n_i | n_{i-1}$ for $2 \leq i \leq k$ and $n_1 > 2$. Let $m = n_1n_2(k-1-\sum_{i=1}^{k}\frac{1}{n_i})$, then $\mathrm{BF}_+(\mathsf{X}_{diag(n_1,\dots,n_k)}) = -\mathbb{Z}/m\mathbb{Z}\oplus\mathbb{Z}/n_3\mathbb{Z}\oplus\cdots\oplus\mathbb{Z}/n_k\mathbb{Z}$.*

Proof. By the arguments above, $\mathsf{X}_{\mathrm{diag}(n_1,\ldots,n_k)}$ is conjugate to an edge shift with adjacency matrix A such that the following transformation can be carried out by row and column addition

$$\mathrm{Id}-A \rightsquigarrow \begin{pmatrix} y & 1 & 1 & \cdots & 1 \\ 0 & n_2 & 0 & \cdots & 0 \\ 0 & 0 & n_3 & \cdots & 0 \\ \vdots & \vdots & \vdots & \ddots & \vdots \\ 0 & 0 & 0 & \cdots & n_k \end{pmatrix} \rightsquigarrow \begin{pmatrix} 0 & 1 & 0 & \cdots & 0 \\ m & 0 & 0 & \cdots & 0 \\ 0 & 0 & n_3 & \cdots & 0 \\ \vdots & \vdots & \vdots & \ddots & \vdots \\ 0 & 0 & 0 & \cdots & n_k \end{pmatrix},$$

where $y = -n_1\left(k-1-\sum_{i=1}^{k} 1/n_i\right)$. It follows that the Smith normal form of $\mathrm{Id}-A$ is $\mathrm{diag}(m,n_3,\ldots,n_k)$, and $\det(\mathrm{Id}-A) < 0$.

Let G be a finite direct sum of finite cyclic groups. Then Proposition 2 shows that G is a subgroup of the Bowen–Franks group of some SFT renewal system, but it is still unclear whether G itself is also the Bowen–Franks group of a renewal system since the term $\mathbb{Z}/m\mathbb{Z}$ in the statement of Proposition 2 is determined by the other terms. Furthermore, the groups constructed in Proposition 2 are all finite. Other techniques can be used to construct renewal systems with groups such as $\mathbb{Z}/(n+1)\oplus\mathbb{Z}$ [7, Ex. 5.54].

The determinants of all the renewal systems with non-cyclic Bowen–Franks groups considered above were negative or zero, so the next goal is to construct a class of SFT renewal systems with positive determinants and non-cyclic Bowen–Franks groups.

Lemma 7. *Let L_d be the generating list of $\mathsf{X}_{diag(n_1,\ldots,n_k)}$ as defined in (9.8), and let $(F_d,\mathcal{L}_d)$ be the left Fischer cover of $\mathsf{X}_{diag(n_1,\ldots,n_k)}$. Let L_m be a left-modular generating list for which $\mathsf{X}(L_m)$ is an SFT with left Fischer cover $(F_m,\mathcal{L}_m)$. For $L_{d+m} = L_d \cup L_m \cup_{i=1}^{k} \{a_i w \mid w \in L_m\}$, $\mathsf{X}(L_{d+m})$ is an SFT for which the left Fischer cover is obtained by adding the following connecting edges to the disjoint union of $(F_d,\mathcal{L}_d)$ and $(F_m,\mathcal{L}_m)$ (sketched in Fig. 9.5):*

- *For each $1 \le i \le k$ and each $e \in F_m^0$ with $r(e) = P_0(L_m)$ draw an edge e_i with $s(e_i) = s(e)$ and $r(e_i) = P_\infty(a_i a_j \ldots)$ labelled $\mathcal{L}_m(e)$.*
- *For each $1 \le i \le k$ and each border point $P \in F_m^0$ draw an edge labelled a_i from $P_\infty(a_i a_j \ldots)$ to P.*

Proof. Let $(F_{d+m},\mathcal{L}_{d+m})$ be the labelled graph defined in the lemma and sketched in Fig. 9.5. The graph is left-resolving, predecessor-separated, and irreducible by construction, so it is the left Fischer cover of some sofic shift X [12, Corollary 3.3.19]. The first goal is to prove that $X = \mathsf{X}(L_{d+m})$. By the arguments used in the proof of Lemma 6, any word of the form $a_{i_0} w_m a_{i_1} a_{i_2}^{l_i} \ldots a_{i_p}^{l_p}$ where $w_m \in L_m^*$, $p \in \mathbb{N}$, $i_j \ne i_{j+1}$ and $l_j < n_{i_j}$ for $1 < j < p$,

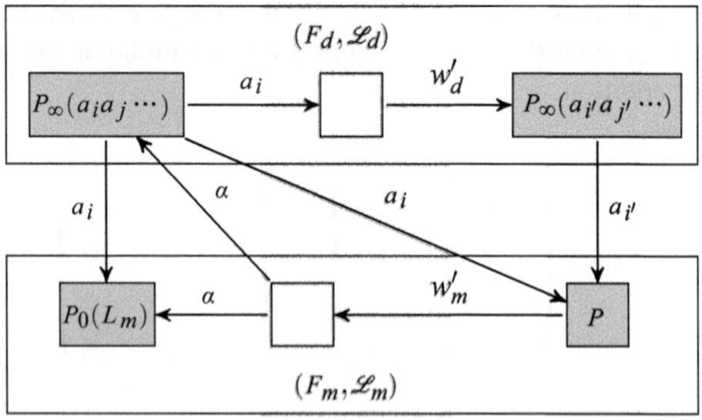

Fig. 9.5 Construction of the left Fischer cover considered in Lemma 7. Here, $w'_m\alpha = w_m \in L_m^*$ and $w'_d a_{i'} = w_d \in \mathcal{B}(\mathsf{X}_{\text{diag}(n_1,\ldots,n_k)})$ with $\text{ll}(w_d) \neq a_i$. Border points are coloured *grey*

and $1 \leq l_p < n_{i_p} - 1$ has a partitioning with empty beginning and end in $\mathsf{X}(L_{d+m})$. Hence, $\mathcal{B}(\mathsf{X}(L_{d+m}))$ is the set of factors of concatenations of words from $\{w_m a_i w_d \mid w_m \in L_m^*, 1 \leq i \leq k, w_d \in \mathcal{B}(\mathsf{X}_{\text{diag}(n_1,\ldots,n_k)}), \text{ll}(w_d) \neq a_i\}$. Since L_m is left-modular, a path $\lambda \in F_m^*$ with $r(\lambda) = P_0(L_m)$ has $\mathcal{L}_m(\lambda) \in L_m^*$ if and only if $s(\lambda)$ is a border point in F_m. Hence, the language recognised by the left Fischer cover $(F_{d+m}, \mathcal{L}_{d+m})$ is precisely the language of $\mathsf{X}(L_{d+m})$.

It remains to show that $(F_{d+m}, \mathcal{L}_{d+m})$ presents an SFT. Let $1 \leq i \leq k$ and let $\alpha \in \mathcal{B}(\mathsf{X}(L_m))$, then any labelled path in $(F_{d+m}, \mathcal{L}_{d+m})$ with $a_i\alpha$ as a prefix must start at $P_\infty(a_i a_j \cdots)$. Similarly, if there is a path $\lambda \in F_{d+m}^*$ with αa_i as a prefix of $\mathcal{L}_{d+m}(\lambda)$, then there must be unique vertex v emitting an edge labelled α to $P_0(L)$, and $s(\lambda) = v$. Let $x \in \mathsf{X}_{(F_{d+m}, \mathcal{L}_{d+m})}$. If there is no upper bound on set of $i \in \mathbb{Z}$ such that $x_i \in \{a_1, \ldots, a_k\}$ and $x_{i+1} \in \mathcal{A}(\mathsf{X}(L_m))$ or vice versa, then the arguments above and the fact that the graph is left-resolving prove that there is only one path in $(F_{d+m}, \mathcal{L}_{d+m})$ labelled x. If there is an upper bound on the set considered above, then a presentation of x is eventually contained in either F_d or F_m. It follows that the covering map of $(F_{d+m}, \mathcal{L}_{d+m})$ is injective, so it presents an SFT.

Example 3. The next step is to use Lemma 7 to construct renewal systems that share features with both $\mathsf{X}_{\text{diag}(n_1,\ldots,n_k)}$ and the renewal systems considered in Example 2. Given $n_1, \ldots, n_k \geq 2$ with $\max_j n_j > 2$, consider the list L_d defined in (9.8) which generates the renewal system $\mathsf{X}_{\text{diag}(n_1,\ldots,n_k)}$, and the list L from (9.6). L is left-modular, and $\mathsf{X}(L)$ is an SFT, so Lemma 7 can be used to find the left Fischer cover of the SFT renewal system X_+ generated by $L_+ = L_d \cup L \cup_{i=1}^k \{a_i w \mid w \in L\}$, and the corresponding symbolic adjacency matrix is

$$A_+ = \left(\begin{array}{c|cccc|ccccc|c|ccccc}
b & \alpha & 0 & b+\beta & a+\alpha & b & 0 & \cdots & 0 & 0 & \cdots & b & 0 & \cdots & 0 & 0 \\ \hline
0 & 0 & \gamma & 0 & 0 & 0 & 0 & \cdots & 0 & 0 & & 0 & 0 & \cdots & 0 & 0 \\
\beta & 0 & 0 & 0 & \beta & \beta & 0 & \cdots & 0 & 0 & & \beta & 0 & \cdots & 0 & 0 \\
0 & 0 & 0 & 0 & \tilde{\alpha} & 0 & 0 & \cdots & 0 & 0 & & 0 & 0 & \cdots & 0 & 0 \\
\gamma & 0 & 0 & \gamma & \gamma & \gamma & 0 & \cdots & 0 & 0 & & \gamma & 0 & \cdots & 0 & 0 \\ \hline
a_1 & 0 & 0 & a_1 & a_1 & 0 & 0 & & 0 & 0 & & a_1 & a_1 & \cdots & a_1 & a_1 \\
0 & 0 & 0 & 0 & 0 & a_1 & 0 & & 0 & 0 & & 0 & 0 & & 0 & 0 \\
\vdots & \vdots & \vdots & \vdots & \vdots & & & \ddots & & & & & & \ddots & & \\
0 & 0 & 0 & 0 & 0 & 0 & 0 & & 0 & 0 & & 0 & 0 & & 0 & 0 \\
0 & 0 & 0 & 0 & 0 & 0 & 0 & & a_1 & 0 & & 0 & 0 & & 0 & 0 \\ \hline
\vdots & & & & & & & & & & \ddots & & & & & \\ \hline
a_k & 0 & 0 & a_k & a_k & a_k & a_k & \cdots & a_k & a_k & & 0 & 0 & & 0 & 0 \\
0 & 0 & 0 & 0 & 0 & 0 & 0 & & 0 & 0 & & a_k & 0 & & 0 & 0 \\
\vdots & \vdots & \vdots & \vdots & \vdots & & & \ddots & & & & & & \ddots & & \\
0 & 0 & 0 & 0 & 0 & 0 & 0 & & 0 & 0 & & 0 & 0 & & 0 & 0 \\
0 & 0 & 0 & 0 & 0 & 0 & 0 & & 0 & 0 & & 0 & 0 & & a_k & 0
\end{array}\right),$$

where $b = a + \alpha + \tilde{\alpha}$. Let Y_+ be a renewal system obtained from X_+ by a fragmentation of a, α, $\tilde{\alpha}$, β, and γ. Then the (non-symbolic) adjacency matrix of the left Fischer cover of Y_+ is obtained from the matrix A_+ above by replacing $a_1, \dots, a_k$ by 1, and replacing a, α, $\tilde{\alpha}$, β, and γ by positive integers. Let B_+ be a matrix obtained in this manner. By doing row and column operations as in the construction that leads to the proof Proposition 2, and by disregarding rows and columns where the only non-zero entry is a diagonal 1, it follows that

$$\mathrm{Id} - B_+ \rightsquigarrow \left(\begin{array}{c|cccc|cccc}
1-b & -\alpha & 0 & -b-\beta & -a-\alpha & -b & -b & \cdots & -b \\ \hline
0 & 1 & -\gamma & 0 & 0 & 0 & 0 & \cdots & 0 \\
-\beta & 0 & 1 & 0 & -\beta & -\beta & -\beta & \cdots & -\beta \\
0 & 0 & 0 & 1 & -\tilde{\alpha} & 0 & 0 & \cdots & 0 \\
-\gamma & 0 & 0 & -\gamma & 1-\gamma & -\gamma & -\gamma & \cdots & -\gamma \\ \hline
-1 & 0 & 0 & -1 & -1 & 1 & 1-n_2 & \cdots & 1-n_k \\
-1 & 0 & 0 & -1 & -1 & 1-n_1 & 1 & & 1-n_k \\
\vdots & \vdots & \vdots & \vdots & \vdots & \vdots & & \ddots & \vdots \\
-1 & 0 & 0 & -1 & -1 & 1-n_1 & 1-n_2 & \cdots & 1
\end{array}\right).$$

Add the third row to the first and subtract the first column from columns $4, \dots, k+4$ as in the proof of Lemma 5 and choose the variables a, α, $\tilde{\alpha}$, β, and γ as in the proof of Theorem 2. Assuming that $n_i | n_{i-1}$ for $2 \le i \le k$, this matrix can be reduced to

$\mathrm{Id} - B_+ \rightsquigarrow$

$$\left(\begin{array}{c|ccccc} x & -1 & -1 & -1 & \cdots & -1 \\ \hline 2x-1 & 0 & -n_2 & -n_3 & \cdots & -n_k \\ 0 & -n_1 & n_2 & 0 & & 0 \\ 0 & -n_1 & 0 & n_3 & & 0 \\ \vdots & \vdots & & & \ddots & \vdots \\ 0 & -n_1 & 0 & 0 & \cdots & n_k \end{array}\right) \rightsquigarrow \left(\begin{array}{cccccc} x & -\sum_{i=1}^{k} \frac{n_1}{n_i} & -1 & 0 & \cdots & 0 \\ 2x-1 & -(k-1)n_1 & 0 & 0 & \cdots & 0 \\ 0 & 0 & n_2 & 0 & & 0 \\ 0 & 0 & 0 & n_3 & & 0 \\ \vdots & \vdots & & & \ddots & \vdots \\ 0 & 0 & 0 & 0 & \cdots & n_k \end{array}\right),$$

where $x \in \mathbb{Z}$ is arbitrary. Hence, the determinant is

$$\det(\mathrm{Id} - B_+) = n_2 \cdots n_k \left((2x-1) \sum_{i=1}^{k} \frac{n_1}{n_i} - x(k-1)n_1 \right), \tag{9.11}$$

and there exists an abelian group G with at most two generators such that the Bowen–Franks group of the corresponding SFT is $G \oplus \mathbb{Z}/n_3\mathbb{Z} \oplus \cdots \oplus \mathbb{Z}/n_k\mathbb{Z}$. For $x = 0$, the determinant is negative and the Bowen–Franks group is $\mathbb{Z}/(\sum_{i=1}^{k} \frac{n_2 n_1}{n_i})\mathbb{Z} \oplus \mathbb{Z}/n_3\mathbb{Z} \oplus \cdots \oplus \mathbb{Z}/n_k\mathbb{Z}$.

This gives the first example of SFT renewal systems that simultaneously have positive determinants and non-cyclic Bowen–Franks groups.

Theorem 3. *Given $n_1, \ldots, n_k \geq 2$ with $n_i | n_{i-1}$ for $2 \leq i \leq k$ there exist abelian groups $G_\pm$ with at most two generators and SFT renewal systems $\mathsf{X}(L_\pm)$ such that* $\mathrm{BF}_+(\mathsf{X}(L_\pm)) = \pm G_\pm \oplus \mathbb{Z}/n_1\mathbb{Z} \oplus \cdots \oplus \mathbb{Z}/n_k\mathbb{Z}$.

Proof. Consider the renewal system from Example 3. Given the other variables, (9.11) shows that x can be chosen such that the determinant has either sign.

The question raised by Adler, and the related question concerning the flow equivalence of renewal systems are still unanswered, and a significant amount of work remains before they can be solved. However, there is hope that the techniques developed in Sect. 9.2 and the special classes of renewal systems considered in Sect. 9.4 can act as a foundation for the construction of a class of renewal systems attaining all the values of the Bowen–Franks invariant realised by irreducible SFTs.

Acknowledgements Supported by VILLUM FONDEN through the experimental mathematics network at the University of Copenhagen. Supported by the Danish National Research Foundation through the Centre for Symmetry and Deformation (DNRF92).

References

1. Berstel, J., Perrin, D.: Theory of codes, *Pure and Applied Mathematics*, vol. 117. Academic Press Inc., Orlando, FL (1985)
2. Fischer, R.: Sofic systems and graphs. Monatsh. Math. **80**(3), 179–186 (1975)
3. Franks, J.: Flow equivalence of subshifts of finite type. Ergodic Theory Dynam. Systems **4**(1), 53–66 (1984). DOI 10.1017/S0143385700002261. URL http://dx.doi.org/10.1017/S0143385700002261
4. Goldberger, J., Lind, D., Smorodinsky, M.: The entropies of renewal systems. Israel J. Math. **75**(1), 49–64 (1991). DOI 10.1007/BF02787181. URL http://dx.doi.org/10.1007/BF02787181
5. Hong, S., Shin, S.: Cyclic renewal systems. Theoret. Comput. Sci. **410**(27–29), 2675–2684 (2009). DOI 10.1016/j.tcs.2009.03.033. URL http://dx.doi.org/10.1016/j.tcs.2009.03.033
6. Hong, S., Shin, S.: The entropies and periods of renewal systems. Israel J. Math. **172**, 9–27 (2009). DOI 10.1007/s11856-009-0060-7. URL http://dx.doi.org/10.1007/s11856-009-0060-7
7. Johansen, R.: On flow equivalence of sofic shifts. Ph.D. thesis, University of Copenhagen (2011), arXiv:1210.3048
8. Johansen, R.: On the structure of covers of sofic shifts. Doc. Math. **16**, 111–131 (2011)
9. Johnson, A.S.A., Madden, K.: Renewal systems, sharp-eyed snakes, and shifts of finite type. Amer. Math. Monthly **109**(3), 258–272 (2002). DOI 10.2307/2695355. URL http://dx.doi.org/10.2307/2695355
10. Jonoska, N., Marcus, B.: Minimal presentations for irreducible sofic shifts. IEEE Trans. Inform. Theory **40**(6), 1818–1825 (1994). DOI 10.1109/18.340457. URL http://dx.doi.org/10.1109/18.340457
11. Krieger, W.: On sofic systems. I. Israel J. Math. **48**(4), 305–330 (1984). DOI 10.1007/BF02760631. URL http://dx.doi.org/10.1007/BF02760631
12. Lind, D., Marcus, B.: An introduction to symbolic dynamics and coding. Cambridge University Press, Cambridge (1995). DOI 10.1017/CBO9780511626302. URL http://dx.doi.org/10.1017/CBO9780511626302
13. Parry, B., Sullivan, D.: A topological invariant of flows on 1-dimensional spaces. Topology **14**(4), 297–299 (1975)
14. Restivo, A.: Codes and local constraints. Theoret. Comput. Sci. **72**(1), 55–64 (1990). DOI 10.1016/0304-3975(90)90046-K. URL http://dx.doi.org/10.1016/0304-3975(90)90046-K
15. Restivo, A.: A note on renewal systems. Theoret. Comput. Sci. **94**(2), 367–371 (1992). DOI 10.1016/0304-3975(92)90044-G. URL http://dx.doi.org/10.1016/0304-3975(92)90044-G. Discrete mathematics and applications to computer science (Marseille, 1989)
16. Weiss, B.: Subshifts of finite type and sofic systems. Monatsh. Math. **77**, 462–474 (1973)
17. Williams, S.: Notes on renewal systems. Proc. Amer. Math. Soc. **110**(3), 851–853 (1990). DOI 10.2307/2047932. URL http://dx.doi.org/10.2307/2047932

Chapter 10
On the Grothendieck Theorem for Jointly Completely Bounded Bilinear Forms

Tim de Laat

Abstract We show how the proof of the Grothendieck Theorem for jointly completely bounded bilinear forms on C^*-algebras by Haagerup and Musat can be modified in such a way that the method of proof is essentially C^*-algebraic. To this purpose, we use Cuntz algebras rather than type III factors. Furthermore, we show that the best constant in Blecher's inequality is strictly greater than one.

Keywords Noncommutative Grothendieck Theorem • Completely bounded bilinear forms • Blecher's inequality • Operator spaces • Cuntz-algebras • KMS-states

Mathematics Subject Classification (2010): 46L07, 47L25, 46L05, 46L55.

10.1 Introduction

In [10], Grothendieck proved his famous *Fundamental Theorem on the metric theory of tensor products*. He also conjectured a noncommutative analogue of this theorem for bounded bilinear forms on C^*-algebras. This *noncommutative Grothendieck Theorem* was proved by Pisier assuming a certain approximability condition on the bilinear form [16]. The general case was proved by Haagerup [11]. Effros and Ruan conjectured a "sharper" analogue of this theorem for bilinear forms on C^*-algebras that are jointly completely bounded (rather than bounded) [9]. More precisely, they conjectured the following result, with universal constant $K = 1$.

T. de Laat (✉)
Department of Mathematical Sciences, University of Copenhagen, Universitetsparken 5, DK-2100 Copenhagen Ø, Denmark
e-mail: tlaat@math.ku.dk

T.M. Carlsen et al. (eds.), *Operator Algebra and Dynamics*, Springer Proceedings in Mathematics & Statistics 58, DOI 10.1007/978-3-642-39459-1_10,

Theorem 1 (JCB Grothendieck Theorem). *Let A, B be C^*-algebras, and let $u : A \times B \to \mathbb{C}$ be a jointly completely bounded bilinear form. Then there exist states f_1, f_2 on A and g_1, g_2 on B such that for all $a \in A$ and $b \in B$,*

$$|u(a,b)| \leq K\|u\|_{jcb}\left(f_1(aa^*)^{\frac{1}{2}}g_1(b^*b)^{\frac{1}{2}} + f_2(a^*a)^{\frac{1}{2}}g_2(bb^*)^{\frac{1}{2}}\right),$$

where K is a constant.

We call this *Grothendieck Theorem for jointly completely bounded bilinear forms on C^*-algebras* the *JCB Grothendieck Theorem*. It is often referred to as the *Effros-Ruan conjecture*.

In [18], Pisier and Shlyakhtenko proved a version of Theorem 1 for exact operator spaces, in which the constant K depends on the exactness constants of the operator spaces. They also proved the conjecture for C^*-algebras, assuming that at least one of them is exact, with universal constant $K = 2^{\frac{3}{2}}$.

Haagerup and Musat proved the general conjecture (for C^*-algebras), i.e., Theorem 1, with universal constant $K = 1$ [12]. They used certain type III factors in the proof. Since the conjecture itself is purely C^*-algebraic, it would be more satisfactory to have a proof that relies on C^*-algebras. In this note, we show how the proof of Haagerup and Musat can be modified in such a way that essentially only C^*-algebraic arguments are used. Indeed, in their proof, one tensors the C^*-algebras on which the bilinear form is defined with certain type III factors, whereas we show that it also works to tensor with certain simple nuclear C^*-algebras admitting KMS states instead. We then transform the problem back to the (classical) noncommutative Grothendieck Theorem, as was also done by Haagerup and Musat.

Recently, Regev and Vidick gave a more elementary proof of both the JCB Grothendieck Theorem for C^*-algebras and its version for exact operator spaces [19]. Their proof makes use of methods from quantum information theory and has the advantage that the transformation of the problem to the (classical) noncommutative Grothendieck Theorem is more explicit and based on finite-dimensional techniques. Moreover, they obtain certain new quantitative estimates.

For an extensive overview of the different versions of the Grothendieck Theorem, as well as their proofs and several applications, we refer to [17].

This text is organized as follows. In Sect. 10.2, we recall two different notions of complete boundedness for bilinear forms on operator spaces. In Sect. 10.3, we recall some facts about Cuntz algebras and their KMS states. This is needed for the proof of the JCB Grothendieck Theorem, which is given in Sect. 10.4 (with a constant $K > 1$) by using (single) Cuntz algebras. We explain how to obtain $K = 1$ in Sect. 10.5. In Sect. 10.6, we show that using a recent result by Haagerup and Musat on the best constant in the noncommutative little Grothendieck Theorem, we are able to improve the best constant in Blecher's inequality.

10.2 Bilinear Forms on Operator Spaces

Recall that an operator space E is a closed linear subspace of $\mathscr{B}(H)$ for some Hilbert space H. For $n \geq 1$, the embedding $M_n(E) \subset M_n(\mathscr{B}(H)) \cong \mathscr{B}(H^n)$ gives rise to a norm $\|.\|_n$ on $M_n(E)$. In particular, C^*-algebras are operator spaces. A linear map $T : E \to F$ between operator spaces induces a linear map $T_n : M_n(E) \to M_n(F)$ for each $n \in \mathbb{N}$, defined by $T_n([x_{ij}]) = [T(x_{ij})]$ for all $x = [x_{ij}] \in M_n(E)$. The map T is called completely bounded if the completely bounded norm $\|T\|_{cb} := \sup_{n\geq 1} \|T_n\|$ is finite.

There are two common ways to define a notion of complete boundedness for bilinear forms on operator spaces. For the first one, we refer to [5]. Let E and F be operator spaces contained in C^*-algebras A and B, respectively, and let $u : E \times F \to \mathbb{C}$ be a bounded bilinear form. Let $u_{(n)} : M_n(E) \times M_n(F) \to M_n(\mathbb{C})$ be the map defined by $([a_{ij}], [b_{ij}]) \mapsto \left[\sum_{k=1}^n u(a_{ik}, b_{kj})\right]$.

Definition 1. The bilinear form u is called *completely bounded* if

$$\|u\|_{cb} := \sup_{n\geq 1} \|u_{(n)}\|$$

is finite. We put $\|u\|_{cb} = \infty$ if u is not completely bounded.

Equivalently (see Sect. 3 of [12] or the Introduction of [18]), u is completely bounded if there exists a constant $C \geq 0$ and states f on A and g on B such that for all $a \in E$ and $b \in F$,

$$|u(a,b)| \leq C f(aa^*)^{\frac{1}{2}} g(b^*b)^{\frac{1}{2}}, \tag{10.1}$$

and $\|u\|_{cb}$ is the smallest constant C such that (10.1) holds.

For the second notion, we refer to [3, 9]. Let E and F be operator spaces contained in C^*-algebras A and B, respectively, and let $u : E \times F \to \mathbb{C}$ be a bounded bilinear form. Then there exists a unique bounded linear operator $\tilde{u} : E \to F^*$ such that

$$u(a,b) = \langle \tilde{u}(a), b \rangle$$

for all $a \in E$ and $b \in F$, where $\langle .,. \rangle$ denotes the pairing between F and its dual.

Definition 2. The bilinear form u is called *jointly completely bounded* if the map $\tilde{u} : E \to F^*$ is completely bounded, and we set

$$\|u\|_{jcb} := \|\tilde{u}\|_{cb}.$$

We put $\|u\|_{jcb} = \infty$ if u is not jointly completely bounded.

Equivalently, if we define maps $u_n : M_n(E) \otimes M_n(F) \to M_n(\mathbb{C}) \otimes M_n(\mathbb{C})$ by

$$u_n\left(\sum_{i=1}^{k} a_i \otimes c_i, \sum_{j=1}^{l} b_j \otimes d_j\right) = \sum_{i=1}^{k}\sum_{j=1}^{l} u(a_i, b_j) c_i \otimes d_j$$

for $a_1, \ldots, a_k \in A$, $b_1, \ldots, b_l \in B$, and $c_1, \ldots, c_k, d_1, \ldots, d_l \in M_n(\mathbb{C})$, then we have $\|u\|_{jcb} = \sup_{n \geq 1} \|u_n\|$.

10.3 KMS States on Cuntz Algebras

For $2 \leq n < \infty$, let $\mathcal{O}_n$ denote the Cuntz algebra generated by n isometries, as introduced by Cuntz in [6], in which one of the main results is that the algebras $\mathcal{O}_n$ are simple. We now recall some results by Cuntz. If $\alpha = (\alpha_1, \ldots, \alpha_k)$ denotes a multi-index of length $k = l(\alpha)$, where $\alpha_j \in \{1, \ldots, n\}$ for all j, we write $S_\alpha = S_{\alpha_1} \ldots S_{\alpha_k}$, and we put $S_0 = 1$. It follows that for every nonzero word M in $\{S_i\}_{i=1}^n \bigcup \{S_i^*\}_{i=1}^n$, there are unique multi-indices μ and ν such that $M = S_\mu S_\nu^*$.

For $k \geq 1$, let $\mathcal{F}_n^k$ be the C^*-algebra generated by $\{S_\mu S_\nu^* \mid l(\mu) = l(\nu) = k\}$, and let $\mathcal{F}_n^0 = \mathbb{C}1$. It follows that $\mathcal{F}_n^k$ is $*$-isomorphic to $M_{n^k}(\mathbb{C})$, and, as a consequence, $\mathcal{F}_n^k \subset \mathcal{F}_n^{k+1}$. The C^*-algebra $\mathcal{F}_n$ generated by $\bigcup_{k=0}^{\infty} \mathcal{F}_n^k$ is a UHF-algebra of type n^∞.

If we write $\mathcal{P}_n$ for the algebra generated algebraically by $S_1, \ldots, S_n$, $S_1^*, \ldots, S_n^*$, each element A in $\mathcal{P}_n$ has a unique representation

$$A = \sum_{k=1}^{N} (S_1^*)^k A_{-k} + A_0 + \sum_{k=1}^{N} A_k S_1^k,$$

where $N \in \mathbb{N}$ and $A_k \in \mathcal{P}_n \cap \mathcal{F}_n$. The maps $F_{n,k} : \mathcal{P}_n \to \mathcal{F}_n$ $(k \in \mathbb{Z})$ defined by $F_{n,k}(A) = A_k$ extend to norm-decreasing maps $F_{n,k} : \mathcal{O}_n \to \mathcal{F}_n$. It follows that $F_{n,0}$ is a conditional expectation.

The existence of a unique KMS state on each Cuntz algebra was proved by Olesen and Pedersen [15]. Firstly, we give some background on C^*-dynamical systems.

Definition 3. A C^*-dynamical system $(A, \mathbb{R}, \rho)$ consists of a C^*-algebra A and a representation $\rho : \mathbb{R} \to \mathrm{Aut}(A)$, such that each map $t \mapsto \rho_t(a)$, $a \in A$, is norm continuous.

C^*-dynamical systems can be defined in more general settings. In particular, one can replace $\mathbb{R}$ with arbitrary locally compact groups.

Let A^a denote the dense $*$-subalgebra of A consisting of analytic elements, i.e., $a \in A^a$ if the function $t \mapsto \rho_t(a)$ has a (necessarily unique) extension to an entire operator-valued function. This extension is implicitly used in the following definition.

Definition 4. Let $(A, \mathbb{R}, \rho)$ be a C^*-dynamical system. An invariant state ϕ on A, i.e., a state for which $\phi \circ \rho_t = \phi$ for all $t \in \mathbb{R}$, is a KMS state if

$$\phi(\rho_{t+i}(a)b) = \phi(b\rho_t(a))$$

for all $a \in A^a$, $b \in A$ and $t \in \mathbb{R}$.

This definition is similar to the one introduced by Takesaki (see [20], Definition 13.1). It corresponds to ϕ being a β-KMS state for ρ_{-t} with $\beta = 1$ according to the conventions of [4] and [15]. In the latter, the following two results were proved (see Lemma 1 and Theorem 2 therein). We restate these results slightly according to the conventions of Definition 4.

Proposition 1 (Olesen-Pedersen). *For all $t \in \mathbb{R}$ and the generators $\{S_k\}_{k=1}^n$ of $\mathcal{O}_n$, define $\rho_t^n(S_k) = n^{it} S_k$. Then ρ_t^n extends uniquely to a $*$-automorphism of $\mathcal{O}_n$ for every $t \in \mathbb{R}$ in such a way that $(\mathcal{O}_n, \mathbb{R}, \rho^n)$ becomes a C^*-dynamical system. Moreover, $\mathcal{F}_n$ is the fixed-point algebra of ρ^n in $\mathcal{O}_n$, and $\mathcal{P}_n \subset (\mathcal{O}_n)^a$.*

Let $\tau_n = \otimes_{k=1}^{\infty} \frac{1}{n}\mathrm{Tr}$ denote the unique tracial state on $\mathcal{F}_n$.

Proposition 2 (Olesen-Pedersen). *For $n \geq 2$, the C^*-dynamical system given by $(\mathcal{O}_n, \mathbb{R}, \rho^n)$ has exactly one KMS state, namely $\phi_n = \tau_n \circ F_{n,0}$.*

For a C^*-algebra A, let $\mathcal{U}(A)$ denote its unitary group. The following result was proved by Archbold [1]. It implies the *Dixmier property* for $\mathcal{O}_n$.

Proposition 3 (Archbold). *For all $x \in \mathcal{O}_n$,*

$$\phi_n(x)1_{\mathcal{O}_n} \in \overline{\mathrm{conv}\{uxu^* \mid u \in \mathcal{U}(\mathcal{F}_n)\}}^{\|.\|}.$$

As a corollary, we obtain the following (well-known) fact (see also [7]).

Corollary 1. *The relative commutant of $\mathcal{F}_n$ in $\mathcal{O}_n$ is trivial, i.e.,*

$$(\mathcal{F}_n)' \cap \mathcal{O}_n = \mathbb{C}1.$$

Proof. Let $x \in (\mathcal{F}_n)' \cap \mathcal{O}_n$. By Proposition 3, we know that for every $\varepsilon > 0$, there exists a finite convex combination $\sum_{i=1}^m \lambda_i u_i x u_i^*$, where $u_i \in \mathcal{U}(\mathcal{F}_n)$, such that $\| \sum_{i=1}^m \lambda_i u_i x u_i^* - \phi_n(x)1_{\mathcal{O}_n} \| < \varepsilon$. Since $x \in (\mathcal{F}_n)' \cap \mathcal{O}_n$, we have $\sum_{i=1}^m \lambda_i u_i x u_i^* = \sum_{i=1}^m \lambda_i x u_i u_i^* = x$. Hence, $\|x - \phi_n(x)1_{\mathcal{O}_n}\| < \varepsilon$. This implies that $x \in \mathbb{C}1$.

Proposition 3 can be extended to finite sets in $\mathcal{O}_n$, as described in the following lemma, by similar methods as in [8], Part III, Chap. 5. For an invertible element v in a C^*-algebra A, we define $\mathrm{ad}(v)(x) = vxv^{-1}$ for all $x \in A$.

Lemma 1. *Let $\{x_1, \ldots, x_k\}$ be a subset of $\mathcal{O}_n$, and let $\varepsilon > 0$. Then there exists a convex combination α of elements in $\{\mathrm{ad}(u) \mid u \in \mathcal{U}(\mathcal{F}_n)\}$ such that*

$$\|\alpha(x_i) - \phi_n(x_i)1_{\mathcal{O}_n}\| < \varepsilon \quad \text{for all } i = 1, \ldots, k.$$

Moreover, there exists a net $\{\alpha_j\}_{j \in J} \subset \mathrm{conv}\{\mathrm{ad}(u) \mid u \in \mathcal{U}(\mathcal{F}_n)\}$ such that

$$\lim_j \|\alpha_j(x) - \phi_n(x)1_{\mathcal{O}_n}\| = 0$$

for all $x \in \mathcal{O}_n$.

Proof. Suppose that $\|\alpha'(x_i) - \phi_n(x_i)1_{\mathcal{O}_n}\| < \varepsilon$ for $i = 1, \ldots, k-1$. By Proposition 3, we can find a convex combination $\tilde{\alpha}$ such that

$$\|\tilde{\alpha}(\alpha'(x_k)) - \phi_n(\alpha'(x_k))1_{\mathcal{O}_n}\| < \varepsilon.$$

Note that $\phi_n(\alpha'(x_k)) = \phi_n(x_k)$ and $1_{\mathcal{O}_n} = \tilde{\alpha}(1_{\mathcal{O}_n})$. By the fact that $\|\tilde{\alpha}(x)\| \leq \|x\|$ for all $x \in \mathcal{O}_n$, we conclude that $\alpha = \tilde{\alpha} \circ \alpha'$ satisfies $\|\alpha(x_i) - \phi_n(x_i)1_{\mathcal{O}_n}\| < \varepsilon$ for $i = 1, \ldots, k$.

Let J denote the directed set consisting of pairs (F, η), where F is a finite subset of $\mathcal{O}_n$ and $\eta \in (0, 1)$, with the ordering given by $(F_1, \eta_1) \preceq (F_2, \eta_2)$ if $F_1 \subset F_2$ and $\eta_1 \geq \eta_2$. By the first assertion, this gives rise to a net $\{\alpha_j\}_{j \in J}$ with the desired properties.

10.4 Proof of the JCB Grothendieck Theorem

In this section, we explain the proof of the Grothendieck Theorem for jointly completely bounded bilinear forms on C^*-algebras. As mentioned in Sect. 10.1, the proof is along the same lines as the proof by Haagerup and Musat, but we tensor with Cuntz algebras instead of type III factors.

Applying the GNS construction to the pair $(\mathcal{O}_n, \phi_n)$, we obtain a $*$-representation π_n of $\mathcal{O}_n$ on the Hilbert space $H_{\pi_n} = L^2(\mathcal{O}_n, \phi_n)$, with cyclic vector ξ_n, such that $\phi_n(x) = \langle \pi_n(x)\xi_n, \xi_n \rangle_{H_{\pi_n}}$. We identify $\mathcal{O}_n$ with its GNS representation. Note that ϕ_n extends in a normal way to the von Neumann algebra $\mathcal{O}_n''$, which also acts on H_{π_n}. This normal extension is a KMS state for a W^*-dynamical system with $\mathcal{O}_n''$ as the underlying von Neumann algebra (see Corollary 5.3.4 of [4]). The commutant $\mathcal{O}_n'$ of $\mathcal{O}_n$ is also a von Neumann algebra, and using Tomita-Takesaki theory (see [4, 20]), we obtain, via the polar decomposition of the closure of the operator $Sx\xi_n = x^*\xi_n$, a conjugate-linear involution $J : H_{\pi_n} \to H_{\pi_n}$ satisfying $J\mathcal{O}_n J \subset \mathcal{O}_n'$.

Lemma 2. *For $k \in \mathbb{Z}$, we have*

$$\mathcal{O}_n^k := \{x \in \mathcal{O}_n \mid \rho_t^n(x) = n^{-ikt}x \forall t \in \mathbb{R}\} = \{x \in \mathcal{O}_n \mid \phi_n(xy) = n^{-k}\phi_n(yx) \forall y \in \mathcal{O}_n\}.$$

The proof of this lemma is analogous to Lemma 1.6 of [21]. Note that $\mathcal{O}_n^0 = \mathcal{F}_n$, and that for all $k \in \mathbb{Z}$, we have $\mathcal{O}_n^k \neq \{0\}$.

Lemma 3. *For every $k \in \mathbb{Z}$, there exists a $c_k \in \mathcal{O}_n$ such that*

$$\phi_n(c_k^* c_k) = n^{\frac{k}{2}}, \qquad \phi_n(c_k c_k^*) = n^{-\frac{k}{2}},$$

and, moreover, $\langle c_k J c_k J \xi_n, \xi_n \rangle = 1$.

The proof is similar to the proof of Lemma 2.1 of [12].

Proposition 4. *Let A, B be C^*-algebras, and let $u : A \times B \to \mathbb{C}$ be a jointly completely bounded bilinear form. There exists a bounded bilinear form $\hat{u}$ on $(A \otimes_{\min} \mathcal{O}_n) \times (B \otimes_{\min} J\mathcal{O}_n J)$ given by*

$$\hat{u}(a \otimes c, b \otimes d) = u(a, b)\langle cd\xi_n, \xi_n \rangle$$

for all $a \in A$, $b \in B$, $c \in \mathcal{O}_n$ and $d \in J\mathcal{O}_n J$. Moreover, $\|\hat{u}\| \leq \|u\|_{jcb}$.

The C^*-algebra $J\mathcal{O}_n J$ is just a copy of $\mathcal{O}_n$. This result is analogous to Proposition 2.3 of [12], and the proof is the same. Note that we use $\|\sum_{i=1}^k c_i d_i\|_{\mathcal{B}(L^2(\mathcal{O}_n, \phi_n))} = \|\sum_{i=1}^k c_i \otimes d_i\|_{\mathcal{O}_n \otimes_{min} J\mathcal{O}_n J}$ for all $c_1, \ldots, c_k \in \mathcal{O}_n$ and $d_1, \ldots, d_k \in J\mathcal{O}_n J$. This equality is elementary, since $\mathcal{O}_n$ is simple and nuclear. In the proof of Haagerup and Musat, one takes the tensor product of A and a certain type III factor M and the tensor product of B with the commutant M' of M, respectively. Note that $J\mathcal{O}_n J \subset \mathcal{O}_n'$.

One can formulate analogues of Lemmas 2.4, 2.5 and Proposition 2.6 of [12]. They can be proved in the same way as there, and one explicitly needs the existence and properties of KMS states on the Cuntz algebras (see Sect. 10.3). The analogue of Proposition 2.6 gives the "transformation" of the JCB Grothendieck Theorem to the noncommutative Grothendieck Theorem for bounded bilinear forms.

Using Lemma 2.7 of [12], we arrive at the following conclusion, which is the analogue of [12], Proposition 2.8.

Proposition 5. *Let $K(n) = \sqrt{(n^{\frac{1}{2}} + n^{-\frac{1}{2}})/2}$, and let $u : A \times B \to \mathbb{C}$ be a jointly completely bounded bilinear form on C^*-algebras A, B. Then there exist states f_1^n, f_2^n on A and g_1^n, g_2^n on B such that for all $a \in A$ and $b \in B$,*

$$|u(a,b)| \leq K(n)\|u\|_{jcb} \left(f_1^n(aa^*)^{\frac{1}{2}} g_1^n(b^*b)^{\frac{1}{2}} + f_2^n(a^*a)^{\frac{1}{2}} g_2^n(bb^*)^{\frac{1}{2}} \right).$$

The above proposition is the JCB Grothendieck Theorem. However, the (universal) constant and states depend on n. This is because the noncommutative Grothendieck Theorem gives states on $A \otimes_{min} \mathcal{O}_n$ and $B \otimes_{min} J\mathcal{O}_n J$, which clearly depend on n, and these states are used to obtain the states on A and B. The best constant we obtain in this way comes from the case $n = 2$, which yields the constant $K(2) = \sqrt{(2^{\frac{1}{2}} + 2^{-\frac{1}{2}})/2} \sim 1.03$.

10.5 The Best Constant

In order to get the best constant $K = 1$, we consider the C^*-dynamical system $(A, \mathbb{R}, \rho)$, with $A = \mathcal{O}_2 \otimes \mathcal{O}_3$ and $\rho_t = \rho_t^2 \otimes \rho_t^3$. It is straightforward to check that it has a KMS state, namely $\phi = \phi_2 \otimes \phi_3$. It is easy to see that $\mathcal{F} = \mathcal{F}_2 \otimes \mathcal{F}_3$ is contained in the fixed point algebra. (Actually, it is equal to the fixed point algebra, but we do not need this.) These assertions follow by the fact that the algebraic tensor product of $\mathcal{O}_2$ and $\mathcal{O}_3$ is dense in $\mathcal{O}_2 \otimes \mathcal{O}_3$. Note that ρ is not periodic.

Applying the GNS construction to the pair (A, ϕ), we obtain a $*$-representation π of A on the Hilbert space $H_\pi = L^2(A, \phi)$, with cyclic vector ξ, such that $\phi(x) = \langle \pi(x)\xi, \xi\rangle_{H_\pi}$. We identify A with its GNS representation. Using Tomita-Takesaki theory, we obtain a conjugate-linear involution $J : H_\pi \to H_\pi$ satisfying $JAJ \subset A'$ (see also Sect. 10.4).

It follows directly from Proposition 3 that $\phi(x)1_A \in \overline{\text{conv}\{uxu^* \mid u \in \mathcal{U}(\mathcal{F})\}}^{\|.\|}$ for all $x \in A$. Also, the analogue of Lemma 1 follows in a similar way, as well as the fact that $\mathcal{F}' \cap A = \mathbb{C}1$.

It is elementary to check that

$$A_{\lambda,k} := \{x \in A \mid \rho_t(x) = \lambda^{ikt} x \forall t \in \mathbb{R}\} = \{x \in A \mid \phi(xy) = \lambda^k \phi(yx) \forall y \in \mathcal{O}_n\}.$$

Let $\Lambda := \{2^p 3^q \mid p, q \in \mathbb{Z}\} \cap (0, 1)$. For all $\lambda \in \Lambda$ and $k \in \mathbb{Z}$, we have $A_{\lambda,k} \neq \{0\}$. This leads, analogous to Lemma 3, to the following result.

Lemma 4. *Let $\lambda \in \Lambda$. For every $k \in \mathbb{Z}$ there exists a $c_{\lambda,k} \in A$ such that*

$$\phi(c_{\lambda,k}^* c_{\lambda,k}) = \lambda^{-\frac{k}{2}}, \qquad \phi(c_{\lambda,k} c_{\lambda,k}^*) = \lambda^{\frac{k}{2}}$$

and

$$\langle c_{\lambda,k} J c_{\lambda,k} J \xi, \xi\rangle = 1.$$

In this way, by the analogues of Lemmas 2.4, 2.5 and Proposition 2.6 of [12], we obtain the following result, which is the analogue of [12], Proposition 2.8.

Proposition 6. *Let $\lambda \in \Lambda$, and let $C(\lambda) = \sqrt{(\lambda^{\frac{1}{2}} + \lambda^{-\frac{1}{2}})/2}$. Let $u : A \times B \to \mathbb{C}$ be a jointly completely bounded bilinear form. Then there exist states f_1^λ, f_2^λ on A and g_1^λ, g_2^λ on B such that for all $a \in A$ and $b \in B$,*

$$|u(a,b)| \leq C(\lambda)\|u\|_{jcb} \left(f_1^\lambda(aa^*)^{\frac{1}{2}} g_1^\lambda(b^*b)^{\frac{1}{2}} + f_2^\lambda(a^*a)^{\frac{1}{2}} g_2^\lambda(bb^*)^{\frac{1}{2}} \right).$$

Note that $C(\lambda) > 1$ for $\lambda \in \Lambda$. Let $(\lambda_n)_{n\in\mathbb{N}}$ be a sequence in Λ converging to 1. By the weak*-compactness of the unit balls $(A_+^*)_1$ and $(B_+^*)_1$ of A_+^* and B_+^*, respectively, the Grothendieck Theorem for jointly completely bounded bilinear forms with $K = 1$ follows in the same way as in the "Proof of Theorem 1.1" in [12].

Remark 1. By Kirchberg's second "Geneva Theorem" (see [14] for a proof), we know that $\mathcal{O}_2 \otimes \mathcal{O}_3 \cong \mathcal{O}_2$. This implies that the best constant in Theorem 1 can also be obtained by tensoring with the single Cuntz algebra $\mathcal{O}_2$, but considered with a different action that defines the C^*-dynamical system. Since the explicit form of the isomorphism is not known, we cannot adjust the action accordingly.

10.6 A Remark on Blecher's Inequality

In [2], Blecher stated a conjecture about the norm of elements in the algebraic tensor product of two C^*-algebras. Equivalently, the conjecture can be formulated as follows (see Conjecture 0.2′ of [18]). For a bilinear form $u : A \times B \to \mathbb{C}$, put $u^t(b,a) = u(a,b)$.

Theorem 2 (Blecher's inequality). *There is a constant K such that any jointly completely bounded bilinear form $u : A \times B \to \mathbb{C}$ on C^*-algebras A and B decomposes as a sum $u = u_1 + u_2$ of completely bounded bilinear forms on $A \times B$, and $\|u_1\|_{cb} + \|u_2^t\|_{cb} \leq K\|u\|_{jcb}$.*

A version of this conjecture for exact operator spaces and a version for pairs of C^*-algebras, one of which is assumed to be exact, were proved by Pisier and Shlyakhtenko [18]. They also showed that the best constant in Theorem 2 is greater than or equal to 1. Haagerup and Musat proved that Theorem 2 holds with $K = 2$ [12, Sect. 3]. We show that the best constant is actually strictly greater than 1.

In the following, let $\mathrm{OH}(I)$ denote Pisier's operator Hilbert space based on $\ell^2(I)$ for some index set I. Recall the noncommutative little Grothendieck Theorem.

Theorem 3 (Noncommutative little Grothendieck Theorem). *Let A be a C^*-algebra, and let $T : A \to \mathrm{OH}(I)$ be a completely bounded map. Then there exists a universal constant $C > 0$ and states f_1 and f_2 on A such that for all $a \in A$,*

$$\|Ta\| \leq C\|T\|_{cb} f_1(aa^*)^{\frac{1}{4}} f_2(a^*a)^{\frac{1}{4}}.$$

For a completely bounded map $T : A \to \mathrm{OH}(I)$, denote by $C(T)$ the smallest constant $C > 0$ for which there exist states f_1, f_2 on A such that for all $a \in A$, we have $\|Ta\| \leq C f_1(aa^*)^{\frac{1}{4}} f_2(a^*a)^{\frac{1}{4}}$. In [12], Haagerup and Musat proved that $C(T) \leq \sqrt{2}\|T\|_{cb}$. Pisier and Shlyakhtenko proved in [18] that $\|T\|_{cb} \leq C(T)$ for all $T : A \to \mathrm{OH}(I)$. Haagerup and Musat proved that for a certain $T : M_3(\mathbb{C}) \to \mathrm{OH}(3)$, the inequality is actually strict, i.e., $\|T\|_{cb} < C(T)$ [13, Sect. 7]. We can now apply this knowledge to improve the best constant in Theorem 2.

Theorem 4. *The best constant K in Theorem 2 is strictly greater than* 1.

Proof. Let A be a C^*-algebra, and let $T : A \to \mathrm{OH}(I)$ be a completely bounded map for which $\|T\|_{cb} < C(T)$. Define the map $V = \overline{T^*JT}$ from A to $\overline{A^*} = \overline{A}^*$, where $J : \mathrm{OH}(I) \to \overline{\mathrm{OH}(I)^*}$ is the canonical complete isomorphism and $T^* : \mathrm{OH}(I)^* \to A^*$ is the adjoint of T. Hence, V is completely bounded. It follows that $V = \tilde{u}$ for some jointly completely bounded bilinear form $u : A \times \overline{A} \to \mathbb{C}$. Moreover, $\|u\|_{jcb} = \|V\|_{cb} = \|T\|_{cb}^2$, where the last equality follows from the proof of Corollary 3.4 in [18]. By Blecher's inequality, i.e., Theorem 2, we have a decomposition $u = u_1 + u_2$ such that $\|u_1\|_{cb} + \|u_2^t\|_{cb} \leq K\|u\|_{jcb}$.

By the second characterization of completely bounded bilinear forms (in the Christensen-Sinclair sense) in Sect. 10.2, we obtain

$$|u_1(a,b)| \leq \|u_1\|_{cb} f_1(aa^*)^{\frac{1}{2}} g_1(b^*b)^{\frac{1}{2}}, \quad |u_2(a,b)| \leq \|u_2^t\|_{cb} f_2(a^*a)^{\frac{1}{2}} g_2(bb^*)^{\frac{1}{2}}.$$

It follows that

$$|u(a,b)| \leq \|u_1\|_{cb} f_1(aa^*)^{\frac{1}{2}} g_1(b^*b)^{\frac{1}{2}} + \|u_2^t\|_{cb} f_2(a^*a)^{\frac{1}{2}} g_2(bb^*)^{\frac{1}{2}}.$$

Let $\overline{g}_i(a) = g_i(\overline{a^*})$ for $i = 1, 2$, and define states

$$\tilde{f} = \frac{\|u_1\|_{cb} f_1 + \|u_2^t\|_{cb}\overline{g}_2}{\|u_1\|_{cb} + \|u_2^t\|_{cb}} \text{ and } \tilde{g} = \frac{\|u_1\|_{cb}\overline{g}_1 + \|u_2^t\|_{cb} f_2}{\|u_1\|_{cb} + \|u_2^t\|_{cb}}.$$

We obtain

$$\begin{aligned}\|T(a)\|^2 = |u(a,\overline{a})| &\leq \|u_1\|_{cb} f_1(aa^*)^{\frac{1}{2}}\overline{g}_1(a^*a)^{\frac{1}{2}} + \|u_2^t\|_{cb} f_2(a^*a)^{\frac{1}{2}}\overline{g}_2(aa^*)^{\frac{1}{2}} \\ &\leq (\|u_1\|_{cb} f_1 + \|u_2^t\|_{cb}\overline{g}_2)(aa^*)^{\frac{1}{2}}(\|u_1\|_{cb}\overline{g}_1 + \|u_2^t\|_{cb} f_2)(a^*a)^{\frac{1}{2}} \\ &\leq (\|u_1\|_{cb} + \|u_2^t\|_{cb})\tilde{f}(aa^*)^{\frac{1}{2}}\tilde{g}(a^*a)^{\frac{1}{2}}.\end{aligned}$$

Hence, $\|u_1\|_{cb} + \|u_2^t\|_{cb} \geq C(T)^2 > \|T\|_{cb}^2 = \|u\|_{jcb}$. This proves the theorem.

Acknowledgements The question if a C^*-algebraic proof of Theorem 1 exists was suggested to me by Uffe Haagerup and Magdalena Musat. I thank them for many useful comments.

The author was supported by the Danish National Research Foundation through the Centre for Symmetry and Deformation (DNRF92).

References

1. Archbold, R.J.: On the simple C^*-algebras of J. Cuntz. J. London Math. Soc. (2) **21**(3), 517–526 (1980). DOI 10.1112/jlms/s2-21.3.517. URL http://dx.doi.org/10.1112/jlms/s2-21.3.517
2. Blecher, D.P.: Generalizing Grothendieck's program. In: Function spaces (Edwardsville, IL, 1990), *Lecture Notes in Pure and Appl. Math.*, vol. 136, pp. 45–53. Dekker, New York (1992)
3. Blecher, D.P., Paulsen, V.I.: Tensor products of operator spaces. J. Funct. Anal. **99**(2), 262–292 (1991). DOI 10.1016/0022-1236(91)90042-4. URL http://dx.doi.org/10.1016/0022-1236(91)90042-4

4. Bratteli, O., Robinson, D.W.: Operator algebras and quantum statistical mechanics. 2, second edn. Texts and Monographs in Physics. Springer-Verlag, Berlin (1997). Equilibrium states. Models in quantum statistical mechanics
5. Christensen, E., Sinclair, A.M.: Representations of completely bounded multilinear operators. J. Funct. Anal. **72**(1), 151–181 (1987). DOI 10.1016/0022-1236(87)90084-X. URL http://dx.doi.org/10.1016/0022-1236(87)90084-X
6. Cuntz, J.: Simple C^*-algebras generated by isometries. Comm. Math. Phys. **57**(2), 173–185 (1977)
7. Cuntz, J.: Automorphisms of certain simple C^*-algebras. In: Quantum fields—algebras, processes (Proc. Sympos., Univ. Bielefeld, Bielefeld, 1978), pp. 187–196. Springer, Vienna (1980)
8. Dixmier, J.: von Neumann algebras, *North-Holland Mathematical Library*, vol. 27. North-Holland Publishing Co., Amsterdam (1981). With a preface by E. C. Lance, Translated from the second French edition by F. Jellett
9. Effros, E.G., Ruan, Z.J.: A new approach to operator spaces. Canad. Math. Bull. **34**(3), 329–337 (1991). DOI 10.4153/CMB-1991-053-x. URL http://dx.doi.org/10.4153/CMB-1991-053-x
10. Grothendieck, A.: Résumé de la théorie métrique des produits tensoriels topologiques. Bol. Soc. Mat. São Paulo **8**, 1–79 (1953)
11. Haagerup, U.: The Grothendieck inequality for bilinear forms on C^*-algebras. Adv. in Math. **56**(2), 93–116 (1985). DOI 10.1016/0001-8708(85)90026-X. URL http://dx.doi.org/10.1016/0001-8708(85)90026-X
12. Haagerup, U., Musat, M.: The Effros-Ruan conjecture for bilinear forms on C^*-algebras. Invent. Math. **174**(1), 139–163 (2008). DOI 10.1007/s00222-008-0137-7. URL http://dx.doi.org/10.1007/s00222-008-0137-7
13. Haagerup, U., Musat, M.: Factorization and dilation problems for completely positive maps on von Neumann algebras. Comm. Math. Phys. **303**(2), 555–594 (2011). DOI 10.1007/s00220-011-1216-y. URL http://dx.doi.org/10.1007/s00220-011-1216-y
14. Kirchberg, E., Phillips, N.C.: Embedding of exact C^*-algebras in the Cuntz algebra $\mathcal{O}_2$. J. Reine Angew. Math. **525**, 17–53 (2000). DOI 10.1515/crll.2000.065. URL http://dx.doi.org/10.1515/crll.2000.065
15. Olesen, D., Pedersen, G.K.: Some C^*-dynamical systems with a single KMS state. Math. Scand. **42**(1), 111–118 (1978)
16. Pisier, G.: Grothendieck's theorem for noncommutative C^*-algebras, with an appendix on Grothendieck's constants. J. Funct. Anal. **29**(3), 397–415 (1978). DOI 10.1016/0022-1236(78)90038-1. URL http://dx.doi.org/10.1016/0022-1236(78)90038-1
17. Pisier, G.: Grothendieck's theorem, past and present. Bull. Amer. Math. Soc. (N.S.) **49**(2), 237–323 (2012). DOI 10.1090/S0273-0979-2011-01348-9. URL http://dx.doi.org/10.1090/S0273-0979-2011-01348-9
18. Pisier, G., Shlyakhtenko, D.: Grothendieck's theorem for operator spaces. Invent. Math. **150**(1), 185–217 (2002). DOI 10.1007/s00222-002-0235-x. URL http://dx.doi.org/10.1007/s00222-002-0235-x
19. Regev, O., Vidick, T.: Elementary proofs of Grothendieck theorems for completely bounded norms (2012), J. Operator Theory (to appear), arXiv:1206.4025
20. Takesaki, M.: Tomita's theory of modular Hilbert algebras and its applications. Lecture Notes in Mathematics, Vol. 128. Springer-Verlag, Berlin (1970)
21. Takesaki, M.: The structure of a von Neumann algebra with a homogeneous periodic state. Acta Math. **131**, 79–121 (1973)

Chapter 11
C^*-Algebras Associated with a-adic Numbers

Tron Omland

Abstract By a crossed product construction, we produce a family of (stabilized) Cuntz-Li algebras associated with the a-adic numbers. Moreover, we present an a-adic duality theorem.

Keywords C^*-dynamical system • Cuntz-Li algebras • a-adic numbers

Mathematics Subject Classification (2010): 46L55, 11R04, 11R56.

11.1 Introduction

In [1] Cuntz introduces the C^*-algebra $\mathcal{Q}_{\mathbb{N}}$ associated with the $ax + b$-semigroup over the natural numbers, that is $\mathbb{Z} \rtimes \mathbb{N}^\times$, where $\mathbb{N}^\times$ acts on $\mathbb{Z}$ by multiplication. It is defined as the universal C^*-algebra generated by isometries $\{s_n\}_{n\in\mathbb{N}^\times}$ and a unitary u satisfying the relations

$$s_m s_n = s_{mn}, \quad s_n u = u^n s_n, \quad \text{and} \quad \sum_{k=0}^{n-1} u^k s_n s_n^* u^{-k} = 1 \quad \text{for } m, n \in \mathbb{N}^\times.$$

Furthermore, $\mathcal{Q}_{\mathbb{N}}$ is shown to be simple and purely infinite and can also be obtained as a semigroup crossed product

$$C(\hat{\mathbb{Z}}) \rtimes (\mathbb{Z} \rtimes \mathbb{N}^\times)$$

T. Omland (✉)
Department of Mathematical Sciences, Norwegian University of Science and Technology (NTNU), NO-7491 Trondheim, Norway
e-mail: tron.omland@math.ntnu.no

T.M. Carlsen et al. (eds.), *Operator Algebra and Dynamics*, Springer Proceedings in Mathematics & Statistics 58, DOI 10.1007/978-3-642-39459-1_11,

for the natural $ax+b$-semigroup action of $\mathbb{Z} \rtimes \mathbb{N}^\times$ on the finite integral adeles $\hat{\mathbb{Z}} = \prod_{p \text{ prime}} \mathbb{Z}_p$ (i.e. $\hat{\mathbb{Z}}$ is the profinite completion of $\mathbb{Z}$). Its stabilization $\overline{\mathscr{Q}_\mathbb{N}}$ is isomorphic to the ordinary crossed product

$$C_0(\mathscr{A}_f) \rtimes (\mathbb{Q} \rtimes \mathbb{Q}_+^\times)$$

where $\mathbb{Q}_+^\times$ denotes the multiplicative group of positive rationals and $\mathscr{A}_f$ denotes the finite adeles, i.e. the restricted product $\prod'_{p \text{ prime}} \mathbb{Q}_p = \prod_{p \text{ prime}} (\mathbb{Q}_p, \mathbb{Z}_p)$. The action of $\mathbb{Q} \rtimes \mathbb{Q}_+^\times$ on $\mathscr{A}_f$ is the natural $ax+b$-action. This crossed product is the minimal automorphic dilation of the semigroup crossed product above (see Laca [8]).

Replacing $\mathbb{N}^\times$ with $\mathbb{Z}^\times$ gives rise to the C^*-algebra $\mathscr{Q}_\mathbb{Z}$ of the ring $\mathbb{Z}$. This approach is generalized to certain integral domains by Cuntz and Li [3] and then to more general rings by Li [10].

In [9] Larsen and Li define the 2-adic ring algebra of the integers $\mathscr{Q}_2$, attached to the semigroup $\mathbb{Z} \rtimes |2\rangle$, where $|2\rangle = \{2^i : i \geq 0\} \subset \mathbb{N}^\times$ acts on $\mathbb{Z}$ by multiplication. It is the universal C^*-algebra generated by an isometry s_2 and a unitary u satisfying the relations

$$s_2 u^k = u^{2k} s_2 \quad \text{and} \quad s_2 s_2^* + u s_2 s_2^* u^* = 1.$$

The algebra $\mathscr{Q}_2$ shares many structural properties with $\mathscr{Q}_\mathbb{N}$. It is simple, purely infinite and has a semigroup crossed product description. Its stabilization $\overline{\mathscr{Q}_2}$ is isomorphic to its minimal automorphic dilation, which is the crossed product

$$C_0(\mathbb{Q}_2) \rtimes (\mathbb{Z}[\tfrac{1}{2}] \rtimes \langle 2 \rangle).$$

Here $\mathbb{Z}[\frac{1}{2}]$ denotes the ring extension of $\mathbb{Z}$ by $\frac{1}{2}$, $\langle 2 \rangle$ the subgroup of the positive rationals $\mathbb{Q}_+^\times$ generated by 2 and the action of $\mathbb{Z}[\frac{1}{2}] \rtimes \langle 2 \rangle$ on $\mathbb{Q}_2$ is the natural $ax+b$-action.

Both $\mathscr{A}_f$ and $\mathbb{Q}_2$ are examples of groups of so-called a-adic numbers, defined by a doubly infinite sequence $a = (\ldots, a_{-2}, a_{-1}, a_0, a_1, a_2, \ldots)$ with $a_i \geq 2$ for all $i \in \mathbb{Z}$. Our goal is to construct C^*-algebras associated with the a-adic numbers and show that these algebras provide a family of examples that under certain conditions share many structural properties with $\mathscr{Q}_2$, $\mathscr{Q}_\mathbb{N}$ and also the ring C^*-algebras of Cuntz and Li.

Our approach is inspired by [5], that is, we begin with a crossed product by a group and use the classical theory of C^*-dynamical systems to prove our results, instead of the generators and relations as in the papers of Cuntz, Li and Larsen. Therefore, our construction only gives analogs of the stabilized algebras $\overline{\mathscr{Q}_\mathbb{N}}$ and $\overline{\mathscr{Q}_2}$.

Even though the C^*-algebras associated with a-adic numbers are closely related to the ring C^*-algebras of Cuntz and Li, they are not a special case of these (except in the finite adeles case). Also, our approach does not fit in general into the framework of [5].

One of the main results in the paper is Theorem 3, which is a general a-adic duality theorem that encompasses the 2-adic duality theorem [9, Theorem 7.5] and the analogous result of Cuntz [1, Theorem 6.5]. In the proof, we only apply crossed product techniques, and not the groupoid equivalence as in [9].

11.2 The a-adic Numbers

Let $a = (\ldots, a_{-2}, a_{-1}, a_0, a_1, a_2, \ldots)$ be a doubly infinite sequence of natural numbers with $a_i \geq 2$ for all $i \in \mathbb{Z}$. Let the sequence a be arbitrary, but fixed.

We use Hewitt and Ross [4, Sects. 10 and 25] as our reference and define the a-adic numbers Ω as the group of sequences

$$\left\{ x = (x_i) \in \prod_{i=-\infty}^{\infty} \{0, 1, \ldots, a_i - 1\} : x_i = 0 \text{ for } i < j \text{ for some } j \in \mathbb{Z} \right\}$$

under addition with carry, that is, the sequences have a first nonzero entry and addition is defined inductively. Its topology is generated by the subgroups $\{\Lambda_j : j \in \mathbb{Z}\}$, where

$$\Lambda_j = \{x \in \Omega : x_i = 0 \text{ for } i < j\}.$$

This turns Ω into a totally disconnected, locally compact Hausdorff abelian group. The group Δ of a-adic integers is defined as $\Delta = \Lambda_0$. It is a compact, open subgroup, and a maximal compact ring in Ω with product given by multiplication with carry. On the other hand, Ω itself is not a ring in general (see (11.4) in Sect. 11.5).

Define the a-adic rationals N as the additive subgroup of $\mathbb{Q}$ given by

$$N = \left\{ \frac{j}{a_{-1} \cdots a_{-k}} : j \in \mathbb{Z}, k \geq 1 \right\}.$$

In fact, all noncyclic additive subgroups of $\mathbb{Q}$ containing $\mathbb{Z}$ are of this form (see Lemma 2 below). There is an injective homomorphism

$$\iota : N \hookrightarrow \Omega$$

determined by

$$\left(\iota\left(\frac{1}{a_{-1} \cdots a_{-k}} \right) \right)_{-j} = \delta_{jk}.$$

Moreover, $\iota(N)$ is the dense subgroup of Ω comprising the sequences with only finitely many nonzero entries. This map restricts to an injective ring homomorphism denoted by the same symbol

$$\iota : \mathbb{Z} \hookrightarrow \Delta$$

with dense range. Henceforth, we will suppress the ι and identify N and $\mathbb{Z}$ with their image in Ω and Δ, respectively.

Now let $\mathscr{U}$ be the family of all subgroups of N of the form $\frac{m}{n}\mathbb{Z}$, where m and n are natural numbers such that m divides $a_0 \cdots a_j$ for some $j \geq 0$ and n divides $a_{-1} \cdots a_{-k}$ for some $k \geq 1$. Then $\mathscr{U}$

1. is downward directed, that is, for all $U, V \in \mathscr{U}$ there exists $W \in \mathscr{U}$ such that $W \subset U \cap V$,
2. is separating, that is,

$$\bigcap_{U \in \mathscr{U}} U = \{e\},$$

3. has finite quotients, that is, $|U/V| < \infty$ whenever $U, V \in \mathscr{U}$ and $V \subset U$,

and the same is also true for

$$\mathscr{V} = \{U \cap \mathbb{Z} : U \in \mathscr{U}\}.$$

In fact, both $\mathscr{U}$ and $\mathscr{V}$ are closed under intersections, since

$$\frac{m}{n}\mathbb{Z} \cap \frac{m'}{n'}\mathbb{Z} = \frac{\operatorname{lcm}(m, m')}{\gcd(n, n')}\mathbb{Z}. \tag{11.1}$$

It is a consequence of (1)–(3) above that the collection of subgroups $\mathscr{U}$ induces a locally compact Hausdorff topology on N. Denote the Hausdorff completion of N with respect to this topology by $\overline{N}$. Then

$$\overline{N} \cong \varprojlim_{U \in \mathscr{U}} N/U.$$

Next, let $U_0 = \mathbb{Z}$ and for $j \geq 1$ define $U_j = a_0 \cdots a_{j-1}\mathbb{Z}$ and set

$$\mathscr{W} = \{U_j : j \geq 0\} \subset \mathscr{V} \subset \mathscr{U}.$$

Note that $\mathscr{W}$ is also separating and closed under intersections. The closure of U_j in Ω is Λ_j, so

$$\Omega/\Lambda_j \cong N/U_j \quad \text{and} \quad \Delta/\Lambda_j \cong \mathbb{Z}/U_j \quad \text{for all } j \geq 0.$$

Next, let

$$\tau_j : \Omega \to N/U_j$$

denote the quotient map for $j \geq 0$, and identify $\tau_j(x)$ with the truncated sequence $x^{(j-1)}$, where $x^{(j)}$ is defined for all $j \in \mathbb{Z}$ by

$$(x^{(j)})_i = \begin{cases} x_i & \text{for } i \leq j, \\ 0 & \text{for } i > j. \end{cases}$$

We find it convenient to use the standard construction of the inverse limit of the system $\{N/U_j, \pmod{a_j}\}$:

$$\varprojlim_{j \geq 0} N/U_j = \left\{ x = (x_i) \in \prod_{i=0}^{\infty} N/U_i : x_i = x_{i+1} \pmod{a_i} \right\},$$

and then the product $\tau : \Omega \to \varprojlim_{j \geq 0} N/U_j$ of the truncation maps τ_j, given by

$$\tau(x) = (\tau_0(x), \tau_1(x), \tau_2(x), \dots) = (x^{(-1)}, x^{(0)}, x^{(1)}, \dots),$$

is an isomorphism.

Furthermore, we note that $\mathscr{W}$ is cofinal in $\mathscr{U}$. Indeed, for all $U = \frac{m}{n}\mathbb{Z} \in \mathscr{U}$, if we choose $j \geq 0$ such that m divides $a_0 \cdots a_j$ then we have $\mathscr{W} \ni U_{j+1} \subset U$.

Therefore,

$$\Omega \cong \varprojlim_{j \geq 0} N/U_j \cong \varprojlim_{U \in \mathscr{U}} N/U \cong \overline{N},$$

and similarly

$$\Delta \cong \varprojlim_{j \geq 0} \mathbb{Z}/U_j \cong \varprojlim_{V \in \mathscr{V}} \mathbb{Z}/V \cong \overline{\mathbb{Z}}.$$

In particular, Δ is a profinite group. In fact, every profinite group coming from a completion of $\mathbb{Z}$ occurs this way (see also Lemma 2).

The following is a consequence of (11.1) and should serve as motivation for our $\mathscr{U}$.

Lemma 1 ([6, Lemma 1.1]). *Every open subgroup of Ω is of the form*

$$\overline{\bigcup_{U \in \mathscr{C}} U}$$

for some increasing chain $\mathscr{C}$ in $\mathscr{U}$. In particular, every compact open subgroup of Ω is of the form $\overline{U}$ for some $U \in \mathscr{U}$.

Whenever any confusion is possible, we write Ω_a, Δ_a, N_a, etc. for the structures associated with the sequence a. If b is another sequence such that $\mathscr{U}_a = \mathscr{U}_b$, we write $a \sim b$. In this case also $N_a = N_b$. It is not hard to verify that $a \sim b$ if and only if there is an isomorphism $\Omega_a \to \Omega_b$ restricting to an isomorphism $\Delta_a \to \Delta_b$. The groups Ω_a and Ω_b can nevertheless be isomorphic even if $a \not\sim b$ (see Example 3 below). In this regard, we have the following result, which is a consequence of Proposition 2.

Theorem 1 ([6, Corollary 5.4]). *We have that $\Omega_a \cong \Omega_b$ if and only if there exists a $(\mathscr{U}_a, \mathscr{U}_b)$-continuous isomorphism $N_a \to N_b$.*

Example 1. Let p be a prime and assume $a = (\dots, p, p, p, \dots)$. Then $\Omega \cong \mathbb{Q}_p$ and $\Delta \cong \mathbb{Z}_p$, i.e. the usual p-adic numbers and p-adic integers.

Example 2. Let $a = (\dots, 4, 3, 2, 3, 4, \dots)$, i.e. $a_i = a_{-i} = i + 2$ for $i \geq 0$. Then $\Omega \cong \mathscr{A}_f$ and $\Delta \cong \hat{\mathbb{Z}}$, because every prime occurs infinitely often among both the positive and the negative tail of the sequence a (see the paragraph after Lemma 2).

Example 3. Let $a_i = 2$ for $i \neq 0$ and $a_0 = 3$, so that

$$N = \mathbb{Z}[\tfrac{1}{2}] \quad \text{and} \quad \mathscr{U} = \{2^i\mathbb{Z}, 2^i 3\mathbb{Z} : i \in \mathbb{Z}\}.$$

Then Ω contains torsion elements. Indeed, let

$$x = (\dots, 0, 1, 1, 0, 1, 0, 1, \dots), \quad \text{so that} \quad 2x = (\dots, 0, 2, 0, 1, 0, 1, 0, \dots),$$

where the first nonzero entry is x_0. Then $3x = 0$ and $\{0, x, 2x\}$ forms a subgroup of Ω isomorphic with $\mathbb{Z}/3\mathbb{Z}$. Hence $\Omega \not\cong \mathbb{Q}_2$ since $\mathbb{Q}_2$ is a field.

Furthermore, let b be given by $b_i = a_{i+1}$, that is, $b_i = 2$ for $i \neq -1$ and $b_{-1} = 3$. Then

$$N_b = \tfrac{1}{3}\mathbb{Z}[\tfrac{1}{2}] \quad \text{and} \quad \mathscr{U}_b = \{2^i\mathbb{Z}, 2^i \tfrac{1}{3}\mathbb{Z} : i \in \mathbb{Z}\}.$$

We have $\Omega_a \cong \Omega_b$, but $a \not\sim b$ since $\Delta_a \not\cong \Delta_b$. Note also that the equation $3x = 1$ has no solution in Ω_a, but two solutions in Ω_b, and these are

$$\tfrac{1}{3} \in N_b \text{ and } y = (\dots, 0, 1, 1, 0, 1, 0, 1, \dots), \text{ where the first nonzero entry is } y_0.$$

11.3 The a-adic Algebras

We now want to define a multiplicative action on Ω, of some suitable subset of N, that is compatible with the natural multiplicative action of $\mathbb{Z}$ on Ω. Let S consist of all $s \in \mathbb{Q}_+^\times$ such that the map $\mathscr{U} \to \mathscr{U}$ given by $U \mapsto sU$ is well-defined and bijective. Clearly, the map $U \mapsto sU$ is injective if it is well-defined and it is surjective if the map $U \mapsto s^{-1}U$ is well-defined. Define a subset P of the prime numbers by

$$P = \{p \text{ prime} : p \text{ divides } a_k \text{ for infinitely many } k < 0 \text{ and infinitely many } k \geq 0\}.$$

It is not hard to see that S coincides with the subgroup $\langle P \rangle$ of $\mathbb{Q}_+^\times$ generated by P. Moreover, S is the largest subgroup of $\mathbb{Q}_+^\times$ that acts continuously on N. Indeed, the action is well-defined since all $q \in N$ belongs to some $U \in \mathscr{U}$. If $q + U$ is a basic open set in N, then its inverse image under multiplication by s, $s^{-1}(q+U) = s^{-1}q + s^{-1}U$, is also open in N as $s^{-1}U \in \mathscr{U}$. By letting S be discrete, it follows that the action is continuous.

We will not always be interested in the action of the whole group S on N, but rather a subgroup of S. So henceforth, let H denote any subgroup of S. Furthermore, let G be the semidirect product of N by H, i.e. $G = N \rtimes H$ where H acts on N by multiplication. This means that there is a well-defined $ax + b$-action of G on N given by

$$(r, h) \cdot q = r + hq \quad \text{for } q, r \in N \text{ and } h \in H.$$

This action is continuous with respect to $\mathscr{U}$, and can therefore be extended to an action of G on Ω, by uniform continuity.

Proposition 1 ([6, Proposition 2.4]). *Assume $P \neq \varnothing$ and let H be a nontrivial subgroup of S. Then the action of $G = N \rtimes H$ on Ω is minimal, locally contractive and topologically free.*

Definition 1. Suppose $P \neq \varnothing$. If H is a nontrivial subgroup of S, we define the C^*-algebra $\overline{\mathscr{Q}} = \overline{\mathscr{Q}}(a, H)$ by

$$\overline{\mathscr{Q}} = C_0(\Omega) \rtimes_{\alpha^{\mathrm{aff}}} G,$$

where

$$\alpha^{\mathrm{aff}}{}_{(n,h)}(f)(x) = f(h^{-1} \cdot (x - n)).$$

Remark 1. The bar-notation on $\overline{\mathscr{Q}}$ is used so that it agrees with the notation for stabilized Cuntz-Li algebras in [1] and [9].

Theorem 2 ([6, Corollary 2.8]). *The C^*-algebra $\overline{\mathscr{Q}}$ is simple and purely infinite. Moreover, $\overline{\mathscr{Q}}$ is a nonunital Kirchberg algebra in the UCT class.*

Example 4. If $a = (\ldots, 2, 2, 2, \ldots)$ and $H = S = \langle 2 \rangle$, then $\overline{\mathscr{Q}}$ is the algebra $\overline{\mathscr{Q}}_2$ of Larsen and Li [9]. More generally, if p is a prime, $a = (\ldots, p, p, p, \ldots)$ and $H = S = \langle p \rangle$, we are in the setting of Example 1 and get algebras similar to $\overline{\mathscr{Q}}_2$.

If $a = (\dots, 4, 3, 2, 3, 4, \dots)$ and $H = S = \mathbb{Q}_+^\times$, then we are in the setting of Example 2. In this case $\overline{\mathcal{Q}}$ is the algebra $\overline{\mathcal{Q}}_\mathbb{N}$ of Cuntz [1].

Both these algebras are special cases of the most well-behaved situation, namely where $H = S$ and $a_i \in H$ for all $i \in \mathbb{Z}$. The algebras arising this way are completely determined by the set (finite or infinite) of primes P, and are precisely the kind of algebras that fit into the framework of [5]. The cases described above are the two extremes, where P consists of either one single prime or all primes.

If $a \sim b$, then $S_a = S_b$ and $\overline{\mathcal{Q}}(a, H) = \overline{\mathcal{Q}}(b, H)$ for all $H \subset S_a = S_b$. Suppose $\Omega_a \cong \Omega_b$. Then $S_a = S_b$ as well, and for all $H \subset S_a = S_b$, we have that $\overline{\mathcal{Q}}(a, H) \cong \overline{\mathcal{Q}}(b, H)$. Indeed, by Theorem 1 there exists an isomorphism $\varphi : \Omega_a \to \Omega_b$ restricting to an isomorphism $N_a \to N_b$. Therefore, the map

$$\varphi_* : C_c(N_a \rtimes H, C_0(\Omega_a)) \to C_c(N_b \rtimes H, C_0(\Omega_b))$$

given by

$$\varphi_*(f)(n, h)(x) = f(\varphi^{-1}(n), h)(\varphi^{-1}(x))$$

determines an isomorphism $\overline{\mathcal{Q}}(a, H) \cong \overline{\mathcal{Q}}(b, H)$.

Example 5. Let a and b be the sequences from Example 3. Then $\overline{\mathcal{Q}}(a, H) \cong \overline{\mathcal{Q}}(b, H)$ for all $H \subset S_a = S_b = \langle 2 \rangle$.

Example 6. If $a = (\dots, 2, 2, 2, \dots)$ and $b = (\dots, 4, 4, 4, \dots)$, then $a \sim b$. Hence, for all nontrivial $H \subset S = \langle 2 \rangle$ we have $\overline{\mathcal{Q}}(a, H) = \overline{\mathcal{Q}}(b, H)$. However, if $H = \langle 4 \rangle$, then $\overline{\mathcal{Q}}(a, S) \ncong \overline{\mathcal{Q}}(a, H)$, as remarked after Question 1.

In light of this example, it could also be interesting to investigate the $ax+b$-action on Ω of other subgroups G' of $N \rtimes S$. It follows from the proof of Proposition 1 that the action of G' on Ω is minimal, locally contractive and topologically free if and only if $G' = M \rtimes H$, where $M \subset N$ is dense in Ω and $H \subset S$ is nontrivial.

Moreover, it can be shown that a proper subgroup M of N is dense in Ω if and only if $M = qN$ for some $q \geq 2$ such that q and a_i are relatively prime for all $i \in \mathbb{Z}$. This property is also invariant under isomorphisms, i.e. if $\Omega_a \cong \Omega_b$ and $q \geq 2$, then qN_a is dense in Ω_a if and only if qN_b is dense in Ω_b (see Sect. 11.5). However, if M is such a subgroup of N that is dense in Ω and $H \subset S$, then

$$C_0(\Omega) \rtimes_{\alpha^{\text{aff}}} (N \rtimes H) \cong C_0(\Omega) \rtimes_{\alpha^{\text{aff}}} (M \rtimes H). \tag{11.2}$$

The reason for the isomorphism (11.2) is the following. If

$$Q = \{p \text{ prime} : p \text{ does not divide any } a_i\},$$

then multiplication by a prime p is an automorphism of Ω if and only if $p \in P \cup Q$. Indeed, if $p \in Q$, then $p\overline{U} = \overline{pU} = \overline{U}$ for all $U \in \mathscr{U}$. Thus, $\frac{1}{p} \in \Omega$ when $p \in Q$ and it is possible to embed the subgroup

$$N_Q = \{\tfrac{n}{q} : n \in N, q \in \langle Q \rangle\} \subset \mathbb{Q}$$

in Ω, where $\langle Q \rangle$ denotes the multiplicative subgroup of $\mathbb{Q}_+^\times$ generated by Q.

We complete this discussion by considering the $ax+b$-action on Ω of potentially larger groups than $N \rtimes S$. The largest subgroup of $\mathbb{Q} \rtimes \mathbb{Q}_+^\times$ that can act on Ω through an $ax+b$-action is $N_Q \rtimes \langle P \cup Q \rangle$. However, the only groups $N \subset M \subset N_Q$ that give rise to the duality theorems in the next section are of the form $M = \frac{1}{q}N$ for $q \in \langle Q \rangle$ (see Remark 2). Moreover, S is the largest subgroup of $\langle P \cup Q \rangle$ that acts on M, and of course, (11.2) also holds for all $H \subset S$ in this case.

Finally, we remark that one may also involve the roots of unity of $\mathbb{Q}^\times$ in the multiplicative action, that is, replace H with $\{\pm h : h \in H\} = \{\pm 1\} \times H$ as in [3]. The associated algebras will then be of the form $\overline{\mathscr{Q}} \rtimes \mathbb{Z}/2\mathbb{Z}$. However, we restrict to the action of the torsion-free part of $\mathbb{Q}^\times$ in this paper.

11.4 The a-Adic Duality Theorem

For any a, let a^* be the sequence given by $a_i^* = a_{-i}$. In particular, $(a^*)^* = a$. We now fix a and write Ω and Ω^* for the a-adic and a^*-adic numbers, respectively.

Let $x \in \Omega$ and $y \in \Omega^*$ and for $j \in \mathbb{N}$ put

$$z_j = e^{2\pi i x^{(j)} y^{(j)}/a_0},$$

where the sequences $x^{(j)}$ and $y^{(j)}$ are treated as their corresponding rational numbers in N. It can be checked that z_j is eventually constant. We now define the pairing $\Omega \times \Omega^* \to \mathbb{T}$ by

$$\langle x, y \rangle_\Omega = \lim_{j\to\infty} e^{2\pi i x^{(j)} y^{(j)}/a_0}.$$

The pairing is a continuous homomorphism in each variable separately and gives an isomorphism $\Omega^* \to \hat{\Omega}$. Indeed, this map coincides with the one in [4, 25.1].

The injection $\iota : N \to \mathbb{R} \times \Omega$ given by $q \mapsto (q, q)$ has discrete range, and N may be considered as a closed subgroup of $\mathbb{R} \times \Omega$. Similarly, N^* may be considered as a closed subgroup of $\mathbb{R} \times \Omega^*$.

Remark 2. Subgroups M of $\mathbb{Q}$ such that $N \subset M \subset N_Q$ also embed densely into Ω. For example, $\mathbb{Q}$ itself can be embedded densely into $\mathbb{Q}_p$ for all primes p. On the other hand, it is not hard to see that the image of the diagonal map $\mathbb{Q} \to \mathbb{R} \times \mathbb{Q}_p$ is not closed in this case. More generally, a subgroup M of $\mathbb{Q}$ embeds densely into Ω

such that the image of the diagonal map $M \to \mathbb{R} \times \Omega$ is closed if and only if M is of the form $\frac{1}{q}N$ for $q \in \langle Q \rangle$.

By applying the facts about the pairing of Ω and Ω^* stated above, the pairing of $\mathbb{R} \times \Omega$ and $\mathbb{R} \times \Omega^*$ given by

$$\langle (u,x),(v,y) \rangle = e^{-2\pi i uv/a_0} \lim_{j\to\infty} e^{2\pi i x^{(j)} y^{(j)}/a_0} = \langle u,v \rangle_{\mathbb{R}} \langle x,y \rangle_{\Omega}$$

defines an isomorphism $\mathbb{R} \times \Omega^* \to \widehat{\mathbb{R} \times \Omega}$ that restricts to an isomorphism $\iota(N^*) \to \iota(N)^{\perp}$. Thus, we get the following theorem.

Theorem 3 ([6, Theorem 3.3]). *We have that*

$$(\mathbb{R} \times \Omega^*)/N^* \cong \widehat{\mathbb{R} \times \Omega/N^{\perp}} \cong \hat{N},$$

where the isomorphism $\omega : (\mathbb{R} \times \Omega^*)/N^* \to \hat{N}$ *is given by*

$$\omega\big((v,y) + N^*\big)(q) = \langle (q,q),(v,y) \rangle \quad \text{for } (v,y) \in \mathbb{R} \times \Omega^* \text{ and } q \in N.$$

Remark 3. In general, note that $P^* = P$ so $S^* = S$. Hence, every subgroup $H \subset S$ acting on N and Ω also acts on N^* and Ω^*. In particular $\overline{\mathscr{Q}}(a,H)$ is well-defined if and only if $\overline{\mathscr{Q}}(a^*,H)$ is.

Theorem 4 ([6, Theorem 4.1]). *Assume that* $P \neq \varnothing$ *and that* H *is a nontrivial subgroup of* S. *Set* $G = N \rtimes H$ *and* $G^* = N^* \rtimes H$. *Then there is a Morita equivalence*

$$C_0(\Omega) \rtimes_{\alpha^{\mathrm{aff}}} G \sim_M C_0(\mathbb{R}) \rtimes_{\alpha^{\mathrm{aff}}} G^*,$$

where the action on each side is the $ax+b$*-action.*

We give an outline of the proof that involves a few classical results in the theory of crossed products. To simplify the notation in the proof, we switch the stars, and seek a Morita equivalence between $C_0(\Omega^*) \rtimes_{\alpha^{\mathrm{aff}}} G^*$ and $C_0(\mathbb{R}) \rtimes_{\alpha^{\mathrm{aff}}} G$. Our strategy is to first find a Morita equivalence

$$C_0(T/\Omega) \rtimes_{\mathrm{lt}} N \sim_M C_0(N\backslash T) \rtimes_{\mathrm{rt}} \Omega,$$

where $T = \mathbb{R} \times \Omega$, that is equivariant for actions α and β of H on $C_0(T/\Omega) \rtimes_{\mathrm{lt}} N$ and $C_0(N\backslash T) \rtimes_{\mathrm{rt}} \Omega$, respectively, and then find isomorphisms

$$\big(C_0(T/\Omega) \rtimes_{\mathrm{lt}} N\big) \rtimes_{\alpha} H \cong C_0(\mathbb{R}) \rtimes_{\alpha^{\mathrm{aff}}} G,$$
$$\big(C_0(N\backslash T) \rtimes_{\mathrm{rt}} \Omega\big) \rtimes_{\beta} H \cong C_0(\Omega^*) \rtimes_{\alpha^{\mathrm{aff}}} G^*.$$

Recall that N and Ω sit inside T as closed subgroups. All the groups are abelian, and therefore, by "Green's symmetric imprimitivity theorem" (for example [12, Corollary 4.11]) we get a Morita equivalence

$$C_0(T/\Omega) \rtimes_{\mathrm{lt}} N \sim_M C_0(N\backslash T) \rtimes_{\mathrm{rt}} \Omega \tag{11.3}$$

via an imprimitivity bimodule X that is a completion of $C_c(T)$. Here N acts on the left of T/Ω by $n \cdot ((t, y) \cdot \Omega) = (n + t, n + y) \cdot \Omega$ and Ω acts on the right of $N\backslash T$ by $(N \cdot (t, y)) \cdot x = N \cdot (t, y + x)$, and the induced actions on C_0-functions are given by

$$\mathrm{lt}_n(f)(p \cdot \Omega) = f\big(-n \cdot (p \cdot \Omega)\big)$$
$$\mathrm{rt}_x(g)(N \cdot p) = g\big((N \cdot p) \cdot x\big)$$

for $n \in N$, $f \in C_0(T/\Omega)$, $p \in T$, $x \in \Omega$, and $g \in C_0(N\backslash T)$.

Moreover, H acts by multiplication on N, hence on Ω, and also on $\mathbb{R}$. Thus H acts diagonally on $T = \mathbb{R} \times \Omega$ by $h \cdot (t, x) = (ht, h \cdot x)$.

One can then show that the Morita equivalence (11.3) is equivariant for the actions α, β, and γ of H on $C_c(N, C_0(T/\Omega))$, $C_c(\Omega, C_0(T\backslash N)$, and $C_c(T)$ given by

$$\alpha_h(f)(n)\big((t, y) \cdot \Omega\big) = f(hn)\big((ht, h \cdot y) \cdot \Omega\big),$$
$$\beta_h(g)(x)\big(N \cdot (t, y)\big) = \delta(h)g(h \cdot x)\big(N \cdot (ht, h \cdot y)\big),$$
$$\gamma_h(\xi)(t, y) = \delta(h)^{\frac{1}{2}}\xi(ht, h \cdot y),$$

where δ is the modular function for the multiplicative action of H on Ω.

The next step is now to show that

$$\begin{aligned}\big(C_0(T/\Omega) \rtimes_{\mathrm{lt}} N\big) \rtimes_\alpha H &\cong \big(C_0(\mathbb{R}) \rtimes_{\mathrm{lt}} N\big) \rtimes_{\alpha'} H\\ &\cong \big(C_0(\mathbb{R}) \rtimes_{\mathrm{lt}} N\big) \rtimes_{\alpha''} H\\ &\cong C_0(\mathbb{R}) \rtimes_{\alpha^{\mathrm{aff}}} (N \rtimes H).\end{aligned}$$

The first isomorphism is induced from $T/\Omega \xrightarrow{\cong} \mathbb{R}$ and then we get the correct α'' by composing α' with the automorphism $h \mapsto h^{-1}$ of H. The last isomorphism is a consequence of a result regarding decomposition of iterated crossed products (see [12, Corollary 3.11]).

The other part requires more work, and the aim is to get through the steps

$$\begin{aligned}\big(C_0(N\backslash T) \rtimes_{\mathrm{rt}} \Omega\big) \rtimes_\beta H &\cong \big(C_0(\widehat{N^*}) \rtimes_{\mathrm{rt}} \Omega\big) \rtimes_{\beta'} H\\ &\cong \big(C_0(\Omega^*) \rtimes_{\mathrm{rt}} N^*\big) \rtimes_{\beta''} H\\ &\cong C_0(\Omega^*) \rtimes_{\alpha^{\mathrm{aff}}} (N^* \rtimes H).\end{aligned}$$

Here, the first isomorphism is induced from the ω in Theorem 3. For the second isomorphism, we need the "subgroup of dual group theorem" (see [6, Appendix A]). Finally, the third isomorphism is, similarly as above, a consequence of the "iterated crossed products decomposition".

Remark 4. The C^*-algebras $C_0(\Omega) \rtimes_{\alpha^{\text{aff}}} G$ and $C_0(\mathbb{R}) \rtimes_{\alpha^{\text{aff}}} G^*$ will actually be isomorphic by Zhang's dichotomy: a separable, simple, purely infinite C^*-algebra is either unital or stable.

If a is defined by $a_i = 2$ for all i and $H = \langle 2 \rangle$, then this result coincides with [9, Theorem 7.5], and if a is the sequence described in Example 2, it coincides with [1, Theorem 6.5].

11.5 Invariants and Isomorphism Results

Let $\mathbb{P}$ be the set of prime numbers. A supernatural number is a function

$$\lambda : \mathbb{P} \to \mathbb{N} \cup \{0, \infty\}$$

such that $\sum_{p \in \mathbb{P}} \lambda(p) = \infty$. Denote the set of supernatural numbers by $\mathbb{S}$. It may sometimes be useful to consider a supernatural number as an infinite formal product

$$\lambda = 2^{\lambda(2)} 3^{\lambda(3)} 5^{\lambda(5)} 7^{\lambda(7)} \cdots .$$

If λ is a supernatural number and p is a prime, let $p\lambda$ denote the supernatural number given by $(p\lambda)(p) = \lambda(p) + 1$ (with the convention that $\infty + 1 = \infty$) and $(p\lambda)(q) = \lambda(q)$ if $p \neq q$. The definition of $p\lambda$ extends to all natural numbers p by prime factorization.

Let λ and ϱ be two supernatural numbers associated with the sequence a in the following way:

$$\lambda(p) = \sup\{i : p^i \text{ divides } a_0 \dots a_j \text{ for some } j \geq 0\} \in \mathbb{N} \cup \{0, \infty\}$$

$$\varrho(p) = \sup\{i : p^i \text{ divides } a_{-1} \dots a_{-k} \text{ for some } k \geq 1\} \in \mathbb{N} \cup \{0, \infty\}$$

Lemma 2. *Let a and b be two sequences. The following hold:*

1. $\Delta_a \cong \Delta_b$ *if and only if* $\lambda_a = \lambda_b$.
2. $N_a = N_b$ *if and only if* $\varrho_a = \varrho_b$.
3. $\mathcal{U}_a = \mathcal{U}_b$ *if and only if both* $\lambda_a = \lambda_b$ *and* $\varrho_a = \varrho_b$.

Indeed, from [4, Theorem 25.16] we have

$$\Delta \cong \prod_{p \in \lambda^{-1}(\infty)} \mathbb{Z}_p \times \prod_{p \in \lambda^{-1}(\mathbb{N})} \mathbb{Z}/p^{\lambda(p)}\mathbb{Z}$$

and hence (1) holds. It is not difficult to see that condition (2) and (3) also hold.

This means that there is a one-to-one correspondence between supernatural numbers and noncyclic subgroups of $\mathbb{Q}$ containing $\mathbb{Z}$, and also between supernatural numbers and Hausdorff completions of $\mathbb{Z}$.

Condition (3) is equivalent to $a \sim b$, and more generally, the following result clarifies when Ω_a and Ω_b are isomorphic.

Proposition 2 ([6, Proposition 5.2]). *Let a and b be two sequences. Then $\Omega_a \cong \Omega_b$ if and only if there are natural numbers p and q such that*

$$(\ldots, a_{-2}, qa_{-1}, pa_0, a_1, \ldots) \sim (\ldots, b_{-2}, pb_{-1}, qb_0, b_1, \ldots).$$

That is, $\Omega_a \cong \Omega_b$ if and only if there are $p, q \in \mathbb{N}$ such that $p\lambda_a = q\lambda_b$ and $q\varrho_a = p\varrho_b$.

Hence, if $\Omega_a \cong \Omega_b$, then $N_a \cong N_b$, $P_a = P_b$ so $S_a = S_b$ and $Q_a = Q_b$.

Corollary 1 ([6, Proposition 5.7]). *The group of a-adic numbers Ω is self-dual if and only if there are natural numbers p and q such that $p\lambda = q\varrho$.*

For two pairs of supernatural numbers (λ_1, ϱ_1) and (λ_2, ϱ_2), we write $(\lambda_1, \varrho_1) \sim (\lambda_2, \varrho_2)$ if there exist natural numbers p and q such that $p\lambda_1 = q\lambda_2$ and $q\varrho_1 = p\varrho_2$. Then the set of isomorphism classes of a-adic numbers coincides with $\mathbb{S}\times\mathbb{S}/\sim$ and the self-dual ones coincide with the diagonal, i.e. are of the form $[(\lambda, \lambda)]$.

Set $\mathscr{U}_P = \{\frac{m}{n}\mathbb{Z} \in \mathscr{U} : n \in S\} = \{U \in \mathscr{U} : U \subset \mathbb{Z}[\{\frac{1}{p} : p \in P\}]\}$. Then the open subgroup

$$R = \overline{\mathbb{Z}[\{\tfrac{1}{p} : p \in P\}]} = \overline{\bigcup_{U \in \mathscr{U}_P} U}$$

in Ω is the maximal open (and closed) ring contained in Ω. In particular, the a-adic numbers Ω can be given the structure of a topological (commutative) ring with multiplication inherited from $N \subset \mathbb{Q}$ if and only if [11, E. Herman, 12.3.35]

$$N = \bigcup_{h \in S} h\mathbb{Z} \quad \left(= \mathbb{Z}\left[\left\{\tfrac{1}{p} : p \in P\right\}\right]\right) \tag{11.4}$$

i.e. if and only if $\Omega = R$.

Moreover, by Theorem 4 and Remark 4, it should be clear that $\overline{\mathscr{Q}}(a, H) \cong \overline{\mathscr{Q}}(b, K)$ if $N_a^* \cong N_b^*$ and $H = K$, although the isomorphism is in general not canonical. Hence, for every sequence a, there is a sequence b such that Ω_b is a ring and $\overline{\mathscr{Q}}(a, H) \cong \overline{\mathscr{Q}}(b, H)$, since one can always pick b so that $\Omega_b = R_a$. (Warning: $\overline{\mathscr{Q}}(b, H)$ is still not a ring algebra in the sense of [10].) If both Ω_a and Ω_b are rings, then $\Omega_a \cong \Omega_b$ as topological rings if and only if $a \sim b$.

Example 7. Let a and b be the sequences of Examples 3 and 5, and let $H = \langle 2 \rangle$. Then $\overline{\mathscr{Q}}(a, H) \cong \overline{\mathscr{Q}}(b, H)$ and these algebras are also isomorphic to $\overline{\mathscr{Q}}_2$, but the isomorphisms are not canonical.

Question 1. Given two sequences a and b and subgroups $H \subset S_a$ and $K \subset S_b$. When is $\overline{\mathcal{Q}}(a,H) \not\cong \overline{\mathcal{Q}}(b,K)$?

To enlighten the question, consider the following situation. Let $a = (n,n,n,\dots)$ and $H = \langle n \rangle$, and note that $H = S$ if and only if n is prime. Then $\mathcal{Q}(a,H) = C(\Delta) \rtimes_{\alpha^{\text{aff}}} G_{\mathbb{Z}}$ (see next section) is the $\mathcal{O}(E_{n,1})$ of [7, Example A.6]. Thus

$$(K_0(\mathcal{Q}(a,H)),[1],K_1(\mathcal{Q}(a,H))) \cong (\mathbb{Z} \oplus \mathbb{Z}/(n-1)\mathbb{Z},(0,1),\mathbb{Z}).$$

Moreover, since all $\mathcal{Q}(a,H)$ are Kirchberg algebras in the UCT class, they are classifiable by K-theory.

In future work we hope to be able to compute the K-theory of $\overline{\mathcal{Q}}(a,H)$ using the following strategy. Since $C_0(\Omega) \rtimes N$ is stably isomorphic to the Bunce-Deddens algebra $C(\Delta) \rtimes \mathbb{Z}$, its K-theory is well-known, in fact

$$(K_0(C(\Delta) \rtimes \mathbb{Z}),[1],K_1(C(\Delta) \rtimes \mathbb{Z})) \cong (N^*,1,\mathbb{Z}).$$

As H is a free abelian group, we can apply the Pimsner-Voiculescu six-term exact sequence by adding the action of one generator of H at a time. For this to work out, we will need to apply Theorem 4 and use homotopy arguments to compute the action of H on the K-groups (see also [2, Remark 3.16]).

11.6 The "Unstabilized" a-Adic Algebras

Fix a sequence a and a nontrivial subgroup $H \subset S$ and set $\overline{\mathcal{Q}} = \overline{\mathcal{Q}}(a,H)$. Let H_+ be the semigroup $H \cap \mathbb{N}^\times$ and for each $U \in \mathcal{U}$, let G_U denote the semigroup $U \rtimes H_+$ with multiplication inherited from G. Moreover, for $n \in N$ let p_{n+U} be the projection in $\overline{\mathcal{Q}}$ corresponding to the projection $\chi_{n+\overline{U}}$ in $C_0(\Omega)$.

Assume $U, V \in \mathcal{U}$ and $V \subset U$, so $U = r\mathbb{Z}$ for some r and set $k = |U/V|$. Then

$$U = \bigsqcup_{j=0}^{k-1} jr + V \quad \text{so that} \quad p_U = \sum_{j=0}^{k-1} p_{jr+V}. \tag{11.5}$$

Proposition 3. *The following hold:*

1. *p_U is a full projection in $\overline{\mathcal{Q}}$.*
2. *The full corner $p_U \overline{\mathcal{Q}} p_U$ is isomorphic to the semigroup crossed product*

$$C(\overline{U}) \rtimes_{\alpha^{\text{aff}}} G_U, \quad \alpha^{\text{aff}}{}_{(n,h)} f(x) = \begin{cases} f(h^{-1} \cdot (x-n)) & \text{if } x \in n + h\overline{U}, \\ 0 & \text{else.} \end{cases}$$

Proof. Note first that if $p_V \in \overline{\mathcal{Q}} p_U \overline{\mathcal{Q}}$ for some $V \in \mathcal{U}$, then $g p_{n+hV} \in \overline{\mathcal{Q}} p_U \overline{\mathcal{Q}}$ for all g and $(n,h) \in G$. Therefore, it suffices to check that $p_V \in \overline{\mathcal{Q}} p_U \overline{\mathcal{Q}}$ for all $V \in \mathcal{U}$.

Pick $V = r\mathbb{Z} \in \mathcal{U}$ and choose $W \in \mathcal{U}$ with $W \subset U \cap V$ (for example $W = U \cap V$). Let $k = |V/W|$, then by (11.5)

$$\begin{aligned} p_V &= \sum_{j=0}^{k-1} p_{jr+W} = \sum_{j=0}^{k-1} (jr,1) p_W (-jr,1) \\ &\in \operatorname{span}\{gpg' : g, g' \in G, p \text{ projection in } \overline{\mathcal{Q}} \text{ with } p \le p_U\} \\ &\subset \operatorname{span} \overline{\mathcal{Q}} p_U \overline{\mathcal{Q}} \end{aligned}$$

as $p p_U p = p$ if $p \le p_U$.

For the second part, we just remark that for $f \in C_0(\Omega)$ and $(n,h) \in G$,

$$p_U f(n,h) p_U = p_{U \cap (n+hU)} f(n,h) = f_{|\overline{U} \cap (n+h\overline{U})}(n,h),$$

which is nonzero only if $n \in U \cup hU$. □

The minimal automorphic dilation of $C(\overline{U}) \rtimes_{\alpha^{\text{aff}}} G_U$ does not necessarily take us back to $\overline{\mathcal{Q}}$. In fact, it gives

$$C_0(\overline{H_+^{-1}U}) \rtimes_{\alpha^{\text{aff}}} (H_+^{-1}U \rtimes H)$$

where

$$H_+^{-1}U = \{\tfrac{n}{h} : n \in U, h \in H_+\} = \bigcup_{h \in H_+} h^{-1}U = \bigcup_{h \in H} hU = \{hn : n \in U, h \in H\}.$$

Therefore, one gets $\overline{\mathcal{Q}}$ back precisely when $N = H_+^{-1}U$. For example, if $U = \mathbb{Z}$ one gets $\overline{\mathcal{Q}}$ back in the settings of Larsen and Li and also Cuntz, since $H = S$ and (11.4) holds in these cases.

In general, however, we get that

$$\overline{\mathcal{Q}} \sim_M C_0(\overline{H_+^{-1}U}) \rtimes_{\alpha^{\text{aff}}} (H_+^{-1}U \rtimes H)$$

which due to Remark 4 means that these are noncanonically isomorphic as well.

Acknowledgements This paper is a survey based on [6]. The work was mainly done while I was visiting Arizona State University in the spring of 2012, and I would like to thank John Quigg and Steve Kaliszewski for their hospitality during this stay. This research was partially supported by the Research Council of Norway (NFR) and by the NordForsk Research Network Operator Algebras and Dynamics (grant #11580).

References

1. Cuntz, J.: C^*-algebras associated with the $ax + b$-semigroup over $\mathbb{N}$. In: K-theory and noncommutative geometry, EMS Ser. Congr. Rep., pp. 201–215. Eur. Math. Soc., Zürich (2008). DOI 10.4171/060-1/8. URL http://dx.doi.org/10.4171/060-1/8
2. Cuntz, J., Li, X.: C^*-algebras associated with integral domains and crossed products by actions on adele spaces. J. Noncommut. Geom. **5**(1), 1–37 (2011). DOI 10.4171/JNCG/68. URL http://dx.doi.org/10.4171/JNCG/68
3. Cuntz, J., Li, X.: The regular C^*-algebra of an integral domain. In: Quanta of maths, *Clay Math. Proc.*, vol. 11, pp. 149–170. Amer. Math. Soc., Providence, RI (2010)
4. Hewitt, E., Ross, K.A.: Abstract harmonic analysis. Vol. I: Structure of topological groups. Integration theory, group representations. Die Grundlehren der mathematischen Wissenschaften, Bd. 115. Academic Press Inc., Publishers, New York (1963)
5. Kaliszewski, S., Landstad, M.B., Quigg, J.: A crossed-product approach to the Cuntz-Li algebras. Proc. Edinb. Math. Soc. (2) **55**(2), 429–459 (2012). DOI 10.1017/S0013091511000046. URL http://dx.doi.org/10.1017/S0013091511000046
6. Kaliszewski, S., Omland, T., Quigg, J.: Cuntz-Li algebras from a-adic numbers (2012), arXiv:1211.4806
7. Katsura, T.: A class of C^*-algebras generalizing both graph algebras and homeomorphism C^*-algebras. IV. Pure infiniteness. J. Funct. Anal. **254**(5), 1161–1187 (2008). DOI 10.1016/j.jfa.2007.11.014. URL http://dx.doi.org/10.1016/j.jfa.2007.11.014
8. Laca, M.: From endomorphisms to automorphisms and back: dilations and full corners. J. London Math. Soc. (2) **61**(3), 893–904 (2000). DOI 10.1112/S0024610799008492. URL http://dx.doi.org/10.1112/S0024610799008492
9. Larsen, N.S., Li, X.: The 2-adic ring C^*-algebra of the integers and its representations. J. Funct. Anal. **262**(4), 1392–1426 (2012). DOI 10.1016/j.jfa.2011.11.008. URL http://dx.doi.org/10.1016/j.jfa.2011.11.008
10. Li, X.: Ring C^*-algebras. Math. Ann. **348**(4), 859–898 (2010). DOI 10.1007/s00208-010-0502-x. URL http://dx.doi.org/10.1007/s00208-010-0502-x
11. Palmer, T.W.: Banach algebras and the general theory of $*$-algebras. Vol. 2, *Encyclopedia of Mathematics and its Applications*, vol. 79. Cambridge University Press, Cambridge (2001)
12. Williams, D.P.: Crossed products of C^*-algebras, *Mathematical Surveys and Monographs*, vol. 134. American Mathematical Society, Providence, RI (2007)

Chapter 12
The Structure of Stacey Crossed Products by Endomorphisms

Eduard Ortega and Enrique Pardo

Abstract We describe simplicity and purely infiniteness (in simple case) of the Stacey crossed product $A \times_\beta \mathbb{N}$ in terms of conditions of the C^*-dynamical system (A, β).

Keywords C^*-crossed products • Purely infinite C^*-algebras • Cuntz-Pimsner algebras

Mathematics Subject Classification (2010): 46L55, 46L35.

12.1 Introduction

In [5], Cuntz defined the fundamental Cuntz algebras $\mathcal{O}_n$ in terms of generators and relations. He also represented these algebras as crossed products of a UHF-algebra by an endomorphism, and in a subsequent paper [6] he realized this construction as a full corner of an ordinary crossed product. However Cuntz did not explain what kind of crossed product by an endomorphism was. Later, Paschke [18] gave an elegant generalization of Cuntz's result, and described the crossed product of a unital C^*-algebra by an endomorphism $\beta : A \to A$, written $A \times_\beta \mathbb{N}$, as the C^*-algebra

E. Ortega (✉)
Department of Mathematical Sciences, Norwegian University of Science and Technology (NTNU), NO-7491 Trondheim, Norway
e-mail: eduard.ortega.esparza@gmail.com

E. Pardo
Departamento de Matemáticas, Facultad de Ciencias, Universidad de Cádiz, Campus de Puerto Real, ES-11510 Puerto Real (Cádiz), Spain
e-mail: enrique.pardo@uca.es

T.M. Carlsen et al. (eds.), *Operator Algebra and Dynamics*, Springer Proceedings in Mathematics & Statistics 58, DOI 10.1007/978-3-642-39459-1_12,

generated by A and an isometry V_∞, such that $V_\infty a V_\infty^* = \beta(a)$. Endomorphisms of C^*-algebras appeared elsewhere (cf. [3, 7] and the references given there), and led Stacey to give a modern description of their crossed products in terms of covariant representations and universal properties [24]. He also verified that the candidate proposed in [5] had the required property (see [2] and [4] for further study and generalization of the Stacey's crossed product).

More recently, constructions such as Exel crossed products raised hopes to extend the scope of crossed products to describe broad classes of C^*-algebras. In this setting it is worth considering the recent work of an Huef and Raeburn [9], who show that:

1. The relative Cuntz-Pimsner algebra of an Exel system is isomorphic to a Stacey crossed product of its core algebra.
2. Any Stacey crossed product is an Exel crossed product.

In particular, they give a presentation of any graph C^*-algebra (over a row-finite graph) $C^*(E)$ as a Stacey crossed product $C^*(E)^\gamma \times_{\beta_E} \mathbb{N}$ by an endomorphism of the core, extending the work of Kwaśniewski on finite graphs [14].

Cuntz's representation of the $\mathcal{O}_n$ as crossed products by an endomorphism aimed to prove the simplicity of these C^*-algebras. Paschke gave conditions on the C^*-algebra A and on the isometry to obtain a simple crossed product [18, Proposition 2.1], later improved in [4, Corollary 2.6]. Finally, Schweizer gave the most powerful result about the simplicity of the Stacey crossed product [23, Theorem 4.1]. Namely, if A is a unital C^*-algebra and β is an injective $*$-endomorphism, then $A \times_\beta \mathbb{N}$ is simple and $\beta(1)$ is a full projection in A if and only if β^n is outer for every $n > 0$ and there are no non-trivial ideals I of A with $\beta(I) \subseteq I$. Certainly, in most cases the simplicity appears in connection with the pure infiniteness property, first introduced by Cuntz in the simple case, and then extended to general C^*-algebras by Kirchberg and Rørdam [12]. Then we can use the Kirchberg-Phillips classification theorems to model Kirchberg algebras as crossed products.

The aim of this work is to study the simplicity of non-unital crossed products, as well as to give sufficient conditions to decide when a simple Stacey crossed product is purely infinite. Our fundamental technique is seeing the Stacey crossed product $A \times_\beta \mathbb{N}$ as a full corner of a crossed product by an automorphism $P(A_\infty \times_{\beta_\infty} \mathbb{Z})P$ (see [6, 24]), where P is a full projection of the multipliers that is invariant under the canonical gauge-action. Therefore, we can define the associated *Connes Spectrum* of the endomorphism in a similar way we do it for an automorphism (see [10, 15, 16]) and construct a parallel Connes spectrum theory for endomorphisms. Hence, following the results of Olesen and Pedersen [16, 17], we characterize simplicity for the Stacey crossed product $A \times_\beta \mathbb{N}$. Secondly, we will deal with the characterization of pure infiniteness for simple Stacey crossed products. For, by using ideas from [10, 21], we give sufficient conditions on A and the endomorphism β in order to guarantee that $A \times_\beta \mathbb{N}$ is simple and purely infinite. The main difference between these results and ours is that we do not ask the C^*-algebra A to be simple.

12.2 Simple Stacey Crossed Product

The pair (A, β), where A is a C^*-algebra and $\beta : A \to A$ an (injective) endomorphism, is called a *C^*-dynamical system.*

Definition 1. We say that (π, V) is a *Stacey covariant representation* of (A, β) if $\pi : A \to \mathscr{B}(\mathscr{H})$ is a non-degenerated representation and V is an isometry of $\mathscr{B}(\mathscr{H})$ such that $\pi(\beta(a)) = V\pi(a)V^*$ for every $a \in A$. We denote by $C^*(\pi, V)$ the C^*-algebra generated by $\{\pi(A)V^n(V^m)^*\}_{n,m\geq 0}$.

Stacey showed in [24] that there exists a C^*-algebra that is generated by a universal Stacey covariant representation (ι_∞, V_∞). We call $A\times_\beta\mathbb{N} := C^*(\iota_\infty, V_\infty)$ the *Stacey crossed product* of A by the endomorphism β.

Remark 1. Observe that, if β is an automorphism then V_∞ is a unitary, and hence $A \times_\beta \mathbb{N}$ is the usual crossed product $A \times_\beta \mathbb{Z}$.

By universality of $A \times_\beta \mathbb{N}$, given $z \in \mathbb{T}$, we define an automorphism in $A \times_\beta \mathbb{N}$ by the rule $\gamma_z(a) = a$ and $\gamma_z(V_\infty) = zV_\infty$ for every $a \in A$. It defines the gauge action $\gamma : \mathbb{T} \to \mathrm{Aut}(A \times_\beta \mathbb{N})$. An ideal I of $A \times_\beta \mathbb{N}$ is said to be *gauge invariant* if $\gamma_z(I) = I$ for every $z \in \mathbb{T}$. We define a canonical faithful conditional expectation $E : A \times_\beta \mathbb{N} \longrightarrow B$ as $E(x) := \int_{\mathbb{T}} \gamma_z(x)dz$ for every $x \in A \times_\beta \mathbb{N}$, where $B := \overline{\mathrm{span}}\{V_\infty^{*n} a V_\infty^n : a \in A, n \geq 0\}$.

We say that the endomorphism $\beta : A \longrightarrow A$ is *extendible* if, given any strictly convergent sequence $\{x_n\}_{n\geq 0} \subset A$, then the sequence $\{\beta(x_n)\}_{n\geq 0}$ converges in the strict topology (i.e., β extends to $\hat{\beta} : M(A) \longrightarrow M(A)$). Observe that, if β is injective, then $\hat{\beta}(a) \in A$ implies that $a \in A$. Indeed, let $\{a_n\}$ be a sequence that converges in the strict topology and such that $\{\beta(a_n)\}$ converges in norm topology. Since β is isometric (β is injective) then $\{a_n\}$ converges in the norm topology too.

We define the inductive system $\{A_i, \varphi_i\}_{i\geq 1}$ given by $A_i := A$ and $\varphi_i = \beta$ for every $i \geq 1$. Let $A_\infty := \varinjlim \{A_i, \varphi_i\}$. For any $i \geq 1$, $\varphi_{i,\infty} : A_i \longrightarrow A_\infty$ denotes the (injective) canonical map. The diagram

$$\begin{array}{ccccccc}
A & \xrightarrow{\beta} & A & \xrightarrow{\beta} & A & \xrightarrow{\beta} & \cdots \\
\downarrow{\scriptstyle\beta} & & \downarrow{\scriptstyle\beta} & & \downarrow{\scriptstyle\beta} & & \\
A & \xrightarrow[\beta]{} & A & \xrightarrow[\beta]{} & A & \xrightarrow[\beta]{} & \cdots
\end{array}$$

gives rise to the dilated automorphism $\beta_\infty : A_\infty \longrightarrow A_\infty$. We call (A_∞, β_∞) the *dilation* of (A, β).

Observe that, if β is an extendible endomorphism, then given any $i \geq 1$ we have that $\varphi_{i,\infty}$ extends to $\widehat{\varphi_{i.\infty}} : M(A) \longrightarrow M(A_\infty)$.

Proposition 1 (cf. [23, Proposition 3.3]). *If A is a C^*-algebra and $\beta : A \longrightarrow A$ is an extendible and injective endomorphism, then $A \times_\beta \mathbb{N} \cong P(A_\infty \times_{\beta_\infty} \mathbb{Z})P$, where $P = \widehat{\varphi_{1,\infty}}(1_{M(A)}) \in M(A_\infty \times_{\beta_\infty} \mathbb{Z})$. Moreover, P is a full projection, so that $A \times_\beta \mathbb{N}$ is strongly Morita equivalent to $A_\infty \times_{\beta_\infty} \mathbb{Z}$.*

The isomorphism given in the above proposition sends V_∞ (the isometry of $A \times_\beta \mathbb{N}$) to $PU_\infty P$ (where U_∞ is the generating unitary of $A_\infty \times_{\beta_\infty} \mathbb{Z}$), and a to $\varphi_{1,\infty}(a)$ for every $a \in A$. Therefore, from now on we will identify $A \times_\beta \mathbb{N}$ with $P(A_\infty \times_{\beta_\infty} \mathbb{Z})P$. If $\gamma' : \mathbb{T} \longrightarrow \text{Aut}\,(A_\infty \times_{\beta_\infty} \mathbb{Z})$ is the canonical gauge action, since $\gamma'_z(P) = P$ for every $z \in \mathbb{T}$, it restricts to the gauge action γ of $A \times_\beta \mathbb{N}$. Thus, we will identify γ with γ'.

Therefore, by Morita equivalence there exists a bijection between the ideals I of $A_\infty \times_{\beta_\infty} \mathbb{Z}$ and the ideals J of $A \times_\beta \mathbb{N}$, given by

$$I \longmapsto PIP \qquad \text{and} \qquad J \longmapsto \overline{(A_\infty \times_{\beta_\infty} \mathbb{Z})J(A_\infty \times_{\beta_\infty} \mathbb{Z})}\,.$$

So, from the above comment the following results comes.

Lemma 1. *If A is a C^*-algebra and $\beta : A \longrightarrow A$ is an extendible and injective endomorphism, then there exists an order preserving bijection between gauge invariant ideals of $A \times_\beta \mathbb{N}$ and $A_\infty \times_{\beta_\infty} \mathbb{Z}$.*

Now, we will describe the gauge invariant ideals in terms of the C^*-dynamical system (A, β).

Definition 2. Let A be a C^*-algebra and let $\beta : A \to A$ an endomorphism. We say that an ideal I of A is *β-invariant* if $I = \beta^{-1}(I)$. A is *β-simple* if there are no non-trivial β-invariant ideals.

Proposition 2. *If A is a C^*-algebra and $\beta : A \longrightarrow A$ is an injective endomorphism, then the map $I \longmapsto I_\infty$, where $I_\infty := \sum_{n\in\mathbb{N}} \varphi_{n,\infty}(I)$, defines an order preserving bijection between the β-invariant ideals of A and the β_∞-invariant ideals of A_∞.*

Proof. Let I be an ideal of A such that $\beta^{-1}(I) = I$. Let us define $I_\infty := \sum_{n\in\mathbb{N}} \varphi_{n,\infty}(I)$, which is an ideal of A_∞. Since $\beta(I) \subseteq I$ we have that $\beta_\infty(I_\infty) \subseteq I_\infty$. Then, given $x \in I$ and $n \in \mathbb{N}$ we have that $\varphi_{n,\infty}(x) = \varphi_{n+1,\infty}(\beta(x)) = \beta_\infty(\varphi_{n+1,\infty}(x))$, so $\beta_\infty(I_\infty) = I_\infty$. Conversely, given K an ideal of A_∞ let us consider the ideal $\varphi_{1,\infty}^{-1}(K)$. Observe that $\varphi_{1,\infty}^{-1}(K) \neq 0$ because β is injective. Given $x \in \varphi_{1,\infty}^{-1}(K)$ we have that $\varphi_{1,\infty}(\beta(x)) = \beta_\infty(\varphi_{1,\infty}(x)) \in K$, and then $\beta(x) \in \varphi_{1,\infty}^{-1}(K)$. Now let $x \in A$ be such that $\beta(x) \in \varphi_{1,\infty}^{-1}(K)$, so $\varphi_{1,\infty}(\beta(x)) \in K$. But then $\beta_\infty^{-1}(\varphi_{1,\infty}(\beta(x))) = \varphi_{1,\infty}(x) \in K$ and hence $x \in \varphi_{1,\infty}^{-1}(K)$.

Finally, since $\beta^{-1}(I) = I$, given $n \in \mathbb{N}$ we have that

$$\varphi_{n,\infty}^{-1}(\varphi_{n+1,\infty}(I)) = \varphi_{n,\infty}^{-1}(\beta^{-1}(I)) = \varphi_{n,\infty}^{-1}(I)$$

the bijection follows.

We would like to remark that our definition of invariant ideal slightly differs from the one given by Adji in [1] for two reasons. First, because we only are interested in actions by injective endomorphisms. And second, because we are just looking for simple crossed products (hence without any gauge invariant ideal), while Adji looks for a characterization of the gauge invariant ideals as another crossed product.

In the following, we will give necessary and sufficient conditions for the simplicity of a Stacey crossed product. The main technical device we use is the Connes spectrum of an endomorphism. This is just a reformulation of the Connes spectrum for automorphisms (see [10, 15]). We will see that for nice endomorphisms (extendible and hereditary image) the Connes spectrum of β and that of the associated automorphism β_∞ coincide. Therefore, we will be able to use results by Olesen and Pedersen to determine the conditions for the simplicity of the Stacey crossed products.

Definition 3. Let A be a C^*-algebra and let $\beta : A \to A$ be an endomorphism. Then we say that:

1. β is *inner* if there exists an isometry $W \in M(A)$ such that $\beta = \operatorname{Ad} W$.
2. β is *outer* if it is not inner.

Recall [8, Definition 2.1] that an automorphism α of a C^*-algebra A is said to be properly outer if for every nonzero α-invariant two-sided ideal I of A and for every unitary multiplier u of I, $\|\alpha_{|I} - \operatorname{Ad}_{u|I}\| = 2$. By [17, Theorem 10.4] the notion of $\alpha^n_{|I}$ being outer for every $n \in \mathbb{N}$ and every α-invariant ideal I is weaker than the properly outer notion. It is known that if the action is properly outer then the automorphism is outer pointwise. However, this was proved by Kishimoto [13] and Olesen and Pedersen [17] in the case that the C^*-algebra is α-simple. It is not known, at least to the knowledge of the authors, if they are equivalent, at least by $\mathbb{Z}$-actions.

Definition 4. Let A be a C^*-algebra, let $\beta : A \to A$ be an extendible injective endomorphism and let $\gamma : \mathbb{T} \longrightarrow \operatorname{Aut}(A \times_\beta \mathbb{N})$ be the gauge action. We define the *Connes spectrum* of β as

$$\mathbb{T}(\beta) := \{t \in \mathbb{T} : \gamma_t(I) \cap I \neq 0 \text{ for every } 0 \neq I \lhd A \times_\beta \mathbb{N}\}.$$

Remark 2. Observe that $\mathbb{T}(\beta)$ is a closed subgroup of $\mathbb{T}$, since γ is strongly continuous. Hence can only be $\{1\}$, $\mathbb{T}$ or a finite subgroup.

This definition of the Connes spectrum coincides with the one given by Olesen [15] and Olesen and Pedersen [16] when β is an automorphism.

Now, using that the bijection between ideals of $A \times_\beta \mathbb{N}$ and those of $A_\infty \times_{\beta_\infty} \mathbb{Z}$, and the fact that the canonical gauge action $\gamma : \mathbb{T} \longrightarrow \operatorname{Aut}(A_\infty \times_{\beta_\infty} \mathbb{Z})$ restricts to the gauge action of $A \times_\beta \mathbb{N}$ (since $\gamma_z(P) = P$ for every $z \in \mathbb{T}$), the following lemma easily follows.

Lemma 2. *If A is a C^*-algebra and $\beta : A \to A$ is an extendible injective endomorphism, then $\mathbb{T}(\beta) = \mathbb{T}(\beta_\infty)$.*

In order to use this results in our context, the following Lemma is essential.

Lemma 3. *Let A be a C^*-algebra, and let $\beta : A \to A$ be an injective extendible endomorphism such that $\beta(A)$ is a hereditary sub-C^*-algebra of A. Then given any β-invariant ideal I of A we have that $I \cong PI_\infty P$.*

Proof. Recall that $P = \widehat{\varphi_{1,\infty}}(1_{M(A)}) = (P_1, P_2, P_3, \ldots) \in M(A_\infty)$, where we define $P_n = \hat{\beta}^{n-1}(1_{M(A)})$ for every $n \in \mathbb{N}$. It is enough to check that given any $n \in \mathbb{N}$ and $a \in I$, then $P\varphi_{n,\infty}(a)P = \widehat{\varphi_{n,\infty}}(P_n a P_n) \in \varphi_{1,\infty}(I)$. Observe that since $\beta(A)$ is a hereditary sub-C^*-algebra of A we have that $\beta^n(A) = \overline{\beta^n(A)A\beta^n(A)}$ for every $n \in \mathbb{N}$. But since $P_n a P_n \in \overline{\beta^{n-1}(A)I\beta^{n-1}(A)} \subseteq \beta^{n-1}(A) \cap I$, and $\beta^{-1}(I) = I$, we have that

$$P\varphi_{n,\infty}(a)P \in \varphi_{n,\infty}(\beta^{n-1}(I)) = \varphi_{1,\infty}(I),$$

as desired.

Remark 3. Combining [16, Lemma 6.1], the bijection stated in Proposition 2, the Morita equivalence between $A \times_\beta \mathbb{N}$ and $A_\infty \times_{\beta_\infty} \mathbb{Z}$, Lemmas 1 and 3, we have a bijection between the β-invariant ideals of A and the gauge invariant ideals of $A \times_\beta \mathbb{N}$ defined by $I \mapsto (A \times_\beta \mathbb{N})I(A \times_\beta \mathbb{N})$, with inverse $K \mapsto K \cap A$.

Remark 4. If (A, β) is a C^*-dynamical system with β extendible and $\beta(A)$ being a hereditary sub-C^*-algebra of A, then it follows that $V_\infty^* A V_\infty \subseteq A$. Indeed, let $a \in A$. By Lemma 3 there exists $b \in A$ such that $V_\infty V_\infty^* a V_\infty V_\infty^* = V_\infty b V_\infty^*$, and hence $V_\infty a V_\infty^* = b$, as desired. Therefore, the conditional expectation can be defined as $E : A \times_\beta \mathbb{N} \to A$.

Now, let us recall a result following from [15].

Theorem 1. *Let A be a C^*-algebra and let $\beta : A \to A$ be an extendible injective endomorphism with $\beta(A)$ being a hereditary sub-C^*-algebra of A. Let us consider the following statements:*

1. *$\mathbb{T}(\beta^n) = \mathbb{T}$ for every $n > 0$.*
2. *Given $a \in \tilde{A}$ (the unitization of A) and any B hereditary sub-C^*-algebra of A, for every $n > 0$ we have that*

$$inf\{\|xa\beta^n(x)\| : 0 \le x \in B \text{ with } \|x\| = 1\} = 0\,.$$

3. *β^n is outer for every $n > 0$.*

Then, (1) $\Rightarrow$ (2) $\Rightarrow$ (3). *Moreover, if A is β-simple, then* (3) $\Rightarrow$ (1) *(and thus they are all equivalent).*

Proof. (1) $\Rightarrow$ (2) This is [17, Theorem 10.4 and Lemma 7.1]. If $\mathbb{T}(\beta^n) = \mathbb{T}$ then $\mathbb{T}(\beta_\infty^n) = \mathbb{T}$ for every $n > 0$, so β_∞^n is properly outer for every $n > 0$. Since any hereditary sub-C^*-algebra B of A is also a hereditary sub-C^*-algebra of A_∞, (see Lemma 3), we can apply [17, Proof of Lemma 7.1] to B. Thus, since $\beta^n_{\infty|A} = \beta^n$, we have the result.

(2) $\Rightarrow$ (3) Suppose that $\beta^n = \operatorname{Ad} W$ for an isometry $W \in M(A)$. Fix $\varepsilon > 0$, and take $b \in A_+$ with $\|b\| = 1$. Set $c := f_\varepsilon(b)$, where $f_\varepsilon(t) : [0,1] \longrightarrow \mathbb{R}_+$ is the continuous function that is $f_\varepsilon(0) = 0$, constant 1 for $t \geq \varepsilon$ and linear otherwise. Then, we have that $xc = cx = x$ for every $x \in \overline{(b-\varepsilon)_+ A (b-\varepsilon)_+}$. Hence, given any $0 \leq x \in \overline{(b-\varepsilon)_+ A (b-\varepsilon)_+}$ with $\|x\| = 1$, we have that

$$\begin{aligned}\|x(cW^*)\beta^n(x)\|^2 &= \|x(cW^*)WxW^*\|^2 = \|xcxW^*\|^2 \\ &= \|x^2W^*\|^2 = \|x^2W^*Wx^2\| = \|x^4\| = \|x\|^4 = 1\,,\end{aligned}$$

which contradicts the hypothesis, since $cW^* \in A$.

Now, suppose that A is β-simple. We will prove that (3) $\Rightarrow$ (1). By [17, Theorem 10.4] we have that $\mathbb{T}(\beta_\infty) = \mathbb{T}$ if and only if $\mathbb{T}(\beta_\infty^n) = \mathbb{T}$ for every $n \in \mathbb{N}$. Let us suppose that $\mathbb{T}(\beta) = \mathbb{T}(\beta_\infty) \neq \mathbb{T}$. Hence, $\mathbb{T}(\beta_\infty)$ is a finite subgroup, and thus the complement $\mathbb{T}(\beta_\infty)^\perp \neq \{0\}$. Therefore, by [17, Theorem 4.5], for every $0 \neq k \in \mathbb{T}(\beta_\infty)^\perp$ we have that $\beta_\infty^k = \operatorname{Ad} U$, where $U \in M(A_\infty)$. But then, by Lemma 3, $V = PUP \in M(A)$ is an isometry such that $\beta^k = \operatorname{Ad} V$, a contradiction.

Corollary 1. *Let A be a C^*-algebra and let $\beta : A \to A$ be an extendible injective endomorphism with $\beta(A)$ being a hereditary sub-C^*-algebra of A. Then $A \times_\beta \mathbb{N}$ is simple if and only if A is β-simple and β^n is outer for every $n > 0$.*

Proof. $A \times_\beta \mathbb{N}$ is simple if and only if $A_\infty \times_{\beta_\infty} \mathbb{Z}$ is simple if and only if A_∞ is β_∞-simple and $\mathbb{T}(\beta_\infty) = \mathbb{T}$ [16, Theorem 6.5] if and only if A is β-simple and $\mathbb{T}(\beta) = \mathbb{T}$. Therefore, by Theorem 1 we have that A is β-simple and $\mathbb{T}(\beta) = \mathbb{T}$ if and only if A is β-simple and β^n is outer for every $n > 0$.

To apply classification results to these crossed products, it will be necessary to compute the K-theory of the crossed product by an endomorphism.

Lemma 4 (cf. [21, Corollary 2.2]). *Let A be a separable C^*-algebra and let $\beta : A \longrightarrow A$ be an injective extendible endomorphism such that $\beta(A)$ is a full hereditary sub-C^*-algebra of A. Then, we have the following six-term exact sequence*

$$\begin{array}{ccccc} K_0(A) & \xrightarrow{id - K_0(\beta)} & K_0(A) & \longrightarrow & K_0(A \times_\beta \mathbb{N}) \\ \uparrow & & & & \downarrow \\ K_1(A \times_\beta \mathbb{N}) & \longleftarrow & K_1(A) & \xleftarrow{id - K_1(\beta)} & K_1(A) \end{array}$$

Proof. First, recall that $A \times_\beta \mathbb{N}$ is strongly Morita equivalent to $A \times_{\beta_\infty} \mathbb{Z}$, so their K-groups are isomorphic. Thus, we can use the Pimsner-Voiculescu six-term exact sequence for $A \times_{\beta_\infty} \mathbb{Z}$. Since $\beta(A)$ is a hereditary sub-C^*-algebra of A, we have that $A \cong PA_\infty P$ by Lemma 3. Moreover, as $\beta(A)$ is a full subalgebra, it follows that P is a full projection, whence A and A_∞ are strongly Morita equivalent. Hence, $K_*(A) \cong K_*(A_\infty)$. Finally, by continuity of the K-theory functor, we have the desired result.

12.3 Purely Infinite Simple Crossed Products

In Theorem 1 we have given necessary and sufficient conditions on the endomorphism β for the simplicity of the C^*-algebra $A \times_\beta \mathbb{N}$. If A is a unital C^*-algebra and $\beta(1) \neq 1$, then $A \times_\beta \mathbb{N}$ contains a proper isometry, and if in addition $A \times_\beta \mathbb{N}$ is simple, we have that it is a properly infinite C^*-algebra. We will see that for a broad class of unital real rank zero C^*-algebras, say A, we have that $A \times_\beta \mathbb{N}$ turns out to be purely infinite. Our results generalize and unify similar results given in [21] and [10].

Recall Cuntz's definition: a unital simple C^*-algebra A is *purely infinite* if given any non-zero element a in A there exist x, y in A such that $xay = 1$. Equivalently, a unital simple C^*-algebra is purely infinite if and only if has real rank zero and every projection is infinite [25].

Lemma 5. *Let A be a unital C^*-algebra, let $\beta : A \longrightarrow A$ be an injective endomorphism, and suppose that does not exist any proper ideal I of A such that $\beta(I) \subseteq I$. Then, given any non-zero $a \in A_+$ there exists $n \in \mathbb{N}$ such that $a + \beta(a) + \cdots + \beta^n(a)$ is a full positive element in A.*

Proof. Consider the ideal $I := \overline{\text{span}}\{x\beta^n(a)y : n \geq 0,\ x, y \in A\} \neq 0$. It clearly satisfies $\beta(I) \subseteq I$ and then, by hypothesis, we have that $I = A$. Therefore we can write

$$1 = \sum_{i=1}^{k} x_i \beta^{n_i}(a) y_i$$

where $x_i, y_i \in A$ and $n_i \in \mathbb{N}$ for every $i \in \{1, \ldots, k\}$. Then, taking $n = \max_i\{n_i\}$, we have the desired result.

Let $T(A)$ be the set of tracial states of A, which is a compact space with the $*$-weak topology. We say that A has *strict comparison* if: (i) $T(A) \neq \emptyset$; (ii) Whenever $p \in \overline{AqA}$ such that $\tau(p) < \tau(q)$ for every $\tau \in T(A)$, we have that $p \lesssim q$. For example, every unital exact and stably finite C^*-algebra that is $\mathcal{Z}$-stable has strict comparison [22, Corollary 4.10].

The following lemma is a slight modification of [21, Lemma 3.2].

Lemma 6 (cf. [21, Lemma 3.2]). *Let A be a unital C^*-algebra that either has strict comparison or is purely infinite. Let $\beta : A \longrightarrow A$ be an injective endomorphism such that $\beta(1) \neq 1$ and let $A \times_\beta \mathbb{N} = C^*(\iota_\infty, V_\infty)$. If there does not exist any proper ideal I of A such that $\beta(I) \subseteq I$, then for every full projection $p \in A$ there exist a partial isometry $u \in A$ and $m \in \mathbb{N}$ such that $(V_\infty^*)^m u^* p u V_\infty^m = (V_\infty^*)^m V_\infty^m = 1$.*

Proof. We need to prove that there exists $m \in \mathbb{N}$ such that $V_\infty^m (V_\infty^m)^* \precsim p$. If that holds, then there exists a partial isometry $u \in A$ such that $u^*u = V_\infty^m (V_\infty^*)^m$ and $uu^* \leq p$. Therefore $(V_\infty^*)^m u^* p u V_\infty^m = (V_\infty^*)^m (V_\infty^m (V_\infty^*)^m) V_\infty^m = 1$, so we are done.

Observe that if A is purely infinite then p is a properly infinite full projection. So, we have that $V_\infty V_\infty^* \in \overline{ApA} = A$. Hence, $V_\infty V_\infty^* \precsim p$, so that $m = 1$ holds.

Now suppose that A has strict comparison. Then $T(A)$ is non-empty and compact. So, given any $k \in \mathbb{N}$ we set

$$\alpha = \inf \{\tau(p) : \tau \in T(A)\} \qquad \text{and} \qquad \gamma_k = \sup \{\tau(V_\infty^k (V_\infty^*)^k) : \tau \in T(A)\}.$$

Observe that, since p is full, we have that $\alpha > 0$. Now, we claim that there exists $n \in \mathbb{N}$ such that $\gamma_n < 1$. Indeed, it is enough to prove that there exists $n \in \mathbb{N}$ such that $1 - V_\infty^n (V_\infty^*)^n$ is a full projection. Let us construct the ideal

$$I := \overline{\operatorname{span}} \{x(V_\infty^l (V_\infty^*)^l - V_\infty^{l+1} (V_\infty^*)^{l+1}) y : l \geq 0,\, x, y \in A\} \neq 0\,.$$

It is clear that $\beta(I) \subseteq I$. Therefore, by Lemma 5, there exists $n \in \mathbb{N}$ such that

$$\begin{aligned}(1 - V_\infty V_\infty^*) + \cdots + \beta^{n-1}(1 - V_\infty V_\infty^*) = (1 - V_\infty V_\infty^*) + \cdots + (V_\infty^{n-1} (V_\infty^*)^{n-1} \\ - V_\infty^n (V_\infty^*)^n) = 1 - V_\infty^n (V_\infty^*)^n\,,\end{aligned}$$

is a full projection. Therefore $\gamma_n < 1$. By the same argument as in the proof of [21, Lemma 3.2], we have that $\tau(V_\infty^{nl} (V_\infty^*)^{nl}) \leq \gamma_n^l$ for every $l \in \mathbb{N}$. Then, there exists $l \in \mathbb{N}$ such that $\tau(V_\infty^{nl} (V_\infty^*)^{nl}) \leq \gamma_n^l < \alpha \leq \tau(p)$. Since A has strict comparison, we have that $V_\infty^{nl} (V_\infty^*)^{nl} \precsim p$.

Lemma 7. *Let A be a C^*-algebra of real rank zero, and let $\beta : A \to A$ be an extendible injective endomorphism with $\beta(A)$ being hereditary such that $\mathbb{T}(\beta) = \mathbb{T}$. Then, given any $a \in A^\sim$ and any B hereditary sub-C^*-algebra of A we have that*

$$\inf\{\|pa\beta(p)\| : p \text{ is a non-zero projection of } B\} = 0\,.$$

Proof. Let $a \in A^+$ and let B be a hereditary sub-C^*-algebra of A. Given $\varepsilon > 0$, by Theorem 1 there exists $x \in B_+$ with $\|x\| = 1$ such that $\|xa\beta(x)\| < \varepsilon/2$. Given $\delta > 0$, let $f_\delta : [0,1] \longrightarrow [0,1]$ be such that $f(t) = 1$ for every $t \in [1 - \delta/2, 1]$ and such that $|f_\delta(t) - t| < \delta$ for every $0 \leq t \leq 1$. Take $\delta > 0$ such that $\|f_\delta(x) a \beta(f_\delta(x))\| < \varepsilon$. Let $C = \{y \in B : f_\delta(x) y = y f_\delta(x) = y\} \neq 0$. Notice that C is a hereditary sub-C^*-algebra of B. Since C has real rank zero, there exists a non-zero projection $p \in C$, and by construction $pf_\delta(x) = f_\delta(x)p = p$. Therefore

$$\|pa\beta(p)\| = \|pf_\delta(x)a\beta(f_\delta(x)p)\| \leq \|f_\delta(x)a\beta(f_\delta(x))\| < \varepsilon .$$

Corollary 2. *Let A be a C^*-algebra of real rank zero, and let $\beta : A \longrightarrow A$ be an extendible injective endomorphism with $\beta(A)$ being hereditary such that $\mathbb{T}(\beta^n) = \mathbb{T}$ for every $n > 0$. Then, given any $\varepsilon > 0$, $a_1, \dots, a_k \in A^\sim$ and $n_1, \dots, n_k \in \mathbb{N}$ and a projection $p \in A$, there exists a projection $q \in pAp$ such that*

$$\|qa_i\beta^{n_i}(q)\| < \varepsilon \qquad \text{for every } i \in \{1, \dots, k\} .$$

Let A be a C^*-algebra. Then, we define the following technical condition:

(†) Given any $n \in \mathbb{N}$ and $p \in A$ there exist $p_1, \dots, p_n \in A$ non-zero pairwise orthogonal subprojections of p with $p \in \overline{Ap_iA}$ for all $i \in \{1, \dots, n\}$.

For example, every $\mathcal{Z}$-stable or purely infinite C^*-algebra of real rank zero satisfies condition (†) [19, Theorem 5.8].

Proposition 3. *Let A be a unital C^*-algebra of real rank zero satisfying (†), let $\beta : A \longrightarrow A$ be an injective endomorphism such that $\beta(A)$ is a hereditary sub-C^*-algebra of A, and let $A \times_\beta \mathbb{N} = C^*(\iota_\infty, V_\infty)$. If does not exist any proper ideal I of A such that $\beta(I) \subseteq I$, then given any non-zero projection $p \in A$ there exist a full projection $q \in A$ and $c \in A \times_\beta \mathbb{N}$ such that $q = cpc^*$.*

Proof. By Lemma 5 there exists $n \in \mathbb{N}$ such that $p + \beta(p) + \cdots + \beta^n(p)$ is a full positive element of A. Since A satisfies (†) there exist non-zero orthogonal projections $p_0, \dots, p_n \in A$ such that $p_0 + \cdots + p_n \leq p$ with $p \in \overline{Ap_iA}$ for all $i \in \{0, \dots, n\}$. Observe that $p + \beta(p) + \cdots + \beta^n(p)$ lies in the ideal generated by $q' := p_0 + \beta(p_1) + \cdots + \beta^n(p_n)$, so q' is also a full positive element of A. Denote $p_i' := \beta^i(p_i)$ for every $i \in \{0, \dots, n\}$. Now we are going to use induction on n to construct a projection $q \in A$ such that $p_0' + \cdots + p_n' \in \overline{AqA}$. The case $n = 0$ is clear. Now, suppose that there exists a projection q_{k-1} such that $p_0' + \cdots + p_{k-1}' \in \overline{Aq_{k-1}A}$.

Using the Riesz decomposition of $V(A)$ [26] we have $p_k' \sim a_k \oplus b_k$ such that $a_k \lesssim q_{k-1}$ and $b_k \lesssim 1 - q_{k-1}$. Let v_k be the partial isometry such that $v_k^* v_k \leq p_k'$ and $v_k v_k^* \leq 1 - q_{k-1}$. If we define the projection $q_k := q_{k-1} + v_k v_k^*$, then we have that $p_1' + \cdots + p_k' \in \overline{Aq_kA}$. Therefore the projection $q := q_n$ is full. If we define $c := p_0 + v_1 V_\infty p_1 + \cdots + v_n V_\infty^n p_n$, then we have that

$$cpc^* = cc^* = p_0 + v_1\beta(p_1)v_1^* + \cdots + v_n\beta^n(p_n)v_n^* = q ,$$

as desired.

Theorem 2. *Let A be a unital C^*-algebra of real rank zero satisfying (†) that has strict comparison, let $\beta : A \longrightarrow A$ be an injective endomorphism such that $\beta(1) \neq 1$ and $\beta(A)$ is a hereditary sub-C^*-algebra of A. If $A \times_\beta \mathbb{N}$ is simple and $\beta(1)$ is a full projection of A, then $A \times_\beta \mathbb{N}$ is purely infinite simple C^*-algebra.*

Proof. It is enough to prove that given a positive element $x \in A \times_\beta \mathbb{N}$ there exist $a, b \in A \times_\beta \mathbb{N}$ such that $axb = 1$. Let $E : A \times_\beta \mathbb{N} \longrightarrow A$ be the canonical faithful conditional expectation. So, $0 \neq E(x) = c \in A_+$. Then, for $\|c\| > \varepsilon > 0$ we have that the hereditary sub-C^*-algebra $\overline{(c-\varepsilon)_+ A (c-\varepsilon)_+} \subseteq c^{1/2} A c^{1/2}$ has real rank zero. Hence, there exists a non-zero projection $p = c^{1/2} y c^{1/2} \in c^{1/2} A c^{1/2}$. Then, $q = y^{1/2} c y^{1/2}$ is a projection, and $E(y^{1/2} x y^{1/2}) = y^{1/2} c y^{1/2} = q$. Thus, we can assume that $E(x) = q$ is a non-zero projection. Given $1/2 > \varepsilon > 0$, there exists $x' = (V^*)^m d_{-m} + \cdots + q + \cdots + d_m V^m$, with $d_j \in A_+$ for every j, such that $\|x - x'\| < \varepsilon$. By Corollary 1, Theorem 1 and Corollary 2, there exists a non-zero projection $p \in qAq$ such that

$$\|p d_i \beta^i(p)\| < \varepsilon/2m \qquad \text{and} \qquad \|\beta^i(p) d_{-i} p\| < \varepsilon/2m$$

for every $i \in \{1, \ldots, m\}$. Therefore

$$\|pxp - p\| \leq \|pxp - px'p\| + \|px'p - p\| \leq \varepsilon + \varepsilon < 1\,.$$

Then, pxp is invertible in $p(A \times_\beta \mathbb{N})p$, whence there exists $y \in p(A \times_\beta \mathbb{N})p$ such that $ypxp = p$. Since we are assuming that $A \times_\beta \mathbb{N}$ is simple and $\beta(1)$ is a full projection, [23, Theorem 4.1] implies that there are no non-trivial ideals I of A such that $\beta(I) \subseteq I$. Thus, by Proposition 3, there exist $c \in A \times_\beta \mathbb{N}$ and a full projection $q \in A$ such that $cpc^* = q$.

By Lemma 6, there exist $m \in \mathbb{N}$ and a partial isometry $u \in A$ with the property that $(V_\infty^*)^m u^* q u V_\infty^m = 1$ and therefore

$$(V_\infty^*)^m u^* (cypxpc^*) u V_\infty^m = (V_\infty^*)^m u^* cpc^* u V_\infty^m = (V_\infty^*)^m u^* q u V_\infty^m = 1\,.$$

Thus, if we set $a := (V_\infty^*)^m u^* c y p$ and $b := p c^* u V_\infty^m$ we have $axb = 1$, as desired.

When A is a purely infinite C^*-algebra, we generalize the result of [10].

Corollary 3. *Let A be a unital purely infinite C^*-algebra of real rank zero, let $\beta : A \longrightarrow A$ be an injective endomorphism such that $\beta(1) \neq 1$ is a full projection and $\beta(A)$ is a hereditary sub-C^*-algebra of A. Then $A \times_\beta \mathbb{N}$ is a simple purely infinite C^*-algebra if and only if $A \times_\beta \mathbb{N}$ is simple.*

Proof. The proof works in the same way as that of Theorem 2, but keeping in mind that Lemma 6 and condition (†) are also satisfied for purely infinite C^*-algebras.

Example 1. This is a generalization of Cuntz's construction of the algebras $\mathcal{O}_n$ [5]. Let $\mathcal{U}_m$ be the m^∞ UHF algebra $\bigotimes_{n=1}^\infty M_m(\mathbb{C})$, and let $B = \mathcal{U}_m \oplus \cdots \oplus \mathcal{U}_m$ be the direct sum of n copies of $\mathcal{U}_m$, that is a nuclear unital C^*-algebra of real rank zero absorbing $\mathcal{Z}$, and hence has strict comparison and satisfies condition (†). Let us consider the endomorphism $\beta : B \longrightarrow B$ given by $\beta(x_1, \dots, x_n) = (P_1 \otimes x_2, P_2 \otimes x_3 \cdots, P_n \otimes x_1)$ for every $(x_1, \dots, x_n) \in B$, where $P_1, \cdots, P_n \in M_m(\mathbb{C})$ are rank 1 projections. Hence, β is injective. Observe that $\beta(1) \neq 1$ is a full projection of B. It is clear that B is β-simple and β^k is outer for any $k > 0$, since B is a unital finite C^*-algebra. Hence, $B \times_\beta \mathbb{N}$ is simple by Theorem 1, and thus applying Theorem 2 it is also a purely infinite C^*-algebra; in particular it is a Kirchberg algebra. Now, we use the modification of the Pimsner-Voiculescu six-term exact given in Lemma 4,

$$\begin{array}{ccccc} K_0(B) & \xrightarrow{Id-K_0(\beta)} & K_0(B) & \longrightarrow & K_0(B \times_\beta \mathbb{N}) \\ \uparrow & & & & \downarrow \\ K_1(B \times_\beta \mathbb{N}) & \longleftarrow & K_1(B) & \xleftarrow{Id-K_1(\beta)} & K_1(B) \end{array}$$

Notice that the induced map $K_0(\beta) : \mathbb{Z}[1/m]^n \longrightarrow \mathbb{Z}[1/m]^n$ is given by

$$K_0(\beta)(x_1, \dots, x_n) = (x_2/m, \dots, x_n/m, x_1/m),$$

for every $(x_1, \dots, x_n) \in \mathbb{Z}[1/m]^n$. Then, we can easily compute $K_0(B \times_\beta \mathbb{N}) = \mathbb{Z}/(m^n - 1)\mathbb{Z}$ and $K_1(B \times_\beta \mathbb{N}) = 0$. Hence, using the Kirchberg-Phillips classification theorems [11, 20], we conclude that $B \times_\beta \mathbb{N}$ is stably isomorphic to the Cuntz algebra $\mathcal{O}_{m^n}$.

Acknowledgements Parts of this work were done during a visit of the second author to the Institute for Matematiske Fag, Norges Teknisk-Naturvitenskapelige Universitet (Trondheim, Norway). The second author thanks the host center for its warm hospitality. This research was supported by the NordForsk Research Network "Operator Algebras and Dynamics" (grant #11580). The second author was partially supported by PAI III grants FQM-298 and P07-FQM-7156 of the Junta de Andalucía. Both authors were partially supported by the DGI-MICINN and European Regional Development Fund, jointly, through Project MTM2011-28992-C02-02 and by 2009 SGR 1389 grant of the Comissionat per Universitats i Recerca de la Generalitat de Catalunya.

References

1. Adji, S.: Semigroup crossed products and the structure of Toeplitz algebras. J. Operator Theory **44**(1), 139–150 (2000)
2. Adji, S., Laca, M., Nilsen, M., Raeburn, I.: Crossed products by semigroups of endomorphisms and the Toeplitz algebras of ordered groups. Proc. Amer. Math. Soc. **122**(4), 1133–1141 (1994). DOI 10.2307/2161182. URL http://dx.doi.org/10.2307/2161182
3. Arveson, W.: Continuous analogues of Fock space. Mem. Amer. Math. Soc. **80**(409), iv+66 (1989)

4. Boyd, S., Keswani, N., Raeburn, I.: Faithful representations of crossed products by endomorphisms. Proc. Amer. Math. Soc. **118**(2), 427–436 (1993). DOI 10.2307/2160319. URL http://dx.doi.org/10.2307/2160319
5. Cuntz, J.: Simple C^*-algebras generated by isometries. Comm. Math. Phys. **57**(2), 173–185 (1977)
6. Cuntz, J.: The internal structure of simple C^*-algebras. In: Operator algebras and applications, Part I (Kingston, Ont., 1980), *Proc. Sympos. Pure Math.*, vol. 38, pp. 85–115. Amer. Math. Soc., Providence, R.I. (1982)
7. Doplicher, S., Roberts, J.E.: Endomorphisms of C^*-algebras, cross products and duality for compact groups. Ann. of Math. (2) **130**(1), 75–119 (1989). DOI 10.2307/1971477. URL http://dx.doi.org/10.2307/1971477
8. Elliott, G.A.: Some simple C^*-algebras constructed as crossed products with discrete outer automorphism groups. Publ. Res. Inst. Math. Sci. **16**(1), 299–311 (1980). DOI 10.2977/prims/1195187509. URL http://dx.doi.org/10.2977/prims/1195187509
9. an Huef, A., Raeburn, I.: Exel and Stacey crossed products, and Cuntz-Pimsner algebras (2011), arXiv:1107.1019v1
10. Jeong, J.A., Kodaka, K., Osaka, H.: Purely infinite simple C^*-crossed products. II. Canad. Math. Bull. **39**(2), 203–210 (1996). DOI 10.4153/CMB-1996-025-2. URL http://dx.doi.org/10.4153/CMB-1996-025-2
11. Kirchberg, E.: Das nicht-kommutative Michael-Auswahlprinzip und die Klassifikation nicht-einfacher Algebren. In: C^*-algebras (Münster, 1999), pp. 92–141. Springer, Berlin (2000)
12. Kirchberg, E., Rørdam, M.: Non-simple purely infinite C^*-algebras. Amer. J. Math. **122**(3), 637–666 (2000). URL http://muse.jhu.edu/journals/american_journal_of_mathematics/v122/122.3kirchberg.pdf
13. Kishimoto, A.: Outer automorphisms and reduced crossed products of simple C^*-algebras. Comm. Math. Phys. **81**(3), 429–435 (1981). URL http://projecteuclid.org/getRecord?id=euclid.cmp/1103920327
14. Kwaśniewski, B.K.: Cuntz-Krieger uniqueness theorem for crossed products by Hilbert bimodules. Preprint
15. Olesen, D.: Inner *-automorphisms of simple C^*-algebras. Comm. Math. Phys. **44**(2), 175–190 (1975)
16. Olesen, D., Pedersen, G.K.: Applications of the Connes spectrum to C^*-dynamical systems. J. Funct. Anal. **30**(2), 179–197 (1978). DOI 10.1016/0022-1236(78)90069-1. URL http://dx.doi.org/10.1016/0022-1236(78)90069-1
17. Olesen, D., Pedersen, G.K.: Applications of the Connes spectrum to C^*-dynamical systems. III. J. Funct. Anal. **45**(3), 357–390 (1982). DOI 10.1016/0022-1236(82)90011-8. URL http://dx.doi.org/10.1016/0022-1236(82)90011-8
18. Paschke, W.L.: The crossed product of a C^*-algebra by an endomorphism. Proc. Amer. Math. Soc. **80**(1), 113–118 (1980). DOI 10.2307/2042155. URL http://dx.doi.org/10.2307/2042155
19. Perera, F., Rørdam, M.: AF-embeddings into C^*-algebras of real rank zero. J. Funct. Anal. **217**(1), 142–170 (2004). DOI 10.1016/j.jfa.2004.05.001. URL http://dx.doi.org/10.1016/j.jfa.2004.05.001
20. Phillips, N.C.: A classification theorem for nuclear purely infinite simple C^*-algebras. Doc. Math. **5**, 49–114 (electronic) (2000)
21. Rørdam, M.: Classification of certain infinite simple C^*-algebras. J. Funct. Anal. **131**(2), 415–458 (1995). DOI 10.1006/jfan.1995.1095. URL http://dx.doi.org/10.1006/jfan.1995.1095
22. Rørdam, M.: The stable and the real rank of $\mathscr{Z}$-absorbing C^*-algebras. Internat. J. Math. **15**(10), 1065–1084 (2004). DOI 10.1142/S0129167X04002661. URL http://dx.doi.org/10.1142/S0129167X04002661
23. Schweizer, J.: Dilations of C^*-correspondences and the simplicity of Cuntz-Pimsner algebras. J. Funct. Anal. **180**(2), 404–425 (2001). DOI 10.1006/jfan.2000.3703. URL http://dx.doi.org/10.1006/jfan.2000.3703

24. Stacey, P.J.: Crossed products of C^*-algebras by $*$-endomorphisms. J. Austral. Math. Soc. Ser. A **54**(2), 204–212 (1993)
25. Zhang, S.: A property of purely infinite simple C^*-algebras. Proc. Amer. Math. Soc. **109**(3), 717–720 (1990). DOI 10.2307/2048211. URL http://dx.doi.org/10.2307/2048211
26. Zhang, S.: A Riesz decomposition property and ideal structure of multiplier algebras. J. Operator Theory **24**(2), 209–225 (1990)

Chapter 13
Quasi-symmetric Group Algebras and C^*-Completions of Hecke Algebras

Rui Palma

Abstract We show that for a Hecke pair (G, Γ) the C^*-completions $C^*(L^1(G, \Gamma))$ and $pC^*(\overline{G})p$ of its Hecke algebra coincide whenever the group algebra $L^1(\overline{G})$ satisfies a spectral property which we call "quasi-symmetry", a property that is satisfied by all Hermitian groups and all groups with subexponential growth. We generalize in this way a result of Kaliszewski et al. (Proc Edinb Math Soc (2) 51(3):657–695, 2008). Combining this result with our earlier results in (Palma, J Funct Anal 264:2704–2731, 2013) and a theorem of Tzanev (J Oper Theory 50(1):169–178, 2003) we establish that the full Hecke C^*-algebra exists and coincides with the reduced one for several classes of Hecke pairs, particularly all Hecke pairs (G, Γ) where G is a nilpotent group. As a consequence, the category equivalence studied by Hall (Hecke C^*-algebras. Ph.D. thesis, The Pennsylvania State University, 1999) holds for all such Hecke pairs. We also show that the completions $C^*(L^1(G, \Gamma))$ and $pC^*(\overline{G})p$ do not always coincide, with the Hecke pair $(SL_2(\mathbb{Q}_q), SL_2(\mathbb{Z}_q))$ providing one such example.

Keywords Hecke algebras • Hecke pairs • Hall's equivalence • C^*-algebras

Mathematics Subject Classification (2010): 46L55, 20C08.

13.1 Introduction

A *Hecke pair* (G, Γ) consists of a group G and a subgroup $\Gamma \subseteq G$, called a *Hecke subgroup*, for which every double coset $\Gamma g \Gamma$ is the union of finitely many

R. Palma (✉)
University of Oslo, P.O. Box 1053, NO-0316 Oslo, Norway
e-mail: ruip@math.uio.no

T.M. Carlsen et al. (eds.), *Operator Algebra and Dynamics*, Springer Proceedings in Mathematics & Statistics 58, DOI 10.1007/978-3-642-39459-1_13,

left cosets. Examples of Hecke subgroups include finite subgroups, finite-index subgroups and normal subgroups. It is many times insightful to think of Hecke subgroups as subgroups which are "almost normal". The *Hecke algebra* $\mathscr{H}(G,\Gamma)$ of a Hecke pair (G,Γ) is a *-algebra of complex-valued functions over the set of double cosets $\Gamma\backslash G/\Gamma$, with suitable convolution product and involution. It generalizes the notion of the group algebra $\mathbb{C}(G/\Gamma)$ of the quotient group when Γ is a normal subgroup.

A natural example of a Hecke pair (G,Γ) is that of a locally compact totally disconnected group G and a compact open subgroup Γ. These type of examples are, in some sense, the general case, since we can always reduce to this case via a construction called the Schlichting completion: given a Hecke pair (G,Γ) we can associate to it a new Hecke pair $(\overline{G},\overline{\Gamma})$ where $\overline{G}$ is locally compact totally disconnected, $\overline{\Gamma}$ is compact and open and the corresponding Hecke algebras $\mathscr{H}(G,\Gamma)$ and $\mathscr{H}(\overline{G},\overline{\Gamma})$ are canonically isomorphic.

For operator algebraists the interest in the subject of Hecke algebras was largely raised by the work of Bost and Connes [2] on phase transitions in number theory and their work has led several authors to study C^*-algebras which arise as completions of Hecke algebras. There are several canonical C^*-completions of a Hecke algebra $\mathscr{H}(G,\Gamma)$ which one can consider (see [17] and [11]): the enveloping C^*-algebra of $\mathscr{H}(G,\Gamma)$, denoted by $C^*(G,\Gamma)$; the enveloping C^*-algebra of the Banach *-algebra $L^1(G,\Gamma)$, denoted by $C^*(L^1(G,\Gamma))$; the canonical corner $pC^*(\overline{G})p$, where p is the characteristic function of $\overline{\Gamma}$; and $C_r^*(G,\Gamma)$, which is the C^*-algebra generated by the left regular representation of $\mathscr{H}(G,\Gamma)$. The question of when does $C^*(G,\Gamma)$ exist and when do some of these completions coincide has been studied by several authors ([2, 6, 11, 14, 17], to name a few).

An important question raised by Hall [6] where C^*-completions of Hecke algebras came to play an important role was if for a Hecke pair (G,Γ) there is a correspondence between unitary representations of G generated by the Γ-fixed vectors and nondegenerate *-representations of $\mathscr{H}(G,\Gamma)$, analogous to the known correspondence between representations of a group and of its group algebra. Whenever such a correspondence holds we say that (G,Γ) satisfies *Hall's equivalence*. It is known that Hall's equivalence does not hold in general [6], and in fact a theorem of Kaliszewski et al. [11] shows that Hall's equivalence holds precisely when $C^*(G,\Gamma)$ exists and $C^*(G,\Gamma)\cong C^*(L^1(G,\Gamma))\cong pC^*(\overline{G})p$, which has been shown to be the case for several classes of Hecke pairs.

The primary goal of this article is to give a sufficient condition for the isomorphism $C^*(L^1(G,\Gamma))\cong pC^*(\overline{G})p$ to hold and to combine this result with the results of [14] in order to establish Hall's equivalence for several classes of Hecke pairs, including all Hecke pairs (G,Γ) where G is a nilpotent group. We will also show that the two C^*-completions $C^*(L^1(G,\Gamma))$ and $pC^*(\overline{G})p$ are in general different, with $(SL_2(\mathbb{Q}_q), SL_2(\mathbb{Z}_q))$ providing an example for which $C^*(L^1(G,\Gamma))\ncong pC^*(\overline{G})p$.

The problem of deciding for which Hecke pairs the completions $C^*(L^1(G,\Gamma))$ and $pC^*(\overline{G})p$ coincide is partially understood. Several properties of the pair (G,Γ) are known to force these two completions to coincide, and in this regard we recall a result by Kaliszewski et al. [11] which states that $C^*(L^1(G,\Gamma)) \cong pC^*(\overline{G})p$ whenever the Schlichting completion $\overline{G}$ is a Hermitian group (meaning that every self-adjoint element $f \in L^1(\overline{G})$ has real spectrum). We will generalize their result in Sect. 13.3 in a way that covers also all Hecke pairs for which G or $\overline{G}$ has subexponential growth. For that we introduce the notion of a *quasi-symmetric* group algebra: a locally compact group G will be said to have a quasi-symmetric group algebra if for any $f \in C_c(G)$ the spectrum of $f^* * f$ relative to $L^1(G)$ is in $\mathbb{R}_0^+$. It follows directly from the Shirali-Ford theorem [16] that Hermitian groups have a quasi-symmetric group algebra and it is a consequence of the work of Hulanicki [7,8] that this is also the case for groups of subexponential growth. We show that $C^*(L^1(G,\Gamma)) \cong pC^*(\overline{G})p$ whenever the Schlichting completion $\overline{G}$ has a quasi-symmetric group algebra.

Besides strictly generalizing Kaliszewski, Landstad and Quigg's result, as there are groups of subexponential growth which are not Hermitian, our result is easier to apply in practice since many times we can use it without any knowledge about the Schlichting completion $\overline{G}$, which is often hard to compute. In fact we will show that if G has subexponential growth then so does $\overline{G}$, which means that knowledge about the original group G is sufficient for applying our result. The relation between Hermitianness and subexponential growth will be discussed in Sect. 13.4.

By combining our result on quasi-symmetric group algebras with the results of [14] and also a theorem of Tzanev [17], we are able to establish in Sect. 13.5 that $C^*(G,\Gamma)$ exists and $C^*(G,\Gamma) \cong C^*(L^1(G,\Gamma)) \cong pC^*(\overline{G})p \cong C_r^*(G,\Gamma)$ for several classes of Hecke pairs, including all Hecke pairs (G,Γ) where G is a nilpotent group. Consequently, it follows that Hall's equivalence holds for all such classes of Hecke pairs.

It is natural to ask if there are examples of Hecke pairs for which we have $C^*(L^1(G,\Gamma)) \ncong pC^*(\overline{G})p$. According to [11], Tzanev claims in private communication with the authors that the Hecke pair $(PSL_3(\mathbb{Q}_q), PSL_3(\mathbb{Z}_q))$ is such that $C^*(L^1(G,\Gamma)) \ncong pC^*(\overline{G})p$, but no proof has been published and no other example seems to be known, as far as we know. We prove in Sect. 13.6 that $C^*(L^1(G,\Gamma)) \ncong pC^*(\overline{G})p$ for the Hecke pair $(PSL_2(\mathbb{Q}_q), PSL_2(\mathbb{Z}_q))$, as suggested in [11], but following a different approach which does not use the representation theory of $PSL_2(\mathbb{Q}_q)$.

The author is thankful to his adviser Nadia Larsen for the very helpful discussions, suggestions and comments during the elaboration of this work.

13.2 Preliminaries

13.2.1 Hecke Pairs and Hecke Algebras

We will mostly follow [12] and [11] in what regards Hecke pairs and Hecke algebras and refer to these references for more details.

Definition 1. Let G be a group and Γ a subgroup. The pair (G,Γ) is called a *Hecke pair* if every double coset $\Gamma g\Gamma$ is the union of finitely many right (and left) cosets. In this case, Γ will be called a *Hecke subgroup* of G.

Given a Hecke pair (G,Γ) we will denote by L and R, respectively, the left and right coset counting functions, i.e.

$$L(g) := |\Gamma g\Gamma/\Gamma| < \infty \qquad \text{and} \qquad R(g) := |\Gamma\backslash\Gamma g\Gamma| < \infty .$$

We recall that L and R are Γ-biinvariant functions which satisfy $L(g) = R(g^{-1})$ for all $g \in G$. Moreover, the function $\Delta : G \to \mathbb{Q}^+$ given by

$$\Delta(g) := \frac{L(g)}{R(g)},$$

is a group homomorphism, usually called the *modular function* of (G,Γ).

Definition 2. The *Hecke algebra* $\mathscr{H}(G,\Gamma)$ is the *-algebra of finitely supported $\mathbb{C}$-valued functions on the double coset space $\Gamma\backslash G/\Gamma$ with the product and involution defined by

$$(f_1 * f_2)(\Gamma g\Gamma) := \sum_{h\Gamma\in G/\Gamma} f_1(\Gamma h\Gamma) f_2(\Gamma h^{-1} g\Gamma),$$

$$f^*(\Gamma g\Gamma) := \Delta(g^{-1})\overline{f(\Gamma g^{-1}\Gamma)}.$$

Remark 1. Some authors, including Krieg [12], do not include the factor Δ in the involution. Here we adopt the convention of [11] in doing so, as it gives rise to a more natural L^1-norm. We note, nevertheless, that there is no loss (or gain) in doing so, because these two different involutions give rise to *-isomorphic Hecke algebras.

Given a Hecke pair (G,Γ), the subgroup $R^\Gamma := \bigcap_{g\in G} g\Gamma g^{-1}$ is a normal subgroup of G contained in Γ. A Hecke pair (G,Γ) is called *reduced* if $R^\Gamma = \{e\}$. As it is known, the pair $(G_r,\Gamma_r) := (G/R^\Gamma, \Gamma/R^\Gamma)$ is a reduced Hecke pair and the Hecke algebras $\mathscr{H}(G,\Gamma) \cong \mathscr{H}(G_r,\Gamma_r)$ are canonically isomorphic. For this reason the pair (G_r,Γ_r) is called the *reduction* of (G,Γ), and the isomorphism of the corresponding Hecke algebras shows that it is enough to consider reduced Hecke pairs, a convention used by several authors. We will not use this convention however, since we aim at achieving general results based on properties of the original Hecke pair (G,Γ), and not its reduction.

A natural example of a Hecke pair (G,Γ) is given by a totally disconnected locally compact group G and a compact open subgroup Γ. It is known that this type of examples are, in some sense, the general case: there is a canonical construction which associates to a given reduced Hecke pair (G,Γ) a new Hecke pair $(\overline{G},\overline{\Gamma})$ with the following properties:

1. $\overline{G}$ is a totally disconnected locally compact group;
2. $\overline{\Gamma}$ is a compact open subgroup;
3. The pair $(\overline{G}, \overline{\Gamma})$ is reduced;
4. There is a canonical embedding $\theta : G \to \overline{G}$ such that $\theta(G)$ is dense in $\overline{G}$ and $\theta(\Gamma)$ is dense in $\overline{\Gamma}$. Moreover, $\theta^{-1}(\overline{\Gamma}) = \Gamma$.

   The pair $(\overline{G}, \overline{\Gamma})$ satisfies a well-known uniqueness property and is called the *Schlichting completion* of (G, Γ). For the details of this construction the reader is referred to [17] and [11] (see also [4] for a slightly different approach). We shall make a quick review of some known facts and we refer to the previous references for all the details.

   Henceforward we will not write explicitly the canonical homomorphism θ, and we will instead see G as a dense subgroup of $\overline{G}$, identified with the image $\theta(G)$. The Schlichting completion $(\overline{G}, \overline{\Gamma})$ of a reduced Hecke pair (G, Γ) satisfies the following additional property:
5. There are canonical bijections $G/\Gamma \to \overline{G}/\overline{\Gamma}$ and $\Gamma\backslash G/\Gamma \to \overline{\Gamma}\backslash\overline{G}/\overline{\Gamma}$ given by $g\Gamma \to g\overline{\Gamma}$ and $\Gamma g\Gamma \to \overline{\Gamma}g\overline{\Gamma}$, respectively.

If a Hecke pair (G, Γ) is not reduced, its *Schlichting completion* $(\overline{G}, \overline{\Gamma})$ is defined as the completion $(\overline{G_r}, \overline{\Gamma_r})$ of its reduction. There is then a canonical map with dense image $G \to \overline{G}$ which factors through G_r, and this map is an embedding if and only if (G, Γ) is reduced, i.e. $G \cong G_r$.

Following [11], we consider the normalized Haar measure μ on $\overline{G}$ (so that $\mu(\overline{\Gamma}) = 1$) and define the Banach *-algebra $L^1(\overline{G})$ with the usual convolution product and involution. We denote by p the characteristic function of $\overline{\Gamma}$, i.e. $p := \chi_{\overline{\Gamma}}$, which is a projection in $C_c(\overline{G}) \subseteq L^1(\overline{G})$. Recalling [17] or [11], we always have canonical *-isomorphisms:

$$\mathscr{H}(G, \Gamma) \cong \mathscr{H}(G_r, \Gamma_r) \cong \mathscr{H}(\overline{G}, \overline{\Gamma}) \cong pC_c(\overline{G})p\,. \tag{13.1}$$

The modular function Δ of a reduced Hecke pair (G, Γ), defined by (13.2.1), is simply the modular function of the group $\overline{G}$ restricted to G.

13.2.2 L^1- and C^*-Completions

There are several ways of defining a L^1-norm in a Hecke algebra. One approach is to simply take the L^1-norm from $L^1(\overline{G})$, since the isomorphisms in (13.1) enables us to see the Hecke algebra as a subalgebra of $L^1(\overline{G})$. The completion of $\mathscr{H}(G, \Gamma)$ with respect to this L^1-norm is isomorphic to the corner $pL^1(\overline{G})p$. Alternatively, one may take the following definition:

Definition 3. The L^1-*norm* on $\mathscr{H}(G, \Gamma)$, denoted $\|\cdot\|_{L^1}$, is given by

$$\|f\|_{L^1} := \sum_{\Gamma g\Gamma \in \Gamma\backslash G/\Gamma} |f(\Gamma g\Gamma)|\, L(g)\,.$$

We will denote by $L^1(G,\Gamma)$ the completion of $\mathscr{H}(G,\Gamma)$ under this norm.

As observed in [17] or [11], the two L^1-norms described above are the same. In fact we have canonical $*$-isomorphisms

$$L^1(G,\Gamma) \cong L^1(\overline{G},\overline{\Gamma}) \cong pL^1(\overline{G})p\,.$$

There are several canonical C^*-completions of $\mathscr{H}(G,\Gamma)$. These are:

- $C_r^*(G,\Gamma)$ – Called the *reduced Hecke C^*-algebra*, it is the completion of $\mathscr{H}(G,\Gamma)$ under the C^*-norm arising from the left regular representation (see [17]).
- $pC^*(\overline{G})p$ – The corner of the full group C^*-algebra $C^*(\overline{G})$.
- $C^*(L^1(G,\Gamma))$ – The enveloping C^*-algebra of $L^1(G,\Gamma)$.
- $C^*(G,\Gamma)$ – The enveloping C^*-algebra (if it exists!) of $\mathscr{H}(G,\Gamma)$. When it exists, it is usually called the *full Hecke C^*-algebra*.

The various C^*-completions of $\mathscr{H}(G,\Gamma)$ are related in the following way, through canonical surjective maps:

$$C^*(G,\Gamma) \dashrightarrow C^*(L^1(G,\Gamma)) \longrightarrow pC^*(\overline{G})p \longrightarrow C_r^*(G,\Gamma)\,.$$

As was pointed out by Hall in [6, Proposition 2.21], the full Hecke C^*-algebra $C^*(G,\Gamma)$ does not have to exist in general. Nevertheless, its existence has been established for several classes of Hecke pairs (see, for example, [6, 11] or [14]).

The question of whether some of these completions are actually the same has also been explored in the literature [2, 11, 14, 17]. We review here some of the main results.

The question of when one has the isomorphism $pC^*(\overline{G})p \cong C_r^*(G,\Gamma)$ was clarified by Tzanev, in [17, Proposition 5.1], to be a matter of amenability. As pointed out in [11], there was a mistake in Tzanev's article (where it is assumed without proof that $C^*(L^1(G,\Gamma)) \cong pC^*(\overline{G})p$ is always true) which carries over to the cited Proposition 5.1. Nevertheless, Tzanev's proof holds if one just replaces $C^*(L^1(G,\Gamma))$ with $pC^*(\overline{G})p$, so that the correct statement of (a part of) his result becomes:

Theorem 1 (Tzanev). $pC^*(\overline{G})p \cong C_r^*(G,\Gamma)$ *if and only if* $\overline{G}$ *is amenable.*

A result concerning the isomorphism $C^*(L^1(G,\Gamma)) \cong pC^*(\overline{G})p$ was obtained by Kaliszewski, Landstad and Quigg in [11, Theorem 6.14], where they showed that this isomorphism holds when $\overline{G}$ is a Hermitian group.

In [14] we established the existence of $C^*(G,\Gamma)$ and also the isomorphism $C^*(G,\Gamma) \cong C^*(L^1(G,\Gamma))$ for several classes of Hecke pairs, recovering also various results in the literature in a unified approach.

Another important result of [11] regarding the existence of $C^*(G,\Gamma)$ and the simultaneous isomorphisms $C^*(G,\Gamma) \cong C^*(L^1(G,\Gamma)) \cong pC^*(\overline{G})p$ will be discussed in the next subsection.

13.2.3 Representation Theory

As it is well-known, for any group G there is a canonical bijective correspondence (i.e. category equivalence) between unitary representations of G and nondegenerate *-representations of the group algebra $\mathbb{C}(G)$. Hall [6] asked whether something analogous was true for Hecke pairs, and the following definition is necessary in order to understand Hall's question:

Definition 4. Let G be a group and $\Gamma \subseteq G$ a subgroup. A unitary representation $\pi : G \to U(\mathscr{H})$ is said to be *generated by its Γ-fixed vectors* if $\overline{\pi(G)\mathscr{H}^{\Gamma}} = \mathscr{H}$, where $\mathscr{H}^{\Gamma} = \{\xi \in \mathscr{H} : \pi(\gamma)\xi = \xi\,, \text{ for all } \gamma \in \Gamma\}$.

The question Hall posed in [6] is the following:

Question 1 (Hall's equivalence). Let (G, Γ) be a Hecke pair. Is there a category equivalence between nondegenerate *-representations of $\mathscr{H}(G, \Gamma)$ and unitary representations of G generated by the Γ-fixed vectors?

Whenever there is an affirmative answer to this question, we shall say the Hecke pair (G, Γ) satisfies *Hall's equivalence*. In the work of Hall [6] and the subsequent work of Glöckner and Willis [4], Hall's equivalence was studied and proven to hold under a certain form of positivity for some *-algebraic bimodules. A more complete approach was further developed by Kaliszewski, Landstad and Quigg in [11], where Hall's equivalence, positivity for certain *-algebraic bimodules, and C^*-completions of Hecke algebras were all shown to be related. We briefly describe here the approach and results of [11] and the reader is referred to this reference for more details.

Let $(\overline{G}, \overline{\Gamma})$ be the Schlichting completion of a Hecke pair (G, Γ). Following [11, Sect. 5], we have an inclusion of two imprimitivity bimodules (in Fell's sense):

$$_{C_c(\overline{G})pC_c(\overline{G})}\big(C_c(\overline{G})p\big)_{\mathscr{H}(\overline{G},\overline{\Gamma})} \qquad \subseteq \qquad _{L^1(\overline{G})pL^1(\overline{G})}\big(L^1(\overline{G})p\big)_{L^1(\overline{G},\overline{\Gamma})}\,,$$

where the left and right inner products, $\langle\rangle_L$ and $\langle\rangle_R$, on these bimodules are given by multiplication within $L^1(\overline{G})$ by

$$\langle f\,,\,g\rangle_L = f * g^*\,, \qquad\qquad \langle f\,,\,g\rangle_R = f^* * g\,.$$

A *-representation π of $\mathscr{H}(G, \Gamma)$ is said to be *$\langle\rangle_R$-positive* if

$$\pi(\langle f\,,\,f\rangle_R) \geq 0\,, \qquad \text{for all } f \in C_c(\overline{G})p\,. \tag{13.2}$$

Similarly, a *-representation π of $L^1(G, \Gamma)$ is said to be *$\langle\rangle_R$-positive* when condition (13.2) holds for all $f \in L^1(\overline{G})p$.

In [11, Corollary 6.19] it is proven that, for a reduced pair (G, Γ), there exists a category equivalence between unitary representations of G generated by the Γ-fixed vectors and the $\langle\rangle_R$-positive representations of $\mathscr{H}(G, \Gamma)$. This is in fact true for non-reduced Hecke pairs (G, Γ) as well, as follows from the following observation:

Proposition 1. *Let (G, Γ) be a Hecke pair and (G_r, Γ_r) its reduction. There exists a category equivalence between unitary representations of G generated by the Γ-fixed vectors and unitary representations of G_r generated by the Γ_r-fixed vectors.*

The correspondence is as follows: a representation $\pi : G_r \to U(\mathscr{H})$ is mapped to the representation $\pi \circ q$, where $q : G \to G_r$ is the quotient map. Its inverse map takes a representation $\rho : G \to U(\mathscr{H})$ to the representation $\tilde{\rho}$ of G_r on the same Hilbert space, given by $\tilde{\rho}([g]) := \rho(g)$.

Proof. First we observe that the assignment $\pi \mapsto \pi \circ q$ does indeed produce a unitary representation of G generated by the Γ-fixed vectors. This is obvious since the spaces of fixed vectors $\mathscr{H}^{\Gamma_r}$ and $\mathscr{H}^{\Gamma}$ for π and $\pi \circ q$, respectively, are the same.

Secondly, for the inverse assignment, we need to check that $\tilde{\rho}$ is well-defined, which amounts to show that $\rho(g) = \rho(gh)$ for any $g \in G$ and $h \in R^{\Gamma}$. For any $s \in G$ and $\xi \in \mathscr{H}^{\Gamma}$ we have

$$\begin{aligned}\rho(gh)\rho(s)\xi &= \rho(g)\rho(s)\rho(s^{-1}hs)\xi \\ &= \rho(g)\rho(s)\xi\,,\end{aligned}$$

because $s^{-1}hs \in R^{\Gamma} \subseteq \Gamma$. Hence, $\rho(gh) = \rho(g)$ on the space $\overline{\pi(G)\mathscr{H}^{\Gamma}}$. Since ρ is assumed to be generated by the Γ-fixed vectors, it follows that $\rho(gh) = \rho(g)$.

It is also easy to see that $\tilde{\rho}$ is generated by the Γ_r-fixed vectors and it is clear from the definitions that these assignments are inverse of one another.

We now have to say a few words about the intertwiners of representations, i.e. the morphisms in the categories we are considering. It follows immediately from the definitions that if we have an intertwiner $V : \mathscr{H}_1 \to \mathscr{H}_2$ between two representations $\pi_1 : G_r \to B(\mathscr{H}_1)$ and $\pi_2 : G_r \to B(\mathscr{H}_2)$, then V itself is an intertwiner between $\pi_1 \circ q$ and $\pi_2 \circ q$ and moreover the composition laws are satisfied. The exact same thing happens for the assignment $\rho \to \tilde{\rho}$, so that we have in fact an isomorphism of categories, and therefore, in particular, a category equivalence. □

In the light of Kaliszewski, Landstad and Quigg's result, for a Hecke pair (G, Γ) for which all *-representations of $\mathscr{H}(G, \Gamma)$ are $\langle\rangle_R$-positive, there exists a category equivalence between unitary representations of G generated by the Γ-fixed vectors and nondegenerate *-representations of $\mathscr{H}(G, \Gamma)$. In other words, Hall's equivalence holds when all *-representations of $\mathscr{H}(G, \Gamma)$ are $\langle\rangle_R$-positive. Furthermore, the authors of [11] show also the following relation between $\langle\rangle_R$-positivity and C^*-completions of Hecke algebras:

Theorem 2 ([11, Corollary 6.11]). *Let (G, Γ) be a Hecke pair.*

1. *Every *-representation of $\mathscr{H}(G, \Gamma)$ is $\langle\rangle_R$-positive if and only if $C^*(G, \Gamma)$ exists and $C^*(G, \Gamma) \cong C^*(L^1(G, \Gamma)) \cong pC^*(\overline{G})p$.*
2. *Every *-representation of $L^1(G, \Gamma)$ is $\langle\rangle_R$-positive if and only if $C^*(L^1(G, \Gamma)) \cong pC^*(\overline{G})p$.*

13.2.4 Groups of Subexponential Growth

Let G be a locally compact group with a Haar measure μ. For a compact neighbourhood V of e, the limit superior

$$\limsup_{n\to\infty} \mu(V^n)^{\frac{1}{n}} \tag{13.3}$$

will be called the *growth rate* of V. Since $0 < \mu(V) \leq \mu(V^n)$ for all $n \in \mathbb{N}$ it is clear that the growth rate of V is always greater or equal to one.

Definition 5. A locally compact group G is said to be of *subexponential growth* if $\limsup_{n\to\infty} \mu(V^n)^{\frac{1}{n}} = 1$ for all compact neighbourhoods V of e. Otherwise it is said to be of *exponential growth*.

The class of groups with subexponential growth is closed under taking closed subgroups [5, Théorème I.2] and quotients [5, Théorème I.3]. We observe that even though in [5] the author is only working with compactly generated groups, the proofs of these results are general and hold for any locally compact group.

It is known that if G has subexponential growth as a discrete group, then it has subexponential growth with respect to any other locally compact topology [8, Theorem 3.1]. The following is a slight generalization of this result, and the proof is done along similar lines:

Proposition 2. *Let H be a dense subgroup of a locally compact group $\overline{H}$. If H has subexponential growth as a discrete group, then $\overline{H}$ has subexponential growth in its locally compact topology.*

Proof. Let $A \subseteq \overline{H}$ be a compact neighbourhood of e. First we claim that $HA = \overline{H}$. Since A is a neighbourhood of $\{e\}$, there is an open set $U \subseteq A$ such that $e \in U$. To show that $HA = \overline{H}$, let $g \in \overline{H}$. Since H is dense in $\overline{H}$ and $g(U \cap U^{-1})$ is open, it follows that there exists $h \in H \cap g(U \cap U^{-1})$. Thus, there exists $s \in U \cap U^{-1}$ such that $h = gs$, or equivalently, $g = hs^{-1}$. Since $s^{-1} \in U \cap U^{-1}$ we then have $g \in hU$, and thus $g \in hA$. Hence $\overline{H} = HA$.

From the previous observation it follows that $\{hA\}_{h\in H}$ is a covering of the compact set AA, and since A has non-empty interior there must exist a finite set $F \subset H$ such that $AA \subseteq FA$. Hence, we have $A^n \subseteq F^{n-1}A$, for all $n \geq 2$. Without loss of generality we can assume that F contains the identity element. Now using the fact that H has subexponential growth we obtain

$$\limsup_{n\to\infty} \mu(A^n)^{\frac{1}{n}} \leq \limsup_{n\to\infty} \mu(F^{n-1}A)^{\frac{1}{n}} \leq \limsup_{n\to\infty} |F^{n-1}|^{\frac{1}{n}} \mu(A)^{\frac{1}{n}} = 1 .$$

□

Corollary 1. *Let (G, Γ) be a discrete Hecke pair. If G (or G_r) has subexponential growth, then so does $\overline{G}$.*

Proof. If G has subexponential growth than so does any of its quotients, so in particular G_r also has subexponential growth. If G_r has subexponential growth then so does $\overline{G}$ by Proposition 2. □

Groups with subexponential growth are always unimodular [15, Proposition 12.5.8] and amenable [15, Sect. 12.6.18].

The class of groups with subexponential growth includes all locally nilpotent groups and all FC^--groups [15, Theorem 12.5.17]. In particular, all abelian and all compact groups have subexponential growth.

13.3 Quasi-symmetric Group Algebras

Given a *-algebra A and an element $a \in A$ we will use throughout this chapter the notations $\sigma_A(a)$ to denote the spectrum of a relative to A, and $R_A(a)$ to denote the spectral radius of a relative to A.

Recall, for example from [15], that a *-algebra A is said to be:

- *Hermitian* if $\sigma_A(a) \subseteq \mathbb{R}$, for any self-adjoint element $a = a^*$ of A.
- *Symmetric* if $\sigma_A(a^*a) \subseteq \mathbb{R}_0^+$, for any $a \in A$.

It is an easy fact that symmetry implies Hermitianness. The two properties are equivalent for Banach *-algebras, as asserted by the Shirali-Ford theorem [16].

Recall also that a locally compact group G is called *Hermitian* if $L^1(G)$ is a Hermitian (equivalently, symmetric) Banach *-algebra. The class of Hermitian groups satisfies some known closure properties, some of which we list below:

1. The class of Hermitian groups is closed under taking open subgroups and quotients [15, Theorem 12.5.18].
2. Let $1 \to H \to G \to G/H \to 1$ be an extension of locally compact groups. If H is Hermitian and G/H is finite, then G is Hermitian [15, Theorem 12.5.18].

The class of groups we are going to consider in this work arises by relaxing the condition of symmetry on the group algebra:

Definition 6. Let G be a locally compact group. We will say that the group algebra $L^1(G)$ is *quasi-symmetric* if $\sigma_{L^1(G)}(f^* * f) \subseteq \mathbb{R}_0^+$ for any compactly supported continuous function f.

Clearly, Hermitian groups have a quasi-symmetric group algebra. Another important class of groups with this property is that of groups with subexponential growth, which comes as a consequence of the work of Hulanicki (for discrete groups this was established in [7]):

Proposition 3. *If G is a locally compact group with subexponential growth, then $L^1(G)$ is quasi-symmetric.*

Proof. Let $\lambda : L^1(G) \to B(L^2(G))$ denote the left regular representation of $L^1(G)$. Hulanicki proved in [8] that if G has subexponential growth then

$$R_{L^1(G)}(f) = \|\lambda(f)\| , \tag{13.4}$$

for any self-adjoint continuous function f of compact support. Moreover, Barnes showed in [1] (a result which he credited to Hulanicki [9]) that if A is a Banach *-algebra, $B \subseteq A$ a *-subalgebra and if $\pi : A \to B(\mathscr{H})$ is a faithful *-representation such that

$$R_A(b) = \|\pi(b)\| ,$$

for all self-adjoint elements $b = b^*$ in B, then $\sigma_A(b) = \sigma_{B(\mathscr{H})}(\pi(b))$ for every $b \in B$.

Considering A and B to be $L^1(G)$ and $C_c(G)$ respectively, we see from (13.4) that by taking π to be λ we immediately get that $\sigma_{L^1(G)}(f^* * f) = \sigma_{B(L^2(G))}(\lambda(f^* * f)) = \sigma_{B(L^2(G))}\big(\lambda(f)^*\lambda(f)\big)$ for any $f \in C_c(G)$. Thus, since $B(L^2(G))$ is a C^*-algebra, we have that $\sigma_{L^1(G)}(f^* * f) \subseteq \mathbb{R}_0^+$ for $f \in C_c(G)$, i.e. $L^1(G)$ is quasi-symmetric. □

The following result is the main result in this section and explains the reason for considering quasi-symmetric group algebras in the context of C^*-completions of Hecke pairs.

Theorem 3. *Let (G, Γ) be a Hecke pair. If $\overline{G}$ has a quasi-symmetric group algebra, then*

$$C^*(L^1(G, \Gamma)) \cong pC^*(\overline{G})p .$$

In particular, there is a category equivalence between *-representations of $L^1(G, \Gamma)$ and unitary representations of G generated by the Γ-fixed vectors.

Lemma 1. *Let (G, Γ) be a Hecke pair and $f \in pL^1(\overline{G})p$. We have that $\sigma_{pL^1(\overline{G})p}(f) \subseteq \sigma_{L^1(\overline{G})}(f)$.*

Proof. Let us denote by $L^1(\overline{G})^\dagger$ the minimal unitization of $L^1(\overline{G})$ and let $\mathbf{1} \in L^1(\overline{G})^\dagger$ be its unit. Let $\lambda \in \mathbb{C}$ and suppose that $f - \lambda\mathbf{1}$ is invertible in $L^1(\overline{G})^\dagger$. We want to prove that $f - \lambda p$ is invertible in $pL^1(\overline{G})p$. Invertibility of $f - \lambda\mathbf{1}$ in $L^1(\overline{G})^\dagger$ means that there exist $g \in L^1(\overline{G})$ and $\beta \in \mathbb{C}$ such that $\mathbf{1} = (f - \lambda\mathbf{1})(g + \beta\mathbf{1})$. Hence we have

$$\begin{aligned} p = p(f - \lambda\mathbf{1})(g + \beta\mathbf{1})p &= (pf - \lambda p)(gp + \beta p) \\ = (fp - \lambda p)(gp + \beta p) &= (f - \lambda p)p(gp + \beta p) \\ = (f - \lambda p)(pgp + \beta p) . & \end{aligned}$$

Hence, $f - \lambda p$ is invertible in $pL^1(\overline{G})p$ and this finishes the proof. □

Proof (Theorem 3). Due to the canonical isomorphism $L^1(G,\Gamma) \cong pL^1(\overline{G})p$, it is enough to prove that $C^*(pL^1(\overline{G})p) \cong pC^*(\overline{G})p$. By [11, Corollary 6.11] we only need to show that every *-representation of $pL^1(\overline{G})p$ is $\langle\rangle_R$-positive. Let $\pi : pL^1(\overline{G})p \to B(\mathscr{H})$ be a *-representation and $f \in L^1(\overline{G})p$. Let $\{g_n\}_{n\in\mathbb{N}}$ be a sequence of functions in $C_c(\overline{G})p$ such that $g_n \to f$ in $L^1(\overline{G})$. Then, we also have $g_n^* * g_n \to f^* * f$ in $L^1(\overline{G})$. It is a standard fact that

$$\sigma_{B(\mathscr{H})}(\pi(g_n^* * g_n)) \subseteq \sigma_{pL^1(\overline{G})p}(g_n^* * g_n)\,,$$

and by Lemma 1 we have $\sigma_{pL^1(\overline{G})p}(g_n^* * g_n) \subseteq \sigma_{L^1(\overline{G})}(g_n^* * g_n)$. Moreover, since $L^1(\overline{G})$ is quasi-symmetric we have that $\sigma_{L^1(\overline{G})}(g_n^* * g_n) \subseteq \mathbb{R}_0^+$. All these inclusions combined give

$$\sigma_{B(\mathscr{H})}(\pi(g_n^* * g_n)) \subseteq \sigma_{pL^1(\overline{G})p}(g_n^* * g_n) \subseteq \sigma_{L^1(\overline{G})}(g_n^* * g_n) \subseteq \mathbb{R}_0^+\,,$$

and therefore $\pi(g_n^* * g_n)$ is a positive operator for every $n \in \mathbb{N}$. Thus, the limit $\pi(f^* * f) = \lim \pi(g_n^* * g_n)$ is also a positive operator. In other words, $\pi(\langle f, f\rangle_R) \geq 0$. □

As a consequence we immediately recover Kaliszewski, Landstad and Quigg's result on Hermitian groups and also that $C^*(L^1(G,\Gamma)) \cong pC^*(\overline{G})p \cong C_r^*(G,\Gamma)$ for Hecke pairs arising from groups of subexponential growth:

Corollary 2 ([11, Theorem 6.14]). *Let (G,Γ) be a Hecke pair. If $\overline{G}$ is Hermitian, then $C^*(L^1(G,\Gamma)) \cong pC^*(\overline{G})p$.*

Corollary 3. *Let (G,Γ) be a Hecke pair. If one of the groups G, G_r or $\overline{G}$ has subexponential growth, then $C^*(L^1(G,\Gamma)) \cong pC^*(\overline{G})p \cong C_r^*(G,\Gamma)$.*

Proof. By Corollary 1, if G or G_r has subexponential growth, then so does $\overline{G}$ in its totally disconnected locally compact topology. Since $\overline{G}$ has subexponential growth, we have that $L^1(\overline{G})$ is quasi-symmetric and therefore $C^*(L^1(G,\Gamma)) \cong pC^*(\overline{G})p$ by Theorem 3. The isomorphism $pC^*(\overline{G})p \cong C_r^*(G,\Gamma)$ follows from Tzanev's theorem (Theorem 1 in the present work), due to the fact that subexponential growth implies amenability of the group $\overline{G}$. □

13.4 Further Remarks on Groups with a Quasi-symmetric Group Algebra

The classes of Hermitian groups and groups with subexponential growth are in general different. On one side, there are examples of Hermitian groups which do not have subexponential growth, such as the affine group of the real line $\mathrm{Aff}(\mathbb{R}) := \mathbb{R} \rtimes \mathbb{R}^*$, with its usual topology as a (connected) Lie group, as shown by Leptin [13]. On the other side, there are examples of groups with subexponential growth which are not Hermitian, such as the Fountain-Ramsay-Williamson group [3], which is the discrete group with the presentation

$$\big\langle \{u_j\}_{j\in\mathbb{N}} \mid u_j^2 = e \text{ and } u_i u_j u_k u_j = u_j u_k u_j u_i \quad \forall i, j < k \in \mathbb{N} \big\rangle .$$

Fountain, Ramsay and Williamson showed that this group is not Hermitian despite being locally finite (thus, having subexponential growth). Another such example was given by Hulanicki in [10].

Using these examples we can show that the class of groups with a quasi-symmetric group algebra is strictly larger than the union of the classes of Hermitian groups and groups with subexponential growth. In that regard we have the following result:

Proposition 4. *Let H be a Hermitian locally compact group with exponential growth and let L be a discrete locally finite group which is not Hermitian. The locally compact group $G := H \times L$ has a quasi-symmetric group algebra, but it is neither Hermitian nor has subexponential growth.*

An example of such a group is given by taking $H :=$ Aff($\mathbb{R}$) and L the Fountain-Ramsay-Williamson group.

Proof. Let us first prove that $G := H \times L$ has a quasi-symmetric group algebra. Given a function $f \in C_c(G)$, the product $f^* * f$ also has compact support, and since L is discrete, the support of $f^* * f$ must lie inside some set of the form $H \times F$, where $F \subseteq L$ is a finite set. Since L is locally finite, F generates a finite subgroup $\langle F \rangle \subseteq G$. Now $H \times \langle F \rangle$ is an open subgroup of G, so that

$$L^1(H \times \langle F \rangle) \subseteq L^1(G) .$$

The group $H \times \langle F \rangle$ is Hermitian, being a finite extension of a Hermitian group, and therefore $\sigma_{L^1(H\times\langle F\rangle)}(f^* * f) \subseteq \mathbb{R}_0^+$. This implies that

$$\sigma_{L^1(G)}(f^* * f) \subseteq \sigma_{L^1(H\times\langle F\rangle)}(f^* * f) \subseteq \mathbb{R}_0^+ ,$$

which shows that G is quasi-symmetric.

This group is not Hermitian, because it has a quotient (L) which is not Hermitian, and it does not have subexponential growth because it has a quotient (H) which does not have subexponential growth. □

Since in the present work we are directly concerned with totally disconnected groups (because of the Schlichting completion), it would be interesting to know if there are examples of totally disconnected groups with a quasi-symmetric group algebra, but which are not Hermitian nor have subexponential growth. We do not know the answer to this question. The example considered in Proposition 4 is of course not totally disconnected since Aff($\mathbb{R}$) is a connected group. But in view of Proposition 4, it would suffice to answer affirmatively the following more fundamental problem:

Question 2. Is there any Hermitian, totally disconnected group, with exponential growth?

As we pointed out above, there are examples of locally compact groups (even connected ones) which are Hermitian and have exponential growth, such as Aff($\mathbb{R}$), but the question of whether this can happen in the totally disconnected setting is, as far as we understand, still open. In the discrete case, Palmer [15] claims that all examples of discrete groups which are known to be Hermitian actually have subexponential growth (even more, polynomial growth).

On the other side, a negative answer to the above question would mean that any Hermitian totally disconnected group necessarily has subexponential growth and is therefore amenable, and thus would bring new evidence for the long standing conjecture that all Hermitian groups are amenable [15], which is known to be true in the connected case [15, Theorem 12.5.18 (e)]. In fact, a negative answer to 2 in the discrete case alone would, through the theory of extensions, imply that all Hermitian groups with an open connected component are amenable.

The fact that we do not know of any totally disconnected group with a quasi-symmetric group algebra which does not have subexponential growth is not a drawback in any way. In fact, the class of groups with subexponential growth is already very rich by itself and will be used to give meaningful examples in Hecke C^*-algebra theory and Hall's equivalence in the next section.

13.5 Hall's Equivalence

Combining the results of [14] on the existence of $C^*(G,\Gamma)$ and the isomorphism $C^*(G,\Gamma) \cong C^*(L^1(G,\Gamma))$, with the results on this paper on groups of subexponential growth and also Tzanev's theorem, we are able to establish that

$$C^*(G,\Gamma) \cong C^*(L^1(G,\Gamma)) \cong pC^*(\overline{G})p \cong C_r^*(G,\Gamma),$$

for several classes of Hecke pairs, including all Hecke pairs (G,Γ) where G is a nilpotent group. As a consequence, [11, Corollary 6.11] (Theorem 2 in the present work) yields that Hall's equivalence is satisfied for all such classes of Hecke pairs.

Proposition 5. *If a group G satisfies one of the following generalized nilpotency properties:*

- *G is finite-by-nilpotent, or*
- *G is hypercentral, or*
- *All subgroups of G are subnormal,*

then for any Hecke subgroup $\Gamma \subseteq G$ we have that $C^*(G,\Gamma)$ exists and

$$C^*(G,\Gamma) \cong C^*(L^1(G,\Gamma)) \cong pC^*(\overline{G})p \cong C_r^*(G,\Gamma).$$

In particular, Hall's equivalence holds with respect to any Hecke subgroup.

Proof. As discussed in [14, Classes 5.8, 5.9, 5.5] for every Hecke pair (G, Γ) where G satisfies one of the aforementioned properties we have that the full Hecke C^*-algebra exists and $C^*(G, \Gamma) \cong C^*(L^1(G, \Gamma))$.

We claim that if G has one of the three properties above, it must have subexponential growth. If G is finite-by-nilpotent, then by definition G is a nilpotent extension of a finite group, and since nilpotent groups have subexponential growth, then so does G. If G is hypercentral or all subgroups of G are subnormal, then it is known that G is locally nilpotent and therefore must have subexponential growth (see [7]). Consequently, by Corollary 3 we must have $C^*(L^1(G, \Gamma)) \cong pC^*(\overline{G})p \cong C_r^*(G, \Gamma)$. □

If we restrict ourselves to finite subgroups $\Gamma \subseteq G$ we get a similar result for other classes of groups:

Proposition 6. *If a group G satisfies one of the following properties:*

- *G is an FC-group, or*
- *G is locally nilpotent, or*
- *G is locally finite,*

then for any finite subgroup $\Gamma \subseteq G$ we have that $C^*(G, \Gamma)$ exists and

$$C^*(G, \Gamma) \cong C^*(L^1(G, \Gamma)) \cong pC^*(\overline{G})p \cong C_r^*(G, \Gamma).$$

In particular, Hall's equivalence holds with respect to any finite subgroup.

Proof. As discussed in [14, Classes 5.10, 5.11, 5.12] for every group G that satisfies one of the aforementioned properties we have that, for any finite subgroup Γ, the full Hecke C^*-algebra exists and we have $C^*(G, \Gamma) \cong C^*(L^1(G, \Gamma))$. Also if G has one of the three properties above, it must have subexponential growth (for FC- and locally nilpotent groups see [7], and for locally finite groups it is obvious). Consequently, by Corollary 3 we must have $C^*(L^1(G, \Gamma)) \cong pC^*(\overline{G})p \cong C_r^*(G, \Gamma)$. □

Remark 2. The results above show that Hall's equivalence holds for any Hecke pair (G, Γ) where G satisfies a certain generalized nilpotency property. An analogous result for the class of solvable groups cannot hold. In [17, Example 3.4] Tzanev gave an example of a Hecke pair (G, Γ) where G is solvable but for which $C^*(G, \Gamma)$ does not exist, and consequently Hall's equivalence does not hold. The example consists of the infinite dihedral group $G := \mathbb{Z} \rtimes (\mathbb{Z}/2\mathbb{Z})$ together with $\Gamma := \mathbb{Z}/2\mathbb{Z}$.

13.6 A Counter-Example

In the previous sections we have established a sufficient condition for the isomorphism $C^*(L^1(G, \Gamma)) \cong pC^*(\overline{G})p$ to hold, namely whenever $\overline{G}$ has a quasi-symmetric group algebra. A natural question to ask is the following: is it even possible that $C^*(L^1(G,\Gamma)) \not\cong pC^*(\overline{G})p$? We will now show that $C^*(L^1(G,\Gamma)) \not\cong pC^*(\overline{G})p$ for the Hecke pair $(PSL_2(\mathbb{Q}_q), PSL_2(\mathbb{Z}_q))$, where q denotes a prime number and $\mathbb{Q}_q$, $\mathbb{Z}_q$ denote respectively the field of q-adic numbers and the ring of q-adic integers. It was already asked in [11, Example 11.8] if $C^*(L^1(G,\Gamma)) \not\cong pC^*(\overline{G})p$ for this Hecke pair and a strategy to achieve this result was designed. Our approach is nevertheless different from the approach suggested in [11] since we make no use of the representation theory of $PSL_2(\mathbb{Q}_q)$.

As we remarked in the introduction, Tzanev has claimed that the Hecke pair $(PSL_3(\mathbb{Q}_q), PSL_3(\mathbb{Z}_q))$ gives another example, but no proof has been published.

Theorem 4. *Let q be a prime number and $\mathbb{Q}_q$ and $\mathbb{Z}_q$ denote respectively the field of q-adic numbers and the ring of q-adic integers. For the Hecke pair $(G,\Gamma) := (PSL_2(\mathbb{Q}_q), PSL_2(\mathbb{Z}_q))$ we have that $C^*(L^1(G,\Gamma)) \not\cong pC^*(G)p$.*

Proof. For ease of reading and so that no confusion arises between the prime number q and the projection p, we will throughout this proof denote the projection p by P. Thus, our goal is to prove that $C^*(L^1(G,\Gamma)) \not\cong PC^*(G)P$.

The pair $(PSL_2(\mathbb{Q}_q), PSL_2(\mathbb{Z}_q))$ coincides with its own Schlichting completion (see [11]) and is the reduction of the pair $(SL_2(\mathbb{Q}_q), SL_2(\mathbb{Z}_q))$. For ease of reading we will work with the pair $(SL_2(\mathbb{Q}_q), SL_2(\mathbb{Z}_q))$ in this proof.

The structure of the Hecke algebra $\mathscr{H}(G,\Gamma)$ is well-known, and for convenience we will mostly refer to Hall [6, Sect. 2.1.2.1] whenever we need to. Letting

$$x_n := \begin{pmatrix} q^n & 0 \\ 0 & q^{-n} \end{pmatrix},$$

it is known ([6, Proposition 2.9]) that every double coset $\Gamma s \Gamma$ can be uniquely represented as $\Gamma x_n \Gamma$ for some $n \in \mathbb{N}$.

For each $0 \le k \le q-1$ let us denote by $y_k \in G$ the matrix

$$y_k := \begin{pmatrix} q & k \\ 0 & q^{-1} \end{pmatrix},$$

and let us take $g \in L^1(G)P$ as the element $g := y_0 P + y_1 P + \cdots + y_{q-1} P$, and $f := P + g$. We then have

$$\begin{aligned} f^* f = (P+g)^*(P+g) &= P + g^* P + P g + g^* g \\ &= P + \sum_{k=0}^{q-1} P y_k^{-1} P + \sum_{k=0}^{q-1} P y_k P + \sum_{i,j=0}^{q-1} P y_i^{-1} y_j P \\ &= (q+1)P + \sum_{k=0}^{q-1} P y_k^{-1} P + \sum_{k=0}^{q-1} P y_k P + \sum_{\substack{i,j=0 \\ i \neq j}}^{q-1} P y_i^{-1} y_j P . \end{aligned}$$

As it is know (see for example [6, Propositions 2.10 and 2.12]), in $\mathscr{H}(G,\Gamma)$ the modular function is trivial and each double coset is self-adjoint. Hence we can write

$$f^*f = (q+1)P + 2\sum_{k=0}^{q-1} Py_kP + 2\sum_{\substack{i,j=0\\ i<j}}^{q-1} Py_i^{-1}y_jP\,.$$

We now notice that, from [6, Proposition 2.9], we have $\Gamma y_k\Gamma = \Gamma x_1\Gamma$, and therefore $Py_kP = Px_1P$. Moreover, for $0 \le i < j \le q-1$, we have that

$$y_i^{-1}y_j = \begin{pmatrix} 1 & (j-i)q^{-1} \\ 0 & 1 \end{pmatrix},$$

and again from [6, Proposition 2.9] we conclude that $Py_i^{-1}y_jP = Px_1P$. Hence, we get

$$\begin{aligned} f^*f &= (q+1)\,P + 2q\,Px_1P + 2\frac{(q-1)q}{2}\,Px_1P \\ &= (q+1)\,P + (q^2+q)\,Px_1P\,. \end{aligned}$$

It is well known that $\mathscr{H}(G,\Gamma)$ is commutative (see for example [6, Sect. 2.2.3.2]) and all of its characters have been explicitly described. Following [11, Example 11.8] the characters of $\mathscr{H}(G,\Gamma)$ are precisely all the functions $\pi_z : \mathscr{H}(G,\Gamma) \to \mathbb{C}$ such that

$$\pi_z(Px_mP) = \frac{1-qz}{(q+1)(1-z)}\left(\frac{z}{q}\right)^m + \frac{q-z}{(q+1)(1-z)}\left(\frac{1}{qz}\right)^m,$$

for a given complex number $z \in \mathbb{C}\backslash\{1\}$ (the expression for π_1 is different and the reader should check [11, Example 11.8] for the correct definition, but we will not need it here). Kaliszewski et al. [11, Example 11.8] have also determined that the characters π_z which extend to *-representations of $L^1(G,\Gamma)$ are precisely those with $z \in [-q,-1/q] \cup [1/q,q]$.

We will now consider the *-representation π_{-q} of $L^1(G,\Gamma)$ and show that $\pi_{-q}(f^*f) < 0$. First we notice that

$$\begin{aligned} \pi_{-q}(Px_1P) &= \frac{1-q(-q)}{(q+1)(1-(-q))}\left(\frac{-q}{q}\right) + \frac{q-(-q)}{(q+1)(1-(-q))}\left(\frac{1}{q(-q)}\right) \\ &= -\frac{1+q^2}{(q+1)^2} - \frac{2}{(q+1)^2q} \\ &= -\frac{q^3+q+2}{(q+1)^2q}\,. \end{aligned}$$

Hence we get

$$\begin{aligned}\pi_{-q}(f^* f) &= \pi_{-q}\big((q+1)\,P + (q^2+q)\,P x_1 P\big)\\ &= q+1-(q^2+q)\frac{q^3+q+2}{(q+1)^2 q}\\ &= q+1-\frac{q^3+q+2}{q+1}.\end{aligned}$$

To prove that $\pi_{-q}(f^* f) < 0$ is then equivalent to show that $(q+1)^2 < q^3+q+2$, or equivalently, $0 < q^3-q^2-q+1$, for any prime number q. This follows from an elementary calculus argument as follows: letting $F(x) = x^3-x^2-x+1$, we have that $F''(x) = 6x-2$ is always greater than 0 for $x \geq 2$ (the first prime number). Hence, $F'(x) = 3x^2-2x-1$ is growing for $x \geq 2$. Since $F'(2) > 0$, it follows that $F'(x)$ is always greater than 0 for $x \geq 2$. Thus, $F(x)$ is growing in this interval, and since $F(2) > 0$, it follows that $F(q) > 0$, for any prime q.

Since $\pi_{-q}(f^* f) < 0$ it then follows that not all representations of $L^1(G,\Gamma)$ are $\langle\rangle_R$-positive and consequently $C^*(L^1(G,\Gamma)) \ncong PC^*(G)P$. □

As a particular consequence of the above theorem, it follows that $PSL_2(\mathbb{Q}_q)$ does not have a quasi-symmetric group algebra. Also, together with Hall's result [6, Proposition 2.21] and the fact that $PSL_2(\mathbb{Q}_q)$ is not amenable, we can say that for this Hecke pair $C^*(G,\Gamma)$ does not exist and $C^*(L^1(G,\Gamma)) \ncong pC^*(\overline{G})p \ncong C_r^*(G,\Gamma)$.

As we have seen in this chapter, the isomorphism $C^*(L^1(G,\Gamma)) \cong pC^*(\overline{G})p$ holds whenever G, G_r or $\overline{G}$ has subexponential growth. We would like know if the same is true or if one counter-example can be found for the class of amenable groups:

Question 3. If $\overline{G}$ is amenable does it follow that $C^*(L^1(G,\Gamma)) \cong pC^*(\overline{G})p$?

Acknowledgements This research was supported by the Research Council of Norway and the NordForsk research network "Operator Algebra and Dynamics" (grant # 11580).

References

1. Barnes, B.A.: When is the spectrum of a convolution operator on L^p independent of p? Proc. Edinburgh Math. Soc. (2) **33**(2), 327–332 (1990). DOI 10.1017/S0013091500018241. URL http://dx.doi.org/10.1017/S0013091500018241
2. Bost, J.B., Connes, A.: Hecke algebras, type III factors and phase transitions with spontaneous symmetry breaking in number theory. Selecta Math. (N.S.) **1**(3), 411–457 (1995). DOI 10.1007/BF01589495. URL http://dx.doi.org/10.1007/BF01589495
3. Fountain, J.B., Ramsay, R.W., Williamson, J.H.: Functions of measures on compact groups. Proc. Roy. Irish Acad. Sect. A **76**(23), 235–251 (1976)

4. Glöckner, H., Willis, G.A.: Topologization of Hecke pairs and Hecke C^*-algebras. In: Proceedings of the 16th Summer Conference on General Topology and its Applications (New York), vol. 26, pp. 565–591 (2001/02)
5. Guivarc'h, Y.: Croissance polynomiale et périodes des fonctions harmoniques. Bull. Soc. Math. France **101**, 333–379 (1973)
6. Hall, R.W.: Hecke C^*-algebras. Ph.D. thesis, The Pennsylvania State University (1999)
7. Hulanicki, A.: On the spectral radius of hermitian elements in group algebras. Pacific J. Math. **18**, 277–287 (1966)
8. Hulanicki, A.: On positive functionals on a group algebra multiplicative on a subalgebra. Studia Math. **37**, 163–171 (1970/71)
9. Hulanicki, A.: On the spectrum of convolution operators on groups with polynomial growth. Invent. Math. **17**, 135–142 (1972)
10. Hulanicki, A.: Invariant subsets of nonsynthesis Leptin algebras and nonsymmetry. Colloq. Math. **43**(1), 127–136 (1981) (1980)
11. Kaliszewski, S., Landstad, M.B., Quigg, J.: Hecke C^*-algebras, Schlichting completions and Morita equivalence. Proc. Edinb. Math. Soc. (2) **51**(3), 657–695 (2008). DOI 10.1017/S0013091506001419. URL http://dx.doi.org/10.1017/S0013091506001419
12. Krieg, A.: Hecke algebras. Mem. Amer. Math. Soc. **87**(435), x+158 (1990)
13. Leptin, H.: Lokal kompakte Gruppen mit symmetrischen Algebren. In: Symposia Mathematica, Vol. XXII (Convegno sull'Analisi Armonica e Spazi di Funzioni su Gruppi Localmente Compatti, INDAM, Rome, 1976), pp. 267–281. Academic Press, London (1977)
14. Palma, R.: On enveloping C^*-algebras of Hecke algebras. Journal of Functional Analysis **264**, 2704–2731 (2013)
15. Palmer, T.W.: Banach algebras and the general theory of $*$-algebras. Vol. 2, *Encyclopedia of Mathematics and its Applications*, vol. 79. Cambridge University Press, Cambridge (2001)
16. Shirali, S., Ford, J.W.M.: Symmetry in complex involutory Banach algebras. II. Duke Math. J. **37**, 275–280 (1970)
17. Tzanev, K.: Hecke C^*-algebras and amenability. J. Operator Theory **50**(1), 169–178 (2003)

Chapter 14
Dynamics, Wavelets, Commutants and Transfer Operators Satisfying Crossed Product Type Commutation Relations

Sergei Silvestrov

Abstract An overview is provided of several recent results, constructions and publications relating dynamical systems, wavelets, transfer operators satisfying covariance commutation relations associated to non-invertible dynamics, defining generalizations of crossed product operator algebras to non-invertible dynamics or actions by semigroups, ideals in the corresponding crossed product type algebras and commutants of elements and subsets in the algebras and in their representations. Some open directions and open problems on this rich interplay motivated by these constructions and results are also indicated.

Keywords Dynamical system • Crossed product algebra • Commutant • Ideal • Wavelet representation • Quadrature mirror filter • Cantor set

Mathematics Subject Classification (2010): 42C40, 28A80, 47L65, 37A30

14.1 Introduction

The interplay between dynamical systems and operator theory and operator algebras is now a well developed subject [13, 17, 22, 42, 72, 74, 86, 87]. The interplay between topological properties of the dynamical system (or more general actions of groups) such as minimality, transitivity, freeness and others on the one hand, and properties of ideals, subalgebras and representations of the corresponding crossed product C^*-algebra on the other hand has been a subject of intensive investigations at least

S. Silvestrov (✉)
Division of Applied Mathematics, The School of Education, Culture and Communication, Mälardalen University, Box 883, SE-721 23 Västerås, Sweden
e-mail: sergei.silvestrov@mdh.se

T.M. Carlsen et al. (eds.), *Operator Algebra and Dynamics*, Springer Proceedings in Mathematics & Statistics 58, DOI 10.1007/978-3-642-39459-1_14,

since the 1960s. In the recent years, substantial efforts are made in establishing broad interplay between C^*-algebras and non-invertible dynamical systems, actions of semigroups, equivalence relations, (semi-)groupoids, correspondences (see for example [1, 3, 4, 10–12, 14–16, 18, 19, 23, 26–28, 31, 37, 39–41, 44, 50–52, 61, 72, 77–79, 88], and references therein).

This interplay and its implications for operator representations of the corresponding crossed product algebras, spectral and harmonic analysis, non-commutative analysis and non-commutative geometry are fundamental for the mathematical foundations of quantum mechanics, quantum field theory, string theory, integrable systems, lattice models, quantization, symmetry analysis and, as it has become clear recently, in wavelet analysis and its applications in signal and image processing (see [11, 22, 48, 58–60, 72, 74, 89] and references therein). In particular, the operator theoretic approach to wavelet theory has been extremely productive [10–12, 19, 43]. The connections between irreducible covariant representations, ergodic shifts on solenoids, fixed points of transfer (or Ruelle) operators, as well as the related investigations on interplay between decompositions or reducible representations, centers and commutants in corresponding crossed product algebras and periodicity and aperiodicity, freeness, minimality, transitivity, ergodicity and related properties of the corresponding topological dynamical system, are of major importance in these contexts.

Wavelets are functions that generate orthonormal bases under certain actions of translation and dilation operators. They have the advantage over Fourier series that they are better localized. More precisely, in the theory of wavelets, orthonormal bases for $L^2(\mathbb{R})$ are constructed by applying dilation and translation operators, in a certain order, to a given vector ψ called the *wavelet*. Thus from the start, in this construction, there are two unitary operators, the dilation operator U and the translation operator T on $L^2(\mathbb{R})$,

$$Uf(x) = \frac{1}{\sqrt{2}} f\left(\frac{x}{2}\right), \quad Tf(x) = f(x-1), \quad (x \in \mathbb{R}, f \in L^2(\mathbb{R})) \tag{14.1}$$

satisfying a *covariance relation* with the action defined by the non-invertible map $z \mapsto z^2$ on the complex plane $\mathbb{C}$ or on the unit circle $\mathbb{T}$ or on the real line $\mathbb{R}$:

$$UTU^{-1} = T^2. \tag{14.2}$$

Since T is a unitary operator, its spectrum is a subset of the unit circle $\mathbb{T}$. Using Borel functional calculus, one can define a representation of $L^\infty(\mathbb{T})$ on $L^2(\mathbb{R})$, by $\pi(f) = f(T)$, $(f \in L^\infty(\mathbb{R}))$, which means in particular that $\pi(z^n) = T^n$ and for polynomials $\pi(\sum_k a_k z^k) = \sum_k a_k T^k$. The representation satisfies the *covariance relation*:

$$U\pi(f)U^{-1} = \pi(f(z^2)), \quad (f \in L^\infty(\mathbb{T})) \tag{14.3}$$

Crossed product W^*-algebras or covariant representations of crossed product C^*-algebras of functions by a group action (invertible dynamics) are defined by such relation except that in the context of Wavelets, the dynamics (the action) on the space of definition of the functions is not invertible, meaning that the action by the group $\mathbb{Z}$ is replaced by the action of a semigroup of non-negative integers put in correspondence with forward iterations of the acting non-invertible map. A natural general way to include the information about the inverse iterations given by inverse branches (pre-images) into such crossed-product algebra structure is to attach to it the transfer operators averaging in some ways the values of the functions over all the pre-images of the acting non-invertible map. This turns out to be vary fruitful and relevant for investigation of dynamics and wavelets using the operator approach, and is also the framework used in the Exel's crossed product algebras by a semigroup.

In the paradigmatic classical example from wavelet analysis described above a wavelet is a function $\psi \in L^2(\mathbb{R})$ with the property that

$$\{2^{j/2}\psi(2^j \cdot -k) : j,k \in \mathbb{Z}\} \tag{14.4}$$

is an orthonormal basis for $L^2(\mathbb{R})$ (see for example Daubechies' classical book [21] for details). Using the operators U and T, the family defined in (14.4) can be written as $\{U^j T^k \psi : j,k \in \mathbb{Z}\}$. The main general technique of constructing wavelets is by a *multiresolution analysis* (multiresolution). A multiresolution analysis or multiresolution is a sequence $(V_n)_{n\in\mathbb{Z}}$ of closed subspaces of $L^2(\mathbb{R})$ with the following properties:

1. $V_n \subseteq V_{n+1}$ for all $n \in \mathbb{Z}$;
2. $UV_{n+1} = V_n$ for all $n \in \mathbb{Z}$;
3. $\cup_n V_n$ is dense in $L^2(\mathbb{R})$ and $\cap_n V_n = \{0\}$;
4. There exists a function $\varphi \in L^2(\mathbb{R})$ called *the scaling function*, such that $\{T^k\varphi : k \in \mathbb{Z}\}$ is an orthonormal basis for V_0.

The subspaces V_n correspond to various resolution levels. Once a multiresolution analysis is given, the wavelet can be found in the *detail space*: $W_0 := V_1 \ominus V_0$. It is a function ψ with the property that $\{T^k\psi : k \in \mathbb{Z}\}$ is an orthonormal basis for W_0. The multiresolution is constructed easily from the scaling function φ. Since $U\varphi$ is in $V_{-1} \subseteq V_0$, it can be written as a combination of translates of φ. This gives the *scaling equation* for the function φ:

$$U\varphi = \sum_{k\in\mathbb{Z}} a_k T^k \varphi. \tag{14.5}$$

Starting from a *quadrature-mirror-filter (QMF)* $m_0 \in L^\infty(\mathbb{T})$ that satisfies the *QMF-condition*

$$\frac{1}{2}\sum_{w^2=z} |m_0(w)|^2 = 1, \quad (z \in \mathbb{T}), \tag{14.6}$$

the *low-pass condition* $m_0(1) = \sqrt{2}$, and perhaps some regularity (Lipschitz, etc.), the *scaling function* φ associated to the QMF m_0 is constructed by an infinite product formula for its Fourier transform

$$\hat{\varphi}(x) = \prod_{n=1}^{\infty} \frac{m_0\left(e^{2\pi i \frac{x}{2^n}}\right)}{\sqrt{2}},$$

where $\hat{f}$ denotes the Fourier transform $\hat{f}(x) = \int_{\mathbb{R}} f(t)e^{-2\pi i t x}\, dt$, $(x \in \mathbb{R})$, of the function f.

The scaling function satisfies the *scaling equation*, which in terms of the representation π can be written as

$$U\varphi = \pi(m_0)\varphi, \tag{14.7}$$

with the low-pass filter $m_0(z) = \sum_{k\in\mathbb{Z}} a_k z^k$, $(z \in \mathbb{T})$, which is the starting point for the construction of the multiresolution analysis. A *multiresolution* associated to φ is generated as a sequence of subspaces V_n, $n \in \mathbb{Z}$:

$$V_0 = \overline{\operatorname{span}}\{T^k\varphi \mid k \in \mathbb{Z}\} = \overline{\operatorname{span}}\{\pi(f)\varphi \mid f \in L^\infty(\mathbb{T})\},$$

$$V_n = U^{-n}V_0, \quad (n \in \mathbb{Z})$$

satisfying the scaling equation $V_n \subseteq V_{n+1}$ and

$$\overline{\bigcup_{n\in\mathbb{Z}} V_n} = L^2(\mathbb{R}). \tag{14.8}$$

With m_0 carefully chosen, one can obtain *orthonormal scaling function* φ, i.e., such that its translates are orthogonal $\langle T^k\varphi, T^l\varphi\rangle = \delta_{kl}$ for $k, l \in \mathbb{Z}$. Equivalently

$$\langle \pi(f)\varphi, \varphi\rangle = \int_{\mathbb{T}} f\, d\mu, \quad (f \in L^\infty(\mathbb{T})). \tag{14.9}$$

Given the orthonormal scaling function and the multiresolution, the wavelet is obtained by considering the *detail space* $W_0 := V_1 \ominus V_0$. Analyzing the multiplicity of the representation π on the spaces V_0 and V_1, one can see that there is a function ψ such that $\{T^k\psi \mid k \in \mathbb{Z}\}$ is an orthonormal basis for W_0. The set $\{U^n T^k\psi \mid n, k \in \mathbb{Z}\}$ is an orthonormal basis for $L^2(\mathbb{R})$, and thus ψ is a *wavelet.* Since one is aiming at scaling functions whose translates are orthogonal, a necessary condition on m_0 is the *quadrature mirror filter (QMF) condition* (14.6).

Wavelet representations were introduced in [24, 31, 49] in an attempt to apply the multiresolution techniques of wavelet theory [21] to a larger class of problems where self-similarity, or refinement is the central phenomenon. They were used to construct wavelet bases and multiresolutions on fractal measures and Cantor sets [29] or on solenoids [25]. Wavelet representations can be defined axiomatically as follows. Let X be a compact metric space and let $r : X \to X$ be a Borel measurable function which is onto and finite-to-one, i.e., $0 < \#r^{-1}(x) < \infty$ for all $x \in X$. Let μ be a *strongly invariant measure* on X, i.e.

$$\int_X f\,d\mu = \int_X \frac{1}{\#r^{-1}(x)} \sum_{r(y)=x} f(y)\,d\mu(x), \quad (f \in L^\infty(X)) \tag{14.10}$$

Let $m_0 \in L^\infty(X)$ be a *QMF filter*, i.e.,

$$\frac{1}{\#r^{-1}(x)} \sum_{r(y)=x} |m_0(y)|^2 = 1 \text{ for } \mu\text{-a.e. } x \in X \tag{14.11}$$

Theorem 1 ([31]). *There exists a Hilbert space $\mathscr{H}$, a unitary operator U on $\mathscr{H}$, a representation π of $L^\infty(X)$ on $\mathscr{H}$ and an element φ of $\mathscr{H}$ such that*

1. *(Covariance) $U\pi(f)U^{-1} = \pi(f \circ r)$ for all $f \in L^\infty(X)$.*
2. *(Scaling equation) $U\varphi = \pi(m_0)\varphi$*
3. *(Orthogonality) $\langle \pi(f)\varphi, \varphi \rangle = \int f\,d\mu$ for all $f \in L^\infty(X)$.*
4. *(Density) $\{U^{-n}\pi(f)\varphi \mid n \in \mathbb{N}, f \in L^\infty(X)\}$ is dense in $\mathscr{H}$.*

Moreover they are unique up to isomorphism.

Definition 2. The quadruple $(\mathscr{H}, U, \pi, \varphi)$ in Theorem 1 is called the *wavelet representation* associated to m_0.

The issues of reducibility, irreducibility and decompositions of wavelet representations are central for analysis and constructions of multiresolutions, wavelets, bases, corresponding harmonic analysis and their applications. The commutant of representation plays important role in these contexts. The paradigmatic classical wavelet representation π on $L^2(\mathbb{R})$, defined via Borel functional calculus by the dilation and translation operators described above, is associated to the map $r(z) = z^2$ on $\mathbb{T}$. The measure μ is the Haar measure on the circle, and m_0 can be any low-pass QMF filter which produces an orthogonal scaling function (see [21]). For example, the Haar filter $m_0(z) = (1 + z)/\sqrt{2}$ produces the Haar scaling function φ. This representation is reducible. The commutant was computed in [43] and the direct integral decomposition was presented in [57]. Some low-pass filters, such as the stretched Haar filter $m_0(z) = (1 + z^3)/\sqrt{2}$ give rise to non-orthogonal scaling functions. In this case super-wavelets appear, and the wavelet representation is realized on a direct sum of finitely many copies of $L^2(\mathbb{R})$. This representation is also reducible and its direct integral decomposition is similar to the one for $L^2(\mathbb{R})$ (see [9,25]). For the QMF filter $m_0 = 1$ the representation can be realized on a solenoid and in this case it is irreducible [25]. The result holds even for more general maps r, if they are ergodic (see [33]).

The general theory of the decomposition of wavelet representations into irreducible components was given in [25], but there is a large class of examples where it is not known whether these representations are irreducible or not.

The wavelet representation associated to the map $r(z) = z^3$ on the unit circle $\mathbb{T}$ with the Haar measure μ and the QMF filter $m_0(z) = (1 + z^2)/\sqrt{2}$ is strongly connected to the middle-third Cantor set $\mathbf{C}$ (see [30]). This representation is reducible [34]. A wavelet representation whose scaling function is the Sierpinski gasket is constructed in [20] by d'Andrea, Merrill and Packer. They also present some numerical experiments showing how this multiresolution behaves under the usual wavelet compression algorithm. In [8, 55, 56] the wavelet representations are given a more operator theoretic flavor. A groupoid approach is presented in [45]. General multiresolution theories are considered in [5–8].

14.2 Wavelet Representations, Solenoids and Symbolic Dynamics

Wavelet representations can be realized on the solenoid associated to the underlying dynamics [31], that is in terms of the symbolic dynamics of the orbit space. The *solenoid* associated to the map r is defined as the set of all inverse iteration paths (backward orbits) for the dynamical system generated by r:

$$X_\infty := \{(x_0, x_1, \ldots) \in X^{\mathbb{N}} \mid r(x_{n+1}) = x_n \text{ for all } n \geq 0\} \tag{14.12}$$

also sometimes being convenient to view as the forward orbits of the iterated function system generated by the pre-image maps (inverse maps) of r. Given the map $r : X \mapsto X$, the map $r_\infty : X_\infty \to X_\infty$ defined by

$$r_\infty(x_0, x_1, \ldots) = (r(x_0), x_0, x_1, \ldots) \text{ for all } (x_0, x_1, \ldots) \in X_\infty \tag{14.13}$$

is a measurable automorphism on X_∞ with respect to the σ-algebra generated by cylinder sets. Let $c(x) := \#r^{-1}(r(x))$ and $W(x) = |m_0(x)|^2/c(x)$ for all $x \in X$. Then

$$\sum_{r(y)=x} W(y) = 1, \quad (x \in X), \tag{14.14}$$

and $W(y)$ can be interpreted as the transition probability from x to one of the roots y of the equation $x = r(y)$ (pre-images of x under r). The path measure P_x on the fibers $\Omega_x := \{(x_0, x_1, \ldots) \in X_\infty \mid x_0 = x\}$ with $x \in X$, defined on cylinder sets for any $z_1, \ldots, z_n \in X$ by

$$P_x(\{(x_n)_{n\geq 0} \in \Omega_x \mid x_1 = z_1, \ldots, x_n = z_n\}) = W(z_1) \ldots W(z_n), \tag{14.15}$$

can be interpreted as the probability of the random walk to go from x to z_n through the points $x_1 = z_1, \dots, x_n = z_n$, and defines the measure μ_∞ on X_∞ via the condition

$$\int f \, d\mu_\infty = \int_X \int_{\Omega_x} f(x, x_1, \dots) \, dP_x(x, x_1, \dots) \, d\mu(x) \tag{14.16}$$

for bounded measurable functions on X_∞. Consider now the Hilbert space $\mathscr{H} := L^2(X_\infty, \mu_\infty)$. Let $\theta_m(x_0, x_1, \dots) = x_m$ for $m \geq 0$ be the projection map $\theta_m : X_\infty \to X$ onto the mth coordinate. Then in particular $\theta_0 : X_\infty \to X$ is the projection map $\theta_0(x_0, x_1, \dots) = x_0$ onto the initial 0th coordinate (x_0-coordinate) and the relation between the maps r, r_∞, θ_0 is described by the commutative diagram

$$\begin{array}{ccc} X_\infty & \stackrel{r_\infty}{\to} & X_\infty \\ \theta_0 \downarrow & & \downarrow \theta_0 \\ X & \stackrel{r}{\to} & X \end{array}, \quad \theta_0 \circ r_\infty = r \circ \theta_0$$

meaning that the projection θ_0 is an intertwining map for the maps r and r_∞. Define the operator

$$U\xi = (m_0 \circ \theta_0) \, \xi \circ r_\infty, \quad (\xi \in L^2(X_\infty, \mu_\infty)) \tag{14.17}$$

and the representation of $L^\infty(X)$ on $\mathscr{H}$

$$\pi(f)\xi = (f \circ \theta_0) \, \xi, \quad (f \in L^\infty(X), \xi \in L^2(X_\infty, \mu_\infty)) \tag{14.18}$$

and let $\varphi = 1$ be the constant function 1 on X_∞. If m_0 is non-singular, i.e., $\mu(\{x \in X \mid m_0(x) = 0\}) = 0$, then the data $(\mathscr{H}, U, \pi, \varphi)$ forms the wavelet representation associated to m_0 (see [31]).

14.3 Commutants and Reducibility of Wavelet Representations and Fixed Points of the Transfer Operators

Irreducibility and reducibility of wavelet representations as well as decomposition theorems (generalized spectral theorems) involve the study of the commutant of the representation. There are actually several equivalent ways to formulate the problem of reducibility or irreducibility of the wavelet representations yielding different approaches and insights. The commutant of the wavelet representations, i.e., the set of operators that commute with both the "dilation" operator U and the "translation" operators $\pi(f)$, has a simple description, and the operators in the commutant are in one-to-one correspondence with bounded fixed points of the transfer operator.

The commutant of the classical wavelet representation on $L^2(\mathbb{R})$ was computed in [19]. The commutant for other choices of filters, such as $m_0 = 1$ or for the wavelet representation associated to the Cantor set and connection to reducibility or irreducibility of wavelet representations have been considered for example in [33]. Some of the results pertaining to irreducibility and reducibility of the wavelet representations and commutant are presented in the next several theorems.

Theorem 3 ([31]). *Suppose m_0 is non-singular. Then there is a one-to-one correspondence between the following data:*

1. *Operators S in the commutant of $\{U, \pi\}$.*
2. *Cocycles, i.e., functions $f \in L^\infty(X_\infty, \mu_\infty)$ such that $f \circ r_\infty = f$, μ_∞-a.e.*
3. *Harmonic functions $h \in L^\infty(X)$ for the transfer operator R_{m_0}, i.e., $R_{m_0} h = h$, where*

$$R_{m_0} f(x) = \frac{1}{\# r^{-1}(x)} \sum_{r(y)=x} |m_0(y)|^2 f(y).$$

The correspondence $1 \leftrightarrow 2$ is given by $S = M_f$ where M_f is the multiplication operator $M_f \xi = f\xi$, $\xi \in L^2(X_\infty, \mu_\infty)$. The correspondence from 2 to 3 is given by

$$h(x) = \int_{\Omega_x} f(x, x_1, \ldots)\, dP_x(x, x_1, \ldots).$$

The correspondence from 3 to 2 is given by

$$f(x, x_1, \ldots) = \lim_{n \to \infty} h(x_n), \text{ for } \mu_\infty\text{-a.e. } (x, x_1, \ldots) \text{ in } X_\infty.$$

Using Theorem 3, the following criteria for irreducibility or reducibility of the wavelet representation has been obtained in [33].

Theorem 4 ([33]). *Suppose that m_0 is non-singular. The following affirmations are equivalent:*

1. *The wavelet representation is irreducible, i.e., the commutant $\{U, \pi\}'$ is trivial.*
2. *The automorphism r_∞ on (X_∞, μ_∞) is ergodic.*
3. *The only bounded measurable harmonic functions for the transfer operator R_{m_0} are the constants.*
4. *There are no non-constant fixed points of the transfer operator $h \in L^p(X, \mu)$, for some $p > 1$ with the property that*

$$\sup_{n \in \mathbb{N}} \int_X |m_0^{(n)}(x)|^2 |h(x)|^p \, d\mu(x) < \infty \tag{14.19}$$

where

$$m_0^{(n)}(x) = m_0(x)m_0(r(x))\dots m_0(r^{n-1}(x)), \quad (x \in X). \tag{14.20}$$

5. *If* $\varphi' \in L^2(X_\infty, \mu_\infty)$*, satisfies the same scaling equation as* φ*, i.e.,* $U\varphi' = \pi(m_0)\varphi'$*, then* φ' *is a constant multiple of* φ*.*

The following theorem, proved in [34], shows that under some mild assumptions the wavelet representations are reducible.

Theorem 5 ([34]). *Suppose* $r : (X, \mu) \to (X, \mu)$ *is ergodic. Assume* $|m_0|$ *is not constant* 1 μ*-a.e., is non-singular, i.e.,* $\mu(\{x \mid m_0(x) = 0\}) = 0$*, and* $\log|m_0|^2$ *is in* $L^1(X)$*. Then the wavelet representation* $(\mathcal{H}, U, \pi, \varphi)$ *is reducible.*

The proof in [34] uses Jensen's inequality, Birkhoff's ergodic theorem, Egorov's theorem and Borel-Cantelli's lemma. As an application of this theorem yields a solution of the problem posed by Judith Packer who formulated the following question: is the wavelet representation associated to the middle third Cantor set described in the introduction irreducible? The answer is that this representation is reducible [34]. Using this result about reducibility of the wavelet representation in combination with results from [33], one can get that there are non-trivial solutions to refinement equations and non-trivial fixed points for transfer operators for m_0 satisfying the conditions in Theorem 5. For m_0 as in Theorem 5 and the associated wavelet representation $(\mathcal{H}, U, \pi, \varphi)$, there exist solutions $\varphi' \in \mathcal{H}$ for the scaling equation $U\varphi' = \pi(m_0)\varphi'$ which are not constant multiples of φ, and there exist non-constant, bounded fixed points for the transfer operator (see [34])

$$R_{m_0} f(x) = \frac{1}{\#r^{-1}(x)} \sum_{r(y)=x} |m_0(y)|^2 f(y), \quad (f \in L^\infty(X), x \in X).$$

In the case $|m_0| = 1$ not covered by Theorem 5, the corresponding representation can be irreducible [35].

Theorem 6. *Let* $m_0 = 1$ *and let* $(L^2(X_\infty, \mu_\infty), U, \pi, \varphi)$ *be the associated wavelet representation. The following affirmations are equivalent:*

1. *The automorphism* r_∞ *on* (X_∞, μ_∞) *is ergodic.*
2. *The wavelet representation is irreducible.*
3. *The only bounded functions which are fixed points for the transfer operator* R_1*, i.e.,*

$$R_1 h(x) := \frac{1}{\#r^{-1}(x)} \sum_{r(y)=x} h(y) = h(x)$$

are the constant functions.

4. *The only $L^2(X,\mu)$-functions which are fixed points for the transfer operator R_1, are the constants.*
5. *The endomorphism r on (X,μ) is ergodic.*

The study of invariant spaces for the wavelet representation $\{U,\pi\}$ is equivalent to the study of the invariant sets for the dynamical system r_∞ on (X_∞,μ_∞). Since the operators in the commutant of $\{U,\pi\}$ are multiplication operators M_g, with $g \in L^\infty(X_\infty,\mu_\infty)$ and $g = g \circ r_\infty$ (see [31]), the orthogonal projection onto a subspace $\mathscr{K}$ which is invariant under U and $\pi(f)$ for all $f \in L^\infty(X)$, is an operator in the commutant and so it corresponds to a multiplication by a characteristic function χ_A, where A is an invariant set for r_∞, i.e., $A = r_\infty^{-1}(A) = r_\infty(A)$, μ_∞-a.e., and $\mathscr{K} = L^2(A,\mu_\infty)$. This can be used for example to show, that under the assumptions of Theorem 5, there are no finite-dimensional invariant subspaces for the wavelet representation (see [35] for the proof).

In [32] a decomposition problem has been investigated for a class of unitary representations associated with wavelet analysis, wavelet representations in a wide framework having applications to multi-scale expansions arising in dynamical systems theory for non-invertible endomorphisms. A direct integral decomposition for the general wavelet representation, and a solution of a question posed by Judith Packer have been obtained, a detailed analysis of the measures contributing to the decomposition into irreducible representations have been performed involving results for associated Martin boundaries, wavelet filters, random walks, as well as classes of harmonic functions. As described previously, with measures on the solenoid (X_∞, r_∞), built from (X,r) the map r_∞ induces unitary operators U on Hilbert space $\mathscr{H}$ and representations π of the algebra $L^\infty(X)$ such that the pair (U, r_∞), together with the corresponding representation π forms a crossed-product in the sense of C^*-algebras, and the traditional wavelet representations fall within this wider framework of $(\mathscr{H},U,\pi)$ covariant crossed products.

With $\tilde{m}_0 = 1$ and $\tilde{m}_n = (m_0 \circ \theta_0) \cdot (m_0 \circ \theta_0 \circ r_\infty) \ldots (m_0 \circ \theta_0 \circ r_\infty^{n-1})$ for $n \geq 1$, and

$$\tilde{m}_n = \frac{1}{(m_0 \circ \theta_0 \circ r_\infty^{-1}) \ldots (m_0 \circ \theta_0 \circ r_\infty^{n})}, \text{ for } n < 0,$$

the function $\tilde{m} : X_\infty \times \mathbb{Z} \to \mathbb{C}^*$ defined by $\tilde{m}(x,n) = \tilde{m}_n(x)$ gives a one-cocycle for the action of $\mathbb{Z}$ on X_∞ determined by r_∞, and U being an isometry yields $\int \xi \, d\mu_\infty = \int |\tilde{m}_n|^2 \xi \circ r_\infty^n \, d\mu_\infty$ for $n \in \mathbb{Z}$ and $\xi \in L^2(X_\infty,\mu_\infty)$. For $z = (z_0, z_1, \ldots)$ in X_∞ consider the Hilbert space $\mathscr{H}_z := \left\{(\xi_n)_{n\in\mathbb{Z}} : \sum_{n\in\mathbb{Z}} |\xi_n|^2 |\tilde{m}_n(z)|^2 < \infty\right\}$, with inner product $\langle \xi, \eta \rangle_{\mathscr{H}_z} := \sum_{n\in\mathbb{Z}} \xi_n \overline{\eta}_n |\tilde{m}_n(z)|^2$. Since m_0 is non-singular, the points $z \in X_\infty$, such that one of the functions $\tilde{m}_n(z) = 0$, form a set of μ_∞-measure zero.

Define the unitary operator

$$U_z(\xi_n)_{n\in\mathbb{Z}} = (m_0 \circ \theta_0 \circ r_\infty^n(z) \xi_{n+1})_{n\in\mathbb{Z}}$$

and the representation π of $L^\infty(X)$:

$$\pi_z(f)(\xi_n)_{n\in\mathbb{Z}} = (f \circ \theta_0 \circ r_\infty^n(z)\xi_n)_{n\in\mathbb{Z}}, \quad (f \in L^\infty(X)).$$

The representation π_z is defined for bounded functions on X, not just essentially bounded. The μ-measure zero sets will affect the individual representations π_z but not their direct integral (see below). For μ_∞ almost every $z \in X_\infty$, the triples $[\mathscr{H}_z, U_z, \pi_z]$ form an irreducible representation [32]. The proof for this fact demonstrates the importance and application of maximal commutativity of the canonical subalgebras in crossed products and significance of the periodic points of the dynamics in this context. Further on, in this work, the maximal commutativity of such canonical subalgebras for crossed product type algebras and C^*-algebras and the interplay with properties of the dynamics connected to periodic and aperiodic points will be addressed again. Meanwhile, returning to the proof, it is a matter of simple computations to check that U_z is unitary, π_z is a representation and that the crossed product type covariance commutation relations $U_z\pi_z(f)U_z^{-1} = \pi_z(f \circ r)$ hold for all $f \in L^\infty(X)$. To see that the representation is irreducible for μ_∞-a.e. z, take z to be non-periodic, i.e., $r_\infty^n(z) \neq z$ for all $n \neq 0$. Then $\{\pi_z(f) : f \in L^\infty(X)\}$ forms a maximal abelian subalgebra with cyclic vector δ_0 (see [85, Corollary III.1.3]), where $\delta_0(n) = 1$ for $n = 0$, and $\delta_0(n) = 0$ otherwise. Then, an operator A that commutes with U_z and π_z has to be of the form $\pi_z(g)$ for some $g \in L^\infty(X)$. Since A commutes with U_z we have $\pi_z(g \circ r) = U_z\pi_z(g)U_z^{-1} = \pi_z(g)$. This implies that g is constant on $\{r_\infty^n(z) : n \in \mathbb{Z}\}$, so A is a multiple of the identity.

A subset $\mathscr{F}$ of X_∞ is called a *fundamental domain* if, up to μ_∞-measure zero:

$$\bigcup_{n\in\mathbb{Z}} r_\infty^n(\mathscr{F}) = X_\infty \quad \text{and} \quad r_\infty^n(\mathscr{F}) \cap r_\infty^m(\mathscr{F}) = \emptyset \text{ for } n \neq m.$$

For any dynamical system or action the question of existence and then construction of fundamental domains are of fundamental importance. The next general theorem, states the existence of such fundamental domain and provides a direct integral decomposition for general wavelet representations into irreducibles in as clean form as is realistically feasible, in particular completely solving a question posed by Judith Packer, see e.g., [5–8, 73].

Theorem 7 ([32]). *In the hypotheses of Theorem 5, there exists a fundamental domain $\mathscr{F}$. The wavelet representation associated to m_0 has the following direct integral decomposition:*

$$[\mathscr{H}, U, \pi] = \int_{\mathscr{F}}^{\oplus} [\mathscr{H}_z, U_z, \pi_z]\, d\mu_\infty(z),$$

where the component representations $[\mathscr{H}_z, U_z, \pi_z]$ in the decomposition are irreducible for a.e., z in $\mathscr{F}$, relative to μ_∞.

In [32], the measures in the decomposition were studied further using p-harmonic functions, Green function or potential functions, trees and sub-trees in the orbit spaces of the non-invertible dynamics, regular and periodic and aperiodic points, transition probabilities and not reversible transition processes and random walks, Martin boundaries, Martin compactification and Martin kernels. The proof of the theorem in [32] is rather long and elaborate and thus is beyond of the scope of this review. What can be however mentioned here about that proof is that it in particular indicates one possible general way to construct the fundamental domains, but this way is not very practical and the fundamental domains obtained in such a way are typically not the most easily describable and not the most convenient for further analysis and computations. Thus the problem of constructing better and easier to handle fundamental domains for wavelet representations is open and is important for gaining further insight into the structure and properties of the corresponding wavelet representations and wavelet bases.

14.4 Maximal Commutativity of Subalgebras, Irreducibility of Representations and Freeness and Minimality of Dynamical Systems

As have been demonstrated in the previous section the property of the maximal commutativity (maximal abelianess) of the canonical commutative subalgebra in the crossed products and their representations play pivotal role in proving that wavelet representations, or in general representations of covariance relations and of crossed product algebras associated to dynamical systems, are irreducible under appropriate conditions on the dynamics closely concerned with the periodicity and aperiodicity in the orbit space of the dynamics [32]. Such properties as ergodicity, minimality or freeness of the dynamics (action) therefore are highly relevant in this context. In [2, 36, 38, 53, 54, 76, 80, 86, 87, 90], it was observed that the property of topological freeness of the dynamics for a homeomorphism, or for more general actions of groups (i.e., reversible dynamics), is equivalent or closely linked to the position of the algebra of continuous functions inside the crossed product C^*-algebra, namely with whether it is a maximal abelian subalgebra or not. Moreover, in these pioneering works the property of topological freeness of the dynamics for a homeomorphism, or for more general actions of groups (i.e., reversible dynamics), has been shown also closely linked with the structure of the ideals in the corresponding crossed product C^*-algebra and in particular with the existence of non-zero intersections between ideals and the algebra of continuous functions embedded as a C^*-subalgebra into the crossed product C^*-algebra. This interplay has been considered both for the universal crossed product C^*-algebra and for the reduced crossed product C^*-algebra, the later providing one of the important insights into the significance of those properties for representations of the crossed product. In one of the novel recent developments, envisioned by the present author, it has been noticed that for reversible dynamical systems and crossed product algebras [81–83] this important interplay holds and can be applied far beyond context of C^*-algebra crossed products, in algebraic, in Banach algebraic and in other contexts possibly with suitable modification of the corresponding properties of the involved dynamics and of subsets of the investigated crossed product. In particular, in these works, first steps were made into approaching this interplay for crossed products associated with non-free dynamics, via studying in detail the relevant commutant of the canonical commutative subalgebra when it is not maximal commutative as well as subalgebras of the relative commutant and properties of dual of the commutative subalgebras in the crossed product algebras and the associated dynamics on the spectra. It has been observed first in these works that the relative commutant has the remarkable intersection property with non-zero two-sided ideals, i.e., it intersects any two-sided ideal non-trivially for any non-invertible dynamics. In a series of follow up works to [81–83], this novel approach, results and ideas have been further explored and substantially expanded, deepened and applied in various directions, for classical crossed product C^*-algebras and Banach $*$-algebras associated to invertible topological dynamics in [46, 47, 84], and for various generalizations of crossed product algebras (strongly graded rings, crystalline graded rings, crossed product type algebras, categorial crossed product, Ore extension rings, etc.) in [62–71].

The author feels that one result from this fast developing direction deserves especially attention of the readers in the context of this review. This is a result (Theorem 10) which extends the classical motivating Theorem 8 about interplay between maximal commutativity, intersection property with two-sided ideals and topological freeness of the dynamical system, from crossed product C^*-algebras associated with actions by $\mathbb{Z}$ of homeomorphisms on topological spaces to crossed product C^*-algebras by semigroup actions of the topological dynamical systems generated by covering maps on topological spaces (a broad class containing many non-invertible maps). Moreover, the connection to certain properties of representations of such generalized crossed products is also introduced in this extended result, showing again clearly the importance of maximal commutativity for investigation of representations of generalized crossed product algebras by non-invertible actions defined using forward action and the transfer operators. The Theorem 10 furthermore implies that in the context of non-invertible maps and associated to them transfer operators the freeness properties of the dynamics as well as the intersections properties of the canonical subalgebra with two-sided ideals are highly relevant to investigation of representations. To present this result in proper historic context, we start by presenting the classic motivating pivotal result for crossed product C^*-algebras by $\mathbb{Z}$ associated to homeomorphism on topological spaces, established in its different parts in [2, 38, 53, 54, 86, 87, 90], and presented in the following clear and convenient formulation first in [87, Theorem 5.4].

Theorem 8. *The following three properties are equivalent for a compact Hausdorff space X and a homeomorphism σ of X:*

1. *The non-periodic points of* (X, σ) *are dense in* X;
2. *Any non-zero closed ideal* I *in the crossed product* C^**-algebra* $C(X) \rtimes_\alpha \mathbb{Z}$ *satisfies* $I \cap C(X) \neq \{0\}$;
3. $C(X)$ *is a maximal abelian* C^**-subalgebra of* $C(X) \rtimes_\alpha \mathbb{Z}$.

Let X be a compact Hausdorff space and let $T : X \to X$ be a covering map, i.e., T is continuous and surjective and there exists for every $x \in X$ an open neighborhood V of x such that $T^{-1}(V)$ is a disjoint union of open sets $(U_\alpha)_{\alpha \in I}$ satisfying that T restricted to each U_α is a homeomorphism from U_α onto V. Let α, L and $\mathscr{L}$ be the maps from $C(X)$ to $C(X)$ given by

$$\alpha(f) = f \circ T,$$

$$\mathrm{L}(f)(x) = \sum_{y \in T^{-1}(x)} f(y),$$

$$\mathscr{L}(f) = \mathrm{L}(1_X)^{-1}\mathrm{L}(f),$$

These are well defined maps of $C(X)$ into $C(X)$ (see [41]). The operator $\mathscr{L}$ is a transfer operator for α. Denote $\alpha(\mathrm{L}(1_X))$ by $\mathrm{ind}(E)$ and for every $k \geq 1$ let $I_k = \mathrm{ind}(E)\alpha(\mathrm{ind}(E)) \cdots \alpha^{k-1}(\mathrm{ind}(E))$. Since $\mathscr{L}$ is a transfer operator for α, one can associate the C^*-algebra $C(X) \rtimes_{\alpha,\mathscr{L}} \mathbb{N}$ to the dynamical system (X, T), where $C(X) \rtimes_{\alpha,\mathscr{L}} \mathbb{N}$ is the crossed-product C^*-algebra associated to the triple $(C(X), \alpha, \mathscr{L})$ according to [40]. In [41] this crossed product C^*-algebra has been characterized as a universal C^*-algebra generated by a copy of $C(X)$ and an isometry s subject to certain relations. Since T is a covering map there exists a finite open covering $\{V_i\}_{i=1}^{l}$ of X such that the restriction of T to each V_i is injective. Let $\{v_i\}_{i=1}^{l}$ be a partition of unit subordinate to $\{V_i\}_{i=1}^{l}$ and let $u_i = (\alpha(\mathrm{L}(1_X))v_i)^{1/2}$.

Theorem 9 ([41, Theorem 9.2]). *The* C^**-algebra* $C(X) \rtimes_{\alpha,\mathscr{L}} \mathbb{N}$ *is the universal* C^**-algebra generated by a copy of* $C(X)$ *and an isometry* s *subject to the relations*

1. $sf = \alpha(f)s$,
2. $s^* fs = \mathscr{L}(f)$,
3. $1 = \sum_{i=1}^{l} u_i ss^* u_i$,

for all $f \in C(X)$.

The following representation turns out to play important role in the context of the Theorem 10. For a compact Hausdorff space X and a covering map $T : X \to X$, let H be a Hilbert space with an orthonormal basis $(e_x)_{x \in X}$ indexed by X. For $f \in C(X)$, define the bounded operators M_f and S on H by

$$M_f(e_x) = f(x)e_x, \quad x \in X,$$

$$S(e_x) = (\mathrm{L}(1_X)(x))^{-1/2} \sum_{y \in T^{-1}(\{x\})} e_y, \quad x \in X.$$

It can be shown [15], that there exists a representation ψ of $C(X) \rtimes_{\alpha,\mathcal{L}} \mathbb{N}$ on H such that $\psi(f) = M_f$ for every $f \in C(X)$ and $\psi(s) = S$, and furthermore, $\ker(\psi) \cap C(X) = \{0\}$. This intersection property of the kernel of the representation makes clear the relevance of this representation to the intersection properties of the ideals with the canonical subalgebra $C(X)$ and thus its important appearance in Theorem 10. The notion of topological freeness for dynamical systems generated by a homeomorphism can in a natural way be extended to possibly non-invertible dynamical systems [41]. The dynamical system (X, T) is said to be *topological free* if for every pair of nonnegative integers (k, l) with $k \neq l$, the set $\{x \in X \mid T^k(x) = T^l(x)\}$ has empty interior. The following Theorem 10 is the promised extension Theorem 8 to possibly non-invertible dynamical systems generated by covering maps on compact Hausdorff spaces and to the corresponding crossed product C^*-algebras $C(X) \rtimes_{\alpha,\mathcal{L}} \mathbb{N}$.

Theorem 10 ([15]). *Let X be a compact Hausdorff space, and let $T : X \to X$ be a covering map. Then the following are equivalent:*

1. *(X, T) is topological free.*
2. *Every nontrivial ideal of $C(X) \rtimes_{\alpha,\mathcal{L}} \mathbb{N}$ has a nontrivial intersection with $C(X)$.*
3. *The representation ψ is faithful.*
4. *$C(X)$ is a maximal abelian C^*-subalgebra of $C(X) \rtimes_{\alpha,\mathcal{L}} \mathbb{N}$.*

Note, that in comparison to Theorem 8, in Theorem 10 there is added a fourth equivalent condition of faithfulness of the representation ψ of $C(X) \rtimes_{\alpha,\mathcal{L}} \mathbb{N}$. We refer the reader to [15] for the details on the definition of this representation. If the space X is infinite, and we consider dynamical systems generated by covering maps, then the class of topologically free systems contains the subclass of irreducible dynamical systems, defined as follows (see [41, Proposition 11.1]). Two points $x, y \in X$ are said to be *trajectory-equivalent* $x \sim y$ (see e.g. [4]) when there are $n, m \in \mathbb{N}$ such that $T^n(x) = T^m(y)$. A subset $Y \subseteq X$ is said to be invariant if $x \sim y \in Y$ implies that $x \in Y$. It is easy to see that Y is invariant if and only if $T^{-1}(Y) = Y$. The covering map T and the dynamical system it generates is said to be *irreducible* when there is no closed (equivalently open) invariant set other than $\emptyset$ and X (see e.g. [4]). Notice that irreducibility is weaker than the condition of minimality defined in [23]. In [41], it was shown that, for dynamical systems generated by covering maps of infinite spaces, irreducibility of the system is equivalent to simplicity of $C(X) \rtimes_{\alpha,\mathcal{L}} \mathbb{N}$. Equivalence of simplicity of crossed product C^*-algebras and minimality for homeomorphism dynamics is a classic result [75]. The most easy and neat conceptually proof of this result known to the author is via specialization of Theorem 8 to minimal dynamical systems. The situation is similar with Theorem 10 and proofs of the described above simplicity criterium for $C(X) \rtimes_{\alpha,\mathcal{L}} \mathbb{N}$. In this sense, Theorem 10 can be also viewed as the result on not necessarily simple C^*-algebras. From the point of view of the problem of description or classification of ideals in non-simple C^*-algebras, Theorem 10 provides explicit conditions on the dynamics, or conditions on the canonical commutative subalgebra which guaranty that it intersects any ideal in a non-empty way, thus providing each ideal with a non-empty ideal of the canonical commutative subalgebra which can be in its turn used for generating and describing properties and elements in the ideal in the crossed product. This correspondence, when it exists, is very fruitful for explicit investigation of dynamical, topological or geometrical structure of ideals and also for investigations of representations and their kernels for the corresponding crossed product algebras. Theorem 10 also answers the question of when there exists ideals without the intersection property. Namely, this happens exactly when the dynamical system generated by the covering map is not topological free, or equivalently when the canonical commutative subalgebra is not maximal commutative. For non-free dynamical systems investigation of ideals as well as extensions of the parts of Theorem 10 via description of the commutants and subalgebras of the commutants and intersection properties of the ideals with the commutants and their subalgebras is a very interesting open problem. Extension of such results, and actually in the first place extensions of Theorem 10 to Banach and normed algebras, and interplay with wavelet analysis via properties of wavelet representations and multiresolutions and detailed spectral analysis of transfer operators and harmonic functions is an open direction of high interest. The results and examples in [32–35] on the commutants of the wavelet representations on fractal sets and solenoids associated to non-invertible dynamics can be viewed as contributions in this direction.

Acknowledgements This research was supported in part by the Swedish Research Council (621-2007-6338), Swedish Foundation for International Cooperation in Research and Higher Education (STINT), Swedish Royal Academy of Sciences, The Royal Physiographic Society in Lund, Crafoord Foundation and the Nordforsk network "Operator algebra and dynamics" (grant # 11580).

References

1. Adji, S., Laca, M., Nilsen, M., Raeburn, I.: Crossed products by semigroups of endomorphisms and the Toeplitz algebras of ordered groups. Proc. Amer. Math. Soc. **122**(4), 1133–1141 (1994). DOI 10.2307/2161182. URL http://dx.doi.org/10.2307/2161182
2. Archbold, R.J., Spielberg, J.S.: Topologically free actions and ideals in discrete C^*-dynamical systems. Proc. Edinburgh Math. Soc. (2) **37**(1), 119–124 (1994). DOI 10.1017/S0013091500018733. URL http://dx.doi.org/10.1017/S0013091500018733
3. Arzumanian, V.A., Vershik, A.M.: Star algebras associated with endomorphisms. In: Operator algebras and group representations, Vol. I (Neptun, 1980), *Monogr. Stud. Math.*, vol. 17, pp. 17–27. Pitman, Boston, MA (1984)
4. Arzumanjan, V.A., Veršik, A.M.: Factor representations of the crossed product of a commutative C^*-algebra with its endomorphism semigroup. Dokl. Akad. Nauk SSSR **238**(3), 513–516 (1978)
5. Baggett, L.W., Furst, V., Merrill, K.D., Packer, J.A.: Generalized filters, the low-pass condition, and connections to multiresolution analyses. J. Funct. Anal. **257**(9), 2760–2779 (2009). DOI 10.1016/j.jfa.2009.05.004. URL http://dx.doi.org/10.1016/j.jfa.2009.05.004
6. Baggett, L.W., Furst, V., Merrill, K.D., Packer, J.A.: Classification of generalized multiresolution analyses. J. Funct. Anal. **258**(12), 4210–4228 (2010). DOI 10.1016/j.jfa.2009.12.001. URL http://dx.doi.org/10.1016/j.jfa.2009.12.001

7. Baggett, L.W., Larsen, N.S., Merrill, K.D., Packer, J.A., Raeburn, I.: Generalized multiresolution analyses with given multiplicity functions. J. Fourier Anal. Appl. **15**(5), 616–633 (2009). DOI 10.1007/s00041-008-9031-3. URL http://dx.doi.org/10.1007/s00041-008-9031-3
8. Baggett, L.W., Larsen, N.S., Packer, J.A., Raeburn, I., Ramsay, A.: Direct limits, multiresolution analyses, and wavelets. J. Funct. Anal. **258**(8), 2714–2738 (2010). DOI 10.1016/j.jfa.2009.08.011. URL http://dx.doi.org/10.1016/j.jfa.2009.08.011
9. Bildea, S., Dutkay, D.E., Picioroaga, G.: MRA super-wavelets. New York J. Math. **11**, 1–19 (2005). URL http://nyjm.albany.edu:8000/j/2005/11_1.html
10. Bratteli, O., Evans, D.E., Jorgensen, P.E.T.: Compactly supported wavelets and representations of the Cuntz relations. Appl. Comput. Harmon. Anal. **8**(2), 166–196 (2000). DOI 10.1006/acha.2000.0283. URL http://dx.doi.org/10.1006/acha.2000.0283
11. Bratteli, O., Jorgensen, P.: Wavelets through a looking glass. The world of the spectrum. Applied and Numerical Harmonic Analysis. Birkhäuser Boston Inc., Boston, MA (2002)
12. Bratteli, O., Jorgensen, P.E.T.: Iterated function systems and permutation representations of the Cuntz algebra. Mem. Amer. Math. Soc. **139**(663), x+89 (1999)
13. Brenken, B., Jorgensen, P.E.T.: A family of dilation crossed product algebras. J. Operator Theory **25**(2), 299–308 (1991)
14. Carlsen, T.M., Silvestrov, S.: C^*-crossed products and shift spaces. Expo. Math. **25**(4), 275–307 (2007). DOI 10.1016/j.exmath.2007.02.004. URL http://dx.doi.org/10.1016/j.exmath.2007.02.004
15. Carlsen, T.M., Silvestrov, S.: On the Exel crossed product of topological covering maps. Acta Appl. Math. **108**(3), 573–583 (2009). DOI 10.1007/s10440-008-9372-6. URL http://dx.doi.org/10.1007/s10440-008-9372-6
16. Carlsen, T.M., Silvestrov, S.: On the K-theory of the C^*-algebra associated with a one-sided shift space. Proc. Est. Acad. Sci. **59**(4), 272–279 (2010). DOI 10.3176/proc.2010.4.04. URL http://dx.doi.org/10.3176/proc.2010.4.04
17. Connes, A.: Noncommutative geometry. Academic Press Inc., San Diego, CA (1994)
18. Cuntz, J., Krieger, W.: A class of C^*-algebras and topological Markov chains. Invent. Math. **56**(3), 251–268 (1980). DOI 10.1007/BF01390048. URL http://dx.doi.org/10.1007/BF01390048
19. Dai, X., Larson, D.R.: Wandering vectors for unitary systems and orthogonal wavelets. Mem. Amer. Math. Soc. **134**(640), viii+68 (1998)
20. D'Andrea, J., Merrill, K.D., Packer, J.: Fractal wavelets of Dutkay-Jorgensen type for the Sierpinski gasket space. In: Frames and operator theory in analysis and signal processing, *Contemp. Math.*, vol. 451, pp. 69–88. Amer. Math. Soc., Providence, RI (2008). DOI 10.1090/conm/451/08758. URL http://dx.doi.org/10.1090/conm/451/08758
21. Daubechies, I.: Ten lectures on wavelets, *CBMS-NSF Regional Conference Series in Applied Mathematics*, vol. 61. Society for Industrial and Applied Mathematics (SIAM), Philadelphia, PA (1992). DOI 10.1137/1.9781611970104. URL http://dx.doi.org/10.1137/1.9781611970104
22. Davidson, K.R.: C^*-algebras by example, *Fields Institute Monographs*, vol. 6. American Mathematical Society, Providence, RI (1996)
23. Deaconu, V.: Groupoids associated with endomorphisms. Trans. Amer. Math. Soc. **347**(5), 1779–1786 (1995). DOI 10.2307/2154972. URL http://dx.doi.org/10.2307/2154972
24. Dutkay, D.E.: Harmonic analysis of signed Ruelle transfer operators. J. Math. Anal. Appl. **273**(2), 590–617 (2002). DOI 10.1016/S0022-247X(02)00284-6. URL http://dx.doi.org/10.1016/S0022-247X(02)00284-6
25. Dutkay, D.E.: Low-pass filters and representations of the Baumslag Solitar group. Trans. Amer. Math. Soc. **358**(12), 5271–5291 (electronic) (2006). DOI 10.1090/S0002-9947-06-04230-9. URL http://dx.doi.org/10.1090/S0002-9947-06-04230-9
26. Dutkay, D.E., Jorgensen, P.E.T.: Hilbert spaces of martingales supporting certain substitution-dynamical systems. Conform. Geom. Dyn. **9**, 24–45 (electronic) (2005). DOI 10.1090/S1088-4173-05-00135-9. URL http://dx.doi.org/10.1090/S1088-4173-05-00135-9

27. Dutkay, D.E., Jorgensen, P.E.T.: Wavelet constructions in non-linear dynamics. Electron. Res. Announc. Amer. Math. Soc. **11**, 21–33 (2005). DOI 10.1090/S1079-6762-05-00143-5. URL http://dx.doi.org/10.1090/S1079-6762-05-00143-5
28. Dutkay, D.E., Jorgensen, P.E.T.: Methods from multiscale theory and wavelets applied to nonlinear dynamics. In: Wavelets, multiscale systems and hypercomplex analysis, *Oper. Theory Adv. Appl.*, vol. 167, pp. 87–126. Birkhäuser, Basel (2006). DOI 10.1007/3-7643-7588-4_4. URL http://dx.doi.org/10.1007/3-7643-7588-4_4
29. Dutkay, D.E., Jorgensen, P.E.T.: Wavelets on fractals. Rev. Mat. Iberoam. **22**(1), 131–180 (2006). DOI 10.4171/RMI/452. URL http://dx.doi.org/10.4171/RMI/452
30. Dutkay, D.E., Jorgensen, P.E.T.: Fourier frequencies in affine iterated function systems. J. Funct. Anal. **247**(1), 110–137 (2007). DOI 10.1016/j.jfa.2007.03.002. URL http://dx.doi.org/10.1016/j.jfa.2007.03.002
31. Dutkay, D.E., Jorgensen, P.E.T.: Martingales, endomorphisms, and covariant systems of operators in Hilbert space. J. Operator Theory **58**(2), 269–310 (2007)
32. Dutkay, D.E., Jorgensen, P.E.T., Silvestrov, S.: Decomposition of wavelet representations and Martin boundaries. J. Funct. Anal. **262**(3), 1043–1061 (2012). DOI 10.1016/j.jfa.2011.10.010. URL http://dx.doi.org/10.1016/j.jfa.2011.10.010
33. Dutkay, D.E., Larson, D.R., Silvestrov, S.: Irreducible wavelet representations and ergodic automorphisms on solenoids. Oper. Matrices **5**(2), 201–219 (2011). DOI 10.7153/oam-05-14. URL http://dx.doi.org/10.7153/oam-05-14
34. Dutkay, D.E., Silvestrov, S.: Reducibility of the wavelet representation associated to the Cantor set. Proc. Amer. Math. Soc. **139**(10), 3657–3664 (2011). DOI 10.1090/S0002-9939-2011-10913-4. URL http://dx.doi.org/10.1090/S0002-9939-2011-10913-4
35. Dutkay, D.E., Silvestrov, S.: Wavelet representations and their commutant. In: K. Åström, L.E. Persson, S.D. Silvestrov (eds.) Analysis for science, engineering and beyond: the tribute workshop in honour of Gunnar Sparr held in Lund, May 8–9, 2008, *Springer proceedings in Mathematics*, vol. 6, pp. 253–265. Springer, Heidelberg (2012)
36. Effros, E.G., Hahn, F.: Locally compact transformation groups and C^*- algebras. Memoirs of the American Mathematical Society, No. 75. American Mathematical Society, Providence, R.I. (1967)
37. Eilers, S.: C^*-algebras associated to dynamical systems. Discrete Contin. Dyn. Syst. **15**(1), 177–192 (2006). DOI 10.3934/dcds.2006.15.177. URL http://dx.doi.org/10.3934/dcds.2006.15.177
38. Elliott, G.A.: Some simple C^*-algebras constructed as crossed products with discrete outer automorphism groups. Publ. Res. Inst. Math. Sci. **16**(1), 299–311 (1980). DOI 10.2977/prims/1195187509. URL http://dx.doi.org/10.2977/prims/1195187509
39. Exel, R.: Crossed-products by finite index endomorphisms and KMS states. J. Funct. Anal. **199**(1), 153–188 (2003). DOI 10.1016/S0022-1236(02)00023-X. URL http://dx.doi.org/10.1016/S0022-1236(02)00023-X
40. Exel, R.: A new look at the crossed-product of a C^*-algebra by an endomorphism. Ergodic Theory Dynam. Systems **23**(6), 1733–1750 (2003). DOI 10.1017/S0143385702001797. URL http://dx.doi.org/10.1017/S0143385702001797
41. Exel, R., Vershik, A.: C^*-algebras of irreversible dynamical systems. Canad. J. Math. **58**(1), 39–63 (2006). DOI 10.4153/CJM-2006-003-x. URL http://dx.doi.org/10.4153/CJM-2006-003-x
42. Furman, A.: Orbit equivalence rigidity. Ann. of Math. (2) **150**(3), 1083–1108 (1999). DOI 10.2307/121063. URL http://dx.doi.org/10.2307/121063
43. Han, D., Larson, D.R.: Frames, bases and group representations. Mem. Amer. Math. Soc. **147**(697), x+94 (2000)
44. an Huef, A., Raeburn, I.: The ideal structure of Cuntz-Krieger algebras. Ergodic Theory Dynam. Systems **17**(3), 611–624 (1997). DOI 10.1017/S0143385797079200. URL http://dx.doi.org/10.1017/S0143385797079200

45. Ionescu, M., Muhly, P.S.: Groupoid methods in wavelet analysis. In: Group representations, ergodic theory, and mathematical physics: a tribute to George W. Mackey, *Contemp. Math.*, vol. 449, pp. 193–208. Amer. Math. Soc., Providence, RI (2008). DOI 10.1090/conm/449/08713. URL http://dx.doi.org/10.1090/conm/449/08713
46. de Jeu, M., Svensson, C., Tomiyama, J.: On the Banach $*$-algebra crossed product associated with a topological dynamical system. J. Funct. Anal. **262**(11), 4746–4765 (2012). DOI 10.1016/j.jfa.2012.03.013. URL http://dx.doi.org/10.1016/j.jfa.2012.03.013
47. de Jeu, M., Tomiyama, J.: Maximal abelian subalgebras and projections in two Banach algebras associated with a topological dynamical system. Studia Math. **208**(1), 47–75 (2012). DOI 10.4064/sm208-1-4. URL http://dx.doi.org/10.4064/sm208-1-4
48. Jorgensen, P.E.T.: Operators and representation theory. Canonical models for algebras of operators arising in quantum mechanics., *North-Holland Mathematics Studies*, vol. 147. North-Holland Publishing Co., Amsterdam (1988). Notas de Matemática [Mathematical Notes], 120
49. Jorgensen, P.E.T.: Ruelle operators: functions which are harmonic with respect to a transfer operator. Mem. Amer. Math. Soc. **152**(720), viii+60 (2001)
50. Jorgensen, P.E.T.: Analysis and probability: wavelets, signals, fractals, *Graduate Texts in Mathematics*, vol. 234. Springer, New York (2006)
51. Kajiwara, T., Watatani, Y.: C^*-algebras associated with complex dynamical systems. Indiana Univ. Math. J. **54**(3), 755–778 (2005). DOI 10.1512/iumj.2005.54.2530. URL http://dx.doi.org/10.1512/iumj.2005.54.2530
52. Kawamura, S.: Covariant representations associated with chaotic dynamical systems. Tokyo J. Math. **20**(1), 205–217 (1997). DOI 10.3836/tjm/1270042409. URL http://dx.doi.org/10.3836/tjm/1270042409
53. Kawamura, S., Tomiyama, J.: Properties of topological dynamical systems and corresponding C^*-algebras. Tokyo J. Math. **13**(2), 251–257 (1990). DOI 10.3836/tjm/1270132260. URL http://dx.doi.org/10.3836/tjm/1270132260
54. Kishimoto, A.: Outer automorphisms and reduced crossed products of simple C^*-algebras. Comm. Math. Phys. **81**(3), 429–435 (1981). URL http://projecteuclid.org/getRecord?id=euclid.cmp/1103920327
55. Larsen, N.S., Raeburn, I.: From filters to wavelets via direct limits. In: Operator theory, operator algebras, and applications, *Contemp. Math.*, vol. 414, pp. 35–40. Amer. Math. Soc., Providence, RI (2006). DOI 10.1090/conm/414/07797. URL http://dx.doi.org/10.1090/conm/414/07797
56. Larsen, N.S., Raeburn, I.: Projective multi-resolution analyses arising from direct limits of Hilbert modules. Math. Scand. **100**(2), 317–360 (2007)
57. Lim, L.H., Packer, J.A., Taylor, K.F.: A direct integral decomposition of the wavelet representation. Proc. Amer. Math. Soc. **129**(10), 3057–3067 (electronic) (2001). DOI 10.1090/S0002-9939-01-05928-7. URL http://dx.doi.org/10.1090/S0002-9939-01-05928-7
58. Mackey, G.W.: Induced representations of groups and quantum mechanics. W. A. Benjamin, Inc., New York-Amsterdam (1968)
59. Mackey, G.W.: The theory of unitary group representations. University of Chicago Press, Chicago, Ill. (1976). Based on notes by James M. G. Fell and David B. Lowdenslager of lectures given at the University of Chicago, Chicago, Ill., 1955, Chicago Lectures in Mathematics
60. Mackey, G.W.: Unitary group representations in physics, probability, and number theory, second edn. Advanced Book Classics. Addison-Wesley Publishing Company Advanced Book Program, Redwood City, CA (1989)
61. Matsumoto, K.: On C^*-algebras associated with subshifts. Internat. J. Math. **8**(3), 357–374 (1997). DOI 10.1142/S0129167X97000172. URL http://dx.doi.org/10.1142/S0129167X97000172
62. Öinert, J.: Simple group graded rings and maximal commutativity. In: Operator structures and dynamical systems, *Contemp. Math.*, vol. 503, pp. 159–175. Amer. Math. Soc., Providence, RI (2009). DOI 10.1090/conm/503/09899. URL http://dx.doi.org/10.1090/conm/503/09899
63. Öinert, J.: Simplicity of skew group rings of abelian groups (2011), arXiv:1111.7214v2

64. Öinert, J., Lundström, P.: Commutativity and ideals in category crossed products. Proc. Est. Acad. Sci. **59**(4), 338–346 (2010). DOI 10.3176/proc.2010.4.13. URL http://dx.doi.org/10.3176/proc.2010.4.13
65. Öinert, J., Lundström, P.: The ideal intersection property for groupoid graded rings. Comm. Algebra **40**(5), 1860–1871 (2012). DOI 10.1080/00927872.2011.559181. URL http://dx.doi.org/10.1080/00927872.2011.559181
66. Öinert, J., Lundström, P.: Miyashita action in strongly groupoid graded rings. Int. Electron. J. Algebra **11**, 46–63 (2012)
67. Öinert, J., Richter, J., Silvestrov, S.: Maximal commutative subrings and simplicity of Ore extensions. J. Algebra Appl. **12**(04), 1250,192 (2013). DOI 10.1142/S0219498812501927
68. Öinert, J., Silvestrov, S., Theohari-Apostolidi, T., Vavatsoulas, H.: Commutativity and ideals in strongly graded rings. Acta Appl. Math. **108**(3), 585–602 (2009). DOI 10.1007/s10440-009-9435-3. URL http://dx.doi.org/10.1007/s10440-009-9435-3
69. Öinert, J., Silvestrov, S.D.: Commutativity and ideals in algebraic crossed products. J. Gen. Lie Theory Appl. **2**(4), 287–302 (2008). DOI 10.4303/jglta/S070404. URL http://dx.doi.org/10.4303/jglta/S070404
70. Öinert, J., Silvestrov, S.D.: On a correspondence between ideals and commutativity in algebraic crossed products. J. Gen. Lie Theory Appl. **2**(3), 216–220 (2008)
71. Öinert, J., Silvestrov, S.D.: Crossed product-like and pre-crystalline graded rings. In: Generalized Lie theory in mathematics, physics and beyond, pp. 281–296. Springer, Berlin (2009). DOI 10.1007/978-3-540-85332-9_24. URL http://dx.doi.org/10.1007/978-3-540-85332-9_24
72. Ostrovskyi, V., Samoilenko, Y.: Introduction to the theory of representations of finitely presented *-algebras. I, *Reviews in Mathematics and Mathematical Physics*, vol. 11. Harwood Academic Publishers, Amsterdam (1999). Representations by bounded operators
73. Packer, J.A.: Projective representations and the Mackey obstruction—a survey. In: Group representations, ergodic theory, and mathematical physics: a tribute to George W. Mackey, *Contemp. Math.*, vol. 449, pp. 345–378. Amer. Math. Soc., Providence, RI (2008). DOI 10.1090/conm/449/08719. URL http://dx.doi.org/10.1090/conm/449/08719
74. Pedersen, G.K.: C^*-algebras and their automorphism groups, *London Mathematical Society Monographs*, vol. 14. Academic Press Inc. [Harcourt Brace Jovanovich Publishers], London (1979)
75. Power, S.C.: Simplicity of C^*-algebras of minimal dynamical systems. J. London Math. Soc. (2) **18**(3), 534–538 (1978). DOI 10.1112/jlms/s2-18.3.534. URL http://dx.doi.org/10.1112/jlms/s2-18.3.534
76. Quigg, J.C., Spielberg, J.: Regularity and hyporegularity in C^*-dynamical systems. Houston J. Math. **18**(1), 139–152 (1992)
77. Renault, J.: A groupoid approach to C^*-algebras, *Lecture Notes in Mathematics*, vol. 793. Springer, Berlin (1980)
78. Renault, J.: Cuntz-like algebras. In: Operator theoretical methods (Timişoara, 1998), pp. 371–386. Theta Found., Bucharest (2000)
79. Renault, J.: Cartan subalgebras in C^*-algebras. Irish Math. Soc. Bull. (61), 29–63 (2008)
80. Spielberg, J.: Free-product groups, Cuntz-Krieger algebras, and covariant maps. Internat. J. Math. **2**(4), 457–476 (1991). DOI 10.1142/S0129167X91000260. URL http://dx.doi.org/10.1142/S0129167X91000260
81. Svensson, C., Silvestrov, S., de Jeu, M.: Dynamical systems and commutants in crossed products. Internat. J. Math. **18**(4), 455–471 (2007). DOI 10.1142/S0129167X07004217. URL http://dx.doi.org/10.1142/S0129167X07004217
82. Svensson, C., Silvestrov, S., de Jeu, M.: Connections between dynamical systems and crossed products of Banach algebras by $\mathbb{Z}$. In: Methods of spectral analysis in mathematical physics, *Oper. Theory Adv. Appl.*, vol. 186, pp. 391–401. Birkhäuser Verlag, Basel (2009). DOI 10.1007/978-3-7643-8755-6_19. URL http://dx.doi.org/10.1007/978-3-7643-8755-6_19

83. Svensson, C., Silvestrov, S., de Jeu, M.: Dynamical systems associated with crossed products. Acta Appl. Math. **108**(3), 547–559 (2009). DOI 10.1007/s10440-009-9506-5. URL http://dx.doi.org/10.1007/s10440-009-9506-5
84. Svensson, C., Tomiyama, J.: On the commutant of $C(X)$ in C^*-crossed products by $\mathbb{Z}$ and their representations. J. Funct. Anal. **256**(7), 2367–2386 (2009). DOI 10.1016/j.jfa.2009.02.002. URL http://dx.doi.org/10.1016/j.jfa.2009.02.002
85. Takesaki, M.: Theory of operator algebras. I. Springer-Verlag, New York (1979)
86. Tomiyama, J.: Invitation to C^*-algebras and topological dynamics, *World Scientific Advanced Series in Dynamical Systems*, vol. 3. World Scientific Publishing Co., Singapore (1987)
87. Tomiyama, J.: The interplay between topological dynamics and theory of C^*-algebras, *Lecture Notes Series*, vol. 2. Seoul National University Research Institute of Mathematics Global Analysis Research Center, Seoul (1992)
88. Watatani, Y.: Index for C^*-subalgebras. Mem. Amer. Math. Soc. **83**(424), vi+117 (1990)
89. Williams, D.P.: Crossed products of C^*-algebras, *Mathematical Surveys and Monographs*, vol. 134. American Mathematical Society, Providence, RI (2007)
90. Zeller-Meier, G.: Produits croisés d'une C^*-algèbre par un groupe d'automorphismes. J. Math. Pures Appl. (9) **47**, 101–239 (1968)

Chapter 15
On a Counterexample to a Conjecture by Blackadar

Adam P.W. Sørensen

Abstract Blackadar conjectured that if we have a split short-exact sequence $0 \to I \to A \to \mathbb{C} \to 0$ where I is semiprojective then A must be semiprojective. Eilers and Katsura have found a counterexample to this conjecture. Presumably Blackadar asked that the extension be split to make it more likely that semiprojectivity of I would imply semiprojectivity of A. But oddly enough, in all the counterexamples of Eilers and Katsura the quotient map from A to $A/I \cong \mathbb{C}$ is split. We will show how to modify their examples to find a non-semiprojective C^*-algebra B with a semiprojective ideal J such that B/J is the complex numbers and the quotient map does not split.

Keywords Semiprojective C^*-algebras • Pullbacks of C^*-algebras • Kirchberg algebras

Mathematics Subject Classification (2010): 46L05, 46L80, 54C56, 55P55.

15.1 Introduction

Semiprojectivity is a lifting property for C^*-algebras. It was introduced in [1] in a successful attempt to transfer some of the power of shape theory for metric spaces to the world of C^*-algebras.

Definition 1. A C^*-algebra A is semiprojective if whenever we have a C^*-algebra B containing an increasing sequence of ideals $J_1 \subseteq J_2 \subseteq \cdots$, and a

A.P.W. Sørensen (✉)
School of Mathematics and Applied Statistics, University of Wollongong, Wollongong, NSW 2522, Australia
e-mail: sorensen@unm.edu.au

T.M. Carlsen et al. (eds.), *Operator Algebra and Dynamics*, Springer Proceedings in Mathematics & Statistics 58, DOI 10.1007/978-3-642-39459-1_15,

$*$-homomorphism $\phi: A \to B/\overline{\cup_k J_k}$, we can find an $n \in \mathbb{N}$ and a $*$-homomorphism $\psi: A \to B/J_n$ such that

$$\pi_{n,\infty} \circ \psi = \phi,$$

where $\pi_{n,\infty}: B/J_n \twoheadrightarrow B/\overline{\cup_k J_k}$ is the natural quotient map.

Pictorially, A is semiprojective if we can always fill in the dashed arrow in the following commutative diagram:

$$\begin{array}{ccc} & & B \\ & & \downarrow \\ & & B/J_n \\ & \overset{\psi}{\nearrow} & \downarrow \\ A & \xrightarrow[\phi]{} & B/\overline{\cup_k J_k}. \end{array}$$

The book [10] is the canonical source for information about semiprojectivity. See also the more recent paper [3], the beginning of which has an expository nature.

Many of the main problems about semiprojectivity are concerned with the permanence properties of semiprojective C^*-algebras. In [1], Blackadar proves that the direct sum of two unital semiprojective C^*-algebras is again semiprojective, and that if A is unital and semiprojective then $M_n(A)$ is also semiprojective. These results were later extended from unital algebras to σ-unital algebras, so in particular to all separable algebras, by Loring in [9]. The results are a little stronger, in fact we have for separable algebras that $A \oplus B$ is semiprojective if and only if both A and B are, and a separable unital algebra D is semiprojective if and only if $M_2(D)$ is. It is still an open problem if a non-unital A must be semiprojective whenever $M_2(A)$ is. It is true if A is commutative, see [15, Corollary 6.9].

For a long time the following conjecture by Blackadar [3, Conjecture 4.5], which was first asked as a question by Loring in [10], was one of the main questions concerning the permanence properties of semiprojective C^*-algebras:

Conjecture 1 (Blackadar). Let

$$0 \to A \to B \to \mathbb{C} \to 0$$

be a split exact sequence of separable C^*-algebras. If A is semiprojective then so is B.

An important partial result was obtained in [5, Theorem 6.2.1]. It was used in [5] to show that all the so called one-dimensional non-commutative CW complexes are semiprojective. Enders [6] has proved a form of converse to Conjecture 1, namely that if $0 \to A \to B \to \mathbb{C} \to 0$ is an exact sequence of separable C^*-algebras with B semiprojective then A is semiprojective.

Recently Eilers and Katsura [4] have found a counterexample to Conjecture 1:

Theorem 1 (Eilers-Katsura). *There exists a split short exact sequence of separable C^*-algebras*

$$0 \to A \to B \to \mathbb{C} \to 0$$

where A is semiprojective but B is not.

The techniques used by Eilers and Katsura come from the world of graph C^*-algebras, and so only leads to split short exact sequences. Their work leaves open the question of whether there is a non-split short exact sequence $0 \to A \to B \to \mathbb{C} \to 0$ with A semiprojective and B not semiprojective. In light of Eilers and Katsura's result we certainly expect such a sequence to exist, and indeed, as we shall see in Theorem 3, it does.

This note is structured as follows: In Sect. 15.2 we prove two propositions that will be our main tools, in Sect. 15.3 we prove the main theorem.

15.2 Toolbox

We will be working with pullbacks. Given two $*$-homomorphisms $\phi\colon A \to D$, $\psi\colon B \to D$, we write, by standard abuse of notation, the pullback of A and B taken over ϕ and ψ as $A \oplus_D B$. That is $A \oplus_D B = \{(a,b) \in A \oplus B \mid \phi(a) = \psi(b)\}$. The pullback is universal for $*$-homomorphisms into A and B that agree after compositions with ϕ and ψ. For a detailed account of the theory of pullbacks (and pushouts) see [12].

Our first tool will let us produce new short exact sequences from old ones. In particular, it gives us a way to alter a split short exact sequence to make it non-split.

Proposition 1. *Suppose we are given two short exact sequences*

$$0 \to I \to A \xrightarrow{\pi} \mathbb{C} \to 0, \tag{15.1}$$

and

$$0 \to J \to B \xrightarrow{\rho} \mathbb{C} \to 0. \tag{15.2}$$

Let P be the pullback of A and B taken over π and ρ. Then the following three sequences are short exact:

$$0 \to I \oplus J \to P \to \mathbb{C} \to 0, \tag{15.3}$$

$$0 \to I \to P \to A \to 0, \quad \text{and,} \tag{15.4}$$

$$0 \to J \to P \to B \to 0. \tag{15.5}$$

Moreover (15.3) is split if and only if both (15.1) and (15.2) are split.

Proof. We begin by proving that (15.4) is exact. The map from P to A is simply projection onto the first coordinate, which is a surjection since both π and ρ are surjections. The kernel consists of pairs $(a,b) \in P$ with $a = 0$, that is pairs $(0,b)$ where $\rho(b) = 0$. Hence the kernel is $0 \oplus I \cong I$. A similar argument shows that (15.5) is exact.

We now consider (15.3). The map from P to $\mathbb{C}$ takes a pair (a,b) and sends it to $\pi(a)(= \rho(b))$. By the surjectivity of π and ρ we see that this is indeed a surjection. The kernel of this map is pairs $(a,b) \in P$ such that $\pi(a) = 0 = \rho(b)$, which is exactly $I \oplus J$.

The universal property of the pullback ensures that if (15.1) and (15.2) both split then (15.3) splits. On the other hand if we have a splitting from $\mathbb{C}$ to P, then simply composing that with the coordinate projections will show that (15.1) and (15.2) both split.

Remark 1. In the form of a diagram we have shown that if we are given sequences (15.1) and (15.2) as in the above proposition, then the following diagram commutes and has exact rows, columns and diagonal.

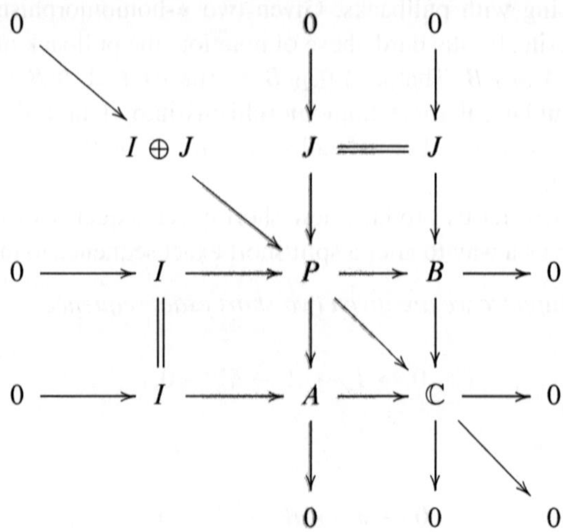

Now that we have a tool to construct non-split extensions from a split and a non-split one, we need a tool to tell us if the new extension is semiprojective.

For this purpose, we recall the definition and a property of compact ideals. We refer to section two of [14] for details on compact ideals

Definition 2. An ideal I in a C^*-algebra A is called compact, if whenever we have an increasing net of ideals $(J_\alpha)_{\alpha\in\Lambda}$ with $I \subseteq \overline{\bigcup_\alpha J_\alpha}$, then there is some $\alpha_0 \in \Lambda$ such that $I \subseteq J_{\alpha_0}$.

It is not hard to see from the definition that an ideal generated by a projection is compact.

Proposition 2 ([14, Corollary 2.2]). *Let A be a C^*-algebra and I an ideal in A. The following are equivalent:*

1. *The ideal I is a compact ideal in A.*
2. *There is a positive $a \in I$ and an $\varepsilon > 0$ such that $(a-\varepsilon)_+$ generates I as an ideal.*

The following proposition is a generalization of [11, Proposition 5.19] (where the ideal has to be the stabilization of a unital C^*-algebra). The proofs are very similar, but since [11] is in German, we include a short proof.

Proposition 3. *Consider a short exact sequence*

$$0 \to I \to A \overset{\rho}{\to} Q \to 0.$$

If I is a compact ideal and A is semiprojective then Q is semiprojective.

Proof. Suppose we are given B, an increasing sequence of ideals (J_k) in B, and a $*$-homomorphism $\phi\colon Q \to B/J$, where $J = \overline{\cup_k J_k}$. For all $k \in \mathbb{N}$, we let $\pi_{k,\infty}\colon B/J_k \to B/J$ be the natural quotient map. By the semiprojectivity of A we can find an $n \in \mathbb{N}$ and a $*$-homomorphism $\psi\colon A \to B/J_n$ such that $\pi_{n,\infty}\circ\psi = \phi\circ\rho$.

Pick a positive $a \in I$ and a $\varepsilon > 0$ such that I is generated as an ideal by $(a-\varepsilon)_+$. We have $\rho(a) = 0$, and therefore we have $(\pi_{n,\infty}\circ\psi)(a) = 0$. Hence, we can use [1, Lemma 2.13] to deduce that there must be some $l \geq n$ such that $\|(\pi_{n,l}\circ\psi)(a)\| < \varepsilon$, where $\pi_{n,l}$ denotes the quotient map from B/J_n to B/J_l. Therefore

$$(\pi_{n,l}\circ\psi)((a-\varepsilon)_+) = ((\pi_{n,l}\circ\psi)(a)-\varepsilon)_+ = 0.$$

So $(\pi_{n,l}\circ\psi)(I) = 0$, and we can conclude that $\pi_{n,l}\circ\psi$ drops to a $*$-homomorphism $\bar{\psi}\colon Q \to B/J_l$ with $\pi_{l,\infty}\circ\bar{\psi} = \phi$. Thus $\bar{\psi}$ and l combine to show that Q is semiprojective.

Our strategy is now the following: Find a non-split short exact sequence

$$0 \to J \to B \to \mathbb{C} \to 0,$$

such that J has a full projection. We will then use the construction in Proposition 1 on that and the Eilers-Katsura example, to produce a new non-split extension, which we can show, using Proposition 3, has the desired properties.

15.3 Constructing a Counterexample

We begin this section by constructing a non-split short exact sequence where the ideal is semiprojective and contains a full projection, and the quotient is the complex numbers. To prove that the constructed sequence is non-split we will use K-theory. In particular, we will show that one of the boundary maps in the six-term exact sequence is non-zero. Since $K_1(\mathbb{C}) = 0$, we need a semiprojective C^*-algebra with non-zero K_1-group. We will use a Kirchberg algebra.

Definition 3. A separable, simple, nuclear, purely infinite C^*-algebra is called a Kirchberg algebra. If it also satisfies the universal coefficient theorem, we call it a UCT Kirchberg algebra.

Definition 4. Denote by $\mathscr{P}_\infty$ the unital UCT Kirchberg algebra with $K_0(\mathscr{P}_\infty) \cong 0$ and $K_1(\mathscr{P}_\infty) \cong \mathbb{Z}$.

Building on the work of Blackadar [3] and Szymanski [17], Spielberg has shown in [16, Theorem 3.12] that any UCT Kirchberg algebra with finitely generated K-theory and torsion-free K_1-group is semiprojective. In particular, we have:

Theorem 2 (Spielberg). *Let $\mathbb{K}$ denote the algebra of compact operators. The Kirchberg algebra $\mathscr{P}_\infty \otimes \mathbb{K}$ is semiprojective.*

We can now construct a non-split sequence with a semiprojective ideal that contains a full projection.

Proposition 4. *There exists a non-split short exact sequence*

$$0 \to J \to E \to \mathbb{C} \to 0,$$

where J is separable, semiprojective, and contains a full projection.

Proof. Put $J = \mathscr{P}_\infty \otimes \mathscr{K}$, which, as the stabilization of the unital algebra $\mathscr{P}_\infty$ contains a full projection. By Theorem 2, it is semiprojective. We will pick E such that the boundary map in K-theory from $K_0(\mathbb{C})$ to $K_1(J)$ is non-zero. Since K-theory is split exact this implies that the sequence does not split.

Let $M(J)$ denote the multiplier algebra of J. We have the following short exact sequence:

$$0 \to J \to M(J) \to M(J)/J \to 0.$$

If we let $\eta\colon K_0(M(J)/J) \to K_1(J)$ be the boundary map in the six-term exact sequence arising from the above extension, then by Blackadar [2, Proposition 12.2.1] η is an isomorphism. In particular

$$K_0(M(J)/J) \cong K_1(J) \cong K_1(\mathscr{P}_\infty) \cong \mathbb{Z}.$$

By Lin and Zhang [8, Theorem 2.2], the corona algebra $M(J)/J$ has a continuous scale and so by Lin [7, Theorem 3.2] it is simple and purely infinite. Since $M(J)/J$ is also unital there is, by Blackadar [2, Corollary 6.11.8], a projection $p \in M(J)/J$ such that the class of p in $K_0(M(J)/J)$ is $1 \in \mathbb{Z}$. Define a $*$-homomorphism $\tau : \mathbb{C} \to M(J)/J$ by $\tau(\lambda) = \lambda p$, and notice that $K_0(\tau)$ is an isomorphism of groups.

Let $E = M(J) \oplus_{M(J)/J} \mathbb{C}$ where the pullback is taken over the quotient map from the multiplier algebra to the corona algebra and τ. We have the following commutative diagram which has exact rows (see [18, Proposition 3.2.9]):

$$\begin{array}{ccccccccc} 0 & \longrightarrow & J & \longrightarrow & E & \longrightarrow & \mathbb{C} & \longrightarrow & 0 \\ & & \| & & \downarrow & & \downarrow{\scriptstyle \tau} & & \\ 0 & \longrightarrow & J & \longrightarrow & M(J) & \longrightarrow & M(J)/J & \longrightarrow & 0 \end{array}$$

Let δ denote the boundary map from $K_0(\mathbb{C})$ to $K_1(J)$ in the six-term exact sequence associated to the short exact sequence on top. By Rørdam et al. [13, Proposition 12.2.1] the following square commutes:

$$\begin{array}{ccc} K_0(\mathbb{C}) & \xrightarrow{\delta} & K_1(J) \\ {\scriptstyle K_0(\tau)}\downarrow & & \| \\ K_0(M(J)/J) & \xrightarrow[\eta]{} & K_1(J) \end{array}$$

Since η and $K_0(\tau)$ are isomorphisms, we must have that δ is an isomorphism. In particular, δ is non-zero, so the sequence

$$0 \to J \to E \to \mathbb{C} \to 0$$

does not split.

We can now prove our main theorem.

Theorem 3. *There exists a non-split short exact sequence*

$$0 \to K \to B \to \mathbb{C} \to 0,$$

such that K is semiprojective but B is not.

Proof. Let

$$0 \to I \to A \xrightarrow{\pi} \mathbb{C} \to 0$$

be a short exact sequence such that I is separable and semiprojective but A is not semiprojective, e.g., one of the extensions constructed by Eilers and Katsura (Theorem 1), and let

$$0 \to J \to E \xrightarrow{\rho} \mathbb{C} \to 0 \tag{15.6}$$

be the non-split extension constructed in Proposition 4.

Put $B = A \oplus_{\mathbb{C}} E$ where the pullback is taken over π and ρ. By Proposition 1, we have the following two short exact sequences:

$$0 \to I \oplus J \to B \to \mathbb{C} \to 0, \quad \text{and,} \tag{15.7}$$

$$0 \to J \to B \to A \to 0. \tag{15.8}$$

Furthermore, (15.7) does not split as (15.6) does not split.

Since J has a full projection it is compact in B and since A is not semiprojective Proposition 3 applied to (15.8) gives us that B is not semiprojective. To complete the proof we put $K = I \oplus J$ and notice that K is semiprojective, as it is the sum of two separable semiprojective C^*-algebras [9, Theorem 4.2].

Acknowledgements The author thanks Takeshi Katsura and Søren Eilers for bringing the problem to his attention. Søren Eilers is also thanked for fruitful discussions concerning the problem. In an earlier version of this paper Proposition 3 assumed the ideal I to be generated by finitely many projections. The author is grateful to Mikael Rørdam for pointing out that the proof works for any compact ideal. This research was supported by the Danish National Research Foundation through the Centre for Symmetry and Deformation (DNRF92).

References

1. Blackadar, B.: Shape theory for C^*-algebras. Math. Scand. **56**(2), 249–275 (1985)
2. Blackadar, B.: K-theory for operator algebras, *Mathematical Sciences Research Institute Publications*, vol. 5, second edn. Cambridge University Press, Cambridge (1998)
3. Blackadar, B.: Semiprojectivity in simple C^*-algebras. In: Operator algebras and applications, *Adv. Stud. Pure Math.*, vol. 38, pp. 1–17. Math. Soc. Japan, Tokyo (2004)
4. Eilers, S., Katsura, T.: Semiprojectivity and properly infinite projections in graph C^*-algebra. In preparation
5. Eilers, S., Loring, T.A., Pedersen, G.K.: Stability of anticommutation relations: an application of noncommutative CW complexes. J. Reine Angew. Math. **499**, 101–143 (1998)
6. Enders, D.: Private communication (2011)
7. Lin, H.: Simple corona C^*-algebras. Proc. Amer. Math. Soc. **132**(11), 3215–3224 (electronic) (2004). DOI 10.1090/S0002-9939-04-07607-5. URL http://dx.doi.org/10.1090/S0002-9939-04-07607-5
8. Lin, H.X., Zhang, S.: On infinite simple C^*-algebras. J. Funct. Anal. **100**(1), 221–231 (1991). DOI 10.1016/0022-1236(91)90109-I. URL http://dx.doi.org/10.1016/0022-1236(91)90109-I
9. Loring, T.A.: Projective C^*-algebras. Math. Scand. **73**(2), 274–280 (1993)
10. Loring, T.A.: Lifting solutions to perturbing problems in C^*-algebras, *Fields Institute Monographs*, vol. 8. American Mathematical Society, Providence, RI (1997)

11. Neubüser, B.: Semiprojektivität und Realisierungen von rein unendlichen C^*-Algebren. Ph.D. thesis, University of Münster (2000)
12. Pedersen, G.K.: Pullback and pushout constructions in C^*-algebra theory. J. Funct. Anal. **167**(2), 243–344 (1999). DOI 10.1006/jfan.1999.3456. URL http://dx.doi.org/10.1006/jfan.1999.3456
13. Rørdam, M., Larsen, F., Laustsen, N.: An introduction to K-theory for C^*-algebras, *London Mathematical Society Student Texts*, vol. 49. Cambridge University Press, Cambridge (2000)
14. Santiago, L.: Reduction of the dimension of nuclear C^*-algebras (2012), arXiv:1211.7159
15. Sørensen, A.P.W., Thiel, H.: A characterization of semiprojectivity for commutative C^*-algebras. Proc. Lond. Math. Soc. (3) **105**(5), 1021–1046 (2012). DOI 10.1112/plms/pdr051. URL http://dx.doi.org/10.1112/plms/pdr051
16. Spielberg, J.: Semiprojectivity for certain purely infinite C^*-algebras. Trans. Amer. Math. Soc. **361**(6), 2805–2830 (2009). DOI 10.1090/S0002-9947-09-04928-9. URL http://dx.doi.org/10.1090/S0002-9947-09-04928-9
17. Szymański, W.: On semiprojectivity of C^*-algebras of directed graphs. Proc. Amer. Math. Soc. **130**(5), 1391–1399 (electronic) (2002). DOI 10.1090/S0002-9939-01-06282-7. URL http://dx.doi.org/10.1090/S0002-9939-01-06282-7
18. Wegge-Olsen, N.E.: K-theory and C^*-algebras. A friendly approach. Oxford Science Publications. The Clarendon Press Oxford University Press, New York (1993)

Chapter 16
The Topological Dimension of Type I C^*-Algebras

Hannes Thiel

Abstract While there is only one natural dimension concept for separable, metric spaces, the theory of dimension in noncommutative topology ramifies into different important concepts. To accommodate this, we introduce the abstract notion of a noncommutative dimension theory by proposing a natural set of axioms. These axioms are inspired by properties of commutative dimension theory, and they are for instance satisfied by the real and stable rank, the decomposition rank and the nuclear dimension.

We add another theory to this list by showing that the topological dimension, as introduced by Brown and Pedersen, is a noncommutative dimension theory of type I C^*-algebras. We also give estimates of the real and stable rank of a type I C^*-algebra in terms of its topological dimension.

Keywords C^*-algebras • Dimension theory • Stable rank • Real rank • Topological dimension • Type I C^*-algebras

Mathematics Subject Classification (2010): 46L85, 46L05, 46L35.

16.1 Introduction

The covering dimension of a topological space is a natural concept that extends our intuitive understanding that a point is zero-dimensional, a line is one-dimensional etc. While there also exist other dimension theories for topological spaces (e.g., small and large inductive dimension), they all agree for separable, metric spaces.

H. Thiel (✉)
Mathematisches Institut, Universität Münster, Einsteinstrasse 62, D-48149 Münster, Germany
e-mail: hannes.thiel@uni-muenster.de

T.M. Carlsen et al. (eds.), *Operator Algebra and Dynamics*, Springer Proceedings in Mathematics & Statistics 58, DOI 10.1007/978-3-642-39459-1_16,

This is in contrast to noncommutative topology where the concept of dimension ramifies into different important theories, such as the real and stable rank, the decomposition rank and the nuclear dimension. Each of these concepts has been studied in its own right, and they have applications in many different areas. A low dimension in each of these theories can be considered as a regularity property, and such regularity properties play an important role in the classification program of C^*-algebras, see [11, 24, 28] and the references therein.

In Sect. 16.3 of this paper we introduce the abstract notion of a non-commutative dimension theory as an assignment $d: \mathscr{C} \to \overline{\mathbb{N}}$ from a class of C^*-algebras to the extended natural numbers $\overline{\mathbb{N}} = \{0, 1, 2, \ldots, \infty\}$ satisfying a natural set of axioms, see Definition 1. These axioms are inspired by properties of the theory of covering dimension, see Remark 1, and they hold for the theories mentioned above. Thus, the proposed axioms do not define a unique dimension theory of C^*-algebras, but rather they collect the essential properties that such theories (should) satisfy.

Besides the very plausible axioms (D1)–(D4), we also propose (D5) which means that the property of being at most n-dimensional is preserved under approximation by sub-C^*-algebras, see 3. This is the noncommutative analog of the notion of "likeness", see 4 and [27, 3.1–3.3]. This axiom implies that dimension does not increase when passing to the limit of an inductive system of C^*-algebras, i.e., $d(\varinjlim A_i) \leq \liminf d(A_i)$, see Proposition 2.

Finally, axiom (D6) says that every *separable* sub-C^*-algebra $C \subset A$ is contained in a *separable* sub-C^*-algebra $D \subset A$ such that $d(D) \leq d(A)$. This is the noncommutative analog of Mardešić's factorization theorem, which says that every map $f: X \to Y$ from a compact space X to a compact, *metrizable* space Y can be factorized through a compact, *metrizable* space Z with $\dim(Z) \leq \dim(X)$, see Remark 1 and [17, Corollary 27.5, p. 159] or [13, Lemma 4].

In Sect. 16.4 we show that the topological dimension as introduced by Brown and Pedersen [6], is a dimension theory in the sense of Definition 1 for the class of type I C^*-algebras. The idea of the topological dimension is to simply consider the dimension of the primitive ideal space of a C^*-algebra. This will, however, run into problems if the primitive ideal space is not Hausdorff. One therefore has to restrict to (locally closed) Hausdorff subsets, and taking the supremum over the dimension of these Hausdorff subsets defines the topological dimension, see Definition 4.

In Sect. 16.5 we show how to estimate the real and stable rank of a type I C^*-algebra in terms of its topological dimension.

Section 16.5 of this article is based on the diploma thesis of the author [26], which was written under the supervision of Wilhelm Winter at the University of Münster in 2009. Sections 16.3 and 16.4 are based upon unpublished notes by the author for the masterclass "The nuclear dimension of C^*-algebras", held at the University of Copenhagen in November 2011.

16.2 Preliminaries

We denote by $\mathscr{C}^*$ the category of C^*-algebras with *-homomorphism as morphisms. In general, by a morphism between C^*-algebras we mean a *-homomorphism.

We write $J \lhd A$ to indicate that J is an ideal in A, and by an ideal of a C^*-algebra we understand a closed, two-sided ideal. Given a C^*-algebra A, we denote by A_+ the set of positive elements. We denote the minimal unitization of A by $\tilde{A}$. The primitive ideal space of A will be denoted by $\mathrm{Prim}(A)$, and the spectrum by $\hat{A}$. We refer the reader to Blackadar's book [1], for details on the theory of C^*-algebras.

If $F, G \subset A$ are two subsets of a C^*-algebra, and $\varepsilon > 0$, then we write $F \subset_\varepsilon G$ if for every $x \in F$ there exists some $y \in G$ such that $\|x - y\| < \varepsilon$. Given elements a, b in a C^*-algebra, we write $a =_\varepsilon b$ if $\|a - b\| < \varepsilon$. Given $a, b \in A_+$, we write $a \ll b$ if b acts as a unit for a, i.e., $ab = a$, and we write $a \ll_\varepsilon b$ if $ab =_\varepsilon a$.

We denote by $\mathbb{K}$ the C^*-algebra of compact operators on an infinite-dimensional, separable Hilbert space, and by $\overline{\mathbb{N}} = \{0, 1, 2, \ldots, \infty\}$ the extended natural numbers.

1. As pointed out in [1, II.2.2.7, p. 61], the full subcategory of commutative C^*-algebras is dually equivalent to the category $\mathscr{SP}_*$ whose objects are pointed, compact Hausdorff spaces and whose morphisms are pointed, continuous maps.

For a locally compact, Hausdorff space X, let αX be its one-point compactification. Let X^+ be the space with one additional point x_∞ attached, i.e., $X^+ = X \sqcup \{x_\infty\}$ if X is compact, and $X^+ = \alpha X$ if X is not compact. In both cases, the basepoint of X^+ is the attached point x_∞.

2. Let X be a space, and let $\mathscr{U}$ be a cover of X. The *order* of $\mathscr{U}$, denoted by $\mathrm{ord}(\mathscr{U})$, is the largest integer k such that some point $x \in X$ is contained in k different elements of $\mathscr{U}$ (and $\mathrm{ord}(\mathscr{U}) = \infty$ if no such k exists). The *covering dimension* of X, denoted by $\dim(X)$, is the smallest integer $n \geq 0$ such that every finite, open cover of X can be refined by a finite, open cover that has order at most $n + 1$ (and $\dim(X) = \infty$ if no such n exists). We refer the reader to Chap. 2 of Nagami's book [17] for more details.

It was pointed out by Morita [16], that in general this definition of covering dimension should be modified to consider only *normal*, finite, open covers. However, for normal spaces (e.g. compact spaces) every finite, open cover is normal, so that we may use the original definition.

The *local covering dimension* of X, denoted by $\mathrm{locdim}(X)$, is the smallest integer $n \geq 0$ such that every point $x \in X$ is contained in a closed neighborhood F such that $\dim(F) \leq n$ (and $\mathrm{locdim}(X) = \infty$ if no such n exists). We refer the reader to [9] and [20, Chap. 5] for more information about the local covering dimension.

It is well-known that $\mathrm{locdim}(X) = \dim(\alpha X)$ for a locally compact, Hausdorff space X. We propose that the natural dimension of a pointed space $(X, x_\infty) \in \mathscr{SP}_*$ is $\dim(X) = \mathrm{locdim}(X \setminus \{x_\infty\})$. Then, for a commutative C^*-algebra A, the natural dimension is $\mathrm{locdim}(\mathrm{Prim}(A))$.

If $G \subset X$ is an open subset of a locally compact space, then $\mathrm{locdim}(G) \leq \mathrm{locdim}(X)$, see [9, 4.1]. It was also shown by Dowker that this does not hold for the usual covering dimension (of non-normal spaces).

3. A family of sub-C^*-algebras $A_i \subset A$ is said to *approximate* a C^*-algebra A (in the literature there also appears the formulation that the A_i "locally approximate" A), if for every finite subset $F \subset A$, and every $\varepsilon > 0$, there exists some i such that $F \subset_\varepsilon A_i$. Let us mention some facts about approximation by subalgebras:

(a) If $A_1 \subset A_2 \subset \ldots \subset A$ is an increasing sequence of sub-C^*-algebras with $A = \overline{\bigcup_k A_k}$, then A is approximated by the family $\{A_k\}$.
(b) If A is approximated by a family $\{A_i\}$, and $J \lhd A$ is an ideal, then J is approximated by the family $\{A_i \cap J\}$. In particular, if $A = \overline{\bigcup_k A_k}$, then $J = \overline{\bigcup_k (A_k \cap J)}$.
   Similarly, A/J is approximated by the family $\{A_i/(A_i \cap J)\}$.
(c) If A is approximated by a family $\{A_i\}$, and $B \subset A$ is a hereditary sub-C^*-algebra, then B might *not* be approximated by the family $\{A_i \cap B\}$. Nevertheless, B is approximated by algebras that are isomorphic to hereditary sub-C^*-algebras of the algebras A_i, see Proposition 4.

4. Let $\mathscr{P}$ be some property of C^*-algebras. We say that a C^*-algebra A is *$\mathscr{P}$-like* (in the literature there also appears the formulation A is "locally $\mathscr{P}$") if A is approximated by subalgebras with property $\mathscr{P}$, see [27, 3.1–3.3]. This is motivated by the concept of $\mathscr{P}$-likeness for commutative spaces, as defined in [15, Definition 1] and further developed in [14].

We will work in the category $\mathscr{S}\mathscr{P}_*$ of pointed, compact spaces, see 1. Let $\mathscr{P}$ be a non-empty class of spaces. Then, a space $X \in \mathscr{S}\mathscr{P}_*$ is said to be $\mathscr{P}$-like if for every finite, open cover $\mathscr{U}$ of X there exists a (pointed) map $f\colon X \to Y$ onto some space $Y \in \mathscr{P}$ and a finite, open cover $\mathscr{V}$ of Y such that $\mathscr{U}$ is refined by $f^{-1}(\mathscr{V}) = \{f^{-1}(V) \mid V \in \mathscr{V}\}$.

Note that we have used $\mathscr{P}$ to denote both a class of spaces and a property that spaces might enjoy. These are just different viewpoints, as we can naturally assign to a property the class of spaces with that property, and vice versa to each class of spaces the property of lying in that class.

For commutative C^*-algebras, the notion of $\mathscr{P}$-likeness for C^*-algebras coincides with that for spaces. More precisely, it is shown in [27, Proposition 3.4] that for a space $(X, x_\infty) \in \mathscr{S}\mathscr{P}_*$ and a collection $\mathscr{P} \subset \mathscr{S}\mathscr{P}_*$, the following are equivalent:

(a) (X, x_∞) is $\mathscr{P}$-like,
(b) $C_0(X \setminus \{x_\infty\})$ is approximated by sub-C^*-algebras $C_0(Y \setminus \{y_\infty\})$ with $(Y, y_\infty) \in \mathscr{P}$.

We note that the definition of covering dimension can be rephrased as follows. Let $\mathscr{P}_k$ be the collection of all k-dimensional polyhedra (polyhedra are defined by combinatoric data, and their dimension is defined by this combinatoric data). Then a compact space X satisfies $\dim(X) \leq k$ if and only if it is $\mathscr{P}_k$-like. This motivates (D5) in Definition 1 below.

5. For the definition of continuous trace C^*-algebras we refer our reader to [1, Definition IV.1.4.12, p. 333]. It is known that a C^*-algebra A has continuous trace if and only if its spectrum $\hat{A}$ is Hausdorff and it satisfies Fell's condition, i.e., for every $\pi \in \hat{A}$ there exists a neighborhood $U \subset \hat{A}$ of π and some $a \in A_+$ such that $\rho(a)$ is a rank-one projection for each $\rho \in U$, see [1, Proposition IV.1.4.18, p. 335].

6. A C^*-algebra A is called a *CCR algebra* (sometimes called a liminal algebra) if for each of its irreducible representations $\pi\colon A \to B(H)$ we have that π takes values inside the compact operators $K(H)$.

A *composition series* for a C^*-algebra A is a collection of ideals $J_\alpha \lhd A$, indexed over all ordinal numbers $\alpha \leq \mu$ for some μ, such that $A = J_\mu$ and:

(a) If $\alpha \leq \beta$, then $J_\alpha \subset J_\beta$,
(b) If α is a limit ordinal, then $J_\alpha = \overline{\bigcup_{\gamma<\alpha} J_\gamma}$.

The C^*-algebras $J_{\alpha+1}/J_\alpha$ are called the successive quotients of the composition series.

A C^*-algebra is called a *type I algebra* (sometimes also called postliminal or GCR algebra) if it has a composition series with successive quotients that are CCR algebras. As it turns out, this is equivalent to having a composition series whose successive quotients have continuous trace.

For information about type I C^*-algebras and their rich structure we refer the reader to Chap. IV.1 of Blackadar's book [1], and Chap. 6 of Pedersen's book [21].

16.3 Dimension Theories for C^*-Algebras

In this section, we introduce the notion of a non-commutative dimension theory by proposing a natural set of axioms that such theories should satisfy. These axioms hold for many well-known theories, in particular the real and stable rank, the decomposition rank and the nuclear dimension, see Remark 2, and this will also be discussed more thoroughly in a forthcoming paper. In Sect. 16.4 we will show that the topological dimension is a dimension theory for type I C^*-algebras.

Our axioms of a non-commutative dimension theory are inspired by facts that the theory of covering dimension satisfies, see Remark 1.

In Definition 2 we introduce the notion of Morita-invariance for dimension theories. If a dimension theory is only defined on a subclass of C^*-algebras, then there is a natural extension of the theory to all C^*-algebras, see Proposition 5. We will show that this extension preserves Morita-invariance.

We denote by $\mathscr{C}^*$ the category of C^*-algebras, and we will use $\mathscr{C}$ to denote a class of C^*-algebras. We may think of $\mathscr{C}$ as a full subcategory of $\mathscr{C}^*$.

Definition 1. Let $\mathscr{C}$ be a class of C^*-algebras that is closed under *-isomorphisms, and closed under taking ideals, quotients, finite direct sums, and minimal unitizations. A *dimension theory* for $\mathscr{C}$ is an assignment $d: \mathscr{C} \to \overline{\mathbb{N}} = \{0, 1, 2, \ldots, \infty\}$ such that $d(A) = d(A')$ whenever A, A' are isomorphic C^*-algebras in $\mathscr{C}$, and moreover the following axioms are satisfied:

(D1) $d(J) \leq d(A)$ whenever $J \lhd A$ is an ideal in $A \in \mathscr{C}$,
(D2) $d(A/J) \leq d(A)$ whenever $J \lhd A \in \mathscr{C}$,
(D3) $d(A \oplus B) = \max\{d(A), d(B)\}$, whenever $A, B \in \mathscr{C}$,
(D4) $d(\tilde{A}) = d(A)$, whenever $A \in \mathscr{C}$.
(D5) If $A \in \mathscr{C}$ is approximated by subalgebras $A_i \in \mathscr{C}$ with $d(A_i) \leq n$, then $d(A) \leq n$.
(D6) Given $A \in \mathscr{C}$ and a separable sub-C^*-algebra $C \subset A$, there exists a separable C^*-algebra $D \in \mathscr{C}$ such that $C \subset D \subset A$ and $d(D) \leq d(A)$.

Note that we do not assume that $\mathscr{C}$ is closed under approximation by sub-C^*-algebra, so that the assumption $A \in \mathscr{C}$ in (D5) is necessary. Moreover, in axiom (D6), we do not assume that the separable subalgebra C lies in $\mathscr{C}$.

Remark 1. The axioms in Definition 1 are inspired by well-known facts of the local covering dimension of commutative spaces, see 2.

Axiom (D1) and (D2) generalize the fact that the local covering dimension does not increase when passing to an open (resp. closed) subspace, see [9, 4.1, 3.1], and axiom (D3) generalizes the fact that $\operatorname{locdim}(X \sqcup Y) = \max\{\operatorname{locdim}(X), \operatorname{locdim}(Y)\}$. Axiom (D4) generalizes that $\operatorname{locdim}(X) = \operatorname{locdim}(\alpha X)$, where αX is the one-point compactification of X.

Axiom (D5) generalizes the fact that a (compact) space is n-dimensional if it is $\mathscr{P}_n$-like for the class $\mathscr{P}_n$ of n-dimensional spaces, see 4. Note also that Proposition 2 generalizes the fact that $\dim(\varprojlim X_i) \leq \liminf_i \dim(X_i)$ for an inverse system of compact spaces X_i.

Axiom (D6) is a generalization of the following factorization theorem, due to Mardešić, see [17, Corollary 27.5, p. 159] or [13, Lemma 4]: Given a compact space X and a map $f: X \to Y$ to a compact, metrizable space Y, there exists a compact, metrizable space Z and maps $g: X \to Z, h: Z \to Y$ such that g is onto, $\dim(Z) \leq \dim(X)$ and $f = h \circ g$. This generalizes (D6), since a unital, commutative C^*-algebra $C(X)$ is separable if and only if X is metrizable.

Axioms (D5) and (D6) are also related to the following concept which is due to Blackadar [1, Definition II.8.5.1, p. 176]: A property $\mathscr{P}$ of C^*-algebras is called *separably inheritable* if:

1. For every C^*-algebra A with property $\mathscr{P}$ and separable sub-C^*-algebra $C \subset A$, there exists a separable sub-C^*-algebra $D \subset A$ that contains C and has property $\mathscr{P}$.
2. Given an inductive system (A_k, φ_k) of separable C^*-algebras with injective connecting morphisms $\varphi_k: A_k \to A_{k+1}$, if each A_k has property $\mathscr{P}$, then does the inductive limit $\varinjlim A_k$.

Thus, for a dimension theory d, the property "$d(A) \leq n$" is separably inheritable.

Axioms (D5) and (D6) imply that $d(A) \leq n$ if and only if A can be written as an inductive limit (with injective connecting morphisms) of separable C^*-algebras B with $d(B) \leq n$. This allows us to reduce essentially every question about dimension theories to the case of separable C^*-algebras.

By explaining the analogs of (D1)–(D6) for pointed, compact spaces, we have shown the following:

Proposition 1. *Let $\mathscr{C}^*_{\mathrm{ab}}$ denote the class of commutative C^*-algebras. Then, the assignment $d: \mathscr{C}^*_{\mathrm{ab}} \to \overline{\mathbb{N}}$, $d(A) := \mathrm{locdim}(\mathrm{Prim}(A))$, is a dimension theory.*

Remark 2. We do not suggest that the axioms of Definition 1 uniquely define a dimension theory. This is clear since the axioms do not even rule out the assignments that give each C^*-algebra the same value.

More interestingly, the following are dimension theories for the class of all C^*-algebras:

1. The stable rank as defined by Rieffel [22, Definition 1.4].
2. The real rank as introduced by Brown and Pedersen [5].
3. The decomposition rank of Kirchberg and Winter [12, Definition 3.1].
4. The nuclear dimension of Winter and Zacharias [29, Definition 2.1].

Indeed, for the real and stable rank, (D1) and (D2) are proven in [10, Théorème 1.4] and [22, Theorems 4.3, 4.4]. Axiom (D3) is easily verified, and (D4) holds by definition. It is shown in [22, Theorem 5.1] that (D5) holds in the special case of an approximation by a countable inductive limit, but the same argument works for general approximations and also for the real rank. Finally, it is noted in [1, II.8.5.5, p. 178] that (D6) holds.

For the nuclear dimension, axioms (D1), (D2), (D3), (D6) and (D4) follow from Propositions 2.5, 2.3, 2.6 and Remark 2.11 in [29], and (D5) is easily verified. For the decomposition rank, (D5) is also easily verified, and axiom (D6) follows from [29, Proposition 2.6] adapted for c.p.c. approximations instead of c.p. approximations. The other axioms (D1)–(D4) follow from Proposition 3.8, 3.11 and Remark 3.2 of [12] for separable C^*-algebras. Using axioms (D5) and (D6) this can be extended to all C^*-algebras.

Thus, the idea of Definition 1 is to collect the essential properties that many different non-commutative dimension theories satisfy. Our way of axiomatizing non-commutative dimension theories should therefore not be confused with the work on axiomatizing the dimension theory of metrizable spaces, see e.g. [18] or [7], since these works pursue the goal of finding axioms that uniquely characterize covering dimension.

Proposition 2. *Let $d: \mathscr{C} \to \overline{\mathbb{N}}$ be a dimension theory, and let $(A_i, \varphi_{i,j})$ be an inductive system with $A_i \in \mathscr{C}$ and such that the limit $A := \varinjlim A_i$ also lies in $\mathscr{C}$. Then $d(A) \leq \liminf_i d(A_i)$.*

Proof. See [1, II.8.2.1, p. 156] for details about inductive systems and inductive limits. For each i, let $\varphi_{\infty,i}\colon A_i \to A$ denote the natural morphism into the inductive limit. Then the subalgebra $\varphi_{\infty,i}(A_i) \subset A$ is a quotient of A_i, and therefore $d(\varphi_{\infty,i}(A_i)) \leq d(A_i)$ by (D2). If $J \subset I$ is cofinal, then A is approximated by the collection of subalgebras $(\varphi_{\infty,i}(A_i))_{i\in J}$. It follows from (D5) that $d(A)$ is bounded by $\sup_{i\in J} d(A_i)$. Since this holds for each cofinal subset $J \subset I$, we obtain:

$$d(A) \leq \inf\{\sup_{i\in J} d(A_i) \mid J \subset I \text{ cofinal}\} = \liminf_i d(A_i),$$

as desired. □

Lemma 1. *Let A be a C^*-algebra, let $B \subset A$ be a full, hereditary sub-C^*-algebra, and let $C \subset A$ be a separable sub-C^*-algebra. Then there exists a separable sub-C^*-algebra $D \subset A$ containing C such that $D \cap B \subset D$ is full, hereditary.*

Proof. The proof is inspired by the proof of [2, Proposition 2.2], see also [1, Theorem II.8.5.6, p.178]. We inductively define separable sub-C^*-algebras $D_k \subset A$. Set $D_1 := C$, and assume D_k has been constructed. Let $S_k := \{x_1^k, x_2^k, \dots\}$ be a countable, dense subset of D_k. Since B is full in A, there exist for each $i \geq 1$ finitely many elements $a_{i,j}^k, c_{i,j}^k \in A$ and $b_{i,j}^k \in B$ such that

$$\|x_i^k - \sum_j a_{i,j}^k b_{i,j}^k c_{i,j}^k\| < 1/k.$$

Set $D_{k+1} := C^*(D_k, a_{i,j}^k, b_{i,j}^k, c_{i,j}^k, i, j \geq 1)$. Then define $D := \overline{\bigcup_k D_k}$, which is a separable sub-C^*-algebra of A containing C.

Note that $D \cap B \subset D$ is a hereditary sub-C^*-algebra, and let us check that it is also full. We need to show that the linear span of $D(D \cap B)D$ is dense in D. Let $d \in D$ and $\varepsilon > 0$ be given. Note that $\bigcup_k S_k$ is dense in D. Thus, we may find k and i such that $\|d - x_i^k\| < \varepsilon/2$. We may assume $k \geq 2/\varepsilon$. By construction, there are elements $a_{i,j}^k, c_{i,j}^k \in D_{k+1}$ and $b_{i,j}^k \in B \cap D_{k+1}$ such that $\|x_i^k - \sum_j a_{i,j}^k b_{i,j}^k c_{i,j}^k\| < 1/k$. It follows that the distance from d to the closed linear span of $D(D \cap B)D$ is at most ε. Since d and ε were chosen arbitrarily, this shows that $D \cap B \subset D$ is full. □

Proposition 3. *Let $d\colon \mathscr{C}^* \to \overline{\mathbb{N}}$ be a dimension theory. Then the following statements are equivalent:*

1. *For all C^*-algebras A, B: If $B \subset A$ is a full, hereditary sub-C^*-algebra, then $d(A) = d(B)$.*
2. *For all C^*-algebras A, B: If A and B are Morita equivalent, then $d(A) = d(B)$.*
3. *For all C^*-algebras A: $d(A) = d(A \otimes \mathbb{K})$.*

Moreover, each of the statements is equivalent to the (a priori weaker) statement where the appearing C^*-algebras are additionally assumed to be separable.

If d satisfies the above conditions, and $B \subset A$ is a (not necessarily full) hereditary sub-C^*-algebra, then $d(B) \leq d(A)$.

Proof. For each of the statements 1., 2., 3., let us denote the statement where the appearing C^*-algebras are assumed to be separable by $1s.$, $2s.$, $3s.$ respectively. For example:

$3s.$ For all separable C^*-algebras A: $d(A) = d(A \otimes \mathbb{K})$.

The implications "1. $\Rightarrow$ $1s.$", "2. $\Rightarrow$ $2s.$", and "3. $\Rightarrow$ $3s.$" are clear. The implication "$2s.$ $\Rightarrow$ $3s.$" follows since A and $A \otimes \mathbb{K}$ are Morita equivalent, and "$1s.$ $\Rightarrow$ $3s.$" follows since $A \subset A \otimes \mathbb{K}$ is a full, hereditary sub-C^*-algebra.

It remains to show the implication "$3s.$ $\Rightarrow$ 1.". Let A be a C^*-algebra, and let $B \subset A$ be a full, hereditary sub-C^*-algebra. We need to show $d(A) = d(B)$. To that end, we will construct separable sub-C^*-algebras $A' \subset A$ and $B' \subset B$ that approximate A and B, respectively, and such that $d(A') = d(B') \leq \min\{d(A), d(B)\}$. Together with (D5), this implies $d(A) = d(B)$.

So let $F \subset A$ and $G \subset B$ be finite sets. We may assume $G \subset F$. We want to find A' and B' with the mentioned properties and such that $F \subset A'$ and $G \subset B'$.

We inductively define separable sub-C^*-algebras $C_k, D_k \subset A$ and $E_k \subset B$ such that:

(a) $C_k \subset D_k$ and $D_k \cap B \subset D_k$ is full,
(b) $D_k \cap B \subset E_k$ and $d(E_k) \leq d(B)$,
(c) $E_k, D_k \subset C_{k+1}$ and $d(C_{k+1}) \leq d(A)$.

We start with $C_1 := C^*(F) \subset A$. If C_k has been constructed, we apply Lemma 1 to find D_k satisfying (a). If D_k has been constructed, we apply (D6) to $D_k \cap B \subset B$ to find E_k satisfying (b). If E_k has been constructed, we apply axiom (D6) to $C^*(D_k, E_k) \subset A$ to find C_{k+1} satisfying (c).

Then let $A' := \overline{\bigcup_k C_k} = \overline{\bigcup_k D_k}$, and $B' := \overline{\bigcup_k (D_k \cap B)} = \overline{\bigcup_k E_k}$. The situation is shown in the following diagram:

$$\begin{array}{ccccccccccc}
C_k & \subset & D_k & \subset & C^*(D_k, E_k) & \subset & C_{k+1} & \subset & \dots & \subset & A' \\
 & & \cup & & \cup & & & & & & \\
 & & D_k \cap B & \subset & E_k & \subset & \dots & & \dots & \subset & B'
\end{array}$$

Let us verify that A' and B' have the desired properties. First, since $d(C_k) \leq d(A)$ for all k, we get $d(A') \leq d(A)$ from (D5). Similarly, we get $d(B') \leq d(B)$. For each k we have that $D_k \cap B \subset D_k$ is a full, hereditary sub-C^*-algebra, and therefore the same holds for $B' \subset A'$. Since A' and B' are separable (and hence σ-unital), we may apply Brown's stabilization theorem [3, Theorem 2.8], and obtain $A' \otimes \mathbb{K} \cong B' \otimes \mathbb{K}$. Together with the assumption $3s.$, we obtain $d(A') = d(A' \otimes \mathbb{K}) = d(B' \otimes \mathbb{K}) = d(B')$. This finishes the construction of A' and B', and we deduce $d(A) = d(B)$ from (D5).

Lastly, if d satisfies condition 1., and $B \subset A$ is a (not necessarily full) hereditary sub-C^*-algebra, then B is full, hereditary in the ideal $J \lhd A$ generated by B. By (D1) and condition 1. we have $d(B) = d(J) \leq d(A)$. □

Definition 2. A dimension theory $d\colon \mathscr{C}^* \to \overline{\mathbb{N}}$ is called *Morita-invariant* if it satisfies the conditions of Proposition 3.

Given positive elements a, b in a C^*-algebra, recall that we write $a =_\sigma b$ if $\|a - b\| < \sigma$. We write $a \ll_\sigma b$ if $ab =_\sigma a$.

Lemma 2. *For every $\varepsilon > 0$ there exists $\delta > 0$ with the following property: Given a C^*-algebra A, and contractive elements $a, b \in A_+$ with $a =_\delta b$, there exists a partial isometry $v \in A^{**}$ such that:*

1. $v(a-\delta)_+ v^* \in \overline{bAb}$.
2. If $d \in A_+$ is contractive with $d \ll_\sigma a$, then $vdv^* =_{4\sigma+\varepsilon} d$.

Proof. To simplify the proof, we will fix $\delta > 0$ and verify the statement for $\varepsilon = \varepsilon(\delta)$ with the property that $\varepsilon(\delta) \to 0$ when $\delta \to 0$.

Fix $\delta > 0$. Let A be a C^*-algebra, and let $a, b \in A_+$ be contractive elements such that $a =_\delta b$. Without loss of generality we may assume that A is unital. It is well-known that there exists $s \in A$ such that $s(a-\delta)_+ s^* \in \overline{bAb}$, see [23, Proposition 2.4]. One could follow the proof to obtain an estimate similar to that in statement 2. It is, however, easier to find $v \in A^{**}$ such that 1. and 2. hold, and for our application in Proposition 4 it is sufficient that v lies in A^{**}.

It follows from $a =_\delta b$ that $a - \delta \leq b$, and hence:

$$(a-\delta)_+^2 = (a-\delta)_+^{1/2}(a-\delta)(a-\delta)_+^{1/2} \leq (a-\delta)_+^{1/2} b (a-\delta)_+^{1/2}.$$

Set $z := b^{1/2}(a-\delta)_+^{1/2}$. Then:

$$|z| = ((a-\delta)_+^{1/2} b (a-\delta)_+^{1/2})^{1/2}, \qquad |z^*| = (b^{1/2}(a-\delta)_+ b^{1/2})^{1/2},$$

and we let $z = v|z|$ be the polar decomposition of z, with $v \in A^{**}$. We claim that v has the desired properties. First, note that $v((a-\delta)_+^{1/2} b (a-\delta)_+^{1/2})v^* = b^{1/2}(a-\delta)_+ b^{1/2} \in \overline{bAb}$, and therefore also $v(a-\delta)_+ v^* \in \overline{bAb}$, which verifies property 1.

For property 2., let us start by estimating the distance from a to z and $|z|$. It is known that there exists an assignment $\sigma \mapsto \varepsilon_1(\sigma)$ with the following property: Whenever x, y are positive, contractive elements of a C^*-algebra, and $x =_\sigma y$, then $x^{1/2} =_{\varepsilon_1(\sigma)} y^{1/2}$, and moreover $\varepsilon_1(\sigma) \to 0$ as $\sigma \to 0$. We may assume $\sigma \leq \varepsilon_1(\sigma)$, and we will use this to simplify some estimates below.

Then, using $(a-\delta)_+ =_\delta a$ and so $(a-\delta)_+^{1/2} =_{\varepsilon_1(\delta)} a^{1/2}$ at the second step,

$$z = b^{1/2}(a-\delta)_+^{1/2} =_{\varepsilon_1(\delta)} b^{1/2}a^{1/2} =_{\varepsilon_1(\delta)} a. \tag{16.1}$$

For $|z|$ we compute, using $(a-\delta)_+^{1/2} b (a-\delta)_+^{1/2} =_{3\varepsilon_1(\delta)} a^2$ at the second step,

$$|z| = ((a-\delta)_+^{1/2} b (a-\delta)_+^{1/2})^{1/2} =_{\varepsilon_1(3\varepsilon_1(\delta))} (a^2)^{1/2} = a. \tag{16.2}$$

Let $d \in A_+$ be contractive with $d \ll_\sigma a$. Then $ada =_{2\sigma} d$, and we may estimate the distance from vdv^* to d as follows:

$$vdv^* =_{2\sigma} vadav^* \overset{(16.2)}{=\!=}_{2\varepsilon_1(3\varepsilon_1(\delta))} v|z|d|z|v^* = zdz \overset{(16.1)}{=\!=}_{4\varepsilon_1(\delta)} ada =_{2\sigma} d.$$

Thus, $\|vdv^* - d\| \leq 4\sigma + 2\varepsilon_1(3\varepsilon_1(\delta)) + 4\varepsilon_1(\delta)$, and this distance converges to 4σ when $\delta \to 0$. □

Proposition 4. *Let A be a C^*-algebra, and let $B \subset A$ be a hereditary sub-C^*-algebra. Assume A is approximated by sub-C^*-algebras $A_i \subset A \otimes \mathbb{K}$. Then B is approximated by subalgebras that are isomorphic to hereditary sub-C^*-algebras of the algebras A_i, i.e., given a finite set $F \subset B$ and $\varepsilon > 0$, there exists a sub-C^*-algebra $B' \subset B$ such that $F \subset_\varepsilon B'$ and B' is isomorphic to a hereditary sub-C^*-algebra of A_i for some i.*

Proof. Let $F \subset B$ and $\varepsilon > 0$ be given. We let $\gamma = \varepsilon/36$, which is justified by the estimates that we obtain through the course of the proof. Without loss of generality, we may assume that F consists of positive, contractive elements.

There exists $b \in B_+$ such that b almost acts as a unit on the elements of F in the sense that $x \ll_\gamma b$ for all $x \in F$. Let $\delta > 0$ be the tolerance we get from Lemma 2 for γ. We may assume $\delta \leq \gamma$, and to simplify the computations below we will often estimate a distance by γ, even if it could be estimated by δ.

By assumption, the algebras A_i approximate A. Thus, there exists i such that there is a positive, contractive element $a \in A_i$ with $a =_\delta b$, and such that for each $x \in F$ there exists a positive, contractive $x' \in A_i$ with $x' =_\delta x$. Then:

$$x'(a-\delta)_+ =_{3\delta} xb =_\gamma x =_\delta x',$$

and so $x' \ll_{5\gamma} (a-\delta)_+$, since $\delta \leq \gamma$. In general, if two positive, contractive elements s, t satisfy $s \ll_\sigma t$, then $s =_{2\sigma} tst \ll_\sigma t$. Thus, if for each $x \in F$ we set $x'' := (a-\delta)_+ x'(a-\delta)_+$, then we obtain:

$$x =_\gamma x' =_{10\gamma} x'' \ll_{5\gamma} (a-\delta)_+. \tag{16.3}$$

Since $a =_\delta b$, we obtain from Lemma 2 a partial isometry $v \in A^{**}$ such that $v(a-\delta)_+ v^* \in \overline{bAb}$. Let $A' := (a-\delta)_+ A_i (a-\delta)_+$, which is a hereditary sub-C^*-algebra of A_i. The map $x \mapsto vxv^*$ defines an isomorphism from A' onto $B' := vA'v^*$. Since B is hereditary, B' is a sub-C^*-algebra of B. Let us estimate the distance from F to B'.

For each $x \in F$, we have computed in (16.3) that $x'' \ll_{5\gamma} (a-\delta)_+$, which implies $x'' \ll_{6\gamma} a$. From statement 2. of Lemma 2 we deduce $vx''v^* =_{25\gamma} x''$. Altogether, the distance between x and $vx''v^*$ is at most 36γ. Since $vx''v^* \in B'$, and since we chose $\gamma = \varepsilon/36$, we have $F \subset_\varepsilon B'$, as desired. □

Proposition 5. *Let $d: \mathscr{C} \to \overline{\mathbb{N}}$ be a dimension theory. For any C^*-algebra A define:*

$$\tilde{d}(A) := \inf\left\{ k \in \mathbb{N} \mid \begin{array}{c} A \textit{ is approximated by sub-}C^*\textit{-algebras } B \\ \textit{with } d(B) \leq k \end{array} \right\},$$

where we define the infimum of the empty set to be $\infty \in \overline{\mathbb{N}}$.

Then $\tilde{d}\colon \mathscr{C}^* \to \overline{\mathbb{N}}$ *is a dimension theory that agrees with* d *on* $\mathscr{C}$.

If, moreover, $\mathscr{C}$ *is closed under stable isomorphism, and* $d(A) = d(A \otimes \mathbb{K})$ *for every (separable)* $A \in \mathscr{C}$, *then* $\tilde{d}$ *is Morita-invariant.*

Proof. If $A \in \mathscr{C}$, then clearly $\tilde{d}(A) \leq d(A)$, and the converse inequality follows from axiom (D5). Axioms (D1)–(D5) for $\tilde{d}$ are easy to check.

Let us check axiom (D6) for $\tilde{d}$. Assume A is a C^*-algebra, and assume $C \subset A$ is a separable sub-C^*-algebra. Set $n := \tilde{d}(A)$, which we may assume is finite. We need to find a separable sub-C^*-algebra $D \subset A$ such that $C \subset D$ and $\tilde{d}(D) \leq n$.

We first note the following: For a finite set $F \subset A$, and $\varepsilon > 0$ we can find a separable sub-C^*-algebra $A(F, \varepsilon) \subset A$ with $d(A(F, \varepsilon)) \leq n$ and $F \subset_\varepsilon A(F, \varepsilon)$. Indeed, by definition of $\tilde{d}$ we can first find a sub-C^*-algebra $B \subset A$ with $d(B) \leq n$ and a finite subset $G \subset B$ such that $F \subset_\varepsilon G$. Applying (D6) to $C^*(G) \subset B$, we may find a separable sub-C^*-algebra $A(F, \varepsilon) \subset B$ with $d(A(F, \varepsilon)) \leq n$ and $C^*(G) \subset A(F, \varepsilon)$, which implies $F \subset_\varepsilon A(F, \varepsilon)$.

We will inductively define separable sub-C^*-algebras $D_k \subset A$ and countable dense subsets $S_k = \{x_1^k, x_2^k, \ldots\} \subset D_k$ as follows: We start with $D_1 := C$ and choose any countable dense subset $S_1 \subset D_1$. If D_l and S_l have been constructed for $l \leq k$, then set:

$$D_{k+1} := C^*(D_k, A(\{x_i^j \mid i, j \leq k\}, 1/k)) \subset A,$$

and choose any countable dense subset $S_{k+1} = \{x_1^{k+1}, x_2^{k+1}, \ldots\} \subset D_{k+1}$.

Set $D := \overline{\bigcup_k D_k} \subset A$, which is a separable C^*-algebra containing C. Let us check that $\tilde{d}(D) \leq n$, which means that we have to show that D is approximated by sub-C^*-algebras $B \in \mathscr{C}$ with $d(B) \leq n$.

Note that $\{x_i^j\}_{i,j \geq 1}$ is dense in D. Thus, if a finite subset $F \subset D$, and $\varepsilon > 0$ is given, we may find k such that $F \subset_{\varepsilon/2} \{x_i^j \mid i, j \leq k\}$, and we may assume $k > 2/\varepsilon$. By construction, D contains the sub-C^*-algebra $B := A(\{x_i^j \mid i, j \leq k\}, 1/k)$, which satisfies $d(B) \leq n$ and $\{x_i^j \mid i, j \leq k\} \subset_{1/k} B$. Then $F \subset_\varepsilon B$, which completes the proof that $\tilde{d}(D) \leq n$.

Lastly, assume $\mathscr{C}$ is closed under stable isomorphism, and assume $d(A) = d(A \otimes \mathbb{K})$ for every separable $A \in \mathscr{C}$. This implies the following: If A is a separable C^*-algebra in $\mathscr{C}$, and $B \subset A$ is a hereditary sub-C^*-algebra, then B lies in $\mathscr{C}$ and $d(B) \leq d(A)$.

We want to check condition 3. of Proposition 3 for $\tilde{d}$. Thus, let a separable C^*-algebra A be given. We need to check $\tilde{d}(A) = \tilde{d}(A \otimes \mathbb{K})$.

If $\tilde{d}(A) = \infty$, then clearly $\tilde{d}(A \otimes \mathbb{K}) \leq \tilde{d}(A)$. So assume $n := \tilde{d}(A) < \infty$, which means that A is approximated by algebras $A_i \subset A$ with $d(A_i) \leq n$. Then

$A \otimes \mathbb{K}$ is approximated by the subalgebras $A_i \otimes \mathbb{K} \subset A \otimes \mathbb{K}$, and $d(A_i \otimes \mathbb{K}) = d(A_i) \leq n$ by assumption. Then $\tilde{d}(A \otimes \mathbb{K}) \leq n = \tilde{d}(A)$.

Conversely, if $\tilde{d}(A \otimes \mathbb{K}) = \infty$, then $\tilde{d}(A) \leq \tilde{d}(A \otimes \mathbb{K})$. So assume $n := \tilde{d}(A \otimes \mathbb{K}) < \infty$, which means that $A \otimes \mathbb{K}$ is approximated by algebras $A_i \subset A \otimes \mathbb{K}$ with $d(A_i) \leq n$. Consider the hereditary sub-C^*-algebra $A \otimes e_{1,1} \subset A \otimes \mathbb{K}$, which is isomorphic to A. By Proposition 4, $A \otimes e_{1,1}$ is approximated by subalgebras $B_j \subset A \otimes e_{1,1}$ such that each B_j is isomorphic to a hereditary sub-C^*-algebras of A_i, for some $i = i(j)$. It follows $d(B_j) \leq n$, and then $\tilde{d}(A) = \tilde{d}(A \otimes e_{1,1}) \leq n = \tilde{d}(A \otimes \mathbb{K})$. Together we get $\tilde{d}(A) = \tilde{d}(A \otimes \mathbb{K})$, as desired. □

16.4 Topological Dimension

One could try to define a dimension theory by simply considering the dimension of the primitive ideal space of a C^*-algebra. This will, however, run into problems if the primitive ideal space is not Hausdorff. Brown and Pedersen [6], suggested a way of dealing with this problem by restricting to (locally closed) Hausdorff subsets of Prim(A), and taking the supremum over the dimension of these Hausdorff subsets. This defines the topological dimension of a C^*-algebra, see Definition 4.

In this section we will show that the topological dimension is a dimension theory in the sense of Definition 1 for the class of type I C^*-algebras. It follows from the work of Brown and Pedersen that axioms (D1)–(D4) are satisfied, and we verify axiom (D5) in Proposition 8. We use transfinite induction over the length of a composition series of the type I C^*-algebra to verify axiom (D6), see Proposition 9.

See 6 for a short reminder on type I C^*-algebras. For more details, we refer the reader to Chap. IV.1 of Blackadar's book [1], and Chap. 6 of Pedersen's book [21].

Definition 3 (Brown and Pedersen [6, 2.2 (iv)]). Let X be a topological space. We define:

1. A subset $C \subset X$ is called *locally closed* if there is a closed set $F \subset X$ and an open set $G \subset X$ such that $C = F \cap G$.
2. X is called *almost Hausdorff* if every non-empty closed subset F contains a non-empty relatively open subset $F \cap G$ (so $F \cap G$ is locally closed in X) which is Hausdorff.

7. We could consider locally closed subsets as "well-placed" subsets. Then, being almost Hausdorff means having enough "well-placed" Hausdorff subsets.

For a C^*-algebra A, the locally closed subsets of Prim(A) correspond to ideals of quotients of A (equivalently to quotients of ideals of A) up to canonical isomorphism, see [6, 2.2(iii)]. Therefore, the primitive ideal space of every type I C^*-algebra is almost Hausdorff, since every non-zero quotient contains a non-zero ideal that has continuous trace, see [21, Theorem 6.2.11, p. 200], and the primitive ideal space of a continuous trace C^*-algebra is Hausdorff.

Definition 4 (Brown and Pedersen [6, 2.2(v)]). Let A be a C^*-algebra. If $\mathrm{Prim}(A)$ is almost Hausdorff, then the *topological dimension* of A, denoted by $\mathrm{topdim}(A)$, is:

$$\mathrm{topdim}(A) := \sup\{\mathrm{locdim}(S) \mid S \subset \mathrm{Prim}(A) \text{ locally closed, Hausdorff}\}.$$

We will now show that the topological dimension satisfies the axioms of Definition 1. The following result immediately implies (D1)–(D4).

Proposition 6 (Brown and Pedersen [6, Proposition 2.3]). *Let* $(J_\alpha)_{\alpha\leq\mu}$ *be a composition series for a* C^**-algebra* A. *Then* $\mathrm{Prim}(A)$ *is almost Hausdorff if and only if* $\mathrm{Prim}(J_{\alpha+1}/J_\alpha)$ *is almost Hausdorff for each* $\alpha < \mu$, *and if this is the case, then:*

$$\mathrm{topdim}(A) = \sup_{\alpha<\mu} \mathrm{topdim}(J_{\alpha+1}/J_\alpha).$$

The following result is implicit in the papers of Brown and Pedersen, e.g. [6, Theorem 5.6].

Proposition 7. *Let* A *be a* C^**-algebra, and let* $B \subset A$ *be a hereditary sub-*C^**-algebra. If* $\mathrm{Prim}(A)$ *is locally Hausdorff, then so is* $\mathrm{Prim}(B)$, *and then* $\mathrm{topdim}(B) \leq \mathrm{topdim}(A)$. *If* B *is even full hereditary, then* $\mathrm{topdim}(B) = \mathrm{topdim}(A)$.

Proof. In general, if $B \subset A$ is a hereditary sub-C^*-algebra, then $\mathrm{Prim}(B)$ is homeomorphic to an open subset of $\mathrm{Prim}(A)$. In fact, $\mathrm{Prim}(B)$ is canonically homeomorphic to the primitive ideal space of the ideal generated by B, and this corresponds to an open subset of $\mathrm{Prim}(A)$.

Note that being locally Hausdorff is a property that passes to locally closed subsets, and so it passes from $\mathrm{Prim}(A)$ to $\mathrm{Prim}(B)$. Further, every locally closed, Hausdorff subset $S \subset \mathrm{Prim}(B)$ is also locally closed (and Hausdorff) in $\mathrm{Prim}(A)$. It follows $\mathrm{topdim}(B) \leq \mathrm{topdim}(A)$.

If B is full, then $\mathrm{Prim}(B) \cong \mathrm{Prim}(A)$ and therefore $\mathrm{topdim}(B) = \mathrm{topdim}(A)$. □

Lemma 3. *Let* A *be a continuous trace* C^**-algebra, and let* $n \in \mathbb{N}$. *If* A *is approximated by sub-*C^**-algebras with topological dimension at most* n, *then* $\mathrm{topdim}(A) \leq n$.

Proof. Since $\mathrm{Prim}(A)$ is Hausdorff, we have $\mathrm{topdim}(A) = \mathrm{locdim}(\mathrm{Prim}(A))$. Thus, it is enough to show that every $x \in \mathrm{Prim}(A)$ has a neighborhood U with $\dim(U) \leq n$. This will allow us to reduce the problem to the situation that A has a global rank-one projection, i.e., that there exists a full, abelian projection $p \in A$, see [1, IV.1.4.20, p. 335], which we do as follows:

Let $x \in \mathrm{Prim}(A)$ be given. Since A has continuous trace, there exists an open neighborhood $U \subset \mathrm{Prim}(A)$ of x and an element $a \in A_+$ such that $\rho(a)$ is a rank-one projection for every $\rho \in U$, see 5. Then there exists a closed, compact neighborhood $Y \subset \mathrm{Prim}(A)$ of x that is contained in U. Let $J \lhd A$ be the ideal corresponding to $\mathrm{Prim}(A) \setminus Y$. The image of a in the quotient A/J is a full, abelian projection. Since A is approximated by subalgebras $B \subset A$ with $\mathrm{topdim}(B) \leq n$, A/J is approximated by the subalgebras $B/(B \cap J)$ with $\mathrm{topdim}(B/(B \cap J)) \leq \mathrm{topdim}(B) \leq n$. If we can show that this implies $\dim(Y) = \mathrm{topdim}(A/J) \leq n$, then every point of $\mathrm{Prim}(A)$ has a closed neighborhood of dimension $\leq n$, which means $\mathrm{topdim}(A) = \mathrm{locdim}(\mathrm{Prim}(A)) \leq n$.

We assume from now on that A has continuous trace with a full, abelian projection $p \in A$. Thus, $pAp \cong C(X)$ where $X := \mathrm{Prim}(A)$ is a compact, Hausdorff space. Assume A is approximated by subalgebras $A_i \subset A$ with $\mathrm{topdim}(A_i) \leq n$. It follows from Proposition 4 that the hereditary sub-C^*-algebra pAp is approximated by subalgebras B_j such that each B_j is isomorphic to a hereditary sub-C^*-algebras of A_i, for some $i = i(j)$. By Proposition 7, $\mathrm{topdim}(B_j) \leq \mathrm{topdim}(A_{i(j)}) \leq n$ for each j.

Thus, $C(X)$ is approximated by commutative subalgebras $C(X_j)$ with $\dim(X_j) = \mathrm{topdim}(C(X_j)) \leq n$. It follows from Proposition 1 that $\dim(X) \leq n$, as desired. □

Proposition 8. *Let A be a type I C^*-algebra, and let $n \in \mathbb{N}$. If A is approximated by sub-C^*-algebras with topological dimension at most n, then* $\mathrm{topdim}(A) \leq n$.

Proof. Let $(J_\alpha)_{\alpha \leq \mu}$ be a composition series for A such that each successive quotient has continuous trace, and assume A is approximated by subalgebras $A_i \subset A$ with $\mathrm{topdim}(A_i) \leq n$.

Then $J_{\alpha+1}/J_\alpha$ is approximated by the subalgebras $(A_i \cap J_{\alpha+1})/(A_i \cap J_\alpha)$, see 3. Since $\mathrm{topdim}((A_i \cap J_{\alpha+1})/(A_i \cap J_\alpha)) \leq \mathrm{topdim}(A_i) \leq n$, we obtain from the above Lemma 3 that $\mathrm{topdim}(J_{\alpha+1}/J_\alpha) \leq n$. By Proposition 6,

$$\mathrm{topdim}(A) = \sup_{\alpha<\mu} \mathrm{topdim}(J_{\alpha+1}/J_\alpha) \leq n,$$

as desired. □

Remark 3. It is noted in [6, Remark 2.5(v)] that a weaker version of the above Proposition 8 would follow from [25]. However, the statement is formulated as an axiom there, and it is not clear that the formulated axioms are consistent and give a dimension theory that agrees with the topological dimension.

We will now prove that the topological dimension of type I C^*-algebras satisfies the Mardešić factorization axiom (D6). We start with two lemmas.

Lemma 4. *Let A be a continuous trace C^*-algebra, and let $C \subset A$ be a separable sub-C^*-algebra. Then there exists a separable, continuous trace sub-C^*-algebra $D \subset A$ that contains C, and such that the inclusion $C \subset D$ is proper, and* $\mathrm{topdim}(D) \leq \mathrm{topdim}(A)$.

Proof. Let us first reduce to the case that A is σ-unital, and the inclusion $C \subset A$ is proper. To this end, consider the hereditary sub-C^*-algebra $A' := CAC \subset A$. Since C is separable, it contains a strictly positive element which is then also strictly positive in A'. Moreover, having continuous trace passes to hereditary sub-C^*-algebras, see [21, Proposition 6.2.10, p. 199]. Thus, A' is σ-unital and $C \subset A'$ is proper. Moreover, $\operatorname{topdim}(A') \leq \operatorname{topdim}(A)$ by Proposition 7.

Thus, by replacing A with CAC, we may assume from now on that A is σ-unital and that the inclusion $C \subset A$ is proper. Set $X := \operatorname{Prim}(A)$. By Brown's stabilization theorem [3, Theorem 2.8], there exists an isomorphism $\Phi\colon A \otimes \mathbb{K} \to C_0(X) \otimes \mathbb{K}$. Let $e_{ij} \in \mathbb{K}$ be the canonical matrix units, and consider the following C^*-algebra:

$$E := C^*\Big(\bigcup_{i,j} e_{1i}\Phi(C \otimes \mathbb{K})e_{j1}\Big) \subset C_0(X) \otimes e_{11}.$$

The following diagram shows some of the C^*-algebras and maps that we will construct below:

$$\begin{array}{ccccccc}
A \otimes e_{11} & \subset & A \otimes \mathbb{K} & \xrightarrow[\cong]{\Phi} & C_0(X) \otimes \mathbb{K} & & \\
\cup & & \cup & & \cup & & \\
D & \subset & \Phi^{-1}(D') & \xrightarrow[\cong]{} & C_0(Z_0) \otimes \mathbb{K} & = & D' \\
\cup & & \cup & & \cup & & \\
C \otimes e_{11} & \subset & C \otimes \mathbb{K} & \xrightarrow[\cong]{} & \Phi(C \otimes \mathbb{K}) & &
\end{array}$$

Note that E is separable and commutative. Thus, there exists a separable sub-C^*-algebra $C_0(Y) \subset C_0(X)$ such that $E = C_0(Y) \otimes e_{11}$. We constructed E such that $\Phi(C \otimes \mathbb{K}) \subset C_0(Y) \otimes \mathbb{K}$.

The inclusion $C_0(Y) \subset C_0(X)$ is induced by a pointed, continuous map $f\colon X^+ \to Y^+$, see 1. Recall that a compact, Hausdorff space M is metrizable if and only if $C(M)$ is separable. Thus, Y^+ is compact, metrizable.

By Mardešić's factorization theorem, see [17, Corollary 27.5, p. 159] or [13, Lemma 4], there exists a compact, metrizable space Z with $\dim(Z) \leq \dim(X)$ and continuous (surjective) maps $g\colon X \to Z$ and $h\colon Z \to Y$ such that $f = h \circ g$. Set $Z_0 := Z \setminus \{g(\infty)\}$, and note that g^* induces an embedding $C_0(Z_0) \subset C_0(X)$. Moreover, $C_0(Z_0)$ is separable, since Z is compact, metrizable.

Consider $D' := C_0(Z_0) \otimes \mathbb{K}$. We have that D' is a separable, continuous trace C^*-algebra such that $\Phi(C \otimes \mathbb{K}) \subset C_0(Y) \otimes \mathbb{K} \subset D'$, and $\operatorname{topdim}(D') = \dim(Z) \leq \dim(X) = \operatorname{topdim}(A)$. We think of C as included in $C \otimes \mathbb{K}$ via $C \cong C \otimes e_{11}$. Set

$$D := (1_{\tilde{A}} \otimes e_{11})(\Phi^{-1}(D'))(1_{\tilde{A}} \otimes e_{11}),$$

which is a hereditary sub-C^*-algebra of $\Phi^{-1}(D') \cong D'$. Hence, D is a separable, continuous trace C^*-algebra with $\operatorname{topdim}(D) \leq \operatorname{topdim}(D') \leq \operatorname{topdim}(A)$. By construction, $C \otimes e_{11} \subset D$, and this inclusion is proper since $D \subset A \otimes e_{11}$ and the inclusion $C \otimes e_{11} \subset A \otimes e_{11}$ is proper. □

Lemma 5. *Let A be a C^*-algebra, let $J \lhd A$ be an ideal, and let $C \subset A$ be a sub-C^*-algebra. Assume $K \subset J$ is a sub-C^*-algebra that contains $C \cap J$ and such that the inclusion $C \cap J \subset K$ is proper. Then K is an ideal in the sub-C^*-algebra $C^*(K, C) \subset A$ generated by K and C. Moreover, there is a natural isomorphism $C^*(K, C)/K \cong C/(C \cap J)$.*

Proof. Set $B := A/J$ and denote the quotient morphism by $\pi\colon A \to B$. Set $D := \pi(C) \subset B$. Clearly, $C^*(K, C)$ contains both K and C, and it is easy to see that the restriction of π to $C^*(K, C)$ maps onto D. The situation is shown in the following commutative diagram, where the top and bottom rows are exact:

$$\begin{array}{ccccccccc}
0 & \longrightarrow & J & \longrightarrow & A & \xrightarrow{\pi} & B & \longrightarrow & 0 \\
 & & \cup & & \cup & & \cup & & \\
 & & K & \longrightarrow & C^*(K,C) & \longrightarrow & D & & \\
 & & \cup & & \cup & & \| & & \\
0 & \longrightarrow & C \cap J & \longrightarrow & C & \longrightarrow & D & \longrightarrow & 0
\end{array}$$

Let us show that K is an ideal in $C^*(K, C)$. Since $C^*(K, C)$ is generated by elements of K and C, it is enough to show that xy and yx lie in K whenever $x \in K$ and $y \in K$ or $y \in C$. For $y \in K$ that is clear, so assume $y \in C$.

Since $C \cap J \subset K$ is proper, for any $\varepsilon > 0$ there exists $c \in C \cap J$ such that $\|cxc - x\| < \varepsilon$. Then $\|xy - cxcy\|, \|yx - ycxc\| < \varepsilon\|y\|$. Moreover, $cxcy \in K$ and $ycxc \in K$ since $cy, yc \in C \cap J \subset K$. For $\varepsilon > 0$ was arbitrary, it follows $xy, yx \in K$. This shows that the middle row in the above diagram is also exact. □

Proposition 9. *Let A be a C^*-algebra, let $J \lhd A$ be an ideal of type I, and let $C \subset A$ be a separable sub-C^*-algebra. Then there exists a separable sub-C^*-algebra $D \subset A$ such that $C \subset D$ and* $\operatorname{topdim}(D \cap J) \leq \operatorname{topdim}(J)$.

Proof. Let $(J_\alpha)_{\alpha \leq \mu}$ be a composition series for J with successive quotients that have continuous trace. To simplify notation, we will write $B[\alpha, \beta)$ for $(B \cap J_\beta)/(B \cap J_\alpha)$ and $B[\alpha, \infty)$ for $B/(B \cap J_\alpha)$ whenever $B \subset A$ is a subalgebra and $\alpha \leq \beta \leq \mu$ are ordinals. In particular, $A[0, \beta) = J_\beta$ and $A[\alpha, \infty) = A/J_\alpha$. We prove the statement of the proposition by transfinite induction over μ, which we carry out in three steps.

Step 1: The statement holds for $\mu = 0$. This follows since J is assumed to have a composition series with length 0 and so $J = \{0\}$ and we can simply set $D := C$.

Step 2: If the statement holds for a finite ordinal n, then it also holds for $n + 1$.

To prove this, assume J has a composition series $(J_\alpha)_{\alpha \leq n+1}$. Let $d := \operatorname{topdim}(J)$. Given $C \subset A$ separable, we want to find a separable subalgebra $D \subset A$ with $C \subset D$ and $\operatorname{topdim}(D[0, n + 1)) \leq d$. The following commutative diagram, whose rows are short exact sequences, contains the algebras and maps that we will construct below:

$$\begin{array}{ccccccccc} 0 & \longrightarrow & A[0,1) & \longrightarrow & A & \longrightarrow & A[1,\infty) & \longrightarrow & 0 \\ & & \cup & & \cup & & \cup & & \\ 0 & \longrightarrow & E[0,1) & \longrightarrow & E & \longrightarrow & E' & \longrightarrow & 0 \\ & & \cup & & \cup & & \cup & & \\ 0 & \longrightarrow & C[0,1) & \longrightarrow & C & \longrightarrow & C[1,\infty) & \longrightarrow & 0 \end{array}$$

Consider $A[1,\infty)$ together with the ideal $A[1,n+1) = J[1,n+1)$. Note that $A[1,n+1)$ has the canonical composition series $(A[1,\alpha))_{1\leq\alpha\leq n+1}$ of length n. By assumption of the induction, the statement holds for n, and so there is a separable sub-C^*-algebra $E' \subset A[1,\infty)$ such that $C[1,\infty) \subset E'$ and $\operatorname{topdim}(E' \cap A[1,n+1)) \leq \operatorname{topdim}(A[1,n+1)) \leq d$. Find a separable sub-$C^*$-algebra $E \subset A$ such that $C \subset E$ and $E[1,\infty) = E'$.

We apply Lemma 4 to the inclusion $E[0,1) \subset A[0,1)$ to find a separable sub-C^*-algebra $K \subset A[0,1)$ containing $E[0,1)$ and such that the inclusion $E[0,1) \subset K$ is proper, and $\operatorname{topdim}(K) \leq \operatorname{topdim}(A[0,1)) \leq d$. Set $D := C^*(K,E) \subset A$, which is a separable C^*-algebra with $C \subset D$. By Lemma 5, D is an extension of E by K, and therefore Proposition 6 gives:

$$\begin{aligned} \operatorname{topdim}(D[0,n+1)) &= \max\{\operatorname{topdim}(D[0,1)), \operatorname{topdim}(D[1,n+1))\} \\ &= \max\{\operatorname{topdim}(K), \operatorname{topdim}(E' \cap A[1,n+1))\} \\ &\leq d. \end{aligned}$$

Step 3: Assume λ is a limit ordinal, and n is finite. If the statement holds for all $\alpha < \lambda$, then it holds for $\lambda + n$.

We will prove this by distinguishing the two sub-cases that λ has cofinality at most ω, or cofinality bigger than ω. We start the construction for both cases together. Later we will treat them separately. Let $d := \operatorname{topdim}(J)$.

We will inductively define ordinals $\alpha_k < \mu$ and sub-C^*-algebras $D_k, E_k \subset A$ with the following properties:

1. $\alpha_1 \leq \alpha_2 \leq \dots$,
2. $D_k \subset E_k$ and $\operatorname{topdim}(E_k[\lambda,\lambda+n)) \leq d$,
3. $E_k \subset D_{k+1}$ and $\operatorname{topdim}(D_{k+1}[0,\alpha_{k+1})) \leq d$.

In both Cases 3a and 3b below, we construct E_k from D_k as follows: Given D_k, consider $D_k[\lambda,\infty) \subset A[\lambda,\infty)$ and the ideal $A[\lambda,\lambda+n) \lhd A[\lambda,\infty)$ which has a composition series of length n. Since $n < \lambda$, we get by assumption of the induction that there exists a separable subalgebra $E'_k \subset A[\lambda,\infty)$ such that $D_k[\lambda,\infty) \subset E'_k$

and $\operatorname{topdim}(E_k' \cap A[\lambda, \lambda+n)) \leq d$. Let $E_k \subset A$ be any separable C^*-algebra such that $D_k \subset E_k$ and $E_k[\lambda, \infty) = E_k'$.

Case 3a: Assume λ has cofinality at most ω, i.e., there exist ordinals $0 = \lambda_0 < \lambda_1 < \lambda_2 < \ldots < \lambda$ such that $\lambda = \sup_k \lambda_k$.

In this case, we let $\alpha_k := \lambda_k$, and we set $D_0 := C$. Given D_k, we construct E_k as described above. Given E_k, we get D_{k+1} satisfying 3. by assumption of the induction.

Case 3b: Assume λ has cofinality larger than ω.

We start by setting $\alpha_0 := 0$ and $D_0 := C$. Given D_k, we construct E_k as described above. Given E_k, we define α_{k+1} as follows:

$$\alpha_{k+1} := \inf\{\alpha \mid \alpha_k \leq \alpha \leq \lambda, \text{ and } E_k[0, \alpha) = E_k[0, \lambda)\}.$$

Since λ has cofinality larger than ω and E_k is separable, we have $\alpha_{k+1} < \lambda$. Hence, we get D_{k+1} satisfying 3. by assumption of the induction.

From now on we treat the Cases 3a and 3b together. Set $D := \overline{\bigcup_k D_k} = \overline{\bigcup_k E_k}$. This is a separable sub-C^*-algebra of A with $C \subset D$. Since $D[\lambda, \lambda+n) = \overline{\bigcup_k E_k[\lambda, \lambda+n)}$ and $\operatorname{topdim}(E_k[\lambda, \lambda+n)) \leq d$ for all k, we get $\operatorname{topdim}(D[\lambda, \lambda+n)) \leq d$ from Proposition 8.

One checks that $D[0, \lambda) = \overline{\bigcup_k D_k[0, \alpha_k)}$. Since $\operatorname{topdim}(D_k[0, \alpha_k)) \leq d$ for all k, we get $\operatorname{topdim}(D[0, \lambda)) \leq d$, again by Proposition 8.

Then Proposition 6 gives:

$$\operatorname{topdim}(D[0, \lambda+n)) = \max\{\operatorname{topdim}(D[0, \lambda)), \operatorname{topdim}(D[\lambda, \lambda+n))\} \leq d.$$

This completes the proof. □

Corollary 1. *The topological dimension of type I C^*-algebras satisfies the Mardešić factorization axiom (D6), i.e., given a type I C^*-algebra A and a separable sub-C^*-algebra $C \subset A$, there exists a separable C^*-algebra $D \subset A$ such that $C \subset D \subset A$ and* $\operatorname{topdim}(D) \leq \operatorname{topdim}(A)$.

This following theorem is the main result of this paper. It follows immediately from the above Corollary 1, Propositions 6 and 8.

Theorem 1. *The topological dimension is a dimension theory in the sense of Definition 1 for the class of type I C^*-algebras.*

8. Let us extend the topological dimension from the class of type I C^*-algebras to all C^*-algebras, as defined in Proposition 5. This dimension theory $\widetilde{\operatorname{topdim}} : \mathscr{C}^* \to \overline{\mathbb{N}}$ is Morita-invariant since $\operatorname{topdim}(A) = \operatorname{topdim}(A \otimes \mathbb{K})$ for any type I C^*-algebra A.

If $\widetilde{\operatorname{topdim}}(A) < \infty$, then A is in particular approximated by type I sub-C^*-algebras. This implies that A is nuclear, satisfies the universal coefficient theorem (UCT), see [8, Theorem 1.1], and is not properly infinite. It is possible that this dimension theory is connected to the decomposition rank and nuclear dimension, although the exact relation is not clear.

Let us show that the (extended) topological dimension behaves well with respect to tensor products. First, if A, B are separable, type I C^*-algebras, then $\mathrm{Prim}(A \otimes B) \cong \mathrm{Prim}(A) \times \mathrm{Prim}(B)$, see [1, IV.3.4.25, p. 390]. This implies:

$$\mathrm{topdim}(A \otimes B) \leq \mathrm{topdim}(A) + \mathrm{topdim}(B).$$

Next, assume A, B are C^*-algebras with $\widetilde{\mathrm{topdim}}(A) = d_1 < \infty$ and $\widetilde{\mathrm{topdim}}(B) = d_2 < \infty$. This means that A is approximated by separable, type I algebras $A_i \subset A$ with $\mathrm{topdim}(A_i) \leq d_1$, and similarly B is approximated by separable, type I algebras $B_j \subset B$ with $\mathrm{topdim}(B_j) \leq d_2$. Then $A \otimes B$ is approximated by the algebras $A_i \otimes B_j$, and we have seen that $\mathrm{topdim}(A_i \otimes B_j) \leq d_1 + d_2$. Thus:

$$\widetilde{\mathrm{topdim}}(A \otimes B) \leq \widetilde{\mathrm{topdim}}(A) + \widetilde{\mathrm{topdim}}(B).$$

Note that we need not specify the tensor product, since $\widetilde{\mathrm{topdim}}(A) < \infty$ implies that A is nuclear.

16.5 Dimension Theories of Type I C^*-Algebras

In this section we study the relation of the topological dimension of type I C^*-algebras to other dimension theories. It was shown by Brown [4, Theorem 3.10], how to compute the real and stable rank of a CCR algebra A in terms of the topological dimension of certain canonical algebras A_k associated to A. We use this to obtain a general estimate of the real and stable rank of a CCR algebra in terms of its topological dimension, see Corollary 2. Using the composition series of a type I C^*-algebra, we will obtain similar (but weaker) estimates for general type I C^*-algebras, see Theorem 3.

Let A be a C^*-algebra. We denote by $\mathrm{rr}(A)$ its real rank, see [5], by $\mathrm{sr}(A)$ its stable rank, and by $\mathrm{csr}(A)$ its connected stable rank, see [22, Definition 1.4, 4.7] We denote by A_k the successive quotient of A that corresponds to the irreducible representations of dimension k.

If t is a real number, we denote by $\lfloor t \rfloor$ the largest integer $n \leq t$, and by $\lceil t \rceil$ the smallest integer $n \geq t$.

Theorem 2 (Brown [4, Theorem 3.10]). *Let A be a CCR algebra with* $\mathrm{topdim}(A) < \infty$. *Then:*

1. *If* $\mathrm{topdim}(A) \leq 1$, *then* $\mathrm{sr}(A) = 1$.
2. *If* $\mathrm{topdim}(A) > 1$, *then* $\mathrm{sr}(A) = \sup_{k \geq 1} \max\{\left\lceil \frac{\mathrm{topdim}(A_k)+2k-1}{2k} \right\rceil, 2\}$.
3. *If* $\mathrm{topdim}(A) = 0$, *then* $\mathrm{rr}(A) = 0$.

4. *If* $\operatorname{topdim}(A) > 0$, *then* $\operatorname{rr}(A) = \sup_{k\geq 1} \max\{\left\lceil \frac{\operatorname{topdim}(A_k)}{2k-1} \right\rceil, 1\}$.

We may draw the following conclusion:

Corollary 2. *Let A be a CCR algebra. Then:*

$$\operatorname{sr}(A) \leq \left\lfloor \frac{\operatorname{topdim}(A)}{2} \right\rfloor + 1, \tag{16.4}$$

$$\operatorname{csr}(A) \leq \left\lfloor \frac{\operatorname{topdim}(A) + 1}{2} \right\rfloor + 1, \tag{16.5}$$

$$\operatorname{rr}(A) \leq \operatorname{topdim}(A). \tag{16.6}$$

Proof. If $\operatorname{topdim}(A) = \infty$, then the statements hold. So we may assume $\operatorname{topdim}(A) < \infty$, whence we may apply [4, Theorem 3.10], see Theorem 2.

Let us show (16.4). If $\operatorname{topdim}(A) \leq 1$, then $\operatorname{sr}(A) = 1 \leq \lfloor \operatorname{topdim}(A)/2 \rfloor + 1$. If $d := \operatorname{topdim}(A) \geq 2$, then we use $\operatorname{topdim}(A_k) \leq d$ to compute:

$$\operatorname{sr}(A) \leq \sup_k \max\{\left\lceil \frac{d + 2k - 1}{2k} \right\rceil, 2\} \leq \max\{\left\lceil \frac{d+1}{2} \right\rceil, 2\} \leq \left\lfloor \frac{d}{2} \right\rfloor + 1.$$

Now (16.5) follows from (16.4) since $\operatorname{csr}(A) \leq \operatorname{sr}(A \otimes C([0, 1]))$ in general, by Nistor [19, Lemma 2.4], and $\operatorname{topdim}(A \otimes C([0, 1])) \leq \operatorname{topdim}(A) + 1$, see 8.

In order to show (16.6), we again use [4, Theorem 3.10], see Theorem 2. If $\operatorname{topdim}(A) = 0$, then $\operatorname{rr}(A) = 0 \leq \operatorname{topdim}(A)$. If $d := \operatorname{topdim}(A) \geq 1$, then we use $\operatorname{topdim}(A_k) \leq d$ to compute:

$$\operatorname{rr}(A) \leq \sup_k \max\{\left\lceil \frac{d}{2k - 1} \right\rceil, 1\} \leq \max\{\lceil d \rceil, 1\} \leq d,$$

which completes the proof. □

Remark 4. What makes type I C^*-algebras so accessible is the presence of composition series with successive quotients that are easier to handle (i.e., of continuous trace or CCR), see 6. They allow us to prove statements by transfinite induction, for which one has to consider the case of a successor and limit ordinal. Let us see that for statements about dimension theories one only needs to consider successor ordinals.

Let $(J_\alpha)_{\alpha\leq\mu}$ be a composition series, and d a dimension theory. If α is a limit ordinal, then $J_\alpha = \overline{\bigcup_{\gamma<\alpha} J_\gamma}$, and we obtain:

$$d(J_\alpha) \leq_{(D5)} \sup_{\gamma<\alpha} d(J_\gamma) \leq_{(D1)} \sup_{\gamma<\alpha} d(J_\alpha),$$

and thus $d(J_\alpha) = \sup_{\gamma<\alpha} d(J_\gamma)$.

Thus, any reasonable estimate about dimension theories that holds for $\gamma < \alpha$ will also hold for α. It follows that we only need to consider a successor ordinal α, in which case $A = J_\alpha$ is an extension of $B = J_\alpha / J_{\alpha-1}$ by $I = J_{\alpha-1}$. By assumption the result is true for I and has to be proved for A (using that B has continuous trace or is CCR). This idea is used to prove the next theorem.

Theorem 3. *Let A be a type I C^*-algebra. Then:*

$$\mathrm{sr}(A) \leq \left\lfloor \frac{\mathrm{topdim}(A)+1}{2} \right\rfloor + 1, \tag{16.7}$$

$$\mathrm{rr}(A) \leq \mathrm{topdim}(A) + 2. \tag{16.8}$$

Proof. Let $(J_\alpha)_{\alpha \leq \mu}$ be a composition series for A such that the successive quotients are CCR algebras. We will prove (16.7) by transfinite induction over μ.

Set $d := \mathrm{topdim}(A)$. Assume the statement holds for some ordinal μ, and let us show it also holds for $\mu + 1$. Consider the ideal $I := J_\mu$ inside $A = J_{\mu+1}$. We obtain the following, where the first estimate follows from [22, Theorem 4.11], and the second estimate follows by assumption of the induction for I and Corollary 2 for the CCR algebra A/I:

$$\begin{aligned} \mathrm{sr}(A) &\leq \max\{\mathrm{sr}(I), \mathrm{sr}(A/I), \mathrm{csr}(A/I)\} \\ &\leq \max\{\left\lfloor \frac{d+1}{2} \right\rfloor + 1, \left\lfloor \frac{d}{2} \right\rfloor + 1, \left\lfloor \frac{d+1}{2} \right\rfloor + 1\} \\ &= \left\lfloor \frac{d+1}{2} \right\rfloor + 1. \end{aligned}$$

Let μ be a limit ordinal, and assume the statement holds for $\alpha < \mu$. This means that $\mathrm{sr}(J_\alpha) \leq \left\lfloor \frac{\mathrm{topdim}(J_\alpha)+1}{2} \right\rfloor + 1$ for all $\alpha < \mu$. As explained in Remark 4, we obtain the desired estimate for μ as follows:

$$\mathrm{sr}(J_\mu) = \sup_{\alpha<\mu} \mathrm{sr}(J_\alpha) \leq \sup_{\alpha<\mu} \left\lfloor \frac{\mathrm{topdim}(J_\alpha)+1}{2} \right\rfloor + 1 = \left\lfloor \frac{\mathrm{topdim}(J_\mu)+1}{2} \right\rfloor + 1.$$

Finally, (16.8) follows from (16.7), using the estimate $\mathrm{rr}(A) \leq 2\mathrm{sr}(A) - 1$, which holds for all C^*-algebras, see [5, Proposition 1.2]. □

Remark 5. It follows from [22, Proposition 1.7], Corollary 2, and Theorem 3 that we may estimate the stable rank of a C^*-algebra A in terms of its topological dimension as follows:

1. $\mathrm{sr}(A) = \left\lfloor \frac{\mathrm{topdim}(A)}{2} \right\rfloor + 1$, if A is commutative.
2. $\mathrm{sr}(A) \leq \left\lfloor \frac{\mathrm{topdim}(A)}{2} \right\rfloor + 1$, if A is CCR.
3. $\mathrm{sr}(A) \leq \left\lfloor \frac{\mathrm{topdim}(A)+1}{2} \right\rfloor + 1$, if A is type I.

This also shows that the inequality for the stable rank in Corollary 2 cannot be improved (the same is true for the estimates of real rank and connected stable rank).

To see that the estimate of Theorem 3 for the stable rank cannot be improved either, consider the Toeplitz algebra $\mathcal{T}$. We have $\mathrm{sr}(\mathcal{T}) = 2$, while $\mathrm{topdim}(\mathcal{T}) = 1$.

Acknowledgements I thank Søren Knudby for valuable comments and feedback. I thank Mikael Rørdam for interesting discussions and valuable comments, especially on the results in Sect. 16.3. I also thank Wilhelm Winter for inspiring discussions on noncommutative dimension theory.

This research was supported by the Danish National Research Foundation through the Centre for Symmetry and Deformation (DNRF92).

References

1. Blackadar, B.: Operator algebras, *Encyclopaedia of Mathematical Sciences*, vol. 122. Springer-Verlag, Berlin (2006). Theory of C^*-algebras and von Neumann algebras, Operator Algebras and Non-commutative Geometry, III
2. Blackadar, B.E.: Weak expectations and nuclear C^*-algebras. Indiana Univ. Math. J. **27**(6), 1021–1026 (1978). DOI 10.1512/iumj.1978.27.27070. URL http://dx.doi.org/10.1512/iumj.1978.27.27070
3. Brown, L.G.: Stable isomorphism of hereditary subalgebras of C^*-algebras. Pacific J. Math. **71**(2), 335–348 (1977)
4. Brown, L.G.: On higher real and stable ranks for CCR C^*-algebras (2007), arXiv:0708.3072
5. Brown, L.G., Pedersen, G.K.: C^*-algebras of real rank zero. J. Funct. Anal. **99**(1), 131–149 (1991). DOI 10.1016/0022-1236(91)90056-B. URL http://dx.doi.org/10.1016/0022-1236(91)90056-B
6. Brown, L.G., Pedersen, G.K.: Limits and C^*-algebras of low rank or dimension. J. Operator Theory **61**(2), 381–417 (2009)
7. Charalambous, M.G.: Axiomatic characterizations of the dimension of metric spaces. Topology Appl. **60**(2), 117–130 (1994). DOI 10.1016/0166-8641(94)00003-4. URL http://dx.doi.org/10.1016/0166-8641(94)00003-4
8. Dadarlat, M.: Some remarks on the universal coefficient theorem in KK-theory. In: Operator algebras and mathematical physics (Constanţa, 2001), pp. 65–74. Theta, Bucharest (2003)
9. Dowker, C.H.: Local dimension of normal spaces. Quart. J. Math. Oxford Ser. (2) **6**, 101–120 (1955)
10. Elhage Hassan, N.: Rang réel de certaines extensions. Proc. Amer. Math. Soc. **123**(10), 3067–3073 (1995). DOI 10.2307/2160661. URL http://dx.doi.org/10.2307/2160661
11. Elliott, G.A., Toms, A.S.: Regularity properties in the classification program for separable amenable C^*-algebras. Bull. Amer. Math. Soc. (N.S.) **45**(2), 229–245 (2008). DOI 10.1090/S0273-0979-08-01199-3. URL http://dx.doi.org/10.1090/S0273-0979-08-01199-3
12. Kirchberg, E., Winter, W.: Covering dimension and quasidiagonality. Internat. J. Math. **15**(1), 63–85 (2004). DOI 10.1142/S0129167X04002119. URL http://dx.doi.org/10.1142/S0129167X04002119
13. Mardešić, S.: On covering dimension and inverse limits of compact spaces. Illinois J. Math. **4**, 278–291 (1960)
14. Mardešić, S., Matijević, V.: $\mathcal{P}$-like spaces are limits of approximate $\mathcal{P}$-resolutions. In: Proceedings of the Tsukuba Topology Symposium (Tsukuba, 1990), vol. 45, pp. 189–202 (1992). DOI 10.1016/0166-8641(92)90004-J. URL http://dx.doi.org/10.1016/0166-8641(92)90004-J
15. Mardešić, S., Segal, J.: ε-mappings onto polyhedra. Trans. Amer. Math. Soc. **109**, 146–164 (1963)

16. Morita, K.: Čech cohomology and covering dimension for topological spaces. Fund. Math. **87**, 31–52 (1975)
17. Nagami, K.: Dimension theory. With an appendix by Yukihiro Kodama. Pure and Applied Mathematics, Vol. 37. Academic Press, New York (1970)
18. Nishiura, T.: An axiomatic characterization of covering dimension in metrizable spaces. In: TOPO 72—general topology and its applications (Proc. Second Pittsburgh Internat. Conf., Pittsburgh, Pa., 1972; dedicated to the memory of Johannes H. de Groot), pp. 341–353. Lecture Notes in Math., Vol. 378. Springer, Berlin (1974)
19. Nistor, V.: Stable range for tensor products of extensions of $\mathcal{K}$ by $C(X)$. J. Operator Theory **16**(2), 387–396 (1986)
20. Pears, A.R.: Dimension theory of general spaces. Cambridge University Press, Cambridge, England (1975)
21. Pedersen, G.K.: C^*-algebras and their automorphism groups, *London Mathematical Society Monographs*, vol. 14. Academic Press Inc. [Harcourt Brace Jovanovich Publishers], London (1979)
22. Rieffel, M.A.: Dimension and stable rank in the K-theory of C^*-algebras. Proc. London Math. Soc. (3) **46**(2), 301–333 (1983). DOI 10.1112/plms/s3-46.2.301. URL http://dx.doi.org/10.1112/plms/s3-46.2.301
23. Rørdam, M.: On the structure of simple C^*-algebras tensored with a UHF-algebra. II. J. Funct. Anal. **107**(2), 255–269 (1992). DOI 10.1016/0022-1236(92)90106-S. URL http://dx.doi.org/10.1016/0022-1236(92)90106-S
24. Rørdam, M.: Structure and classification of C^*-algebras. In: International Congress of Mathematicians. Vol. II, pp. 1581–1598. Eur. Math. Soc., Zürich (2006)
25. Sudo, T.: A topological rank for C^*-algebras. Far East J. Math. Sci. (FJMS) **15**(1), 71–86 (2004)
26. Thiel, H.: One-dimensional C^*-algebras and their K-theory. Diplom thesis, University of Münster (2009). URL http://www.math.ku.dk/~thiel
27. Thiel, H.: Inductive limits of projective C^*-algebras (2011), arXiv:1105.1979
28. Winter, W.: Nuclear dimension and $\mathcal{Z}$-stability of pure C*-algebras. Invent. Math. **187**(2), 259–342 (2012). DOI 10.1007/s00222-011-0334-7. URL http://dx.doi.org/10.1007/s00222-011-0334-7
29. Winter, W., Zacharias, J.: The nuclear dimension of C^*-algebras. Adv. Math. **224**(2), 461–498 (2010). DOI 10.1016/j.aim.2009.12.005. URL http://dx.doi.org/10.1016/j.aim.2009.12.005

Index

T.M. Carlsen et al. (eds.), *Operator Algebra and Dynamics*, Springer Proceedings in Mathematics & Statistics 58, DOI 10.1007/978-3-642-39459-1,

L

M

N

O

P

Q

R

S

MIX
Papier aus verantwortungsvollen Quellen
Paper from responsible sources
FSC® C105338

If you have any concerns about our products,
you can contact us on
ProductSafety@springernature.com

In case Publisher is established outside the EU,
the EU authorized representative is:
Springer Nature Customer Service Center GmbH
Europaplatz 3, 69115 Heidelberg, Germany

Printed by Libri Plureos GmbH
in Hamburg, Germany